Mycorrhizal Biology

Mycorrhizal Biology

Edited by

K. G. Mukerji

B. P. Chamola
University of Delhi
Delhi, India

and

Jagjit Singh
Environmental Buildings Solution, Ltd.
Dunstable, United Kingdom

Kluwer Academic / Plenum Publishers
New York, Boston, Dordrecht, London, Moscow

Library of Congress Cataloging-in-Publication Data

Mycorrhizal biology / edited by K.G. Mukerji, B.P. Chamola, and Jagjit Singh.
   p. cm.
  Includes bibliographical references.
  ISBN 0-306-46294-X
   1. Mycorrhizas. I. Mukerji, K. G. II. Chamola, B. P. III. Singh, Jagjit, 1956-

QK604.2.M92 M96 1999
571.2'95178--dc21

99-049587

Cover illustration by K. G. Mukerji

ISBN: 0-306-46294-X

© 2000 Kluwer Academic / Plenum Publishers, New York
233 Spring Street, New York, New York 10013

http://www.wkap.nl/

10 9 8 7 6 5 4 3 2 1

A C.I.P. record for this book is available from the Library of Congress

Printed in the United States of America

LIST OF CONTRIBUTORS

1. Adholeya, Alok
   Tata Energy Research Institute
   Habitat Place, D.S. Block
   Lodi Road
   New Delhi - 110003, INDIA
   Tel. : 91-11-4601550, 4622246
   Fax : 91-11-4621770, 4632609

2. Annapurna, K.
   Division of Microbiology
   Indian Agricultural Research Institute
   New Delhi - 110012, INDIA
   Tel. : 91-11-5787649, 5731270
   Fax : 91-11-5751717, 5766420

3. Bansal, Manju
   Department of Botany
   Arbindo College
   University of Delhi
   Delhi - 110007, INDIA

4. Chamola, B.P.
   Applied Mycology Laboratory
   Department of Botany
   University of Delhi
   Delhi-110007, INDIA
   Tel. : 91-11-7257573
   Fax : 91-11-7257830

5. Dixon, R.K.
   U.S. Country Studies Program
   1000 Independence Avenue, SW
   Washington, DC, 20585
   USA
   Tel. : 202-586-3288
   Fax : 202-586-3486
   E-mail : csmt@igc.apc.org

6. Douds, David D.
   USDA-ARS ERRC;
   600E Mermaid Lane
   Wyndmoor, PA, 19038-8598
   USA
   Fax : 215-233-5681
   E-mail : douds@arserrc.gov

7. Gadkar, Vijay
   Tata Energy Research Institute
   Habitat Place, D.S. Block
   Lodi Road
   New Delhi - 110003, INDIA
   Tel. : 91-11-4601550, 4622246
   Fax : 91-11-4621770, 4632609

8. Garg, Sandeep
   Department of Microbiology
   University of Delhi, South Campus
   Benito Juarez Road
   New Delhi-110021, INDIA
   Tel. : 91-11-6882231, 6882503
   Fax : 91-11-6886425, 68852270

9. Gupta, Rajni
   Applied Mycology Laboratory
   Department of Botany
   University of Delhi
   Delhi - 110007, INDIA
   Tel. : 91-11-7257573
   Fax : 91-11-7257830

10. Gupta, R.K.
    Department of Environmental Science
    G.B. Pant University of Agriculture
        and Technology
    Pantnagar, U.P., INDIA
    Tel. : 91-05948-33321, 33374
    Fax : 91-05948-33473, 33608

11. Gupta, Vandana
    Department Microbiology
    University of Delhi, South Campus
    Benito Juarez Road
    New Delhi-110021, INDIA
    Tel. : 91-11-6882231, 6882503
    Fax : 91-11-6886427, 6885270
    E-mail : micro@dusc.ernet.in

12. Kumar, Pradeep
    Department of Plant Pathology
    G.B. Pant University of Agriculture
        and Technology
    Pantnagar, U.P., INDIA
    Tel. : 91-05948-33321, 33374
    Fax : 91-05948-33473, 33608

13. Lakhanpal, T.N.
    Department of Biosciences
    Himachal Pradesh University
    Summar Hill
    Shimla-171005, H.P., INDIA
    Tel. : 91-0177-230946

14. Mandelbaum, Carol I.
    Department of Botany
    University of Texas at Austin
    Austin, Texas 78713-7460
    USA
    Tel. : 512-471-5858
    Fax : 512-471-3878
    E-mail : cim@uts.cc.utexas.edu

15. Manoharachary, C.
    Department of Botany
    Osmania University
    Hyderabad-500007
    A.P., INDIA
    Tel. : 91-040-7018176
    Fax : 91-040-7019020

16. Mukerji, K.G.
    C-29, Probyn Road
    University of Delhi
    Delhi-110007, INDIA
    Tel. : 91-11-7257502
    Fax : 91-11-7257830, 7256708

17. Pardha Saradhi, P.
    Plant Physiology and Biotechnology Laboratory
    Department of Biosciences
    Jamia Millia Islamia
    New Delhi-110025, INDIA
    Fax   : 91-11-6916275
    E-mail : pardh.bi@jmi.ernet.in

18. Piche, Yves
    Centre de recherche en biologie forestiere
    Faculté de foresterie et de géomatique
    Pavillon C.E. Marchand
    Université Leval, Ste. Foy
    Québec G1K7P4
    CANADA

19. Puthur, Jos T.
    Plant Physiology and Biotechnology Laboratory
    Department of Biosciences
    Jamia Millia Islamia
    New Delhi-110025, INDIA
    Fax   : 91-11-6916275
    E-mail : pardh.bi@jmi.ernet.in

20. Raina, Shalini
    Applied Mycology Laboratory
    Department Botany
    University of Delhi
    Delhi-110007, INDIA
    Tel. : 91-11-7257573
    Fax : 91-11-7257830

21. Reddy, Jagan Mohan P.
    Department of Botany
    Osmania University
    Hyderabad-500007
    A.P., INDIA
    Tel. : 91-040-7018176
    Fax : 91-040-7019020

22. Sharmila, P.
    Plant Physiology and Biotechnology Laboratory
    Department of Biosciences
    Jamia Millia Islamia
    New Delhi-110025, INDIA
    Fax   : 91-11-6916275
    E-mail : pardh.bi@jmi.ernet.in

23. Sarwar, Nikhat
    Applied Mycology Laboratory
    Department Botany
    University of Delhi
    Delhi-110007, INDIA
    Tel. : 91-11-7257573
    Fax : 91-11-7257830

24. Satyanarayana, T.
    Department of Microbiology
    University of Delhi, South Campus
    Benito Juarez Road
    New Delhi-110021, INDIA
    Tel. : 91-11-6882231, 6882503
    Fax : 91-11-6886427, 6885270
    E-mail : micro@dusc.ernet.in

25. Saxena, A.K.
    Division of Microbiology
    Indian Agricultural Research Institute
    New Delhi - 110012, INDIA
    Tel. : 91-11-5787649, 5731270
    Fax : 91-11-5751717, 5766420

26. Singh, Archana
    School of Life Sciences
    Jawaharlal Nehru University
    New Delhi-110067, INDIA
    Tel. : 91-11-6170016, 6107676, 6167557/2571
    Fax : 91-11-6165886, 6167338

27. Singh, Reena
    Tata Energy Research Institute
    Habitat Place, D.S. Block
    Lodi Road
    New Delhi - 110003, INDIA
    Tel. : 91-11-4601550, 4622246
    Fax : 91-11-4621770, 4632609

28. Tilak, K.V. B.R.
    Division of Micobiology
    Indian Agricultural Research Institute
    New Delhi - 110012, INDIA
    Tel. : 91-11-5787649, 5731270
    Fax : 91-11-5751717, 5766420

29. Varma, Ajit
    School of Life Sciences
    Jawaharlal Nehru University
    New Delhi-110067, INDIA
    Tel. : 91-11-6170016, 6107676, 6167557/2571
    Fax : 91-11-6165886, 6167338

30. Vyas, Sameer
    Department of Plant Pathology
    J.N. Agricultural University
    Indore - 482004, M.P., INDIA

31. Vyas, S.C.
    Department of Plant Pathology
    J.N. Agricultural University
    Indore - 482004, M.P., INDIA
    Tel. : 91-0761-344232

PREFACE

Sustainability in agriculture, forestry, and range management requires balanced functional microbial ecosystems. The association of plant roots with mycorrhizal fungi is a key factor in the below ground network essential to ecosystem function; these associations are known to benefit plants under conditions of nutritional and water stress and pathogen challenge. Molecular and genetic tools are, and will be used increasingly, to explore the structural and regulatory genes in both fungus and plant and permit mycorrhiza formation. Our understanding of the genetic loci that govern mycorrhiza formation will aid in our ability to maintain sustainability.

However, the mycorrhizal status in roots of plants is extremely varied. Mycorrhizal formation is associated with many diverse fungi, and the types of mycorrhizae that are formed differ in structure and efficacy. Of the seven types of mycorrhizae, the two prevalent mycorrhizal types are the ectomycorrhizae common with woody species and related to forestry, and the endomycorrhizae more often associated with herbaceous plants with relevance to range, horticultural, and agronomic plants.

Agriculture has a dominant role in the economy of a developing country providing employment to about 65% of the working population, accounting for a sizeable share of the country's foreign exchange earnings. After the green revolution of 1960s and 1970s, a general decline and stagnation in soil productivity was observed during 1980s. Much of this decline was due to improper use and poor management and exploitative farming practices, resulting in land degradation by wind and water erosion, nutrient depletion and overall decline in soil productivity. Land degradation problems and the associated loss of soil productivity, and the decline in soil quality continue to be the subject of environmental concern attracting attention of the world's scientific community. It is a major concern for at least two reasons. First, soil degradation undermines the productive capacity of an ecosystem. Second, it affects global climate through alteration in water and energy balances and disruptions in cycles of carbon, nitrogen, sulphur, and other elements. Through its impact on agricultural productivity and environment, soil degradation leads to political and social instability, enhanced rate of deforestation, intensive use of marginal and fragile lands, accelerated run-off and soil erosion, pollution of natural waters and emission of greenhouse gases into the atmosphere.

In India, a large proportion of land area shows clear evidence of soil degradation which in turn is affecting the country's productive resource base. Large parts of land are wasted due to various reasons such as salinity and alkalinity, soil erosion, water logging, etc. All these lands are characterised by some kind of deficiency or non-availability of nutrients. The socioeconomic and ecological consequences of land degradation are far reaching, affecting over 50% of the total geographical area of the country. It has been estimated that a total of more than 5000 million tonnes of top soil is permanently getting lost to the sea. Out of the total geographical area of 329 m ha about 187 m ha (57%) are suffering from different soil degradation problems. Water erosion is the major problem causing loss of topsoil (in 132 m ha) and terrain deformation (in 164 m ha). Wind erosion is dominant in the Western region causing loss of top soil and terrain deformation in 13 m ha. The human induced chemical deterioration is observed in 14m ha causing salination in 10 m ha and loss of nutrient and organic matter in about 4 m ha. In India, the problem of soil degradation assumes greater significance because of burgeoning population (over 887 million) and decreasing per capita land holding. Overpopulation, harsh climatic conditions, overexploitation, improper use of soil resources, deforestation, etc., not only induce soil-food-population inbalance, but also render most of tropical and subtropical ecosystems extremely vulnerable to soil erosion and erosion-induced land degradation. Globally, land degradation affects of about one-sixth of the world's population, 70% of all drylands, amounting to 3.6 billion ha and one-quarter of the total land area of the world. In Africa the situation is so deteriorated the term "desertification" has been used to describe the creation of human-induced desert-like condition. The problem is not confined only to the developing world. Even advanced countries, like the USA, face serious problems of erosion resulting in loss of nearly six billion tons of soil each year.

The fundamental problem which the world faces today, is the rapidly increasing pressure of population on the limited resources of the land. To meet the ever increasing demand of expanding, population, agriculture production has been raised through the abundant use of inorganic fertilizers, adopting multicropping system and liberal application of chemical pesticides (fungicides, bactericides, etc.). Though the use of chemicals has increased the yield dramatically, it has also resulted in rapid deterioration of land and water resources apart from wastage of scarce resources. This has adversely affected the biological balance and lead to the presence of toxic residues in food, soil, and water in addition to imposing economic constrains to the developing countries. India is among the top in fertilizer consuming countries in the world and its use is increasing rapidly. Use of agrochemicals results in increase in nitrates in water, accumulation of heavy metals, cadmium, and phosphates. Nutrient enrichment, eutrophication, and deterioration of surface water quality due to transportation of nutrients applied through fertilizers via leaching and/or run off and sediment erosion are other problems.

It is becoming increasing clear that the present agricultural production system has been responsible for degradation of earth's environment. For sustainable development, a clear priority has to be given to productive agriculture, essential to efficiently manage the agricultural inputs for sustaining high crop productivity on a long term basis with minimum damage to ecological and socioeconomic environment. To reduce cost of agrochemicals and harm rendered by them biofertilizers are used. These systems minimise the use purchased inputs by substitution of farm-generated inputs like green manures or exploit biological systems such as symbiotic N_2-fixation to increase soil fertility and potentially through management of mycorrhizal fungi (EM and VAM) to improve the efficiency of P use.

Variability in whether a plant will or will not permit mycorrhizal formation occurs naturally, and also as a result of laboratory manipulations with both endo- and ectomycorrhizal fungi. As discussed although only a small percentage of plants are non-mycorrhizal, the actual number is certainly in the tens of thousands of species (over 90% of known plants). With ectomycorrhizal fungi, some are generalists whereas others are highly specific and form functional mycorrhizae with only certain host species. The endomycorrhizal fungi are unable to colonise certain hosts although the majority of plants are compatible. In the field, preferences between endomycorrhizal fungi are seen for a host and differences in functional efficiency are measured for various host fungal combinations. Virtually nothing is known about the specific genes that regulate mycorrhizal symbiosis. However, it is possible to use the variation that exists, between different hosts and fungal genotype to gain clues relevant understanding the genetic regulation of the symbiosis.

The association creates an intimate link between plant roots and the soil, and plays a pivotal role in the acquisition of mineral nutrients. The ability of the association to enhance plant growth and development has stimulated research, and the recent application of biochemical, genetic, and molecular approaches is providing new insight into the symbiosis. Vesicular-arbuscular mycorrhizal (VAM) symbiosis is the most common and has been estimated to occur in more than 80% of flowering plants. These associations are extremely ancient and VAM fungi have even been identified in fossils of early Devonian land plants. It could be that they assisted plants in their colonization of land.

Improved growth, health, and stress resistance of mycorrhizal plants are widespread, particularly for plants growing in nutrients limiting conditions. Increased resistance to plant pathogens has been noted, this may be mediated by factors other than mineral nutrition.

In contrast to the 'gene for gene' specificity observed in many plant-pathogen interactions, the VAM symbiosis is essentially non-specific, and a single species of VAM fungus has the capacity to colonize many plant species. Although extensive colonization of the root cortex occurs, host defence responses, if elicited are weak and transient. Thus the association is an example of extreme compatibility, where mechanisms have evolved to enable coordinate cellular development of both organisms to achieve a functional metabolic state.

They help plants acquire mineral nutrients from the soil, especially immobile elements such as P, Zn, Cu, but also mobile ions such as S, Ca, K, Fe, Mg, Mn, Cl, Br, and N. They also help in increasing the extent of soil particle aggregation.

Mycorrhizal association is known to increase drought resistance of plants by means of several mechanisms. Mycorrhizal fungi improved host nutrition particularly, which increases P delivery to roots and plants hydraulic conductivity in comparison to P deficient non-mycorrhizal plants. It may however be stated that interest in this phenomenon has escalated dramatically in recent years, partly because of what we have learnt about the benefits of mycorrhizae and partly because of economic and geopolitical events. Also because most of the economically important plants have been found to be mycorrhizal, the subject is currently attracting much attention in agricultural, horticultural, and forestry research. Furthermore due to their unique ability to increase the uptake of phosphorus by plants, mycorrhizal fungi have the potential for utilization as a substitute for phosphatic fertilizers.

The editors express their appreciation to all authors for the quality of the work and cooperation during the preparation of the text. The articles are original and some have been written for the first time. Since these chapters have been written by independent authors there is the possibility of a slight overlap or repetition of certain statements, but this is difficult to avoid assignment like this.

It is our hope that this book will be useful to all students and researchers in microbial biotechnology, microbial ecology, and applied mycology.

The editors are thankful to Dr. Atimanav Gaur (South Africa) and Miss Mandeep Kaur for collecting and scanning literature for the book.

A word of appreciation is also due to M/s. Neelam Graphics for active cooperation in preparing the neat electronic copies of the text of the chapters.

K.G. Mukerji
B.P. Chamola
30-5-1999 **J. Singh**

CONTENTS

EVOLUTION OF MYCORRHIZA

Shalini Raina, B.P. Chamola, and K.G.Mukerji

Applied Mycology Laboratory
Department of Botany
University of Delhi
Delhi-110007, INDIA

1.   INTRODUCTION

Symbiosis means living together. This association has evolved and has been inherited over the generation as it provides selective advantage to the super-organism. The symbionts in general are nutritional symbionts show a kind of mutualistic association in which the plant provides the carbohydrate skeletons in turn the symbiont enables the plant to have access to the sparsely available nutrients and also stress resistance. The most obvious example of symbiosis in the plant kingdom is that of fungus-green plant (Frank, 1885), bacteria-green plant (Beijerinck, 1888) and actionomycetes - green plants (Callaham et al., 1978), their abundance being in order of their mention.

The fungal - plant symbiosis is the most common type of symbiosis encountered in the plant kingdom. This is observed from the lowest level of organization i.e. from alge to the angiosperms. Furthermore, this type of association is also observed in the extinct genera of Pteridophytes and Gymnosperms. The fungus-algae symbiosis helps in synthesis of the lichenoid plants which are better suited than either of the two components-the mycobiont or the phycobiont in colonizing virgin land scape.

Similarly, it is supported by fossil evidence that the fungal-primitive land plants association existed and might have played important role in the colonization of the land by the primitive land plants. The Ectomycorrhiza and Endomycorrhiza, the two types of root biotrophic associations known in plant compensate for the poor growth of the roots in the plants and brings about enhanced absorption of the nutrients, water and minerals from the soil, thereby providing selective advantage to the plant and obviously better survivability (survival ability). And today, the stage has advanced so much that there are certain plants (e.g. orchids) which have an obligate requirement of root association with mycorrhizal fungi. Obviously this type of complex association did not exist in the early geologic eras (Devonian and Silurian) when the virgin land was first encroached upon by the primitive land plant (Mukerji and Sharma, 1996).

It is further believed that this type of association might have originated as a result of accidental association of roots with fungi which later became important for vigorous growth and the case today we encounter in certain genera are where this has become obligatory for establishment.

In the present review we have tried to synthesize the various available literature till date

Mycorrhizal Biology, edited by Mukerji et al.
Kluwer Academic/Plenum Publishers, 2000

on the origin and evolution of mycorrhiza. The observations have led to the stage where very interesting inferences can be made. The most outstanding of these is that the evolution and diversification of land plants has been in some way helped by the symbiotic or near symbiotic association with microbes in the early geological history.

The diversification of plant was also related to the degree of complexity that arose within the plant-symbiont relationship and in certain cases even got genetically fixed. Further, the present day wide spread distribution of flora under diverse conditions owe to the same association which have played a major role in the colonization of the land by the ancestors of the present day land plants.

2.   THE CONCEPT OF SYMBIOSIS

In 1887, de Bary coined the term symbiosis, which he defined in the phrase '...*des Zusammelebens ungleichnamiger Organismen...*,' the living together of differently named organisms. He considered two classes of symbiosis - parasitic (in which one organism benefits to the detriment of other members of the association) and mutualistic (in which all the organisms involved are believed to derive benefits). Too often now, especially to botanists, symbiosis means mutualistic symbiosis only. However, Stainer *et al.* (1977) pointed out whether a particular symbiotic association is parasitic or mutualistic can only be evaluated by comparing the fitness of the two members when living independently and also when living in association. Furthermore, the nature of a particular symbiosis can shift under changing environmental conditions, so that a relationship that starts out as mutualistic may become parasitic or *vice versa*.

2.1.   The Origin and Evolution of Modern Symbiosis

Symbiosis can be considered to have evolutionary potential in that it enables an organism to 'acquire' novel characteristic in the form of properties of its partner. Symbiosis was of crucial importance in the evolution of eukaryotes, with particular reference to mitochondria and plastids.

The most significant events in the terrestrialization of plants was the evolution of biotrophic root inhabiting symbioses. The invasion of the land by the ancestors of the vascular plants clearly seems to have been facilitated by the origin of symbiotic associations between these plants and fungi, bacteria and other microorganisms. The progression of life on land is linked to the ability of limited quantities of photosynthetically fixed carbon and nitrogen. Competition for this resource necessitates more or less intimate relationships between heterotrophic and autotrophic organisms, and has been the central selective force in the coevolutionary development of life. Of the heterotrophs that associate with fungus directly few have a longer history or have formed the intimate relationships than fungi. Foremost among these are-fungus association and almost universal are biotrophic symbioses in which fungi (mycobionts) partially function as absorbing organs for plants to benefit from what has been called a short-circuiting of carbon direct from photosynthesis (Harley, 1975).

Although the biology of mycotrophic symbioses is largely uninvestigated, the biological significance of symbiosis generally can be realised from the studies of unicellular partnerships (Taylor, 1979). Symbiosis provides a means of exchanging genetic information between the partners, despite the restriction imposed by sex. This thereby, helps the partners to exploit the available gene pool and achieve evolutionary success.

Modern plants manifest several kinds of mycotrophic symbiosis that arose at different times and in different eras, between different members of both groups of organisms. Each new partnership being presumably better equipped to cope with harsher environment, thus representing an evolutionary 'leap'.

There are three events involving mycotrophic symbioses which have been landmark events in the progression of plant life on earth (Pirozynski, 1981), (1) the evolution of endotrophic symbiosis, (2) the evolution of ectotrophic symbiosis and (3) the evolution of independence from mycotrophic symbiosis.

2.2.   The Evolution of Endotrophic Symbiosis

The evolution of endotrophic symbiosis appears to be the first, and the most important event in the history of land plants i.e. the event of their origin. The hypothesis of symbiotic origin of land plants, was reformulated by and postulated as a partnership between two basically aquatic protists, a green alga and a "phycomycetous" fungus (Pirozynski and Malloch, 1975).

If marginal habitats are successfully exploited by symbiotic systems, it is not unreasonable to assume that the first land plants could have benefited from adaptations lacking in any of the basically aquatic progenitors individually, but which could result from an intimate association of different organisms.

A fungus belonging to Endogonaceae was discovered earlier, during the classical study of the Rhynie chert flora which was found within the tissues of *"Rhynia"* and *"Asteroxylon"* and a symbiotic relationship was suspected. The inferred existence of mycotrophism in representatives of two lines of early Devonian land plants suggests that the association may have arisen earlier (Pirozynski, 1981).

There are subsequent fossil records which are punctuated by sporadic reports of the occurrence of apparently similar fungi in the subterranean organs of Paleozoic, Mesozoic and Coenozoic plants. Today, the vast majority of living plants harbour identical intracellular fungi in their roots or rhizomes and it is becoming increasingly obvious that their mutual dependence render independent existence of each partners unlikely or impossible, especially in ancient or primitive plants.

Within pteridophytes, the case of symbiosis has been cited but it decreases progressively from obligate in Psilotaceae, Tmesipteridaceae and Lycopodiaceae, to constant in Eusporangiate ferns, to faculative in Leptosporangiate ferns (Boullard, 1979).

The mycobiont plays the important role of an absorptive organ or its extension, so it is possible that this function and its localization would have influenced the course of evolution of the phytobiont. It is assumed that the mycobiont could have exerted pressure, both physical and selective, for the evolution of a polarized, negatively geotopic growth habitat, and also of conducting and strengthening elements in the phytobiont (Raven, 1977).

Vascularization of prothalli of some modern lycopods and ferns is initiated in response to the external chemical environment (Boullard, 1979) which under natural conditions is exploited by the mycobiont. This environment contains boron (Lewis, 1980), to be implicated in the biosynthesis of lignin. If the evolution of vascular plants is linked with mycotrophism and if involvement of boron was indeed a prerequisite for the evolution of xylem and the concomitant lignification, the assimilation of boron should in some way be mediated by the mycobiont.

The conspicuous features of the endomycobiont are :

1.   Its morphological stability since early Paleozoic time. These morphologically conspecific strains are capable of colonization of plants as divergent as pteridophytes, gymnosperms and angiosperms. It would thus appear that as a subordinate endosymbiont the fungus was not selected upon for the morphological or physiological diversity but it is phytobiont that responded to selective pressures.

2.   The whatsoever little morphological innovation there has been, was not towards elaboration of adaptation promoting active disposal, but only to enhance the survival *in situ*.

3.   The fungus is associated with perhaps 80% of living terrestrial plants. Ubiquitous "host" range and Pangean pattern of distribution are unusual fare for an obligate endophyte, unless the symbiont coevolved with plants throughout their history.

As far as known characteristics of the endosymbionts indicate that all endotrophic plants from *Rhynia* and *Asteroxylon* through *Lepidodendron* and *Coenopteris* to *Psilotum*, *Sequoia* and *Taraxacum*, appears to be merely steps in a monophyletic progression.

2.3.   The Evolution of Ectotrophic Symbiosis

The evolution of ectotrophic symbiosis predictably had very different consequences because the peculiarities of ectomycobionts contrast sharply with those of endomycobionts :

1.   There are atleast 5000 species of ectomycorrhizal fungi which are morphologically and physiologically distinct suggesting that the selective pressure has been on the ectosymbiont which, in this case, is the mycobiont forming a sheath on the outside of roots.

2.   However, ectotrophism has evolved only in some 2000 species of plants, chiefly in Pinaceae, Fagaceae, Betulaceae, Salicaceae and within Dipterocarpaceae, Caesalpiniaceae and Myrtaceae.

3.   The evolution of ectomycorrhizal symbiosis may be a relatively recent event, because the ectotrophs are selective in more extreme environments. The fossil record of ectotrophs extends back to the second half of the mesozoic and the geographical disjunctions of ectrophic communities reflect tectonic events of the same interval (Pirozynski, 1981; Taylor, 1995).

In ectotrophic forests the diversity at the root-soil interface appears to be provided by the mycobionts. The dominant trees can select as many as 2000 different species of mycobionts according to topography, soil type or growth phase. Thus, individual trees behave as physiologically different organisms, as a result they do not seem to compete directly with each other and this may allow them to grow in close proximity. The difference in the composition of forests have a mycotrophic basis and the peculiarities of each community may have evolved in consequences.

2.4.   The Evolution of Independence from Mycotrophic Symbiosis

The monotonous ectotrophic systems appears to be on the rise but the endotrophism is beginning to show its age. Progressive weakening of symbiosis, leading to facultative mycotrophism or to the nonmycorrhizal condition may be correlated with unusual metabolic pathways in the phytobiont and accompanied by reduction in size and lignification, by shortening of the life cycle, by evolution of vegetative propagation and a variety of storage organs, and by elaboration of more extensive root systems equipped with root hairs (Baylis, 1975).

The trend towards the evolution of phytobiont self sufficiency, with all its correlated characteristic, is shown in flowering and nonflowering plants alike. In pteridophytes it is seen as the progressive reduction in extent and duration of endotrophic symbiosis in what are taken to be more advanced groups, ending in nonmycorrhizal conditions in some Pteridaceae, as well as in the aquatic families of pteridophytes and Isoetaceae. Interestingly, endotrophic symbiosis in *Equisetum* is rare and when present atypical, though it was well represented in the plants carboniferous arborescent progenitors. Among flowering plants only a few families or groups of thier members are normally nonmycorrhizal. Like ectotrophs, the nonmycorrhizal groups of plants have expanded into more marginal environment since mid-Mesozoic (Pirozynski, 1981). However, we can not ignore the importance of plant-fungus associations. Their respective mode of life make close interaction mandatory.

3.   ORIGIN OF BIOTROPHY

The two concepts in fungal nutrition has been rendered from the recent researches in fungal physiology, physiological plant pathology and mutualisic symbiosis and these are symbiosis and obligate parasitism. The origin of biotrophy, the condition in which fungi derive nutrients from their hosts with only minimal tissue damage and which is probably, evolutionarily, the most advanced form of fungal nutrition. The dependence of mutualistic symbiosis on biotrophy with particular reference to the assimilation of carbon, nutritional interrelationship in various types of mycorrhizae are examined in the light of fungal behaviour. As the fungi are a heterogenous group of organisms of diverse phylogeny, where generalization involving taxonomically unrelated fungi are possible, this can be interpreted as the result of parallel or convergent evolutionary trends.

Fungi are chemoheterotrophic as they derive their carbon compounds from either nonliving organic material i.e., as saprophytes, or from living tissues i.e., as parasites. Their relationship with their hosts is that of a common life, a symbiosis. Thus the bilateral exchange of nutrient is the essential feature of mutualism (Quispel, 1943). As the fungus cannot supply major organic nutrients to its partner since it cannot synthesize them from inorganic precursors, it is a further feature of most mutualistic symbiosis involving fungi that the fungus must be in contact with a supply of inorganic nutrients which are passed to its partner symbiont. Such a supply is from the soil for mycorrhizal fungi and in surface water for lichen fungi. In parasitic symbiosis, however, the movement of major nutrients is essentially one way, from host to parasite. Simultaneously with this bi-or uni-lateral passage of major nutrients, there is probably bilateral exchange of minor organic metabolites such as vitamins, hormones and as yet unidentified grwoth factors. These movements may be quantitatively minor but their effects are great and undoubtedly control the nature of the symbiosis itself.

Fungi have been divided into many groups based on their nutritive adaptations by de Bary (1887), Brian (1967) and Thrower (1966). The classification of de Bary and Brian are based essentially on ecological behaviour and culturability whereas those of Thrower, although retaining culturability, disregard ecology in favour of nutritional criteria (Table 1).

It should be noted that just as environmental conditions can determine whether an association is parasitic or mutualistic, they may also determine whether the nutrition of the fungal partner is biotrophic or necrotrophic. However, it is possible that environmentally induced shifts from biotrophy to necrotrophy convert mutualistic symbioses into parasitic ones. The necrotrophy and biotrophy are only applicable to symbiotic existence and saprotrophy to the saprophytic habit and that facultative saprotrophy is merely a form of facultative necrotrophy or biotrophy. However, it is clear that obligate and facultative refers to ecological behaviours only. With this provision in mind, following five categories of fungal behaviour can be recognized :

(i)   Obligate saprotrophs
(ii)   Facultative necrotrophs
(iii)   Obligate necrotrophs
(iv)   Facultative biotrophs
(v)   Obligate biotrophs

Brian (1967) characterized six features of host-parasitic interaction to delimit the higher plants and unculturable fungi. These are - intracellular penetration; minimal tissue damage; highly developed physiological specialization and relatively restricted host range; morphological disturbances in the host plant and nuclear disturbances. If biotrophic is substituted for 'unculturable' Brains's list will become even more valid since fewer reservations are then required and some of these features are also characteristic of interactions in lichens and some mycorrhizas. Although direct evolution of biotrophy from saprotrophy is possible (Harley 1948; Lewis 1974), which also permits facultative biotrophy and evolution from necrotrophy may also have occurred. The significant clues to the origin of biotrophy showed that specialized symbiosis (as found with biotrophic fungi) is evolutionarily more advanced than unspecialized symbiosis (Garrett, 1970).

Patterns of translocation in healthy plants can be influenced by exogenous hormones and there is circumstantial evidence that, in biotrophic infections, hormones produced by fungi alter patterns of translocation to the advantage of the fungus (Brian, 1967; Smith *et al.,* 1969). It is thus possible that, in these host-parasite combinations, the fungi have a relatively inefficient capacity to alter patterns of transolcation. But by producing an overall increased retention of metabolites in leaves or additonal import and application of kinetin enables this inefficiency to be overcome, resulting in enhanced development of these biotrophs.

Table 1. Classification of fungi based on their nutritive adaptations

A. de Bary (1887)

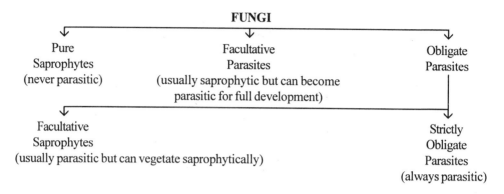

B. BRIAN (1967)

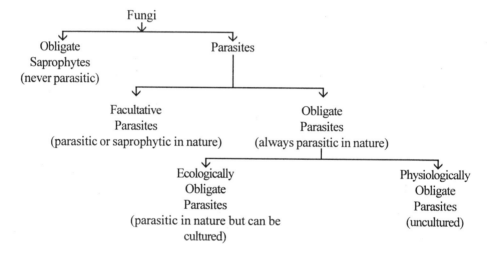

C. THROWER (1966)

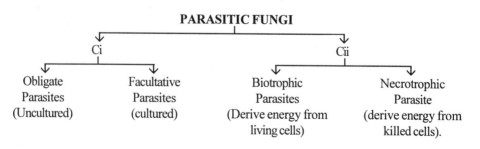

Under low light intensity or darkness, the normally biotrophic bacteria of the *Rhizobium*-legume symbiosis become necrotrophic resulting in a degeneration of existing nodules (Thornton, 1930). The effect of low light intensity and other adverse conditions on established symbioses involving fungi have been less studied, but several, normally biotrophic, lichen fungi do show a tendency towards necrotrophy under such conditions (Ben-Shaul *et al.*, 1969; Galun *et al.*, 1970;

James, 1970; Tschermak, 1943). With regard to development of symbiotic unions in both ectotrophic and vesicular arbuscular mycorrhizae the level of colonisation has commonly been found to be greater under high light intensity. (Harely, 1969; Slankis, 1971).

The difference between biotrophic and nectrotrophic pathogens is in the amount and kind of hydrolytic enzymes secreted by them into the host tissue. The latter produce copious amounts of enzymes of varied specificity resulting in a degradation of tissue, characteristic of the pathogen (Wood, 1967), wheras minimal tissue damage is the hallmark of biotrophy. However, some biotrophic fungi of ectomycorrhizas can utilize some complex polysaccharides. Palmer and Hacskaylo (1970) and Lundeberg (1970) showed that some strains of some mycorrhizal fungal species produce pectinases, cellulases, hemicellulases or polyphenol oxidases in culture. Although such a distinction may be general, but it is also possible that some mycorrhizal fungi, which are capable of producing such enzymes in culture, only do so to a limited and controlled extent in the symbiotic state. Melin (1953) and Norkrans (1950) suggested that such mycorrhizal fungi only produce cellulase when soluble sugars in the root are depleted. Inhibition of enzyme synthesis, and therefore activity, by high sugar levels is the classic example of the phenomenon of catabolite repression, which is well documented for bacteria, yeasts and some filamentous fungi (Anderson and Wood, 1969; Magasanik, 1961; Nisizawa et al., 1972; Paigen and Williams, 1970).

From the evidence of these diverse aspects of host-parasite physiology changes in patterns of translocation, fungal production of hormones, environmental control of fungal development, endogenous sugar levels in infected host tissues and enzyme production in culture. Following events in the evolution of biotrophy from necrotrophy can be envisaged.

(i)   Catabolite repression of degradative enzymes (e.g. the hydrolytic pectinases, hemicellulases, cellulases, proteinase are responsible for cell wall breakdown and polyphenol oxidases are responsible for degradation of lignins) limits there activity. This is only possible in "high sugar" tissues.

(ii)   Concurrently with (1), localized production of hormones by infecting fungi and transmission of the appropriate 'message' from the relatively undamaged tissue of the infected area to other plant parts permit infected tissues to become importers of photosynthetic products.

(iii)   Continued provision of simple metabolites after the changes in pattern of translocation maintains catabolite repression while serving the needs of the new biotrophic symbionts.

(iv)   Genetic loss of the capacity to produce the enzymes responsible for cell collapse and death occurs in some groups. Here, catabolite repression is no longer relevant, but the fungi will not only have become ecologically obligate symbionts owing to reduced competitive saprophytic ability but also have become dependent on living tissues rich in metabolites.

The catabolite repression remains an important control in the symbiosis particularly in those biotrophic associations which have not yet reached stage iv, so that low light intensity induces necrotrophy via reduced photosynthesis, reduced sugar levels and increased fungal synthesis of degradative enzymes.

Biotrophy can thus be seen as a sacrifice of versatility of carbon source for continued provision of the major sugars from the translocation stream of the host. It may be pertinent, therefore, to seek clues to the specificity of many fungal infections in the capacity of the fungi to exploit successfully the hormonal balance of their hosts.

The salient features of biotrophy - a lack of massive destruction of host cells or tisues, an altered physiology of the host so that a continous supply of simple soluble substances to infected areas occurs, enhanced longevity and photosynthetic capacity of cells or tisues under the influence of infection and a tendency to greater host specificity - are the essential prerequisites upon which mutualistic symbiosis involving fungi are based. Lichens, Ectomycorrhiza and Vesicular-arbuscular mycorrhizas have the further feature that the fungi are in contact with a source from which they are potentially capable of directly abstracting nutrients that can be passed to their autographic partners. However, tight nutrient cycling is a characteristic of many mutualistic symbiosis (Barlett and Lewis, 1972; Lewis and Smith, 1971). In Ectomycorrhiza, enhanced lateral exploitation of the litter layer for nutrients instead of more vertical exploitation of the nutrient-poor, lower soil horizons is possibly favoured by increased diageotrophy brought about by fungal destruction of the root-cap, the site of perception of the gravitational stimulus (Juniper *et al.,* 1966). From this, it is concluded that whether the assocaition is mutualistic or parasitic depends on the prevailing environmental conditions and by controlling weather and nutrition of the fungus it can be made to adopt biotrophic or necrotrophic mode of nutrition respectively.

4. EVOLUTION OF THE SYMBIOTIC SYSTEMS

4.1. Mycorrhiza

The word mycorrhiza comes from Greek words *mike* and *rrhiza* meaning fungus and root. It was first coined by Frank (1885). A mycorrhiza is a symbiotic, non-pathogenic, permanent association between a plant root and specialized fungus both in natural environment and in cultivation. Mycorrhizal fungi may be involved in: improved uptake of macro and micronutrients, increased tolerance to stresses (by affecting water relations and pathogen resistance) and beneficial alternations of plant growth regulators (PGRs).

Although it was Frank who gave the name "Mycorrhiza" to permanent associations of roots with hyphal fungi, there are earlier reports of occurrence of such associations. Reissek (1847) described hyphae in the cells of various angiosperms, especially in the Orchidaceae. In 1881, Kamienski published an account on *Monotropa*, describing its assocaition with fungal hyphae and showing that a complete fungal layer was formed around the roots. He was the first person to point out that whatever was absorbed from the soil must pass through the roots.

The three basic functional components of the mycorrhizal symbiosis are : (i) fungal mycelium that explores large volumes of soil and helps in retrieving mineral nutrients; (ii) the fungus-plant interface where the nutrient transfer occurs; and (iii) plant tissues which produce and store carbohydrate.

4.1.1. Classification of Mycorrhizas

Frank (1885) distinguished two main types of mycorrhiza : (i) Ectotrophic - which has well defined external fungal sheath around the root and (ii) Endotrophic - which has no fungal sheath around the root but has intercellular and intracellular penetration of the host by the fungus. However, these terms were later rejected and now named as ectomycorrhiza and Endomycorrhiza respectively.

There have been repeated classifications of mycorrhizas made in recent times by several workers (Bhandari and Mukerji, 1993; Harley and Smith 1983; Lewis 1975, 1976; Read, 1982). The types of mycorrhiza were earlier divided on the basis of fungal partners into those formed with aseptate, zygomycetous endophytes and those formed with septate endophytes with more affinity to ascomycetes or basidiomycetes. The autobionts in mycorrhiza are so many, and taxonomically so diverse that primary classification cannot be based on that.

The classification that is most commonly followed is that proposed by Harley and Smith, 1985; who have divided mycorrhiza into following 7 categories on the basis of their morphology, morphogenate and physiological features (i) Ectomycorrhiza, (ii) Endomycorrhiza (Vesicular-arbuscular mycorrhiza), (iii) Ectendomycorrhiza, (iv) Arbutoid mycorrhiza, (v) Monotropoid mycorrhiza, (vi) Ericoid mycorrhiza, (vii) Orchidoid mycorrhiza (Fig.1).

4.1.2. Vesicular - Arbuscular Mycorrhiza

The vesicular - arbuscular mycorrhiza are geographically ubiquitous in distribution and occurs over a broad ecological range. The VAM forms mutualistic association with almost all natural

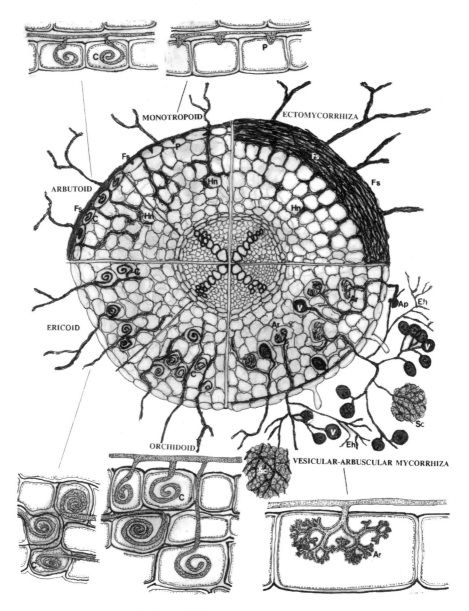

Fig. 1 . Different types of Mycorrhizae.

and man made vegetation. Ninety per cent (90%) of the plants ranging from thallophytes to angiosperms has this association. This association is not restricted to the roots of plants only but is also found in all those organs of higher plants, which are concerned with the absorption of substances from the soil (Srivastava *et al.*, 1996).

The VAM fungi have a long evolutionary history and have been reported from early Devonian era (Pirozynski and Dalphe, 1989). These are members of Zygomycotina which are obligate biotrophs thus cannot be cultured. Therefore, not much is known about their life cycle. At present, the mode of spore formation and sub-cellular structure of spores are used for their classification (Mukerji, 1996).

4.1.3. Fossil Vesicular Arbuscular Mycorrhiza and Primitive Land Plants

The widespread geographic and biologic distribution of extant arbuscular mycorrhizae suggested to some the antiquity of the symbiosis (Trappe, 1987) while others hypothesized that this kind of mutualism was pivotal in the origin of the terrestrial flora. More recently, the morphological similarity of fossil spores with the chlamydospores of living genus *Glomus* Tulasne & Tulasne (Pirozynski and Dalpe, 1989) and sequence divergence data based on a molecular clock model (Simon *et al.*, 1993) provide additional evidence for this hypothesis.

The adaptations necessary to transform a characean green algae into a land plant have been discussed by Raven (1977, 1984, 1985, 1986). The observation that early vascular land plants had Endomycorrhizas (Harley, 1969; Nicolson, 1975; Pirozynski, 1976) led Pirozynski and Malloch (1975) to conclude that their very origin was intimately dependent on the mycorrhizal habit (Malloch *et al.*, 1980; Pirozynski, 1981; Raven, 1977; Raven *et al.*, 1978).

Although several authors have speculated on the occurrence of endomycorrhizae associated with the roots and underground organs of fossil plants (Malloch *et al.*, 1980; Nicolson, 1981), the only evidence to date includes nonseptate mycelium, coiled hyphae, thin-walled, slightly elongated spores interpreted as vesicles and chlamydospores (Kidston and Lang, 1921; Sharma *et al.*, 1993; Wanger and Taylor, 1984). Until recently (Remy *et al.*, 1994) the oldest structure that convincingly can be considered an arbuscule, the *sine quine non* of an endomycorrhizal symbiosis comes from the Triassic of Antarctica (Stubblefield *et al.*, 1987).

Although there are several reports of Paleozoic arbuscules (Halket, 1930; Osborn, 1909), all have been demonstrated to be either artifacts or condensed cytoplasmic cell contents (Cridland, 1962). One of the most often cited examples of fossil endomycorrhizae is *Palaeomyces*, a fungi from the Rhynie chert (Kidston and Lang, 1921). They illustrated that within the axes of the early land plants, there were hyphae, variety of vesicles and chlamydospores which resembles those formed by extant endogenous fungi. However, most definitive morphological indication of a VAM association, the arbuscule has not been demonstrated from the fossil records. Earlier reports of fossil arbuscule actually refer to nonhyphal aggregations of cytoplasm or ergastic substances.

Structurally preserved plant remains from early to middle Triassic of Antarctica yielded a wealth of fungal remains. As in most extant VA mycorrhizal fungi are most abundant in the central cortex and do not occupy the central cylinder. The distribution of the fungi in the fossil roots is consistent with the pattern typical of extant mycorrhizal roots (Harely and Smith, 1983; Kessler, 1966; Kinden and Brown, 1975a, b).

Several types of fungal remains resembling modern day VAM fungi including those colonizing cycads (Gorbunova, 1958; Stanczak-Boratynska, 1954) occur in the colonized fossil roots. The nonseptate hyphae fall within the size range of their modern counterparts and sometimes show S and Y junctions like those described in *Glomus* (Abbott and Robson, 1979). Multiple branching of the hyphae of VA mycorrhiza has also been found in modern cycads (Gorbunova, 1958). The bending, looping or coiling of hyphae in cortical cells that occurs to various degrees in extant endogenous fungi is also observed in the Triassic roots (Andrew and Lenz, 1943; Brown and King, 1982; Cox and Sanders, 1974; Kinden and Brown, 1975 a, b; Taber and Trappe, 1982; Weiss,

1904). Furthermore, the penetration of cortical cell wall by hyphae in the fossil roots resembles that exhibited by VAM fungi in *Acer saccharum* (Kesseler, 1966) in which there is no constriction in the penetrating hypha.

The Triassic fungus from Antarctica is more completely known than any other reputed fossil VA mycorrhizal fungus. An extensively branched structure has been found in the *Antracticycas*. This structure is organized like the arbuscules of modern day VA mycorrhizae found in present day cycads (Asai, 1934; Gorbunova, 1958; Mada, 1954; Stanczak-Boratynska, 1954; Von Tubeuf, 1896). They are found closely associated with fungus that shows a pattern of colonization, vesicles and chlamydospore that are identical with those of extant VA mycorrhize. Thus, Stubblefield *et al.* (1987) believe that the Triassic roots indeed had vesicular - arbuscular mycorrhizae.

The roots of *Antracticycas* (Stubblefield *et al.,* 1987) which are generally radial and diarch with broad parenchymatous cortex separated from a central vascular zone by a distinct band of cells filled with a dark, probably ergastic substance. An extensive colonization may develop without evident damage to cortex i.e. both intra and intercellular hyphae occur. Hyphae are nonseptate sparsely within the outer cortex. They are also characterized by bends or loops and by short, truncated projections. Hyphae may branch without the formation of septa as a result Y-shaped or swollen S-shaped junctions occur. Hyphae further penetrate the walls of adjacent cells but are not necessarily constricted at the point of penetration.

Hypha dichotomizes repeatedly, forming a much branched, three-dimensional structure that nearly fills the host cell. Septate hyphae also occur in central cortex. Both dichotomous and trichotomous branching occur. The septa appear as dark encircling rings on outer hyphal surfaces. Several types of inflated structures occur within roots. Many are subtended by hyphae of varying diameter, and are ellipsoidal to spherical and thick-walled or thinwalled. A variety of other spherical structures also may be present. Some appear to be sporangia filled with sporangiospores, while others include hyphae or undefined globose structures (Stubblefield *et al.*, 1987).

The occurrence of chlamydospores and vesicles in conjunction with hyphae in plant tissue suggests the presence of fungi similar to the modern Endogonaceae. However, these structures alone are not conclusive proof for Paleozoic mycorrhizae since extant endogonaceous fungi are not all known to be mycorrhizal (Trappe and Schenck, 1982). Perhaps the most reliable structural indication of this types of physiological relationship is the arbuscule since it is through this organ that metabolic exchange occurs. Among modern VAM fungi, the arbuscule is a highly branched structure, usually formed within the cortex of roots (Kinden and Brown 1975 a, b; Mukerji and Sharma, 1996). Arbuscules are transistory, and eventually break down to form a mass of material in the centre of the cortical cells. Reports of fossil arbuscules are few and even more problematic. Halket (1930), Obson (1909), reported arbuscules in *Amyelon radians.*

Statistically valid data on the occurrence of chlamydospore in particular genera or group of vascular plants are lacking with respect to fossils and even qualitative reports are often contradictory. For example, Pirozynski (1981) commented that endotrophic symbiosis are well represented in the arborescent Carboniferous ancestors of the modern Equisetales, while some workers suggest that the lack of mycorrhizae in these plants may have been responsible for their evolutionary demise. Similarly, chlamydospores are also found in the subterranean portions of certain Pennsylvanian coal ball plants such as *Psaronius* and *Stigmaria*. Furthermore, they are present in the aerial axes of more primitive plants such as *Psilophyton* which do not yet show histological differentiation between roots and stems (Stubblefield and Banks, 1983). Endogonaceous fungi, once believed to be restricted to roots, are now known from the xylem of *Tradescantia virginiana* and the rhizomes, scale-like leaves and xylem of ginger (Mago *et al.,* 1993; Taber and Trappe, 1982). However, as their role in these portions of the plant is unclear, it is not possible to evaluate their presence in aerial parts of fossil plants. While the occurrence of chlamydospores in Paleozoic stems clearly does not preclude symbiotic relationship, a consistent association with the underground parts of specific plants would be expected if mycorrhizae have played a major role in evolution.

The fungi in *Psilophyton dawnsonii* also show considerable similarity to chlamydospores described from the roots, stems and leaves of several premineralized Pennsylvanian plants (Wanger and Taylor, 1982 a,b). These chlamydospores are predominantly spherical and exhibit two-or three-layered wall. They are often attached to subtending hyphae and there is variability in hyphal attachment. Among extant fungi, the organism in *Psilophyton* most closely resemble members of Oomycetes and the endogonaceous Zygomycetes. The presence of multiple wall layers is often not pronounced suggesting the occurrence of splitting and shrinking. The fungal bodies in *Psilophyton* may also be chlamydospores related to the Endogonaceae, possibly of *Glomus* (Stubblefield and Banks, 1983). However, despite of morphological similarities, there is little evidence to suggest a mycorrhizal relationship. First of all, unlike the modern Endogonaceae which are restricted to subterranean organs, the fungi in *Psilophyton* are found within aerial axes. Secondly, since the inner cortex is not preserved there is no evidence that these fungi were intracellular. Finally, the lack of inner cortical tissue raise the possibility that the fungus was actually saprophyte responsible for the decay of the plant. However, the overwhelming majority of axes lack these tissues, fungi are present only in a small percentage of them. Furthermore, the absence of hyphae suggests that perhaps thin-walled cortical cells are missing not because they were destroyed by fungal activity, but simply because they were delicate in contrast to the well-preserved thick-walled fungal bodies, tracheids and cell of the outer cortex (Stubblefield and Banks, 1983).

The fossil fungus consisting of hyphae, chlamydospores and arbuscules is also reported from the aerial stems and rhizomes of *Aglaophyton major* Edwards (Edwards, 1986). *Aglaophyton major* remains an enigmatic land plant of which systematic affinities are poorly developed. Historically, it has been regarded as a vascular plant but recent studies suggest its closer affinity with a bryophyte level of evolution. Endomycorrhizae have been reported in bryophytes (Gourret and Strullu, 1979; Mago *et al.,* 1992) and pteridophytes (Boullard, 1979; Mandeep, 1995).

The extant arbuscular mycorrhizae shows considerable hyphal polymorphism (Friese and Allen, 1991). Powell (1976) reported that the hyphae from spores of *Glomus* germinate on agar plates buried in the soil, were thick walled, aseptate and rarely branched. As hyphae grew closer to roots there was a pronounced change in hyphal morphology resulting in spetate, highly branched segments. These septate hyphae are believed to represent the initial phases of colonisation since they are generally absent when the endophytes are well established in the host. The same pattern of hyphal morphology is present in *Glomites* where aseptate and septate hyphae are found in close association. Moreover, the fossils also show evidence of small angular projections on some hyphae that have been interpreted as the sites of ephemeral, septate hyphae that have deteriorated (Powell, 1976).

In *Aglaophyton* cell type and organization of the tissue system appears to greatly influence the distribution of the fungus. On the surface of *Aglaophyton major* axes there are numerous, randomly distributed bulges. These bulges represent the sites of rhizoid development. Between the rhizoids, extramatrical hyphae enters the plant. The extraradical mycelium consists of thick-walled, parallel hyphae that form almost cord-like units that contain upto five hyphae. Hyphae infrequently branch and loops are formed in a few. The hyphal wall consist of an outer transparent layer and inner, more opaque zone. Appressorium, an opaque swelling on the surface of the rhizoidal bulge is present immediately above a large, thick-walled intraradical hypha in the hypodermis. Some hyphae give rise to narrower, thin-walled branches that occasionally contain septations. Both the bi-layered wall and presence of septations are the features which are also found in the extraradical hyphal phase of modern endophytes (Bonfante Fasolo, 1984; Holleg and Peterson, 1979).

Axes of *Aglaophyton* consist of a cutinized, uniseriate epidermis overlying a one-to four-cell-thick hypodermis. In the hypodermis hyphae are thick-walled and highly branched, with dichotomies. Extending out from these thick-walled hyphal branches are narrow hyphae, some of these possess H-shaped branches. The hyphae are septate at this stage.

The spores of *Aglaophyton* resembles chlamydospores both morphologically and in the multilayered organization of the wall and also occur in the inner cortical tissues. They vary from nearly elongate to globose and are produced terminally, although sometimes more than one spore is present at the end of a hyphae. The thin-walled intercellular hyphae that produce spores also give rise to branches that ultimately form arbuscule-like structures. These are produced in a specific zone of thin-walled cells that delimits the outer edge of the cortex.

Penetration of the cortical cell by the arbuscule-forming hyphae may take place anywhere along the cell. Inside the host cell, the arbuscule trunk dichotomizes repeatedly to form a dense aggregation that give the structure a bush-like appearance. Typically there is single arbuscule per cell but a few cells appear to contain multiple arbuscules which is similar to the pattern in extant arbuscular mycorrhizae (Kinden and Brown, 1975c). Often the distal ends of the arbuscule branches appear slightly swollen. A few appears as if they bifurcate at the tip. Many of the cortical cells filled with granular material that are partially collapsed, since they are similar to the amorphous mass that results when extant arbuscules deteriorate (Scannerini and Bonfante-Fosolo, 1983). A few of the bifurcating terminal hyphae also show signs of collapse. Although these cortical cell fungi occur associated with cells containing arbuscules, they represent mycoparasites that are now known to have been common in the Rhynie Chert plants (Hass *et al.*, 1994).

The rhizome of *Aglaophyton* is a structurally unique organ possessing stomata on all surfaces and rhizoidal bulges and conducting elements that are similar to the leptoids and hydroids of bryophytes than to the tracheids of vascular plants. As a result of this anatomy the rhizome may not have been a very efficient absorbing structure but this may explain, the extensive development of *Glomites* arbuscules.

Aglaophyton and *Rhynia* are the only plants in the Rhynie chert that possess a zone of palisade-like cells beneath the hypodermis in which arbuscules develop. Many of the cells in this layer contain remnants of arbuscules. These host cells also appear collapsed and disassociated while arbuscule free cell look normal. It means that the arbuscule-containing zone may be meristematic, perhaps throughout the life of the plant, to produce new cells that provide sites for the entry of VAM fungi. This appears to be unlike the situation in modern VAM colonisation (Peterson and Frquahar, 1994) where nonfunctioning arbuscules are absorbed by the host cells. According to Remy *et al.* (1994) the meristematic zone in *Aglaophyton* may represent a tissue directed response that predates the capability of individual cells to absorb arbuscules.

It is generally assumed that members of the true fungi diverged about 1 billion years ago. The oldest fossils believed to represent true fungi are nonseptate hyphae from Cambrian sediments and hyphae and septate spores from the Silurian (Sherwood-Pike and Gray, 1985). In both instances nothing is known about the organization of these fungi or the type of interactions they had with other organisms. At present, all known Silurian plants are preserved as either impressions or compressions making it impossible to determine whether endophytic fungi were present or not. Therefore, the earliest land plant in which mycorrhizae has been identified are from the Rhynie chert.

The discovery of arbuscules in *Aglaophyton* establishes the existence of VAM fungi by early Devonian period but throws no additional light on the history and affinities of these fungi. However, the fossil evidence supports the time of origin of VAM as recently deduced from the sequence data from small subunit rRNA obtained from spores of 12 living species of representative VAM fungi (Simon *et al.,* 1993). According to these molecular data, VAM fungi originated between 462 and 353 million-years ago, well with the approximate 400-million-year time frame of the lower Devonian fossils (Lewis, 1991).

Wilson (1993) has recently suggested that certain types of chemical plant defenses may be the result of endophyte infection and as a result it might have contributed to the coevolutionary relationships between fungi and plants. Since these chemical interrelationships can not be documented from the fossil record, the realization that saprophytic, parasitic and mutualistic fungi are well established by early Devonian time underscores the fact that such interactions accompanied the rise of the terrestrial flora.

The first VA mycorrhizal fungi that was isolated form the roots of poplar was *Rhizophagus populinus*. Peyronel (1923, 1924) recognized VAM fungi as *Endogone* sp. (Link, 1809). The family Endogonaceae is a diverse group of soil borne fungi found worldwide. It is classified in the order Endogonales, class Zygomycetes (Benjamin, 1979).

The classification of the arbuscular mycorrhizal fungi was first presented in modern terms by Gerdemann and Trappe (1974, 1975) as a "temporary taxonomic solution". Seven genera were included in the family Endogonaceae viz: *Acaulospora, Endogone, Gigaspora, Glaziella, Glomus, Modicella* and *Sclerocystis*. The genus *Entrophospora* and *Scutellospora* were added in family endogonaceae (Walker and Sanders, 1986). Subsequently, *Glaziella* and *Modicella* were transferred out of Endogonaecae into Ascomycetes and Mortierellaceae (Trappe and Schenck, 1982) respectively. The systematics of the mycorrhizal fungi has moved at a rapid pace in recent years and the VAM fungi have received particular attention (Morton, 1988, 1990, a, b; Morton and Benny, 1990; Morton and Bentivenga, 1994; Mukerji, 1996; Walker, 1992). Revised classification of Endogonales was given by Morton and Benny (1990) to arrange the groups of fungi according to pattern of common descent, spore ontogeny and mode of spore germination. A new order Glomales was proposed to include all soil fungi and it is further subdivided into two suborders Glominaeae and Gigasporineae. Members of Glomineae form both arbuscules and vesicles while Gigasporineae form only arbuscules (Fig. 1).

It is known that the natural (Monophyletic) groups were defined by shared characters (synapomorphs). Result of such study support the hypothesis that different symbiotic fungal groups (e.g. arbuscular, arbutoid, ectotrophic) are polyphyletic (Harley and Smith, 1983). Development of the arbuscule and the obligate symbiosis united all the endomycorrhizal fungi in a monophyletic group, but no other synapomorphies are available to establish an evolutionary connection with other symbiotic taxon (Morton, 1990 a,b). The phylogenetic studies indicate that *Acaulospora, Entrophospora, Glomus* and *Sclerocystis* evolved from the same common ancestor as : (i) all members share intraradical vesicle development; (ii) members of these groups share many common parallelism; and (iii) a primitive *Acaulospora* is dimorphic with *Glomus* status. However, commonality of the *Gigaspora* and *Scutellospora* lineage is more difficult to establish with a *Glomus*-like hypothetical ancestor. Characters important in spore formation of *Gigaspora* and *Scutellospora*, such as sporogenous cell and a persistent thin unit wall enclosing the structural wall of the spore are unique evolutionary innovations (Morton, 1990a,b).

There is a small amount of divergence among taxa in *Glomus* and *Sclerocystis*. Walker (1987) suggested that *Glomus* and *Sclerocystis* should be combined in the same genus. First, the degree of divergence between sporocarpic *Glomus* and *Sclerocystis* species was low, secondly putative dimorphic *Glomus* sp. provided evidence of an evolutionary linkage between spore aggregation and sporocarpic development. Third, *Sclerocystis* species were positioned as sister taxa in the same monophyletic group with several *Glomus* species. Further it is expected that the classification of this group will progress only with phenetic and ontogenetic studies to complement the cladistic and paleontological data.

Further Walker (1984, 1987) proposed that *Acaulospora* and *Entrophospora* are sister groups. However, both genera are much more closely related to *Glomus*. The importance of intraradical vesicles as an evolutionary character uniting these divergent lineages has gone largely unrecognized, considering that all arbuscular fungi often are lumped together as "Vesicular-arbuscular" or "VAM". Most of the characters found in *Acaulospora* evolved in parallel in *Entrophospora*. This evidence suggests that evolution of both genera are driven by similar selection pressures.

Walker and Sanders, (1986) gave some evidences which indicates that *Gigaspora* is composed of ancestral species existing as contemporaries with descendant *Scutellospora* species.

No autapomorphic characters are expected to be found in *Gigaspora* which define it as a divergent lineage from *Scutellospora*. One of the most important results of this study was the discovery of the extent of phylogenetic isolation between the *Gigaspora* and *Glomus*. Morton and Benny (1990) explained that *Gigaspora* and *Scutellospora* should be placed in a family separate from *Glomus, Acaulospora* and *Entrophospora*.

The evolutionary ties between *Endogone* and other genera of arbuscular mycorrhizal fungi have never been addressed adequately, even when the Glomaceae was proposed as a new family (Pirozynski and Dalpe, 1989). Thus the placement of *Endogone* with any group of arbuscular mycorrhizal fungi is artificial. The zygospores of *Endogone* clearly are not homologous with putative chlamydospores of *Glomus* and *Sclerocystis*. And also the presence of septal perforations in hyphae of some species of *Endogone* and arbuscular species (Gibson *et al.,* 1986) cannot be used to unite the two groups because this character has evolved independently in other zygomycetous fungi.

Thus morphological characters are valuable in comparing extant species with paleontological records and fossils establish a chronology of evolutionary events in a phylogeny (Stuessy, 1987), even though the fossil record is extensive for arbuscular fungi, only *Glomus* and *Sclerocystis* like spores are represented (Pirozynski and Dalpe, 1989). Morefrequent appearance of *Glomus* probably is due, in part, to the ability of some members to form abundant intraradical spores.

Striking similarities have been found between the fossilized specimen and *Glomus intraradices* (Wanger and Taylor, 1982 a,b), particulalry in the aggregation of spores in the roots and the separation of laminae of the structural spore wall. *Glomus intraradices* and other species with similar characters possibly are contemporary ancestral species. Their position in the phylogenetic tree suggests that they and more primitive *Glomus* sp. probably originated as early as the carboniferous period and the morphology of the spores has not changed appreciably since then. The fossil record does not provide a reliable interpretation of the chronology of evolution in *Acaulospora, Entrophospora, Gigaspora* and *Scutellospora*, as preserved spores have not been found. Although differences in life cycles account for their absence, it is more likely that morphological divergence occurred rapidly during times of environmental upheavals when fossil record was less apt to occur. This hypothesis was supported by evidence of extensive divergence of *Acaulospora* and *Entrophospora* from the *Glomus* lineage and of a sexual stage in *Gigaspora* (an advanced character). Extrinsic environmental factors (Meglitsch, 1954) as well as characteristics of the symbiosis (Law, 1985) are important in origination and preservation of identity of asexual species. Environment and genotype of plant hosts must be among the most important selection pressures on fungal symbionts, since the mycorrhizal symbiosis is obligatory and is localized for the most part within the host (Law, 1985). Mechanisms of coevolution may be hidden in historical rather than recent interactions. Relationships between distributional patterns of host fungus associations and major geological changes could provide valuable indirect evidence of coevolutionary processes.

4.1.4. Development of VAM

Nicolson (1967), Mukerji *et al.* (1984) emphasized that VAM though designatd as endotropic, is composed of two phase mycelium system:- (i) an internal mycelium within the cortex of the mycorrhizal root, and (ii) an external mycelium in soil which varies considerably in extent in different hosts. There is no organized fungal growth on the root surface. The VAM association does not cause alterations in the root morphology.

The vesicular-arbuscular mycorrhizae are the most complex group of mycorrhizae which forms interradical structures (i) intracellular hyphae forming coils, often found in the outer layers of cortical parenchyma; (ii) the intercellular hyphae; (iii) the intracellular hyphae with numerous ramifications i.e., the arbuscules; (vi) the inter or intracellular hypertrophied hyphae i.e., the vesicles.

Table 2. Events in establishment of VAM mycorrhiza (modified from Peterson and Farqutar, 1994)

Event	Process
Fungal hyphae in soil around root	Chemotropism
Contact of hyphae with root surface	Recognition
Adhesion of hyphae	Compatibility
Alteration in fungal hyphae i.e., the appressoria formation	Changes in fungal cytoskeleton
Entry of hyphae into or between root cells	Production of enzymes
Further alteration in hyphae i.e., arbuscule and vesicle formation	Further changes in fungal cytoskeleton
Alterations in root cells and root morphology, Establishment of nutrient exchange interface	Hormone production

The mycelial system surrounding the roots is dimorphic (Mosse 1959 a,b; Nicolson, 1959, 1967): (i) with coarse thickwalled irregular non-septate hyphae, and (ii) smaller, thin walled ephemeral lateral branches. The thickwallled hyphae penetrate the host root and cause internal infection. At the entry point, the penetrating hyphae form appressoria in the host plants (Gianinazzi-Pearson *et al.,* 1991; Givoannetti *et al.,* 1991; 1993). The penetrating, infecting hyphae spread inter and intracellularly in the host root cortex. Characteristic H and Y connections are seen in the cortex. The highly branched structures called arbuscules are usually formed in the inner cortex. These are ephemeral structures formed by the repeated dichotomous branching and completes their development in 4 or 5 days (Brundrett *et al.*, 1985). Arbuscules are the key sites for nutrient exchange and remain active only for 4 to 15 days (Carling and Brown, 1982; Cox and Tinker, 1976).

Many but not all endomycorrhizal fungi which forms arbuscules later also form terminal or intercalary vesicles in the root cortex. These are expanded, thin-walled structures which are not delimited by a septum but often contain a large quantity of lipids. They may be spherical, oval or lobed and may become thick walled and resemble resting spores. They serve as the endophytic storage organs and are rich in lipids.

4.2. Ectomycorrhiza

The fungus which does not penetrate living cells in the roots but, instead, only surround them is known as ectomycorrhiza. The extensive mycelium of ectomycorrhizal fungi extends out into the soil and may function in transferring nutrients directly from the decaying leaves, especially in nutrient-poor tropical soils (Meyer, 1974; Suvercha *et al.,* 1991).

Ectomycorrhizal roots are characterized by (i) a fungal sheath or mantle which encloses the root in a fungal tissue, and (ii) a hartig's net which is a plexus of fungal hyphae between epidermal and cortical cells.

The fungal sheath appears to be of either one layered, constructed of coherent hyphae or pseudoparenchymatous throughout, often of two layers, outer usually more dense, compact than the inner which is frequently hyphal and loosely aggregated. The variations in structure of the sheath, the ornamentation of its surface by hyphae, rhizomorphs, setae or cystidium-like structures and the colour have been used, together with chemical, immunological and other tests and the structure of the hartig's net to classify ectomycorrhizas (Chilvers, 1968 a; Dominik, 1956; Melin, 1927; Zak, 1971; 1973).

The term Hartig's net was first described by Strullu (1976). It consists of complicted fan-like or labyrinthine branch systems which provide a very large surface of contact between cells of the two symbionts.

The fungi forming ectomycorrhiza belong to many families of Basidiomycetes, some Ascomycetes with hypogeous or bulky fruit bodies and sterile fungi imperfecti. These represent the most advanced groups of true fungi. They coevolved with plants on land and utilize a diet of complex organic substrates.

In ectomycorrhiza, the mycobionts are very diverse and the phytobionts often form stands that are monotonous and uniform. In most cases, symbiosis is obligatory, but the ectomycobiont is often specific to one or only a few kinds of phytobionts. The phytobionts themselves have the capacity of forming consortia with a wide range of mycobionts, usually simultaneously (Trappe, 1977; Trappe and Fogel, 1977), so that individuals occuring side by side may avoid direct competition with one another.

Furthermore, ectomycorrhizal phytobionts appear to take up mycobionts selectively, according to developmental phase, ecological conditions and possibly climatic fluctuations (Meyer, 1973). Seedlings often have different mycobionts than established plants and these mycobionts are replaced as plant matures (Bowen and Theodorov, 1973). In such communities, the diversity is below the ground where various mycobionts on roots of the same species of phytobionts form symbiotic associations that may not compete directly with each other for the same nutrients at the same time.

In the tropics, ectomycorrhiza are characteristic of marginal conditions, both at high elevations and on very poor soils (Janzer, 1974). In temperate regions, they have been noted as efficient colonizers on black wastes from anthracite mining where endotrophs do not survive. Clearly, ectomycorrhizal associatons, have selective value in extreme environments, perhaps from their direct role in breaking down leaf litter and more specialized and controlled recycling of nutrients to the plants concerned. The mycobionts may have the ability, lacking in the phytobionts, to utilize organic nitrogen taken directly from decaying leaves or ammonia-rich soils.

Ectomycorrhizae are unknown in monocots except of a doubtful record from *Pandanus*. Among the gymnosperms, they are characteristic of Pinaceae, some Cupressaceae and *Gnetum* as well. Among the dicots, ectotrophic mycorrhizae are probably characteristics of all members of Fagaceae, Betulaceae, Salicaeae, Dipterocarpecae and Myrtaceae.

4.2.1. Origin and Evolution of Ectomycorrhizas

Extant ectomycorrhizas are characteristic of mull and moder soils. However, although such soils had developed by the early carboniferous, the ectomycorrhizal habit probably had not, only really coming into its owns during cretaceous (Malloch *et al.*, 1980; Pirozynski and Malloch, 1975). Nonetheless, it was during the carboniferous that the stage was set for these symbiosis. The massive deposits of carbon in this and later eras were due to low rates of decay (Schopf, 1952). This suggests either lack of appropriate degradative organisms or deficiency of oxygen, the co-substrate for rapid decary, or both.

The basidiomycetes were thought to have originated in the carboniferous (Pirozynski, 1976). However, Stubblefield *et al.* (1985) have provided evidences for the existence of wood-inhabiting fungi, likely to be basidiomycetes, in the late Devonian, so appropriate organisms for rapid decay of wood existed at the start of the carboniferous. While much deposition of carbon undoubtedly took place under acidic waterlogged conditions, it is not clear what the precise atmospheric concentration of oxygen was during the period when the basidiomycetes emerged.

The white rot basidiomycetes degrade lignin most rapidly (Reddy, 1984). So to understand the relevance of this, it is necessary to digress to the mechanism by which lignin is degraded. This is because the process is highly oxidative one. As ligninase catalyse polymerisation of fragments, stabilization of these by oxidation catalysed by phenoloxidases utilizing molecular oxygen is may be an important adjunct to the initial degradative steps. The oxidases, which also catalyse further degradation, have a low affinity for oxygen so that the network of reactions, as well as the synthesis of the enzymes themselves, is highly aerobic. The ability to degrade lignin is a feature of secondary metabolism, a metabolic phase induced by low availability of carbon, sulphur or especially, nitrogen, a situation especially relevant in naturally nitrogen-poor wood (Reddy, 1984).

Holland, (1984) and Raven (1985, 1986) suggested that atmospheric oxygen concentration was not less than 0.3-0.6 of the extant value when basidiomycetes first appeared. From this, it is clear that basidiomycetes capable of destroying wood either as saprotrophs or necrotrophs were present by the start of carboniferous. However, basidiomycetes were not `tamed' into the ectomycorrhizal state until much late.

Exquisitely preserved ectomycorrhizae have been found in association with the roots of *Pinus* from the middle Eocene Princeton chert. The fungi and host tissues are preserved as permineralizations. The *Pinus* root show a morphology typical of those with ectomycorrhizal associations including numerous dichotomies that form coralloid mass and lack root hairs. The fungus produces a hartig's net of densely-packed, interdigitate hyphae between and around the cortical cells in young roots which extends through the cortex to the endodermis. A pseudo-parenchymatous mantle and extramatrical hyphae lacking clamp connections which are also characteristic of some extant ectomycorrhizae are also present on these roots. Fungal morphology is similar to the genus *Rhizopogon*, an ectomycorrhizal colonist of living *Pinus*.

The origin of ectomycorrhizal associations is also a significant landmark in the evolution and diversification of terrestrial ecosystems. Molecular data suggests that the ectomycorrhizal fungi originated by the Early Cretaceous. These fossils demonstrate that such association were well established at least 50 million years ago and represent the first evidence of this important type of symbiosis from the fossil record. So, it is concluded that the ectomycorrhizal plants are geologically more recent than their fungi from the fact that fungal genera are common to both Northern and Southern hemisphere plants (Trappe, 1962). This is consistent with establishments of the fungi before the split of the Pangean land mass (Pirozynski, 1981) and with the propensity of ectomycorrhizal symbiosis to be characteristic of the higher latitudes.

5.   CONCLUSIONS

Life and evolution are dependent on each other. Since the beginnings of life on Earth they have been the most intimate of symbiotic partners and they continue to coexist in our world today.. By means of evolution, living systems change through time and the alterations happening are in response to environmental changes. In its widest sense of change driven by natural selection, evolution can operate in non-living systems in interacting sites of chemical reactions for instance. But it is at its most supreme when it operates on the complex self-copying systems of living things. Over a period of more than 3 billion years of accumulating changes this intermeshing of evolution and the life of our plant has produced an extraordinary variety of life forms-organisms ranging from the tiniest bacteria to the hightiest dinosaurs and whales. That epic of change has been charted through fossil evidence and is manifest in the diversity of animals and plants that live now.

It is now well established that the primitive life originated in the oceans in the Cambrian period and the primitive plants originated around one billion years ago. The earlier form of life originated and diversified in water. With the passage of time these primitive plants invaded land, which was still unoccupied by any other living organism, to avoid the fierce competition that had set in water. The colonization of land proved to be the turning point in the history of earth and provided the impetus for the evolution and diversification of the plants. Their subsequent evolution as revealed by the testimony of fossils and the groups of plants that dominate the world today reached, dramatic climax, when they conquered the multitude of habitats available on land.

The terrestrialization of the land by the plants was not an easy task rather a tough ordeal, because the conditions on land were not at all favourable. The colonization of the land by these primititve plants was helped to a great extent by the symbiotic, near - symbiotic and the mutualistic association which they had with the microbes i.e. fungi and bacteria.

The invasion of land by the ancestors of the present day vascular plants clearly seems to have been facilitated by the origin of symbiotic associations between these plants and certain zygomycetous fungi, similar to those that are involved in the endotrophic mycorrhizae in the present time. Fossil evidence from the later Ordovician and Silurian suggest that the association of the plant roots with the microbes is very similar to which we observed today.

The association between the roots of the primitive plants and the microbes helped them to derive the nutrients from soil which are not easily available in free form and this provided competitive advantage over those which lacked this. The nature of association changed from accidental to non obligatory to obligatory mutualism. This change in the nature of the association over the time was accompanied by the diversification of symbiotic partners and has led to the evolution of the present day highly complex plant-microbe symbiotic system, where the association in some cases has become obligatory for the successful survival of either of the partners. This is substantiated by the fact that more than ninty percent of the present day land plant from bryophytes to angiosperms show association with root inhabiting biotrophs. The evolution of mycorrhizae has occurred with the land plants and the complexity and diversification of the taxas have also occurred simultaneously (Pirozynski and Malloch, 1975).

REFERENCES

Abbott, L.K. and Robson, A.D. 1979, A quantitative study of the spores and anatomy of mycorrhizas formed by species of *Glomus*, with references to taxonomy, *Aust. J. Bot.* 27: 363-375.

Abbott, L.M. 1982, Comparative anatomy of vesicular arubsucular mycorrhizas formed on subterranean clover, *Aus. J. Bot.* 30: 485-499.

Ahmadjian, V. 1986, *Symbiosis*, University Press of New England, Hanover, New Hampshire, 212 pp.

Anderson, R.L. and Wood, W.A. 1969, Carbohydrate metabolism in microorganism, *A. Rev. Microbiol.* 23:539-578.

Andrews, H.N. and Lenz, L.W. 1943, A mycorrhizome from the Carboniferous of Illinois, *Bull. Torrey Bot. Club* 70:120-125.

Asai, T. 1934, Uber das Vokommen und die Bedeutung der Wurzelpilze in den Landpflanzen, *Jap. J. Bot.* 7:107-150.

Bartlett, E.M. and Lewis, D.H. 1972, Surface phosphatase activity of mycorrhizal roots of beech, *Soil Biol. Biochem.*4:

Baylis, G.T.S. 1975, The magnolioid mycorrhiza and mycotrophy in root systems derived from it, in : *Endomycorrhizas*, F.E. Sanders, B. Mosse and P.B. Tinker, eds., Academic Press, London, pp.373-389.

Beijerinck, M.W. 1888, Die Bacterian der Papilionaceen-knollchen, *Botanische Zeitung*, 46:797-804.

Benjamin, R.K.1979, Zygomycetes and their spores, in : *The Whole Fungus*, B.Kendrick, ed., Nat. Museums of Con. Ottawa pp. 573-622.

Ben-Shaul, Y., Paran, N. and Galun, M. 1969, The ultrastructure of the association between phycobiont and mycobiont in three ecotypes of the lichen, *Caloplaca aurantia* var. *aurantia, J. Microscopie* 8:415-422.

Bhandari, N.N. and Mukerji, K.G. 1993, `The Haustorium', Research Studies Press Ltd., England pp. 308.

Bonfante-Fasolo, P.1984. Anatomy and morphology of mycorrhizae, in:*VA mycorrhizae*, C.L. Powell and D.J. Bagyaraj, eds., CRC Press, Baco Raton, Florida, pp.5-33.

Boullard, B.1979,Considerations surla symbiose fongique chezles Pteriodophytes, Syllogeous No. 19.Natl. Mus.Nat.Sci., Ottawa, Canada.

Bowen, G.D. and Theodorou, C.1973, In : *Ectomycorrhizae : Their Ecology and Physiology*, G.C. Marks, and T.T. Kozlowsk, eds., Academic Press, New York pp.107-150.

Brian, P.W. 1967, Obligate parasitism in fungi, *Proc.R.Soc.B.* 168:101-118.

Brown, M.F. and King, E.J. 1982, Morphology and histology of vesicular-arbuscular mycorrhizae : Anatomy and cytology, in : *Methods and Principles of Mycorrhizal Research*, N.C. Schenck, ed., Amer Phytopathol.Soc., St.Paul, Minnesota.

Brundrett, M.C., Piche, Y. and Peterson, R.L. 1985, A developmental study of the early stages in vesicular arbuscular mycorrhiza formation, *Can.J.Bot*.63.194.

Callaham, D., Del Tredici, P. and Torrey, J.G. 1978, Isolation and cultivation *in vitro* of the actinomycetes causing root nodulation in *Compotonia, Science* 199:899-902.

Carling, D.E. and Brown, M.F. 1982, Anatomy and physiology of vesicular-arbuscular and nonmycorrhizal roots, *Phytopath*.72:1108-1114.

Chilvers, G.A. 1968a, Some distinctive types of eucalypt mycorrhiza, *Aust.J.Bot*.26:49-70.

Cox, G. and Sanders, F. 1974, Ultrastructure of the host-fungus interface in a vesicular-arbuscular mycorrhiza, *New Phytol*.73:901-912.

Cox., G. and tinker, P.B. 1976, Translocation and transfer of nutrients in vesicular arbuscular mycorrhizas, 1, The arbuscule and phosphorus transfer: a quantitative ultrastructural study, *New Phytol*.77:371-378.

Cridland, A.A. 1962, Fungi in cordiaitean roots, *Mycologia* 54:230-234.

De Bary, A.1887, Comparative morphology and biology of the fungi, mycetozoa and bacteria, Oxford University Press, Oxford.

Dominik, T.1956, Vorschlag einer neuer Klassification der ectotrophen Mykorrhizon auf morphologishch-anatomishen Merbmalen begrundet, roozen Mauk, *Lesn*.14:223-245.

Edwards, D.S. 1986, *Aglaophyton major*, a non-vascular land plant from the Devonian Rhynie Chert, *Bot.J.Linn.Soc*.93:173-204.

Frank, A.B. 1885, Uber die auf Wurzelymbiose beruhende Ernahrung Gewisser Baume durch unterirdische Pilze, *Ber. dt. bot*.3:128-145.

Friese, C.F. and Allen, M.F. 1991, The spread of VA mycorrhizal fungal hyphae in the soil:inoculum types and external hyphal architecture, *Mycologia*. 83:409.

Galun, M., Paran, N. and Ben-Shaul, Y.1970, An ultrastructural study of the fungus-alga association in *Lecanora radiosa* growing under different environmental conditions, *J. Microscopie* 9:801-806.

Garrett, S.D.1970, *Pathogenic Root Infecting Fungi*, Cambridge University Press, Cambridge.

Gerdemann, J.W. and Trappe, J.M. 1974, The Endogonaceae in the Pacific Northwest, *Mycologia Memoir No. 5*, pp. 65.

Gerdemann, J.W. and Trappe, J.M. 1975, Taxonomy of the Endogonaecae, in *Endomycorrhizas*, F.E. Sanders, B. Mosse and P.B. Tinkes, eds., Academic Press, London and New York, pp.35-51.

Gianinazzi-Pearson, V., Smith, S.E.,Gianinazzi, S. and Smith, F.A. 1991, Enzymatic studies on the metabolim of Vesicular arbuscular mycorrhizas V, Is H⁺-ATP-hydroloysing enzyme a component of ATP- hydrolysing activities in palnt-fungus interfaces, *New Phytol*.117:61-74.

Gianinazzi-Pearson, V., Gianinazzi, S., Guillemin, J.P., Trouvelot, A. and Duc, G. 1991, Genetic and cellular analysis of the resistance to vesicular-arbuscular (VA) mycorrhizal fungi in pea mutants, in : *Advances in Molecular Genetics of Plant-Microbe Interactions*, H. Hennecke and D.P.S. Verma, eds., Kluwer Academic Publishers, pp. 336-342.

Gibson, J.L., Kombrough, J.K. and Benny, G.L. 1986, Ultrastructural observations on Endogonaceae (Zygomycetes) : II Glaziellales ord. nov. and Glaziellaceae fam. nov. new taxa based upon light and electron microscopie observations of *Glaziella aurantiaca, Mycologia* 78:941-954.

Gibson, J.L., Kombrough, J.K. and Benny, G.L. 1986, Ultrastructural observations on Endogonaceae (Zygomycetes):*Endogone pisiformis, Mycologia* 79:543-553.

Giovannetti, M., Sbrana, C., Avio, L. and Citernesi, A.S. 1991, Appressorium formation in VAM fungi in presence of host and non-host plants, (Abs), *Fungal Cell Biology:Cytology and Ultrastructure*, April 1991, Portsmouth, U.K.

Giovannetti, M., Sbrana, C., Avio, L., Citernesi, A.S. and Logi, C.1993, Differential hyphal morphogenesis in arbuscular mycorrhizal fungi during preinfection stages, *New Phytol*. 125:587-593.

Gorbunova, N.P. 1958, Mikoriza Encephalartos hildebrandtii Arb. et Bauche i nekotorye soobrazheniya o vzaimoolnosheniyakh griba-endofita i vysshego rasteniya v endotrofnykh, *Byul.Mosk. Obshch. Ssp. Otd. Biol.* 63:123-134.

Gourret, J.P. and Strullu, D.S. 1979, Etude cytophysiologique et ecologique des symbioses racinaires:nodules fixateurs d'azote et mycorhizes, Compte-Rendu Scientifique de I'A.T.P.28-20 CNRS Paris.

Halket, A.C. 1930, The rootlets of *Amyelon radicans* Will., their anatomy, their apices and their endophytic fungus, *Ann. Bot.* (London) 44:865-905.

Harley, J.L. 1948, Mycorrhiza and soil ecology, *Biol. Rev.* 23:127-58.

Harley, J.L. 1968, Fungal symbiosis, *Trans Br. Mycol. Soc.* 51: I-II.

Harley, J.L. 1970, Mycorrhiza and nutrient uptake in forest trees, in : *Physiology of Tree Crops,* L.C. Luckwill and C.V. Cutting, eds., Academic Press, London, pp. 163-167.

Harley, J.L. 1975, Problems of mycotrophy, in : *Endomycorrhizas,* F.E. Sanders, B. Mosse and P.B. Tinker, eds., Academic Press, London, pp.1-24.

Harley, J.L. and Smith, S.E. 1983, *Mycorrhizal Symbiosis,* Academis Press, New York.

Hass, H., Taylor, T.N. and Remy, W. 1994, Fungi from the Lower Devonian Rhynie chert:mycoparasitism, *Amer.J.Bot.* 81:29-37.

Holland, H.D. 1984, *The Chemical Evolution of the Atmosphere and Oceans* : Princeton University Press, Princeton.

Holleg, J.D. and Peterson, R.L. 1979, Development of vesicular - arbuscular mycorshizal fungi in bean roots, *Can.J. Bot.* 57:1960-1978.

James, P.W. 1970, The lichen flora of shaded acid rock crevices and overhangs in Britain, *Lichenologist* 4:309-322.

Janzen, D.H. 1974, *Biotrophica* 6:60-103.

Jeffrey, C.1962, The origin and differentiation of the acheogoniate land plants, *Bot.Not.*115:446-454.

Juniper, B.E., Goves, S., Landau-Schacher, B. and Addus, L.J. 1966, Root cap and the perception of gravity, *Nature,* 209:93-104.

Kamienski, F.1881, Die Vegetationsorgane der *Monotropa hypopitys* L., *Bot. Ztg.* 29:458.

Kuar, Mandeep 1995, Vesicular-arbuscular mycorrhizae in certain pteridophytes, M.Phil. Diss., Univ. of Delhi.

Keeler, K.H. 1985, Cost benefit models of mutualism, in : *The Biology of Mutualism:Ecology and Evolution,* D.H. Boucher, ed., Beckenham, Croom Helm, pp.100-127.

Kessler, K.J. 1966, Growth and development of mycorrhizae of sugar maple (*Acer saccharum* Marsh.), *Can.J.Bot.*44:1413-1425.

Kidston, R. and Lang, W.H. 1921, On old red sand stone plants showing structure from the Rhynne Chert Bed, Aberdeenshire, Part V, The Thallophyta occuring in the peat-bed; the succession of the plants through a vertical section of the bed, and the conditions accumulation and preservation of the deposit, *Trans R. Soc.* Edinburgh 52:855-902.

Kinden, D.A. and Brown, M.F. 1975a, Electron microscopy of vesicular-arbuscular mycorrhizae of yellow poplar, I, Characterization of endophyte structures by scanning electron microscopy, *Can. J. Microbiol.* 21:989-993.

Kinden, D.A. and Brown, M.F. 1975b, Electron microscopy of vesicular-arbuscular mycorrhizae of yellow poplar, IV, Host-endophyte interactions during arbuscular deterioration, *Can. J. Microbiol.*22:64-75.

Kinden, D.A. and Brown, M.F. 1975c, Electron microscopy of vesicular-arbuscular mycorrhizae of yellow poplar, III, Host-endophyte interactions during arubscular development, *Can. J. Microbiol.* 21:1930-1939.

Law, R. 1985, Evolution in a mutualistic environment, in : *The Biology of Mutualisms,* D.H. Boucher, ed., Oxford Univ. Press, New York, pp.29-39.

Lewis, D.H. 1975, Comparative aspects of the carbon nutrition of mycorrhizas, in : *Endomycorrhizas,* F.E. Sanders, B. Mosse and P.B. Tinker, Academic Press, London and New York, pp. 199-148.

Lewis, D.H. 1976, Interchange of metabolites in biotrophic symbiosis between angiosperms and fungi, in: *Prespectives in Experimental Biology,* Vol. 2; Sutherland, ed., Pergamon Press, Oxford, pp. 207-219.

Lewis, D.H. 1980, Boron, lignification and the origin of vascular plants-a united hypothesis, *New Phytol.* 84:209-229.

Lewis, D.H. 1985, Symbiosis and mutualism: crisp concepts and soggy semahlics, in : *The Biology of Mutualism: Ecology and Evolution,* D.H. Boucher, ed., Croom-Helm, London, pp. 29-39.

Lewis, D.H. 1986, Inter-relationships between carbon nutrition and morphogenesis in mycorrhizas, in : *Proceedings of the First European Symposium of Mycorrhizae,* Gianinazzi-Perason, V. and Gianinazzi, S., INRA, Dijon, pp. 85-100.

Lewis, D.H. 1991, Mutualistic symbioses in the origin and evolution of land plants, in : *Symbiosis as a Source of Evolutionary Innovation,* L. Margulis, and R. Fester, eds., MIT Press, Cambridge, Massachusetts, pp. 288-300.

Lewis, D.H. and Smith, D.C. 1971, The autotrophic nutrition of symbitic marine colenterates with special reference to hermatypic corals, I, Movement of photosynthetic products between the symbionts, *Proc. R. Soc. B.* 178: 111-129.

Lundeberg, G. 1970, Utilization of various nitrogen sources, in particular bound soil nitrogen, by mycorrhizal fungi, *Studia Forestalia Suecica* No. 97:1-95.

Mada, M. 1954, The meaning of mycorrhiza in regard to systematic botany, Kumamoto *J. Sci., Ser. B.* 1954:57-84.

Magasanik, B. 1961. Catabolite repression, Cold Spring Harb. Symp. quanti. Biol. 26:249-254.

Mago, P., Agnes, C.A. and Mukerji, K.G. 1992, VAMycorrhizal status of some Indian bryophytes, *Phytomorphology* 42: 231-239.

Mago, P., Pasricha, P. and Mukerji, K.G. 1993, Survey of vesicular-arbuscular mycorrhizal fungi in scale leaves of modified stem, *Phytomorphology* 43:81-85.

Malloch, D.W., Pirozynski, K.A. and Raven, P.H. 1980, Ecological and evolutionary significance of mycorrhizal symbioses in vascular plants (a review), *Proc. Natl. Acad. Sci.* USA, 77:2113-2118.

Marx, D.H. 1991, The practical significance of ectomycorrhizae in forest establishment, in : *Ecophysiology of Evomycorrhizue of Forest Tress,* Vol. 7, M. Wallenberg Foundation, Stockholm, Sweaden, pp. 34-40.

Maslin, T.P. 1952, Morphological criteria of phylogenetic relationships, *Sysn. Zool.* 1:49-70.

Meglitsch, P.A. 1954, On the nature of the species, *Syst. Zool* 3:49-65.

Melin, E. 1927, Studies over barrtradsplatns utveckling i rahmus, II, Mykorrhizans utbildning hos tallplantan i olika rahumus former, *Meddeln St. Skogsforks Inst.* 23:433-494.

Melin, E. 1953, Physiology of mycorrhizal relations in plants, *A. Rev. Pl. Physiol.* 4:325-346.

Meyer, F.H. 1973, Distribution of ectomycorrhizae in native and man made forests, in : *Ectomycorrhizne,* G.C. Marks and T.T. Kozlowski, eds., Accdemic Press, New York and London, pp. 79-105.

Morton, J.B. 1988, Taxonomy of VA mycorrhizal fungi: classification, nomenclature and identification, *Mycotaxon* 32:267-324.

Morton, J.B. 1990a, Species and clones of arbuscular mycorrhizal fungi (Golmales, Zygomycetes): their role in macro-and microevolutionary processes, *Mycotaxon* 37:493-515.

Morton, J.B. 1990b, Evolutionary relationship among arbuscular mycorrhizal fungi in the Endogonaceae, *Mycologia* 82:192-207.

Morton, J.B. and Benny, G.L. 1990, Revised classification of arbuscular mycorrhizal fungi (Zygomycetes): a new order, Glomales, two new suborders, Glomineae and Gigasporineae, and two new families, Acaulosporaceae and Gigasporaceae with an emendation of Glomaceae, *Mycotaxon* 37:471-491.

Morton, J.B. and Bentivenga, S.P. 1994, Levels of diversity in endomycorrhizal fungi (Glomales, Zygomycetes) and their role in defining taxonomic and non-taxonomic group *Plant. Soil.* 159:47-60.

Mosse, B.1959a, The regular germination of resting spores and some observations on the growth requirements of an *Endogone* sp. causing vesicular-arbuscular mycorrhiza, *Trans. Br. Mycol. Soc.* 42:273-286.

Mosse, B. 1959b, Observations on the extramatrical mycelium of a vesicular-arbuscular endophyte, *Trans. Br. Mycol. Soc.* 42:439-448.

Mukerji, K.G. 1996, Taxonomy of endomycorrhizal fungi, in : *Advances in Botany,* K.G. Mukerji, B. Mathur, B.P. Chamola and P.Chitralekha, eds., A.P.H. Publishing Corporation, New Delhi, pp. 212-219.

Mukerji, K.G., Bhattacharjee, M. and Tewari, J.P. 1983, New species of vesicular-arbuscular mycorrhizal fungi, *Trans. Br.Mycol.Soc.*81:641-643.

Mukerji, K.G., Sabharwal, A., Kochar, B. and Ardey, J.1984, Vesicular-arbuscular mycorrhizae:Concept and advances, in : *Progress in Microbial Ecology,* K.G. Mukerji, V.P. Agnihotri and R.P. Singh, eds., Print House, Lucknow, India, pp.489-525.

Mukerji, K.G. and Sharma, M. 1996. Mycorrhizal relationships in forest ecosystem, in : *Forests: A Global Perspective,* S.K. Mazumdar, E.W. Miller and F.J. Bremmer, eds., The Pennsylvania Acad. Sci. USA, pp.95-125.

Nicolson, T.H. 1959, Mycorrhizae in the Gramineae, I, Vesicular-arbuscular endophytes, with special reference to the external phase, *Trans. Brit. Mycol.* Soc. 42:421-438.

Nicolson, T.H. 1967, Vesicular-arbuscular mycorrhiza-a universal plant symbiosis, *Sci. Prog., Oxford* 55:561-581.

Nicolson, T.H. 1975, Evolution of Vesicular-arbuscular mycorrhizas, in : *Endomycorrhizas,* F.E. Sandres, B. Mosse and P. Tinker, eds., Acadmic Press, New York, pp. 25-34.

Nicolson, T.H. 1981, Palaeobotanical evidence for mycorrhizas, *13th Intl. Bot. Cong.,* Sydney, Australia, p.187.

Norkrans, B.1950, Studies in growth and cellulytic enzymes of *Tricholoma* with special reference to mycorrhiza formation, *Symb. Bot. Upsal,* II: 1-126.

Nisizawa, T., Suzuki, H. and Nisizawa, K. 1972, Catabolite repression of cellulase formation in *Trichoderma viride, F. Biochem.* 71:999-1008.

Osborn, T.G.B. 1909, The lateral roots of *Amyelon radicans,* Will. and their mycorrhiza, *Ann. Bot.* 23:603-611.

Paigen, K. and Williams, B.1970, Catabolite repression and other control mechanisms in carbohydrate utilization, *Adv. Microb. Physiol.* 4:251-324.

Palmer, J.G. and Hacskaylo, E. 1970, Ectomycorrhizal fungi in pure culture, I, Growth on single carbon sources, *Physiol. Pl.*2:1187-97.

Peterson, R.L. and Farquhar, M.L. 1994, Mycorrhizas-integrated development between roots and fungi, *Mycologia* 86:311-326.

Peyronel, B.1923, Fructification de I' endophyte a arbuscules et a vesicules des mycorhizes endotrophes, *Bull. Soc. Mycol.* France, 39:119-126.

Peyronel, B. 1924, Prime ricerche sulle micorize endotrophe e sulla microhizes radiciola normide delle famerogame, *Memorie R. Staz. Palol. Veg.* 1-61.

Pirozynski, K.A. 1976, Fossil fungi, *Ann. Rev. Phytopath.* 14:237-246.

Pirozynski, K.A. 1981, Interactions between fungi and plants through the ages, *Can. J. Bot.* 59:1824-1827.

Pirozynski, K.A. and Malloch., D.W. 1975, The origin of land plants: a matter of microtrophism, *Bio. System* 6:153-164.

Pirozynski, K.A. and Dalpe, Y. 1989. Geological history of the Glomaceae with particular reference to mycorrhizal symbiosis, *Symbiosis* 7:1-36.

Powell, C.L. 1976, Development of mycorrhizal infections from *Endogone* spores and infected root segments, *Trans. Brit. Mycol. Soc.* 66:439-445.

Quispel, A. 1943, The mutual relations between algae and fungi in lichens,*Rec. Trav. Bot, Neerl.*40.

Raven, J.A. 1977, The evolution of vascular plants in relation to superacellular transport processes, *Adv. Bot. Res.* 5:153-219.

Raven, J.A. 1984, Physiological correlates of the morphology of early vascular plants, *Botanical Journal of the Linnean Society* 88:105-126.

Raven, J.A. 1985, Comparative physiology of plant and arthopod land adaptaton. *Philosophical Transactions of the Royal Soceity,* series B. 309:273-288.

Raven, J.A. 1986, Evolution of plant life forms, in : *On the Economy of Plant Form and Function,* T. Givnish, ed., Cambridge Uni. Press, pp. 421-492.

Raven, J.A., Smith, S.E. and Smith, F.A. 1978, Ammonium assimilation and the role of mycorrhizas in climax communities in Scoltland, *Trans. Bot. Soc.,* Edinburgh, 43: 27-35.

Read, D.J. 1982, The biology of mycorrhiza in the Ericales, *Proc. 5th NACOM.*

Reddy, C.A. 1984, Physiology and biochemistry of lignin degradation, in : *Current Perspectives in Microbial Ecology,* J. Khug and C.A. Reddy, eds., American Society for Microbiology, Washington, DC pp. 558-571.

Remy. W., Taylor, T.N. Hass, H. and Kerp, H. 1994, 400 million year old vesicular arbuscular mycorrhizae (VAM), *Proc. Nat. Acad. Sci. USA* 91:11841-11843.

Scannerini, S. and Bellando, M. 1968, Sullutrastruttura delle micorrize endotrofiche di *Ornithogalum umbellatum* L. inattivata vegetativa, *Atti. Accad. Sci. Torino Cl. Sci. Fis. Mat. Nat.* 102:795-809.

Scannerini, S. and Bonfante-Fasolo, P. 1983, Comparative ultrastructural analysis of mycorrhizal association, *Can. J. Bot.* 61:917-943.

Schilling, G. 1988, Hellriegel and Wilfarth and their discover of nitrogen fixation at Bernburg, in: *Nitrogen Fixation:Hundred Years After Stuttagart*: H. Bothe, F.J. de Bruijn and W.E. Newton, eds., Gustav Fischer, pp. 13-19.

23

Schopf, J.M. 1952, Was decay importnt in origin of Coal? *Journal of Sedimentary Petrology* 22:61-69.

Schopf, J.M. 1970, Antarctic collections of plant fossils, 1967-1970, *Antarct.J.U.S.*5:89.

Schopf, J.M. 1978, An unusual osmundaceous specimen from Antarctica, *Can. J.Bot.*56:3083-3095.

Sharma, B.D., Bohra, N. and Harsh, R. 1993, Vesicular arbuscular mycorrhizae association in Lower Devonian plants of the Rhynie chert, *Phytomorphology* 43:105-110.

Shaw, M. and Samrorski, D.J. 1956, The physiology of host-parasite relatins, I, The accumulation of radio-active substances at infections of facultative and obligate parasites including tobacco mosaic virus, *Can.J.Bot.*34:389-405.

Sherwood-Pike, M.A. and Gray, J. 1985, Silurian fungal remains:probable records of the class Ascomycetes, *Lethaia* 18:-20.

Sherwood-Pike, M.A. 1991, Fossils as keys to evolution in fungi, *Bio Systems* 25:121-129.

Simon, L., Bosquet, Levesque, R.C. and Lalonde, M. 1993, Origin and diversification of endomycorrhizal fungi and coincidence with vascular land plants, *Nature* 363:67-69.

Slankis, V. 1971, Formation of ectomycorrhizae of forest trees in relation to light, carbohydrates and auxins, in : *Mycorrhizae,* E. Hacskaylo, ed., Proc. 1st N. Am. Conf. Mycorrhizae, pp. 151-67.

Smith, D.C., Muscatine, Li and Lewis, D.H. 1969, Carbohydrate movement from autotrophs to heterotrophs in parasitic and mutualistic symbioses, *Biological Reviews* 44:19-90.

Smith, D.C. 1981, The role of nutrient exchange in recognition between symbionts, Berichte Deutsche Botanische Gesellschaft, 94 (suppl.) : 517-528.

Srivastava, D., Kapoor, R., Srivastava, S.K. and Mukerji, K.G. 1996, Vesicular arbuscular mycorhiza - an overview, in : *Concepts in Mycorrhizal Research,* K.G. Mukerji ed., Kluwer Academic Publishers, Netherland, pp. 1-39.

Staneir, R.Y., Adelberg, E.A. and Ingraham, J.L. 1977, *General Micrbiology* (fourth edition), MacMillan, London.

Stanczak-Boratynska, W.1954, Badania anatomiczne mykorhizy egzotycznych roslin Palmiarni Poznan-skiej, *Ann. Univ Maric Curie Skoldowsa,* Sec C. 9:1-60.

Strullu, D.G. 1976, Recherches de biologie et de micrbiologie forestieres, Etude des relations nutrition, development et cytologie des mycorrhizes chez le douglas (*Pseudotsuga menziesii* Mirb.) et les abietacees, These. Univ. Rennes 1976.

Stubblefield, S.P. and Banks, H.P. 1983, Fungal remains in the Devonian trimerophyte *Psilophyton dawsonii, Amer. J. Bot.* 70:1258-1261.

Stubblefied, S.P., Taylor. T.N. and Beck. C.B. 1985, Studies on Paleozoic fungi, IV, Wood-decaying fungi in *Callixylon newberryi* from the Upper Devonian*, Amer. J. Bot.* 72:1765-1774.

Stubblefied, S.P., Taylor, T.N. and Trappe, J.M. 1987, Vesicular-arbuscular mycorrhizae from the Triassic of Antarctica, *Amer. J. Bot.* 74:1904-1911.

Stuessy, T.F. 1987, Explicity approches for evolutionary classification, *Syst. Bot.* 12:251-262.

Suvercha, Mukerji, K.G. and Arora, D.K. 1991, Ectomycorrhizae, in : *Hand Book of Applied Mycology*, Vol. I, D.K. Arora, B. Rai, K.G. Mukerji and G.R. Knudson, eds., Marcel Dekker Inc., New York, U.S.A., pp. 187-217.

Taber, R.A. and Trappe, J.M.1982, Vesicular-arbuscular mycorrhiza in rhizomes, scale-like leaves, roots, and xylem of ginger, *Mycologia* 74:156-161.

Taylor, F.J.R. 1979, Symbionticism revisited:a discussion on the evolutionary impact of intracellular symbiosis, *Proc. R. Soc.* London, Ser. B. 204:268-286.

Thomton, H.G. 1930, The infleunce of the host plant in inducing parasitism in lucerne and elover nodules. *Proc. R. Soc.* B 106:110-122.

Thrower, L.B. 1966, Terminology for plant parasites, *Phytopath.Z.*56:258-254.

Trappe, J.M. 1962, Fungus associates of ectomycorrhizae, *Bot. Rev.* 28:538-606.

Trappe, J.M. 1987, Phylogenetic and ecological aspects of mycorrhiza in the angiosperms from an evolutionary standpoint, in : *Ecophysiology of VA mycorrhiza plants,* G. Safir, ed., CRC Press, Boco Raton, Florida, pp. 5-25.

Trappe, J.M. and Fogel, R.C. 1977, Range Science Department Science Series (Colorado State Univ, Fort Collins, CO), Vol. 26, pp.205-214.

24

Trappe, J.M. and Schenck, N.C. 1982, Taxonomy of the fungi forming endomycorrhizae, A, Vesicular - arbuscular mycorrhiza fungi (Endogonales), in : *Methods and Principles of Mycorrhizal Research,* N.C. Schenck, ed., The American Phytopathologica Society, St. Paul, Minnesota, pp.1-9.

Tschermark, E. 1943, Weitere Untersuchungen zur Frage des Zusammenlebens von Pilz und Alge in den Flechten, *Ost. Bot. Z.* 92:15-24.

Von Tubeuf, C.1896, Die Haarbildungen der Coniferen, *Forstl. -Naturiv. Zeitschr.* 5:173-193.

Wagner, C.W. and Taylor, T.N. 1982, Fungal chlamydospores from the Pennsylvanian of North America, *Rev. Palaeobot. Paynol.* 37:317-328.

Wagner, C.A. and Taylor, T.N. 1982a, Evidence endomycorrhizae in Pennsylvanian age plant fossil, *Science* 212:562-563.

Wagner, C.A. and Taylor, T.N. 1982b, Fungal chlamydospores from the Pennsylvanian of North America, *Rev Palaeobo. Paynol.* 37:317-328.

Walker, C.1984, Taxonmy of the Endogonaceae, in : *Proceedings of the 6th North American Conference on Mycorrhizae,* R. Molina, ed., Forest Research Lab., Corvallis, Oregon, pp. 193-199.

Walker, C.1987, Current concepts in the taxonony of the Endogonaceae, in : *Mycorrhizae in the Next Decade, Paracitical Applications and Research Priorities,* D.M. Sylvia, L.L. Hung. and J.H. Graham, eds., IFAS, University of Florida, Gainsville, pp. 300-302.

Walker, C.M. 1992, Systematics and taxonomy of the arbuscular endomylorolizul fungi (Glomales) - a possible wing forward, *Agronomie Agrmomie,* 12:887-897.

Walker, C. and Sanders, F.E. 1986, Taxonomic concepts in Endogonaceae III, Separation of *Scutellospora* gen. nov. from *Gigaspora*, Gerdemann and Trappe, *Mycotaxon* 27:169-182.

Weiss, F.E. 1904, A mycorrhiza from the Lower Coal Measures, *Ann. Bot.* (London) 18:255-267.

Wood, R.K.S. 1967, *Physiologica Plant Pathology,* Oxford University Press, Oxford.

Zak, B. 1971, Characterization and identification of Douglas fir mycorrhizae, in: *Mycorrhizae,* E. Haeskaylo, ed., *Proc. 1st NACOM,* US Government Printing Office, Washington, DC, pp. 38-53.

Zak, B.1973, Classification of Ecomycorrhizae, in : *Ectomycorrhizae*, G.C. Marks and T.T. Kozlowski, eds., Academic Press, New York and London, pp. 43-78.

GENERAL ASPECTS OF MYCORRHIZA

Vandana Gupta, T. Satyanarayana, and Sandeep Garg

Department of Microbiology
University of Delhi, South Campus
New Delhi - 110021, INDIA

1. INTRODUCTION

The mycorrhizal association appears to have evolved as survival mechanism for both the fungi and the higher plants, allowing each to survive in the existing environments of low temperatures, soil fertility, periodic drought, diseases, exterene temperatures and other natural stresses. Mycorrhiza appears to be the first line of biological defense against stress, for trees. The following benefits can be derived from this relationship:

(i)   Increased nutrient and water absorption through improved absorbance area (Boyd, 1987)
(ii)   Increased nutrient mobilization through biological weathering.
(iii) Increase in feeder root longevity by provision of biological deterrent to root infection by soil-borne pathogens (Duchesne *et al.,* 1989; Marx, 1973).
(iv)   Accumulation of elements such as nitrogen, phosphorus, potassium, calcium and zinc and their translocation to the host tissue.
(v)   Ectomycorrhizal hyphae completely permeate the F and H horizons of forest floor and thus minerals mobilized in these zone can be absorbed before they reach the subsoil system.
(vi)   Some EM fungi can degrade complex minerals and organic substances in soil and make essential elements available to the host plant.
(vii) Mycorrhizal fungi afford protection to delicate root tissue from attack by pathogenic fungi through strategies such as the use of surplus carbohydrates, provision of a physical barrier and secretion of antibiotics (Duchesne *et al.,* 1989; Garrido *et al.,* 1982; Tsantrizos *et al.,* 1991).
(viii) Provide host plant with growth hormones like auxins, cytokinins, gibberellins and growth regulators such as B vitamins (Gopinathan and Raman, 1992; Ho, 1987; Kraigher *et al.,* 1991).
(ix)   Important contributions to organic matter turnover and nutrient cycling in forest ecosystems and mycorrhizal fungal biomass can account for as much as 15% of the net primary production and a majority of it enter the soil organic matter pool annually.
(x)   Increase in the tolerance of the plant to adverse conditions including water stress, pH stress, temperature stress, heavy metal and toxin stress (Dixon *et al.,* 1994; Gardner and Malajczuk, 1988; Marx and Artman, 1979; Osonuki *et al.,* 1991; Peiffer and Bloss, 1988).

Molecular (Simon *et al.,* 1993a) and fossil (Remy *et al.,* 1994; Taylor, 1986) evidences together indicate that symbiotic interactions between fungi and plant roots developed very early in the process of colonization of the terrestrial environments. Evolutionary considerations suggest that the relationship between the host and fungus has been inherently of 'mutualistic' kind, selection has favored its persistence in the overwhelming majority of species to the present day when over 90% of the world's vascular plants are mycorrhizal (Trappe, 1987). Fungi of the order Glomales that form 'arbuscular' (AM) or 'vesicular arbuscular' (VAM) mycorrhiza colonize most of these plants.

On the basis of calibrated molecular clock in which rates of substitution of gene sequences in the small subunit rRNA are used to provide dates, VAM fungi appear to have originated between 463 and 353 million years ago, when Glomales diverged from Endogonales (Simon *et al.,* 1993a). The members of families such as Brassicaceae are devoid of functional mycorrhiza due to the presence of glucosinolates and their hydrolysis products, isothiocyanates in and around roots (Glenn *et al.,* 1988), which are toxic to the growth of fungi. Since land plants and insects evolved simultaneously, the effects of leaf-eating insects became more and more severe. This selection pressure favored those plants that contained unpalatable or toxic substances such as tannins, phenols and resins. If these substances are spread throughout the plant as in Pinaceae and Myrtaceae, they could be inhibitory to the continued presence of endomycorrhizal fungi. As a result, the less intimate association involved in ectomycorrhizas may well have evolved to fill the void left by VAM fungi which were effectively expelled from some of their hosts. Subsequently selection led to distinct types of mycorrhizal symbiosis such as ectomycorrhiza (ECM) formed between trees and basidiomycetes or ascomycetes. Beautifully preserved fossil ectomycorrhizas formed by a *Rhizopogon* like fungus have been found in chert of the 50 million old Eocene period (Le Page *et al.,* 1996). The ericoid endomycorrhizal association between ericaceous shrubs and Ascomycetes and the orchid mycorrhiza formed by basidiomycetes are of more recent origin (Read, 1996).

Seven types of mycorrhizal associations are known namely - vesicular arbuscular mycorrhiza, ectomycorrhiza, ectendomycorrhiza, arbutoid, monotropoid, ericoid and orchidoid mycorrhiza. Types of mycorrhizas are categorized on the basis of taxonomic group of fungi and plants involved and the alteration in the morphogenesis of fungi and roots, which occur during the development of the new structure that is mycorrhiza (Harley and Smith, 1987). Mycorrhizal effectiveness is being governed by the interaction among a number of factors including fungal genome, plant genome, soil type and edaphic factors.

Enormous amount of work has been done on various aspects of mycorrhizal associations such as role of mycorrhizal fungi in nutrient uptake, stress tolerance, wasteland reclamation, reforestation and afforestation and in sustainable agriculture management. But still gaps remain in the understanding of molecular mechanism of mycorrhizal symbiosis, field identification of mycorrhizas from its natural habitat and use of recombinant DNA technology for improving the efficiency and competitive ability of mycorrhizal fungi. Considering the importance of these fungi it is necessary to work extensively on inoculum production and application technology. In this chapter an attempt has been made to provide an account of each type of mycorrhiza, current status of the understanding of fungus-root interaction during the establishiment of mycorrhizal association and molecular aspects.

2. MYCORRHIZAL TYPES

2.1. Endomycorrhiza

Characteristic feature of the endomycorrhizas is intracellular penetration of root cortical cells by the symbiotic fungi and formation of highly branched haustoria like structures called arbuscules. Arbuscules are the site of nutrient exchange between the two symbionts. In most of the cases round to oval thick walled spores are formed within the cells of plant roots called vesicles. This type of endomycorrhiza is known as Vesicular Arbuscular Mycorrhiza' or 'VAM', but in some cases no vesicles are formed and they are known as 'Arbuscular Mycorrhiza" or 'AM'

Endomycorrhiza is the most extensively studied type of mycorrhiza because it involves the majority of vascular plants including most crops and horticultural species (Peterson *et al.,* 1984). Very little specificity is shown between mycorrhizal fungi and the higher plants (Molina *et al.,* 1992). Host plants derive benefits from the associations such as better tolerance of nutrient deficiencies of potassium, phosphorus, calcium and other minerals, water and also soil-borne pathogens including fungi and nematodes. Fungi derive carbon and energy source from the host plant, which it synthesizes. During the vegetative growth period, before seed or fruit setting most higher plants are not carbon or energy limited, as shown by the starch or lipid accumulation. Thus VAM infection results mostly in a profit for the plant (the exchange of its 'excess' against that of which it has too little).

Endomycorrhizal fungi belong to class Zygomycetes of the order Glomales. Order Glomales is further subdivided in two suborders, namely Glomineae and Gigasporineae. Glomineae suborder has two families, Glomaceae to which genera *Glomus* and *Sclerocystis* belong and Acaulosporaceae which includes genera *Acaulospora* and *Entrophospora*. A single family Gigasporaceae with genera *Gigaspora* and *Scutellospora* has been included in the suborder Gigasporineae. All these fungi are obligate symbionts and do not grow *in vitro*. With few exceptions, species from all angiosperm families can form endomycorrhizal associations. A few gymnosperms such as *Taxas* and *Sequoia* also show infection. Phylogenetically endomycorrhiza are oldest symbionts infecting Bryophytes as well as Pteridophytes.

Advantages conferred on plants by VAM infection are enormous including increased phosphorus uptake, protection from root pathogens (nematodes and fungi), tolerance to toxic heavy metals, tolerance to adverse conditions such as temperature, high salinity, high or low pH, better performance during transplantation shock, aggregation of soil particles, increase in rhizospheric microflora and enhanced uptake of nutrients and water. But these fungi pose problems to their large scale application because of their obligate symbiotic nature. There is a need to evolve certain approaches like large scale inoculum production strategies, controlled inoculum trials and mycorrhizal effectiveness so that VAM can be commercialized. Inoculum production technologies for VAM fungi include expanded clay as substratum, nutrient film technique (hydroponics and aeroponics), surface disinfected plant roots, inorganic carrier based inoculum, root organ culture and Ri T-DNA transformed roots.

2.2. Ectomycorrhiza

Ectomycorrhiza is characterized by an outer shealth of fungal hyphae surrounding the host root called 'Mantle', a layer of hyphae between the root rind cells called 'Hartig's net' and hyphae and rhizomorphs extending outside in soil. Fungi infect secondary or tertiary roots. After the establishment of infection there is a reduction in root hair formation. Roots become coated with fungi and there is reduced growth of secondary and tertiary roots lengthways and they are, therefore often celled short roots. Fungi induce extensive branching in the roots and branching pattern is dependent on the infecting fungi, therefore this is one of the identification criteria. Fungal penetration is restricted to the outermost cell layers. Hyphae grow between the cells, by forcing their way mechanically or by excreting pectinases.

Plant forming ectomycorrhiza include 140 genera of seed plants belonging to families Betulaceae, Fagaceae, Pinaceae, Rosaceae, Mimosaceae and Salicaceae. Though the number of ectomycorrhizal plant species is much less than endomycorrhizal plants, the significance of the ectomycorrhizal symbiosis at the global level lies in the fact that while it is found in a relatively small number of plant families, these contain the dominant species of some of the World's most important terrestrial ecosystems (Read, 1991). The members of the Pinaceae, Fagaceae and Myrtaceae, which dominate boreal, temperate and many sub-tropical forests, respectively are largely made up of species that are ectomycorrhizal (Alexander, 1989). Ectomycorrhizas are absent from monocots. The fungi involved in this symbiosis are almost exclusively basidiomycetes and ascomycetes except for genus *Endogone* which belongs to Zygomycetes. Common genera belonging to Basidiomycetous fungi include both hypogeous and epigeous genera such as *Amanita, Boletus, Leccinium, Suillus, Hebeloma, Gomphidius, Paxillus, Clitopilus, Lactarius, Russula, Laccaria, Thelephora, Rhizopogon, Pisolithus, Scleroderma* and others.

2.3. Ectendomycorrhiza

Ectendomycorrhiza (EEM) are characterized by characters similar to ectomycorrhiza but in addition to intercellular penetration of epidermal and cortical cells, there is also intracellular penetration of the cells. Development of EEM is determined by plant species involved and it is restricted primarily to *Pinus* and *Picea* and to a lesser extent to *Larix. Pinus* and *Larix* become ectendomycorrhizal with a fungal strain E of *Wilcoxina* sp., while all other tree species become ectomycorrhizal with the same mycobiont (Laiho, 1965). Initial contact area between the symbionts is epidermis. Mantle hyphae soon become embedded in large quantities of mucigel resulting in a smooth root surface without an apparent mantle. Mucigel is of root origin and signals from this mucigel are important in the initiation of hyphal branching near the root surface (Scales and Peterson, 1991).

Intracellular penetration of healthy root cortical and epidermal cells which occurs five to six cells behind the apical meristem is through appressorium formation (Wilcox, 1971), or hypha induced cortical cell distention inwards. Piche *et al.* (1986) hypothesized that nutrient exchange may occur during intracellular penetration phase when both hyphae and root cells are healthy and compatible.

2.4. Ericoid Mycorrhizas

Plants belonging to the families Ericaceae, Empetraceae and Epacridaceae are involved in ericoid mycorrhizal associations. Mycorrhiza structure varies from ectotypes through ectendo to pure endotypes. Ericoid mycorrhizas are unique in certain characters such as there is more specificity shown between plant and fungal species (Harley and Smith, 1983) than in other types, often plants with this type of colonization occur in extreme soil conditions (Read, 1983; Read and Kerley, 1995). The role of symbiosis in nature is not well understood because the plant species forming this type of mycorrhiza have unusually fine roots that function to increase the absorptive surface of the plant (Read, 1992). Mycorrhiza formation in these roots doesn't really help in increasing the absorptive surface, rather the fungal partner play an important role in releasing enzymes and other exudates into the substrate to make recalcitrant substance available to the plant (Read, 1992). The fungal partner of ericaceous plants are either Ascomycetes or Basidiomycetes (Stoyke and Currah, 1991), the best studied being the ascomycetes *Hymenoscyphus ericae* (Read) which is associated with a number of ericoid species. Its involvement in the nutrition of the host plant and in the detoxification of its rooting environment has been extensively documented over the last twenty years (Read, 1996).

As in all mycorrhizal associations, recognition between the two symbionts is followed by root penetration by hyphae with the help of an appressorium. From the appressorium thin septate hypha originates and penetrates the epidermal cell wall. Once within cytoplasm of the epidermal cell, the host cell plasma membrane and an interfacial matrix material encase the penetrating hypha. Penetrating hypha grows and coils inside the cell (Bonfante-Fasolo and

Gianinazzi-Pearson, 1982). Presumably nutrient exchange occurs between hyphal coils and epidermal cell cytoplasm.

2.5. Orchid Mycorrhiza

By definition orchid mycorrhiza involves association between the roots of orchid species and fungi, but the association between fungi and protocorm (this structure intercalates between the embryo and seedling) is often included in this type of mycorrhiza (Hadley, 1982). This association has received the most attention because it plays an imprtant role in seedling establishment and the fungus supplies a source of carbon to the achlorophyllous structure. *In vitro* studies on colonization process reveal that seeds of many orchid species can be germinated either in the presence of various carbon sources or a complex carbon source such as cellulose with an appropriate fungal species. There is very little information available on the first contact of hyphae with protocorm and there is no evidence of the presence of chemotrophic substances being produced by the protocorm (Williamson and Hadley, 1970). The entry point for the fungus could be either the suspensor end of the embryo (Clement, 1988; Peterson and Currah, 1990; Richardson *et al.,* 1992) or the epidermal hairs (Williamson and Hadley, 1970). Burgeff (1959) and Currah *et al.* (1988) in their published monographs depicted the entry points for the fungus.

After entering into developing protocorms, the hyphae colonizes parenchyma cells and assume a coiled configuration referred to as 'pelotons'. Nutrient exchange may occur through peloton-parenchymatous cell protoplasm. Pelotons resemble arbuscules of VA mycorrhizas since they are digested by the host cells shortly after their formation. Under certain culture conditions, endophytic fungi become parasitic on the protocorm and this leads to degradation of this structure (Hadley, 1982).

2.6. Arbutoid Mycorrhizas

This type of mycorrhiza is restricted to phytobionts from Ericales, particularly the genera *Arbutus, Pyrola* and *Arctostaphylos.* There is a little or no specificity between the phytobiont and mycobionts (Molina and Trappe, 1982). Fungal species forming mycorrhizas with arbutoid plant species are frequently ectomycorrhizal in association with other phytobionts (Molina and Trappe, 1982), these mycobionts belong to Ascomycetes and Basidiomycetes (Molina and Trappe, 1982; Zhu *et al.,* 1988). In arbutoid mycorrhizas the root morphogenesis frequently involves precocious lateral root formation (Massicotte *et al.,* 1993) giving rise to a 'cruciform' branching pattern (Fuscini and Bonfante-Fasolo, 1984), followed by subsequent development of a pinnate cluster or tubercle (Massicotte *et al.,* 1993; Rivett, 1924).

Anatomical details reveal the presence of mantle of differing thickness (mantle is absent sometimes) (Robertson and Robertson, 1985), and intracellular penetration of the epidermal cells by Hartig's Net or by mantle similar to that in ericoid (Bonfante-Fasolo and Gianinazzi-Pearson, 1982) and ectendomycorrhizas (Scales and Peterson, 1991). Intracellular hyphae are a potential site for the nutrient exchange or for communication between the partners. The extensively branched intracellular hypha is surrounded by host plasma membrane and matrix material (Robertson and Robertson, 1982). Root cell cytoplasm degenerates before the degeneration of intracellular fungus, indicating that lysis of the fungal component is not a feasible method for nutrient exchange to the root (Read, 1992).

2.7. Monotropoid Mycorrhizas

This type of association is limited to plants from a small sub family Monotropoidae that belongs to Ericales. Here the phytobiont is achlorophillous and has requirement for carbon compounds and other nutrients, which it receives from green plants through a common ectomycorrhizal association rather than by direct parasitic connection. The molecular identification methods (Cullings *et al.,* 1996) have shown that some monotropoides are highly specific in their fungal associations and at least one species, *Pterospora andromedea* is specialized on a single species within the genus *Rhizopogon.* Phylogenetic analysis of the Monotropoideae shows that specialization has been derived through narrowing of fungal associations within the lineage containing *P. andromedea* during the evolution.

Colonization initiates in a manner similar to ectomycorrhizas in that rhizosphere hyphae contact the root epidermis and develop a mantle of multilayered tightly interwoven hyphae (Robertson and Robertson, 1982). The most distinctive and unique feature of this association is the epidermal and hyphal cell wall modifications that occur during intracellular penetration stage. The protrusion by a single fungal peg is concurrent with extension of the epidermal wall to surround the invading hypha, suggesting the maintenance of a principally intercellular fungal association (Duddridge and Read, 1982).

3. MOLECULAR ASPECTS

3.1. Molecular Methods in Taxonomy of Mycorrhizal Fungi

Molecular techniques have been used in taxonomic studies of fungi. Two major applications of molecular methods are systematics and the elucidation of evolutionary relationships and the delineation of species concept. Molecular methods employed in the classification of fungi include isozyme analysis, immunochemical methods, polypeptide analysis, RNA coding DNA sequence analysis and RFLP and RAPD analysis. Systematics and methods of characterizing and classifying mycorrhizas are needed because it is always very difficult to define and identify each mycorrhizas as a function of both a known tree species and a known fungus. The different kinds of mycorrhizas are a confusion of colors, sizes and shapes (Zak, 1971). A few attempts have been made to identify ectomycorrhizal fungi by the protein pattern of different species (Burgess *et al.,* 1995b; Mouches *et al.,* 1981). Burgess *et al.* (1995b) compared and classified 85 Australian and 15 non-Australian *Pisolithus* isolates according to basidiospore and basidiome morphology, culture characteristics and separation of polypeptides using one dimensional SDS-PAGE. They found that one cluster was composed of all the non-Australian isolates collected beneath *Pinus,* whilst within Australia, isolates from the eastern, southern and western seaboards fell into distinct clusters. They concluded that the analysis of polypeptide pattern could povide a meaningful classification system to assist in isolate selection for future experiments.

3.2. Immunochemical Approach

Serological techniques have been used to characterize various fungi including ectomycorrhizal fungi. Cleyet-Marel *et al.* (1989) purified an acid phosphatase from *Pisolithus tinctorius* and used it to prepare polyclonal antibodies. These antibodies were applied in homologous and heterologous tests on the mycelium and mycorrhizas of several fungi. They concluded that immunochemical approach seems to be a way to characterize and to detect fungi provided a particular protein is used as a marker.

Immunochemical methods facilitate the study of difficult systematics of the arbuscular mycorrhiza (AM) forming fungi beloging to the order Glomales. Though the use of polyclonal antibodies rarely reach below the generic level, but monoclonal antibodies (mAb) can differentiate AM fungal spores to the species and strain level. But these serological techniques need to be combined with an improved enrichment procedure for spores and hyphae of AM fungi, so that the risk of unspecific binding is reduced. The production of specific antibodies assumes that fungal species or isolates contain different antigens or groups of antigens, which can be extracted. Polyclonal antisera varied greatly in specificity from discriminating between VAM and non-VAM fungi (Wilson *et al.*, 1983) and also in distinguishing between VAM fungi at the generic level (Adwell *et al.*, 1985; Kough *et al.*, 1983). Wright *et al.* (1987) successfully attempted to raise mAb against soluble antigens from spores of *Glomus occultum*. These mAbs had a high specifcity for their homologous antigens and gave significantly weaker signals with 29 other specific mycorrhizal and 5 non-mycorrhizal fungal isolates when tested in an ELISA. These mAbs were even able to differentiate their antigens from isolates of the same species from different locations (Oramas-Shirey and Morton, 1990). Scanders *et al.* (1992) raised polyclonal antisera to soluble spore fractions of VAM fungi, *Gigaspora margarita* and *Acaulospora laevis*. They found a major cross reactivity between *A. laevis* and *G. margarita* and also to some extent with the heterologous antigens from spores of the other VAM fungi tested in a dot immunobinding assay. While an indirect ELISA gave more specific reactions. Fries and Allen (1991) have shown by using polyclonal Abs coupled to a fluorescent stain that inoculated AM fungi can survive in the field for upto two years, even in competition with indigenous AM fungi. Thingstrup *et al.* (1995) detected *Scutellospora heterogama* within the roots using polyclonal antisera.

Isozyme analysis is now used by mycologists to resolve taxonomic disputes, identify unknown fungal taxa, 'fingerprinting' patentable lines, analyze the amount of genetic variability in population, trace the origin of pathogens, follow the segregation of loci and identify ploidy levels throughout the life cycle of an organism (Burdon and Masrhall, 1983; Micales *et al.*, 1986). The technique is based on analyzing the variation in isozyme electrophoretic mobilities encoded by different alleles or separate genetic loci. Such variations are the result of variations in the amino acid content of the molecule, which is dependent on the sequence of nucleotides in the DNA.

Cameleyre and Olivier (1983) studied variability between twelve isolates of the EM fungus *Tuber melanosporum* using isoelectric focussing of four enzymes including acid and alkaline phosphatases, phosphoglucomutases and esterases. Authors observed intraspecific variation resulting in the separation of isolates into several groups. Zhu *et al.* (1988) studied the genetic variability of isozymes of eight different enzymes of various strains of *S. tomentosus* showing geographical clustering of isolates. Ho and Trappe (1987) have shown that the isolates mobilities representing six species and three host sections of *Rhizopogon* had acid phosphatase isozyme that differed greatly in different species but were more conserved with in a species. The isolates of *L. laccata* were divided into three host related groups based on similar analysis (Ho, 1987). While Sen and Hepper (1986) used selective enzyme staining following polyacrylamide gel electrophoresis to characterize VAM fungi like *Glomus*, and showed that the isozymic analysis allowed at least the characterization of species. Hepper *et al.* (1986) could identify the mycelium of the VAM fungi in roots of leak and maize on the basis of isozyme analysis. Hepper *et al.* (1988) also showed that electrophoretic analysis of isozymes can differentiate isolates of *Glomus clarum, G. monosporum* and *G. mosseae* into several clusters. Host genotype and fungal diversity was characterized in scot pine ectomycorrhiza from natural humus microcosms using isozyme analysis (Timonen *et al.*, 1997)

3.3. Analysis of Conserved DNA Sequences

Molecular biology finds wide applications in the identification of genetic variability in the mycorrhizal fungal genome by DNA fingerprinting. This technique has the potential to identify species and isolates by observing DNA polymorphism resulting from mutations and chromosomal rearrangements. Restriction Fragment Length Polymorphism (RFLP) analysis which detects length mutations and alterations in base sequences has been used for constructing DNA fingerprints of *L. bicolor* and *L. laccata* (Martin *et al.*, 1991), *Hebeloma* (Marmeisse *et al.*, 1992), *Cenococcum geophilum* (Lobugio *et al.*, 1991) and *Tuber* species (Hanrion *et al.*, 1994) and to distinguish EM fungi (Armstrong *et al.*, 1989). The PCR amplification of targeted genomic sequences from microorganisms followed by RFLP, allele specific hybridization, direct sequencing or single strand conformation polymorphism is increasingly used to detect and characterize rhizospheric microbes, including mycorrhizal fungi in natural ecosystems (Bruns *et al.*, 1991; Erland, 1995; Gardes *et al.*, 1991; Henrion *et al.*, 1992; Karenet *et al.*, 1997; Pritsch and Buscot, 1996; Pritsch *et al.*, 1997, Simon *et al.*, 1992a, b; Timonen *et al.*, 1997). The PCR primers have been designed based on highly conserved regions of rDNA to amplify the internal transcribed spacers (ITS) and the intergenic spacers (IGS) which are highly polymorphic non-coding regions providing a useful tool for taxonomic and phylogenetic studies (Gardes and Bruns, 1993). The technique involves enzymatic amplification of DNA fragments spanning between terminal sequences recognized by specific oligonucleotide primers. Amplified products are then digested by restriction endonucleases giving rise to DNA fingerprints. PCR probes with high specificity and high sensitivity have been developed based on DNA sequences with low degree of conservation and which are highly repetitive. PCR-RFLP analysis of ITS regions have been used successfully by Sweeney *et al.* (1996), for *L. proxima* isolates, Gandeboeuf *et al.* (1997a) for *Tuber melanosporum*, and Anderson *et al.* (1998) for *Pisolithus* isolates.

The use of different target sites such as mitochondrial DNA and tRNA genes and different endonuclease combinations would extend the scored RFLPs. The small size of the mtDNA and its mode of inheritance independent of the nuclear genome make this molecule attractive for studies of fungal biology, taxonomy and evolution (Taylor *et al.*, 1995). The RFLP analysis of mtDNA has been used to estimate the relationships among and within the species of many groups of fungi including Basidiomycetes, Ascomycetes and Oomycetes (Gardes *et al.*, 1991), for isolate identification of *Laccaria, Tuber aestivum* (Guillemaud *et al.*, 1996). Primers designed for microsatellite DNA were also found to be reliable, sensitive and technically simple tools for assaying genetic variability in mycorrhizal fungi, and can be used to discriminate mycorrhizal symbionts with different taxonomic features (Bonjante *et al.*, 1997; Longato and Bonfante, 1997).

Randomly Amplified Polymorphic DNA (RAPD) is an alternative way of obtaining polymorphisms based on PCR (Tommerup, 1992; Tommerup *et al.*, 1992). Short oligonucleotide primers of arbitratry sequences anneal to complementary sequences occuring randomly in the genome. Amplified templates are separated by electrophoresis to reveal DNA fingerprints. This technique is very rapid and does not require target sequence information and can provide markers in genome regions inaccessible to PCR/RFLP analysis. Successful distinction of isolates of various ectomycorrhizal fungi has been achieved using RAPD technique (Anderson *et al.*, 1998; Doudrick *et al.*, 1995; Gandeboeuf *et al.*, 1997b; LePage *et al.*, 1996; Tommerup, 1992; Tommerup *et al.*, 1992; Wyss and Bonfante, 1992). According to Tommerup *et al.* (1995), RAPD fingerprinting of *Laccaria, Hydnanagium* and *Rhizoctonia* are reproducible in any laboratory providing the same set of reaction and thermocycle conditions are used. Identification of ectomycorrhizal fungi using molecular techniques has been extensively reviewed by Gupta and Satyanarayana (1996).

Since endomycorrhizal fungi are unable to grow in pure culture and because the spores that can be identified are produced in the soil outside the colonized roots, the identity of the endophyte inside the field collected colonized roots is difficult to determine with any degree of certainty. The development of DNA based methods could allow the identification of endomycorrhizal fungi in a sample and provide a useful tool to complement the expertise of fungal taxonomists. Recently, sequences of the gene coding for the small subunit rRNA (SSU) were obtained from 12 endomycorrhizal fungal species and comparisons with other 18S sequences showed that the Glomales formed a phylogenetically coherent group that originated about 400 million years ago (Bruns *et al.*, 1991; Simon *et al.*, 1993a). Twelve nuclear genes coding for SSU were sequenced from different species of VAM fungi (Simon *et al.,* 1992 a,b; 1993b). These were aligned and regions showing potential for the design of taxon specific probes or primers were identified by Simon *et al.* (1993b) and four taxon specific primers (VAACAU, VAGIGA, VAGLO and VALETC) were designed and synthesized. Simon *et al.* (1992a,b) used these in conjunction with Glomale specific VANS 1 primer described. These taxon specific primers were used to amplify portions of the nuclear gene coding for the small subunit rRNA. By coupling the sensitivity of the PCR and the specificity afforded by taxon-specific primer, a variety of samples can be analyzed including small amounts of colonized roots. The amplified products are then subjected to single-strand conformation polymorphism analysis to detect sequence differences. Advantage of this approach is the possibility of directly identifying the fungi inside field collected roots, without having to rely on the fortuitous presence of spores (Simon *et al.,* 1993b), Simon *et al.* (1993b) devised a rapid quantitation method of endomycorrhizal fungus *Glomus vasiculiferum* colonizing leak roots (*Allium porum*). Diversity of the ribosomal internal tranctribed sequences within and among isolates of *Glomus mosseae* and related mycorrhizal fungi was explored by Lloyd-Macgilp *et al.* (1996) in order to various isolates. A highly repeated sequence from the genome of *Scutellospora castanea* was isolated and characterized and it was shown to be *Scutellospora* specific probe in Southern and dot blot hybridizations (Zeze *et al.,* 1996).

3.4. Genetic Transformation of Mycorrhizal Fungi

The transformation systems described so far have been developed for fungi that are well characterized and can be cultured in the laboratory in defined media. Since the extraction of DNA and cloning of certain genes in *E. coli* appears to be possible for a wide variety of fungi, theoretically it should be possible to develop systems in any fungus that can be readily cultured *in vitro*. This makes development of gene cloning systems in the culturable EM fungi possible. The most obvious drawback to developing such systems for VAM fungi is the lack of a defined culture medium and problems in obtaining amounts of pure fungal material for nucleic acid extraction. Though there is a possibility of introducing DNA into germinating spores and the presence of an introduced gene can be selected for by the limited growth occurring on defined medium. Since the spores are multinucleated a strong selection for the introduced gene would be necessary to ensure its maintenance during subsequent propagation of the fungus in association with a suitable plant.

The applications of gene cloning in mycorrhizal fungi include the possibility of introducing novel genes determining products that would enhance the growth or health of the plant partner, introduction of fungicide resistance in the mycorrhizal fungi and it also provides the information about the fundamental biology of mycorrhizal associations.

Prospects and procedures for DNA mediated transformation of EM fungi has been reviewed earlier by Barrett (1992), Lemke *et al.* (1991), Martin *et al.* (1994) and Tigano-Milani *et al.* (1995). The very first report of genetic transformation of a mycorrhizal fungus came from Barrett *et al.* (1990). They successfully transformed *L. laccata* protoplasts at a frequency of 5 transformants per μg of DNA. The promoter and terminator signals used were of *Aspergillus nidulans* (Ascomycetous) origin and worked well in *L. laccata* which is a basidiomycetous fungus. Another evidence for transformation of ectomycorrhizal fungus *Hebeloma cylindrosporum* came from the study of Marmeisse *et al.* (1992b). They transformed *H. cylindrosporum* protoplasts with *E. coli* hygromycin phosphotransferase (hph) gene along with tryptophan biosynthesis gene and /or NADP-glutamate dehydrogenase gene from *Coprinus cinereus*. Stable transformants were obtained and the transformants retained their ability to form mycorrhiza. Bills *et al.* (1995) described transformation of EM fungus *Paxillus involutus* by particle bombardment. Transformation was determined using the *hph* gene as the selectable marker and the β-glucuronidase gene (GUS) as a reporter gene. The hygromycin resistant transformants were obtained which were mitotically stable and maintained both the *hph* and GUS genes in the fungal genome in integrated form. Synthesis experiments showed that transformed *P. involutus* can form ectomycorrhiza with *Pinus resinosa*. These provided the first report of a successful transformation of an EM fungus using particle bombardment. Occurrence of two mitochondrial, linear double stranded DNA plasmids have been reported in *Hebeloma circinans* (Schrunder *et al.*, 1991). This suggests that there is a possibility of identifying sequences of origin of replication from these plasmids and to construct transformation vectors utilizing those sequences.

4. FUNGUS ROOT INTERACTIONS

The interaction between mycorrhizal fungi and roots involves a series of events ultimately leading to an integrated, functional structure, and these events are not understood as yet at molecular level.

4.1. Endomycorrhizal Associations

Endomycorrhizas are the most widespread and most ancient type of plant-fungus association. Majority of terrestrial plants have evolved with compatibility systems towards the fungal symbionts. Cellular interactions leading to mycorrhiza formation are complex. Some events are shared with nodulation processes but molecular modifications specific to arbuscular mycorrhiza formation are also found. It has been suggested that defense gene expression in plants is controlled during establishment of a successful symbiosis. Modifications such as changes in wall metabolism and protein expression are also induced in fungal symbiont. Since endomycorrhizal fungi cannot be grown in cloning of DNA from these fungi opens the possibility of identifying genes involved in symbiosis establishment.

4.2. Fungal Induced Plant Responses

Host root show little cellular response to invasion by an arbuscular mycorrhizal fungus until arbuscule formation occurs in the cortical cells. Which results in the induction of increased abundance of cell organelles, enhanced nuclear acitvity, extensive proliferation of membrane systems and accumulation of a number of host derived molecules in the resulting interface formed between the symbiont cells (Bonfante-Fasolo, 1994; Gianinazzi-Pearson, 1995). It has been observed that very weak activation of plant defense genes occurs which is transient and very localized during mycorrhiza establishment (Bonfante-Fasolo, 1994; Gianinazzi-Pearson *et al.*, 1992; Harrison and Dixon, 1994; Lambais and Mehdy, 1995). It has been proposed that in mycorrhizal plants specific host genes may play a regulatory role towards defense genes, so that their expression is suppressed or maintained at a low level.

Some events in symbiosis establishment by AM fungi and rhizobia are determined by commom plant genes, that is why modifications in similar molecular components (glycoconjugates and nodulins) occur during the two types of root infections (Perotto *et al.*, 1994; Wyss *et al.*, 1990). Now the presence of symbiosis specific plant genes has been confirmed because of : (i) The detection of new proteins or polypeptides (endomycorrhizins) indicating *de novo* gene expression, (ii) mycorrhiza induced modifications in gene expression as confirmed by analysis of mRNA translation products and differential RNA display banding pattern and (iii) disappearance of certain root polypeptides but their precise role in mycorrhiza development and functioning is still unknown.

4.3. Changes Induced in Fungal Behaviour

Spore germination and hyphal elongation is stimulated by factors secreted by host plants, whilst root exudates of non-host plants do not have effect (Giovannetti *et al.*, 1994; Glenn *et al.*, 1988). Root exudates also elicit morphogenetic modification in the mycelium growth leading to hyphal branching and appressoria formation (Giovannetti *et al.*, 1994; Glenn *et al.*, 1988). When the hyphae reach the cortical cells, extensive proliferation of mycelium and differentiation of arbuscules is induced, accompanied by modifications in cell wall structure and molecular composition. Investigation into the structure and regulation of fungal genes require the establishment of representative genomic libraries of AM fungi.

In conclusion development of the mycorrhizal association involves cosiderable morphogenetic and physiological changes in both symbionts which are triggered by a complex cascade of signals. Rapid growth in molecular techniques is facilitating the possibility of analysing temporal and spatial gene expression in the two partners.

4.4. Ectomycorrhizal Associations

Physiological, morphological and anatomical studies have demonstrated that the metabolism of free-living plants and EM fungi are altered while entering into symbiosis (Harley and Smith, 1983). However, the molecular events associated with the establishment and maintenance of these associations are poorly understood. Compatibility in EM associations is genetically controlled in plants (Walker *et al.*, 1986) and in fungi (Kropp and Fortin, 1987). A significant contribution to this field is the observation that EM root formation leads to differential gene expression (Hilbert *et al.*, 1991a) The protein profile of *Eucalyptus globulus* and EM fungus *Pisolithus tinctorius* association are different than the protein of these two organisms living axenically. Specific proteins called 'ectomycorrhizins' were uniquely associated with EM roots (Hilbert *et al.*, 1991a,b). Duchesne (1989) obtained similar results while studying the protein profiles of *Pinus resinosa* and *Paxillus involutus* association. He concluded that metabolism of *Pinus resinosa* roots and *P. involutus* change even prior to physical contact between these two organisms indicating gene expression in EM plants and fungi is altered by the onset of the symbiotic stage.

The EM association between *Eucalyptus* and *Pisolithus tinctorius* has been a useful model for the study of many diverse, basic aspects of the biology of symbiosis such as early stages of EM formation and the discovery of new gene products involved in establishment of the symbiosis and its morphogenesis (Martin and Tagu, 1995), because of its amenability to manipulation *in vitro* (Malajczuk *et al.*, 1990). Symbiosis induced gene expression is well studied for *Eucalyptus-Pisolithus* associations (Martin and Tagu, 1995; Nehls and Martin, 1995). Changes in gene expression leads to appearance and disappearance of new protein and changes in metabolic organization in fungal and plant cells (Burgess *et al.*, 1995a; Hilbert and Martin, 1988). The role and function of several symbiosis related (SR) proteins have recently been characterized by molecular and biochemical approaches. The SR proteins correspond to pathogenesis-related proteins, products of auxin-induced genes, enzymes (Nehls and Martin, 1995) and fungal hydrophobins (Tagu *et al.*, 1996).

The key event in mycorrhiza formation is the adhesion of hyphae which involves cellular differentiation including dramatic changes in the structure and composition of the partner cell walls. The knowledge of these processes is mostly based on cytological observations but several cell wall proteins (CWP) and genes coding for CWP nad their regulation have recently been described (Tagu and Martin, 1996).

4.5. Morphological Alterations during early Stages of Ectomycorrhiza Formation

Initial changes in host plant include stimulation of lateral root formation by fungal auxins, radial elongation of epidermal cells and arrest of cell division of ensheathed roots (Peterson and Bonfante, 1994). Changes in fungi include, shortening of hyphal cells, with few clamp connections and branching leading to pseudoparenchymatous mode of growth (Horan and Chilvers, 1990). Hyphae in the mantle are highly branched, glued together, and embedded in an abundant extracellular matrix composed of polysaccharides, cysteine rich proteins and glycoproteins. Mycelium in the Hartig's net has multilobed hyphal fronts with a complex fan like structure (Scheidegger and Brunner, 1995). The development of interface between the two partners is remarkably uniform regardless of the species of host or fungus, which contains a complex matrix of polysaccharides and proteins. Insight into the cell wall formation and alteration during the symbiosis shows a high level of up regulated 32 kDa symbiosis related acidic polypeptides and hydrophobins and decreased content of mannoprotein gp95, which simultaneously regulates the molecular architecture of protein network in a manner to allow new development fates for both fungal cell cohesion and root colonization by the fungus (Tagu and Martin, 1996).

4.6. Molecular Cloning of Symbiosis Related Fungal Genes

Since it is well documented that there is a shift in the expression of a set of developmentally related genes during ectomycorrhiza formation both in fungal as well as plant symbiont. Several clones were obtained by differential screening of a cDNA library of *E. globulus - P. tinctorius* mycorrhiza using cDNA probes prepared either from mRNA isolated from free-living mycelium or *P. tinctorius* colonized roots (Tagu *et al.,* 1993), that represent about one third of the cDNA clones screened confirming early contentions (Hilbert and Martin, 1988) based on protein analysis, that early development steps of ectomycorrhiza formation induce major shift in fungal gene expression. This is confirmed by the analysis of mRNA populations, and the drastic decreased expression of plant genes is noteworthy.

5. CONCLUSIONS

Mycorrhizal associations are extremely important in nature, since majority of plants occurring in ecosystem are mycorrhizal. Mycorrhiza appear to confer various advantages on to the host plant including protection against heavy metals and soil toxins, plant pathogens, adverse temperature, adverse soil pH, high salinity; stimulation of plant growth by elaborating plant growth promoting hormones, transport and solubilization of minerals by siderophore and organic acid production. Seven types of mycorrhizal forms have been recognized including endomycorrhiza, ectomycorrhiza, ectendomycorrhiza, arbutoid, ericoid, monotropoid and orchid mycorrhiza. Molecular methods such as polypeptide analysis, isozyme analysis, immunochemical approach, and analysis of conserved DNA sequences including PCR-RFLP of rRNA sequences and RAPD have been employed in the taxonomy, and to study geographical distribution of these fungi. Attempts have been made to transform some of mycorrhizal fungi so as to improve their efficacy as mycorrhizal symbiont, and also to understand the molecular mechanism of mycorrhiza formation.

REFERENCES

Adwell. F.E.B., Hall, I.R. and Smith, J.M.B. 1985, Enzyme linked immunosorbant assay as an aid to taxonomy of the Endogonaceae, *Trans. Brit. Mycol. Soc.* 84: 399-402/

Albrecht, C., Laurent, P. and Lapeyrie, F. 1994, Eucalyptus root and shoot chitinases induced following root colonisation by pathogens versus ectomycorrhizal fungi, compared on one-and two-dimensional activity gels, *Plant Sci.* 100: 157-164.

Alexander, I.J. 1989, Mycorrhizas in topical forest, in: *Mineral Nutrients in Tropical Forest and Savanna Ecosystems*, J. Proctor, ed., Oxford : Blackwell Scientific Publishers, pp.169-188

Anderson, I.C., Chambers, S.M. and Cairney, J.W.G. 1998, Molecular determination or genetic vriation in *Pisolithus* isolates from a defined region in New South Wales, Australia, *New Phytol.* 138:151-162.

Armstrog, J.L., Fowles, N.L. and Rygiewiz, P.T. 1989, Restriction fragmen length polymorphisms distringuish ectomycorrhizal fungi, *Plant Soil* 116:1-7.

Barrett, V., Dixon, R.K. and Lemke, P.A. 1990, Genetic transformation from selected species of ectomycorrhizal fungi, *App. Microbiol. Biotechnol.* 33:313-316.

Barrett, V. 1992, Ectomycorrhizal fungi as experimental organisms, in: *Hand Book of Applied Mycology.* Vol.l. Soil and Plants, D.K. Arora, B Rai, K.G. Mukerji and G.R. Knudsen, eds., Marcel Dekker Inc., USA, pp.217-229.

Bills, S.N., Richter, D.L. and Podila, G.K. 1995, Genetic transformation of the ectomycorrhizal fungus *Paxillus involutus* by particle bombardment, *Mycol. Res.* 99: 557-561.

Bonfante, P., Lamfrance, L.O., Cometti, V. and Gure, A. 1997, Inter-and intraspecific variability in strains of the ectomycorrhizal fungus *Suillus* as revealed by molecular techniques, *Microbiol. Res.* 152:287-292.

Bonfante-Fasolo, P. 1994, Ultrastructural analysis reveals the complex interactions between root cell and arbuscular mycorrhizal fungi, in: *Impact of Arbuscular Mycorrhizas on Sustainable Agriculture and Natural Ecosystems,* S. Gianinazzi and H. Schuepp, eds., Birkhauser, Basle pp. 73-87.

Bonfante-Fasolo, P. and Gianinazzi-Pearson, V. 1982, Ultrastructural aspects of endomycorrhiza in the Ericaceae. III. Morphology of dissociated symbionts and modifications occurring during their re-association in axenic culture, *New Phytol.* 91: 691-704.

Boyd, R. 1987, The role of ectomycorrhizas in the water relations of plants, Ph.D. thesis, University of Sheffield.

Bruns, T.D., White, T. J. and Taylor , J.W. 1991, Fungal molecular systematics, *Ann. Rev. Ecol. Systemat.* 22: 525-564.

Burdon, J.J. and Marshall, D.R. 1983, The use of isozymes in plant disease research, in: *Isozymes in Plant Genetics and Breeding,* S.D. Tanksley and T.J. Orton, eds., Vol. A, Elsevier New York, pp. 401-412.

Burgeff, H. 1959, Mycorrhiza of orchids, in: *The Orchids : A Scientific Survey,* C.L. Withner, ed., Ronald Press Co., New York, pp. 361-395.

Burgess, T., Laurent, P., Dell, B., Malajczuk, N. and Martin, F. 1995a, Effect of the fungal isolate aggrresivity on the biosynthesis of symbiosis-related polypeptides in differentiating eucalypt ectomycorrhiza, *Planta* 196: 408-417.

Burgess, T., Malajczuk, N. and Dell, B. 1995b, Variaion in *Pisolithus* based on basidiome and basidiospore morphology; Cultural characteeristics and analysis of polypeptides using ID SDS PAGE, *Mycol. Res.* 99: 1-13.

Cameleyre, I. and Olivier, J.M.1983, Evidence for intraspecific isozyme variations among French isolates of *Tuber melanosporum* (Vitt,), *FEMS Microbiol. Lett.* 110: 159-162.

Clements, M.A.1988, Orchid Mycorrhizal Associations, Lindeyana 3: 73-86.

Cleyet-Marel, J.C., Bausquet, N. and Mousain, D. 1989, The immunochemical approach for the characterization of ectomycorrhizal fungi, *Agri. Ecosy. Environ.* 28: 79-83.

Cullings, K.W., Szaro, T.M. and Bruns, T.D. 1996, Evolution of extreme specialization within a lineage of ectomycorrhizal epiparasites, *Nature* 379: 63-60.

Currah, R.S., Hambelton, S. and Smreciu, A. 1988, Mycorrhizae and Mycorrhizal fungi of *Calypso bilbosa,* *Amer. J. Bot.* 75: 739-752.

Dixon, R.K., Rao, M.V. and Garg, V.K. 1994, *In situ* and *in vitro* response of mycorrhizal fungi to salt stress, *Mycorrhiza News* 5: 6-8.

Doudrick, R.L., Raffle, V.L., Nelson, C.D. and Furnier, G.R. 1995, Genetic analysis of homokaryones from a basidiome of *Laccaria bicolor* using random amplified polymorphic DNA (RAPD) markers, *Mycol. Res.* 99: 1361-1366.

Duchesne, L.C. 1989, Protein synthesis in *Pinus resinosa* and the ectomycorrnizal fungus *Paxillus involutus* prior to ectomycorrhiza formation, *Trees* 3: 73-77.

Duchesne, L.C., Peterson, R.L. and Ellis, B.E. 1989, The future of ectomycorrhizal fungi as biological control agents, *Phytoprotec.* 70: 51-58.

Duddridge, J.A. and Read, D.J. 1982, An ultrastructural analysis of the development of mycorrhizas in *Monotropa hypopitys* L., *New Phytol.* 92: 203-214.

Erland, S. 1995, Abundance of *Tylospora fibrillosa* ectomycorrhizas in a south Swedish spruce forest measured by RFLP analysis of the RCR- amplified rDNA ITS region, *Mycol. Res.* 99: 1425-1428.

Fries, C.F. and Allen, M.F. 1991, Tracking the fates if exotic and local VA mycorrhizal fungi: methods and patterns, *Agric. Ecosys. Environ.* 34: 87-96.

Fusconi, A. and Bonfante-Fasolo, P. 1984, Ultrastructural aspects of host-endophyte relationships in *Arbutus unedo* L. mycorrhizas, *New Phytol.* 96: 397-410.

Gandeboeuf, D., Dupre, C., Roeckel-Drevet, P., Nicolas, P. and Chevalier, G. 1997a, Typing *Tuber* ectomycorrhizas by polymerase chain amplification of the internal transcribed spacer of rDNA and the sequence characterized amplified region markers, *Can. J. Microbiol.* 43: 723-728.

Gandeboeuf, D., Dupre, C., Roeckel-Drevet, P., Nicolas, P. and Chevalier, G. 1997b, Grouping and identification of *Tuber* using RAPD markers, *Can. J. Bot* 75: 36-45.

Gardes, M., White, T.J., Fortin, J.A.,Bruns,T.D.and Taylor, J.W.1991, Identification of indigenous and introduced symbiotic fungi in ectomycorrhizas by amplification of nuclear and mitochondrial ribosomal DNA, *Can. J. Bot.* 69: 180-190.

Gardner, J.H. and Malajczuk, N. 1988, Recolonization of rehabilitated boaxite mine sites in Western Australia by mycorrhizal fungi, *For. Ecol. Management* 24: 27-42.

Garrido, N., Becerra, I., Manticorea, C., Oehrens, E., Silva, M. and Horak, E. 1982, Antibiotic properties of ectomycorrhizas and saprophytic fungi growing on *Pinus radiata.* D. Don I. *Mycopathol.* 77: 93-98.

Gianinazzi-Pearson, V. 1995, Morphological compatibility in interactions between roots and arbuscular endomycorrhizal fungi : molecular mechanisms, genes and gene expression, in: *Pathogenesis and Host Parasite Specificity in Plant Disease,* Vol.II, K. Kohmots, R.P. Singh, and U.S. Singh, eds., Pergamon Press, Elsevier Science, Oxford.

Gianinazzi-Pearson, V., Tahiri-Alaoni, A., Antoniw, J.F., Gianinazzi, S. and Dumas Gandot, E. 1992, Weak expression of the pathogenesis related PR-bl gene and localization of related proteins during symbiotic endomycorrhizal interaction in tobacco roots, *Endocytobiosis Cell Res.* 8: 177-185.

Giovannetti, M., Sharana, C., Citternesi, A.S., Avio, L., Gollotte, A., Gianinazzi-Pearson, V. and Gianinazzi, S. 1994, Recognition and infection process, basis for host specificity of arbuscular mycorrhizal fungi, in: *Impact of Arbuscular Mycorrhizas on Sustainable Agriculture and Natural Ecosystems,* S. Gianinazzi, and H. Schuepp, eds., Birkhauser, Basel. pp.61-72.

Glenn, M.G., Chew, F.S. and Williams, P.H. 1988, Influence of glucosinolate content of *Brassica* (Cruciferae) roots on growth of vesicular-arbuscular mycorrhizal fungi, *New Phytol.* 110:217-225.

Gopinathan, S. and Raman. N. 1992, Indole 3-acetic acid production by ectomycorrhizal fungi, *Ind. J. Exper. Biol.* 30: 142-143.

Guillemaud, T., Raymond, M., Callot, G., Cleyet-Marel, J.C. and Fernandez, D. 1996, Variability of nuclear and mitochondrial ribosomal DNA of a truffle species (*Tuber aestivum*), *Mycol. Res.* 100:547-550.

Gupta, V. and Satyanarayana, T. 1996, Molecular and general genetics of Ectomycorrhizal fungi. in: *Concepts in Mycorrhizal Research,* K.G. Mukerji, ed., Kluwer Academic Publishers, Netherlands, pp. 343-361.

Hadley, G. 1982, Orcid mycorrhia, in: *Orchid Biology : Reviews and Perspectives*, Cornell University Press, Ithaca, New York, pp. 83-188.

Harley, J.L. and Smith, S.E. 1983, *Mycorrhizal Symbiosis*, Academic Press, New York.

Harrison, M.J. and Dixon, R.A. 1994, Spatial patterns of expression of flavonoid/isoflavonoid pathway genes during interactions between roots of *Medicago trunculata* and the mycorrhizal fungus *Glomus versiforme, Plant* J. 6: 9-20.

Henrion, B., Chevalier, G. and Martin, F. 1994, Typing truffle species PCR amplification of the ribosomal DNA spacers, *Mycol. Res.* 98: 37-43.

Henrion, B., Letacon, F. and Martin, F. 1992, Rapid identification of genetic variation of ectomycorrhizal fungi by amplification of ribosomal RNA genes, *New Phytol.* 122: 289-298.

Hepper, C.M., Sen, R. and Maskall, C.S. 1986, Identification of vesicular arbuscular mycorrhzal fungi in roots of leak (*Allium porrum* L.) and maize (*Zea mays* L.) on the basis of enzyme mobility during polyacrylamide gel electrophoresis, *New Phytol.* 102: 529-539.

Happer, C.M., Sen, R., Azcon-Aguiliar, C. and Grace, C. 1988, Variation in certain isozymes among different geographical isolates of the vesicular-arbuscular mycorrhizal fungi *Glomus clarum, Glomus monosporum* and *Glomus mosseae, Soil Biol. Biochem.* 20: 51-59.

Hilbert, J.L. and Martin, F. 1988, Regulation of gene expression in ectomycorrhizas 1, Protein changes and the presence of ectomycorrhiza specific polypeptides in the *Pisolithus-Eucalytus* symbiosis, *New Phytol.* 110: 339-436.

Hilbert, J.L., Costa, G. and Martin, F. 1991a, Regulation of gene expression in ectomycorhizas, Early ectomycorrhizas and polypeptide cleansing in eucalypt ectomycorrhizas, *Plant Physiol.* 97: 977-984.

Hilbert, J.L., Costa, G. and Martin. F. 1991b, Ectomycorrhizin synthesis and polypeptide changes during the early stage of eucalypt mycorrhiza development, *Plant Physiol.* 97: 977-984.

Ho, I. 1987, Enzyme activity and phytohormone production of a mycorrhizal fungus *Laccaria laccata, Can J. For. Res.* 17: 855-858.

Ho, I. and Trappe, J.M. 1987, Enzymes and growth substances of *Rhizopogon* species in relation to mycorrhizal hosts and infrageneric taxonomy, *Mycologia* 79:553-558.

Horan, D.P. and Chilvers, G.A. 1990, Chemotropism : the key to ectomycorrhizal formation, *New Phytol.* 116: 297-301.

Karen, O., Hogberg, N., Dahlberg, A., Jonsson, L. and Nylund, J. 1997, Inter-and intraspecific variation in the ITS region of rDNA of ectomycorrhizal fungi in Fennoscandia as detected by endonuclease analysis, *New Phytol.* 136: 313-325.

Kough, J., Malajczuk, N. and Linderman, R.G. 1983, Use of an indirect immunofuorescent technique to study the vesicular arbuscular fungus *Glomus epigaeum* and other *Glomus* species, *New Phytol.* 94: 57-62.

Kraigher, H., Grayling, A., Wang, T.L. and Hanke, D.F. 1991, Cytokinin production by two ectomycorrhizal fungi in liquid culture, *Phytochem.* 30: 2249-2254.

Kropp, B.R. and Fortin, J.A. 1987, The incompatibility system and ectomycorrhizal performace of monokaryons and reconstituted dikaryons of *Laccaria bicolor, Can. J. Bot.* 66: 289-294.

Laiho, O. 1965, Further studies on the ectendotrophic mycorrhiza, *Acta Forest Fenn.* 79: 1-35.

Lambais, M.R. and Mehdy, M.C. 1995, Differential expression of defence-related genes in arbuscular mycorrhiza, *Can. J. Bot.* 73: S533-S540.

Lanfranco, L., Wyss, P., Marzachi, C. and Bonfante, P. 1993, DNA probes for identification of the ectomycorrhizal fungus *Tuber magnatum* Pico, *FEMS Microbiol. lett.* 114: 245-252.

Lepage, B.A., Currah, R.S., Stockey, R.A. and Rothwell, G.W. 1996, Fossil ectomycorrhizas of Eocene *Pinus* roots, *Proc. Natl. Acad. Sci.* USA.

Lemke, P.A., Barrett, B. and Dixon, R.K. 1991, Procedures and prospects for DNA mediated transformation of ectomycorrhizal fungi, in: *Methods in Mocrobiology*, Vol. 23, J.R. Norris, D.J. Read and A.K. Verma, eds., Academic Press, London, pp. 281-293.

Lloyd-Magilp, S.A., Chambers, S.M., Dodd, J.C., Filter, A.H., Walker, C. and Young, J.P.W. 1996, Diversity of the ribosomal internal transcribed spacers within and among isolates of *Glomus mosseae* and related mycorrhizal fungi, *New Phytol.* 133: 103-111.

Lo Buglio, K.F., Rogers, S.O. and Wang, C.J.K. 1991, Variation in ribosomal DNA among isolates of the mycorrhizal fungus *Cenococcum geophilum, Can. J. Bot.* 69: 2331-2343.

Longato, S. and Bonfante, P. 1997, Molecular identification of mycorrhizal fungi by direct amplification of microsatellite regions, *Mycol. Res.* 101: 425-432.

Malajczuk, N., Garbay, J. and Lapeyrie, F. 1990, Infectivity of pine and eucalyptus isolates of *Pisolithus tinctorius* on roots of *Eucalytus urophilla in vitro*, 1, Mycorrhiza formation in Model Systems, *New Phytol.* 111: 627-631.

Marmeisse, R., Debaud, J.C. and Casselton, L.A. 1992, DNA probes for species identification in the ectomycorrhizal fungus *Hebeloma, Mycol. Res.* 96: 161-165.

Marmeisse, R., Gay, G., Debaud, J.C. and Casselton, L.A. 1992, Genetic transformation of the ectomycorrhizal fungus *Hebeloma cylindrosporum, Curr. Genetics* 228: 41-45.

Martin, F, and Tagu, D. 1995, Ectomycorrhiza development:a molecular perspective, in: *Mycorrhiza Structure, Function, Molecular Biology and Biotechnology*, A.K. Varma and B. Hock, eds., Springer-Verlag, Heidelberg : Berlin, pp. 29-58.

Martin, F., Laurent, P., DeCarvalho, D., Burgess, T., Murphy, P., Nehls, U. and Tagu, D. 1995, Fungal gene expression during ectomycorrhiza formation, *Can. J. Bot.* 73: S541-S547.

Martin, F., Tommerup, I.C. and Tagu, D. 1994, Genetics of ecotmycorrhizal fungi: progress and prospects, *Plant Soil* 159: 159-170.

Martin, F., Zaiou, M., Le Tacon, G. and Rygiewicz, P. 1991, Strain specific difference in ribosomal DNA from the ectomycorrhizal fungi *Laccaria bicolor* (Maire) Orton and *Laccaria laccata* (Scop exfr), *Bri. Ann. Sci. For.* 48: 133-142.

Marx, D.H. 1973, Mycorrhizae and feeder root disease, in: *Ectomycorrhizas*, G.C. Marks. and T.T. Kazlowsky, eds., Academic Press, London, pp. 351-382.

Marx, D.H. and Artman, J.D. 1979, *Pisolithus tinctorius* ectomycorrhizas improve survival and growth of pine seedlings on acid coal spoils in Kentucky and Virginia, *Reclam. Rev.* 2: 23-31.

Massicotte, H.B., Melville, L.H., Molina, R.and Peterson, R.L. 1993, Structure and histochemistry of mycorrhizas synthesised between *Arbutus menziesii* (Ericaceae) and two basidiomycetes, *Pisolithus tinctorius* (Pisolithaceae) and *Piloderma bicolor* (Cortacaceae), *Mycorrhiza* 3: `1-11.

Micales, J.A., Bonde, M.R. and Peterson, G.L. 1986, The use of isozyme analysis in fungal taonomy and genetics, *Mycotaxon* 27: 404-449.

Molina, R. and Trappe, J. 1982, Lack of mycorrhizal specificity by the ericaceous hosts *Arbutus menziesii* and *Arctostaphylos uva-ursi, New Phytol.* 90: 495-509.

Molina, R., Massicotte, H.B. and Trappe, J.M. 1992, Specificity phenomena in mycorrhizal symbiosis: community-ecological consequences and practical implications, in: *Mycorrhizal Functioning : An Integrative Plant Fungal Process*, M.F. Allen, ed., Chapman and Hall, New York and London.

Mouches, C.P., Duthil, N., Poitou, J., Delmas and Bove, J.M. 1981, Characterisation des especes truffieres par analyse de leurs protienes en gel depolyacrylamide et application de ces techniques a la taxonomie des champignons, *Mushroom Sci.* 11: 819-831.

Nehls, U. and Martin, F. 1995, Changes in root gene expression in ectomycorrhiza, in: *Biotechnology of Ectomycorrhizas: Molecular Approaches,* V. Stocchi, P. Bonfante, and M. Nuti, eds., Plenum Press, New York and London, pp. 125-137.

Oramas-Shirey, M. and Morton, J.S. 1990, Immunological stability among different geographical isolates of the arbuscular mycorrhizal fungus *Glomus occultum*, Abstracts of the 90th Annual meeting of American Society for Microbiology, Washington, pp. 311.

Osonuki, O.K., Mulongoy, O.O., Atayesa, M.O. and Okari, D.U.U. 1991, Effects of ectomycorrhizal and VAM fungi on drought tolerance of four leguminous woody seedlings, *Plant Soil* 136: 131-143.

Peiffer, C.M. and Bloss, H.E. 1988, Growth and nutrition of guayule (*Parthenium aggentatum*) in saline soil as influenced by VAM and phosphorus fertilization, *New Phytol.* 108: 351-361.

Perotto, S., Brewin, N.J. and Bonfante-Fasolo, P. 1994, Colonization of pea roots by the mycorrhizal fungus *Glomus versiforme* and by *Rhizobium*: immunological comparison using monoclonal antibodies as probes for cell surface components, *Mol. Plant Microbe Interaction* 7: 91-112.

Peterson, C.A. and Currah, R.S. 1990, Synthesis of mycorrhizas between protocorms of *Goodyera repens* (Orchidaceae) and *Ceratobasidium cereale, Can J. Bot.* 68: 1117-11125.

Peterson, R.L., Piche, Y. and Plenchette, C. 1984, Mycorrhiza and their potential use in the agricultural and forestry industries, *Biotechnol. Advan.* 2: 101-120.

Piche, Y., Ackerley, C.A. and Peterson, R.H. 1986, Structural characteristics of ectendomycorrhizas synthesised between roots of *Pinus resinosa* and E-strain fungus, *Wilcoxina mikolae* var. *mikolae, New Phytol.* 104: 447-452.

Pritsch, K. 1996, Untersuchungen tur Diversitat and Okologie von Mycorhizen der Schwarzele (*Alnux glutinous* (L.) Gaestin), Dissertation, University of Tuebingen, pp. 1-197.

Pritsch, K. and Buscot, F. 1996, Biodiversity of ectomycorrhizas from morphotypes to species, in: *Mycorrhizas in Integrated Systems from Genes to Plant Development*, C. Azcon-Aguilar and J.M. Barea, eds., Proceedings of the Fourth European Symposium, on Mycorrhizas, Europian Commission report 16728, Luxembour : Office for Official Publications of the Europian Communities, pp. 9-31.

Pritsch, K., Boyle, J.C., Munch, J.C. and Buscot, F. 1997, Characterization and identification of black alder ectomycorrhizas by PCR/RFLP analysis of the rDNA internal transcribed spacers (ITS), *New Phytol.* 137: 357-369.

Read, D.J. 1983, The biology of mycorrhiza in the Ericale, *Can. J. Bot.* 61: 985-1004.

Read. D.J. 1991, Mycorrhizas in ecosystems-nature's response to the law of the minimum, in: *Frontiers in Mycology,* D.L. Hawksworth, ed., CAB International, Wallingford, U.K. pp. 101-130.

Read, D.J. 1992, The ectomycorrhizal mycelium, in: *Mycorrhizal Functioning*, M.F. Allen, ed., Chapman and Hall, New York, U.S.A. pp. 102-133.

Read, D.J. 1996, The nature and extent of mutualism in the mycorrhizal symbiosis, in: *A Century of Mycology*, B. Sutton, ed., Cambridge University Press, U.K. pp. 255-292.

Read, D.J. and Kerley, S. 1995, The status and functions of ericoid mycorrhizal systems, in: *Mycorrhiza: Structure, Function, Molecular Biology and Biotechnology*, A. Varma, and B. Hock, eds., Spinger Verlag, Berlin. pp. 499-520.

Remy, W., Taylor, T.N., Hass , H. and Kerp, H. 1994, Four hundred million year old vesicular-arbuscular mycorrhizas, *Proc. Nat. Acad. Sci.* 91:11841-11849.

Richardson, K.A., Peterson, R.L. and Currah, R.S. 1992, Seed reserves and early symbiotic protocorm development of *Platanthera hyperlorea* (Orchidaceae), *Can J. Bot.* 70:291-300.

Rivett, M.F. 1924, The root-tubercles in *Arbutus unedo*, *Ann. Bot.* 38:661-667.

Robertson, D. and Robertson, J. 1982, Ultrastructure of *Petrospora andromedea* Nutall and *Sarcodes sanguinea* Torrey mycorrhizas, *New Phytol.* 92:529-551.

Robertson, D. and Robertson, J. 1985, Ultrastructural aspects of *Pyrola* mycorrhizas, *Can.J. Bot.* 63:1089-1098.

Scales, P. and Peterson, R.L. 1991, Structure and development of *Pinus banksiana* Wilcox ectendomycorrhizas, *Can. J. Bot.* 69:2135-2145.

Scanders, I.R., Ravolanirina, F., Gianinazzi-Pearson, V., Gianinazzi, S. and Lemoine, M.C. 1992, Detection of specific antigens in the vesicular arbuscular mycorrhizal fungi *Gigaspora margerita* and *Acaulospora laevis* using polonal antibodied to soluble spore fractions, *Mycol. Res.* 96:477-480.

Scheidegger. C. and Brunner, I. 1995, Electron microscopy of ectomycorrhiza: methods, applications and findings, in : *Mycorrhiza: Structure, Function, Molecular Biology and Biotechnology*, A. Verma, and B. Hock, eds., Springer Verlag, Heidelberg Berlin.

Schrunder, J., Debaud, J.C. and Mainhardt, F. 1991, Adenoviral like genetic elements in *Hebeloma circinans*, in: *Mycorrhizas in Ecosystems-Structure and Function*, (Abstracts), 3rd ESM Sheffield, U.K.

Sen, R. and Hepper, C.M. 1986, Characterization of vesicular arbuscular mycorrhizal fungi (*Glomus* spp.) by selective enzyme staining following polyacrylamide gel electrophoresis, *Soil Biol. Biochem.* 18:29-34.

Simon, L., Lalonde, M. and Bruns, T.D. 1992a, Specific amplification of 18S fungal ribosomal genes from vesicular arbuscular endomycorrhizal fungi colonizing roots, *App. Environ. Microbiol.* 58:291-295.

Simon, L., Levesque, R.C. and Lalonde, M. 1992b, Rapid quantitation by PCR of endomycorrhizal fungi colonizing roots, *PCR Methods Application* 2:76-80.

Simon, L., Bousquet, J., Levesque, R.C. and Lalonde, M. 1993a, Origin and diversification of endomycorrhizal fungi and coincidence with vascular land plants, *Nature* 363:67-69.

Simon, L., Levesque, R.C. and Lalonde, M. 1993b, Identification of endomycorrhizal fungi colonizing roots by fluorescent single-strand conformation polymorphismpolymerase chain reaction, *App. Environ. Microbiol.* 59:4211-4215.

Stoyke, G. and Currah, R.S. 1991, Endophytic fungi from the mycorrhizas of alpine ericoid plants, *Can. J. Bot.* 69:347-352.

Sweeney, M., Harmey, M.A. and Mitchell, D.T.1996, Detection and identification of *Laccaria* species using a repeated DNA sequence from *Laccaria proxima*, *Mycol. Res.* 100:1515-1521.

Tagu, D. and Martin, F. 1996, Molecular analysis of cell wall proteins expressed during the early steps of ectomycorrhiza development, *New Phytol.* 133:73-85.

Tagu, D., Nasse, B. and Martin, F. 1996, Cloning and characterization of hydrophobins-encoding cDNAs from the ectomycorrhizal basidiomycete *Pisolithus tinctorius, Gene* 168:93-97.

Tagu, D., Python, M., Cretin, C. and Martin, F. 1993, Cloning symbiosis-related cDNAs from eucalypt ectomycorrhizas by PCR assisted differential screening, *New Phytol.* 125:339-343.

Taylor, J.W. 1995, Fungal evolutionary biology and mitochondrial DNA, *Exper. Mycol.* 10:259-269.

Thingstrup, I., Rozycha, M., Jeffries, P., Rosendahl, S. and Dodd, J.C. 1995, Detection of the arbuscular mycorrhizal fungus *Scutellospora heterogama* within roots using polyclonal antisera, *Mycol. Res.* 99:1225-1232.

Tigano-Milani, M.S., Samson, R.A., Martin, I. and Sobral, B.W.S. 1995, DNA markers for differentiating isolates of *Paecilomyces lilacinus, Microbiol.* 141:239-245.

Timonen, S., Tammi, H. and Sen, R. 1997, Characterization of the host genotype and fungal diversity in Scots pine ectomycorrhizal from natural humus microcosms using isozyme and PCR-RFLP analyses, *New Phytol.* 135:313-323.

Tommerup, I.C. 1992, Genetics of eucalypt ectomycorrhizal fungi, in: *International Symposium on Recent Topics in Genetics, Physiology and Tehnology of Basidiomycetes*, M. Miyaji, A. Suzuki, and K. Nishimura, eds., Chiba University, Chiba, Japan, pp. 74-79.

Tommerup, I.C., Barton, J.E. and O'Brian, P.P. 1992, RAPD fingerprinting of *Laccaria, Hydnangium* and *Rhizoctonia* isolates, in: *The International Symposium on Management of Mycorrhizas in Agriculture, Horticulture and Forestry*, the University of Westen Australia, Nadlands (Abstracts), pp. 161.

Tommerup. I.C., Barton, J.E. and O'Brian. P.P. 1995, Reliability of RAPD fingerprinting of three basidiomycete fungi, *Laccaria, Hydnangium* and *Rhizoctonia, Mycol. Res.* 99:179-186.

Trappe, J.M. 1987, Phylogenetic and ecological aspects of mycotrophy in the angiosperms from an evolutionary standpoint, in: *Ecophysiology of VA Mycorrhizal Plants*, G.R. Safir, ed., CRC, Boca Raton, pp. 2-25.

Tsantrizos, Y.S., Kope, H.H., Fortin, J.A. and Ogilive, K.K. 1991, Antifungal antibiotics from *Pisolithus tinctorius, Phytochem.* 30:1113-1118.

Walker, C., Biggin, P. and Jardine, D.C.1986, Differences in mycorrhizal status among clones of Sitka spruce, *For. Eco. Manag.* 14:275-283.

Wilcox, H.E.1971, Morphology, of ectendomycorrhiza in *Pinus resinosa,* in: *Mycorrhiza,* E. Hacskaylo, ed., USDA Miscellaneous Publications, pp. 54-68.

Wilson, J.M., Trinick, M.J. and Parker, C.A. 1983, The identification of vesicular arbuscular mycorrhizal fungi using immunofluorescence, *Soil Biol. Biochem. 15:*439-445.

Wright, S.F., Morton, J.B. and Sworobuk, J.E. 1987, Identification of a vesicular arbuscular mycorrhizal fungus by using monoclonal antibodies in an enzyme linked immunosorbent assay, *App. Environ. Mircorbiol.* 53:2222-2225.

Wyss, P. and Bonfante, P. 1992, Identication of mycorrhizal fungi by DNA fingerprinting using short aribtrary primers, in: *The International Symposium on Management of Mycorrhizas in Agriculture, Horticulture and Forestry,* The University of Western Australia, Nedlands (Abstract), pp. 154.

Wyss, P., Mellor, R.B. and Wiemken, A. 1990, Vesicular-arbuscular mycorrhizas of wild type soybean and non-nodulating mutants with *Glomus mosseae* contain symbiosis-specific polypeptids (mycorrhizins), immunologically cross-reactive with nodulins, *Planta* 182:22-26.

Zak, B. 1971, Characterization and identification of Douglas-Fir Mycorrhizas, in: *Mycorrhizas,* E. Hacskaylo, eds., U.S.Government Printing Office, Washington, pp. 38-53.

Zeze, A., Hosmy, M., Gianinazzi-Pearson, V. and Dulieu, H. 1996, Characterization of a highly repetitive DNA sequence (SCI) from the arbuscular mycorrhizal fungus *Scutellospora castanea* and its detection in plants, *Appl. Environ. Microbiol.* 62:2443-2448.

Zhu, H., Higginbotham, K.O., Dancik, B. and Navratil, S. 1988, Intraspecific genetic variability of isozymes in the ectomycorrhizal fungus *Suillus tomentosus, Can. J. Bot.* 66:588-594.

MOLECULAR APPROACHES TO MYCORRHIZAL ECOLOGY

A.K.Saxena, K.Annapurna, and K.V.B.R. Tilak

Division of Microbiology
Indian Agricultural Research Institute
New Delhi -110012, INDIA

1.   INTRODUCTION

Mycorrhizas, the term used to denote association of certain soil fungi with plant roots, have been an integeral part of terrestrial ecosystems since the invasion of land by plants. More than 95 % of plant taxa form mycorrhizal associations with soil fungi. Mycorrhizas are usually divided into three morphologically distinct groups depending on whether or not there is fungal penetration of the root cells : endomycorrhizas, ectomycorrhizas and ectendomycorrhizas. Endomycorrhizas are more prevalent and forms association with 90% of plant species. Of the wide array of endomycorrhizas, the arbuscular mycorrhizal (AM) association formed between vascular plants and zygomycetous fungi in the Glomales, are certainly the most ancient and widespread mycorrhizal symbiosis. Ericoid endomycorrhizas are formed by few fungal species (e.g. *Hymenoscyphus ericae*) and orchid endomycorrhizae by number of basidomycetes (e.g. *Armillaria, Tulasnella*) or imperfect fungi (e.g. *Rhizoctonia*).

Ectomycorrhizas are formed by few temperate tree species of angiosperms and gymnosperms. Fungi belonging to basidiomycetes (e.g. *Amanita, Boletus, Laccaria, Paxillus, Russula, Rhizopogon, Scleroderma, Suillus*), ascomycetes (such as *Balsamia, Cenococcum, Tuber*) and zygomycetes (*Endogone*) are involved in symbiosis.

Ectendomycorrhizas are formed by few plant families and fungi involved belongs to ascomycetes (e.g. *Phialophora* and *Chloridium*).

Studies on taxonomy, ecology and genetic diversity of mycorrhizal fungi is hampered by the use of conventional techniques primarily based on morphological, anatomical and physiological characters. These characters loose their significance in the identification of the fungus during its symbiotic phase. AM fungi are usually classified on the basis of the morphology of their resting spores and several comprehensive descriptions (Gerdemann and Trappe, 1974) and dichotomous and synoptic taxonomic keys (Hall, 1984; Hall and Fish, 1979; Trappe, 1982) are available. Other characters which have been proposed as an aid to identification are anatomical differences shown by the hyphae when growing within host tissue (Abbott, 1982; Morton, 1985), form of the vesicles borne on external mycelium (Hall, 1984), immunological diagnosis (Aldwell *et al.,* 1983; 1985; Kough, 1985, Sanders *et al.,*

1992; Wright and Morton, 1989) and biochemical characteristics like isozyme analysis and fatty acid analyses (Morton *et al.*, 1995).

Identification of fungal symbiont in ectomycorrhizae has always been difficult (Danielson, 1984; Harley and Smith, 1983; Rogers *et al.*, 1989) because morphological and structural characters of ectomycorrhizae do not vary greatly, several species of ectomycorrhizal symbionts cannot be isolated in pure culture and very few ectomycorrhiza form fruiting bodies in pure culture (Gardes *et al.*, 1991b).

The study of the community structure of ectomycorrhizal fungi has been based largely on surveys of the species forming sporocarps, whose appearance is dependent on weather conditions and other abiotic or external biotic factors (Arnolds, 1981; 1992; Vogt *et al.*,. 1992). Sporocarp sampling does not reveal true structure of community and moreover, at the species level, sporocarps cannot be used to differentiate between individual strains. Extensive morphological descriptions and analysis of mycorrhiza enable the identification of several species (Agerer, 1987;1993; Gronbach, 1988; Ingleby *et al.*, 1990). However many morphophytes remain unidentified. Mating behaviour, enzyme polymorphisms, immunological diagnosis and somatic incompatibility have all been used to identify strains of mycorrhizal fungi (Dahlberg and Stenlid, 1990; Fries, 1987; Fries and Neumann, 1990; Sen, 1990a,b; Zhu *et al.*, 1988). All these techniques are an aid to identification of genera or species (Jacobson *et al.*, 1993; Rizzo *et al.*, 1998; Sen, 1990a,b) but are not suitable for identifying isolates (Dupre *et al.*, 1992), to detect unknown fungi in a natural community and to provide phylogenetic information. In addition these different approaches are not sensitive enough to identify the fungus on a single mycorrhizal tip (Henrion *et al.*, 1994).

In the last one decade, there has been a sudden spurt in the development of molecular markers, particularly those based on DNA, to evaluate levels of genetic diversity and phylogenetic relationships within and between species, to identify new species and biological species, to construct genetic map, to track the fate of introduced fungi and to determine their survival, growth and dissemination within a microbial community. Recently it has been shown that specific probe can be used for estimation of fungal or plant rRNA in the symbiotic tissues and to determine whether an mRNA is down - or up - regulated in mycorrhiza (Carnero - Diaz *et al.*, 1997).

Different techniques have been employed by various workers to generate DNA fingerprints or to develop species-specific or taxon-specific probes and include restriction fragment length polymorphism (RFLP) of whole DNA followed by probing with nuclear or mitochondrial rDNA (Armstong *et al.*, 1989; Buglio *et al.*, 1991; Gardes *et al.*, 1991a; Lo *et al.*, 1991; Marmeisse *et al.*, 1992); PCR amplification of targeted genomic sequences followed by RFLP (Agerer *et al.*, 1996; Amicucci *et al.*, 1996; Gardes *et al.*, 1991a; Henrion, *et al.*, 1992; 1994; Hughes, *et al.*, 1998; Sanders *et al.*, 1996; Simon *et al.*, 1993; Wyss and Bonfante 1993); random amplified polymorphic DNA (RAPD) PCR (Abbas *et al.*, 1996; Doudrick *et al.*, 1995; Junghans *et al.*, 1998; Lan Franco *et al.*, 1993; Tommerup, 1992; Wyss and Bonfante 1993); amplified fragment length polymorphism (AFLP) (Majer *et al.*, 1996) and microsatellites (Sastry *et al.*, 1995; Zeze *et al.*, 1997). All these molecular approaches employed to study ecology of ectomycorrhizal and AM fungi are discussed in the present review.

2. ECTOMYCORRHIZA

DNA techniques have been developed as an aid to identification of species and genera of EM fungi particularly during its symbiotic phase, before sporocarp development. Mitochondrial DNA polymorphism have been used as taxonomic tool to delimit *Laccaria* species (Gardes *et al.*, 1991a). Ribosomal DNA (rDNA) has been shown to be remarkably conserved between different organisms and the coding sequences cloned from one species generally hybridize even under full stringency conditions to the homologous sequences of other related species (De Long *et al.*, 1989). The observed polymorphisms result mainly from divergences in non-coding spacer sequences [internal transcribed sequence (ITS) and intergenic sequence (IGS)] and have been used to make distinctions between species (Marmeisse *et al.*, 1992). Using the rDNA restriction fragment phenotypes, four species of *Laccaria* (*L. bicolor, L. laccata, L. proxima* and *L. amethystina*) as well as four biological species within *L. laccata* were distinguishable (Gardes *et al.*, 1990). Armstrong *et al.* (1989) could distinguish five genera of basidiomycetous fungi, two saprophytes and three ectomycorrhizae through RFLP analysis of genomic DNA followed by hybridization with a radioactively labelled probe (pCc 1) from *Coprinus cinereus*. Randomly cloned genomic sequences of the EM fungus *Hebeloma cylindrosporum* have been used as hybridization probes to digests of genomic DNA of ten *Hebeloma* species and two sequences (Clones RC3 and RC11) could identify unique RFLP patterns in all species tested and thus be used for species differentiation. Another clone RCI contained an 18 kb insert which hybridized only to *H. cylindrosporum* and is thus considered as species - specific probe (Marmeisse *et al.*, 1992). The random amplified polymorphic DNA (RAPD) technique was used to develop DNA probes for the identification of EM fungi belonging to the genus *Tuber*. All six species tested showed very different banding patterns after amplification with random primers. In addition a species specific band of about 1.5 kb appeared in the amplification products of *T. magnatum* with primer OPA-18. The 1.5 kb insert was cloned and used as a probe in southern blots and could hybridize specifically with *T. magnatum* (Lanfranco *et al.*, 1993).Specific primers were developed for the identification of *T. borchii* isolates. The methodology followed is again based on RAPD-PCR. A RAPD fragment constant in *T. borchii* isolates and polymorphic among other species of *Tuber* was selected, sequenced and a pair of primers were synthesized which amplified an internal fragment only within *T. borchii* (Bertini *et al.*, 1998). Additional markers to distinguish species of truffle were obtained following PCR-RFLP on ITS and IGS region (Henrion *et al.*, 1994; Melo *et al.*, 1996). Specific *T. borchii* markers was selected by using upstream and downstream sequences of the *tb* 11.9 gene overexpressed in *T. borchii* fruit body and cloned from the mycelium of this species. Using the sequence, primer pair EST1/EST2 was designed which could amplify a product of 699 bp and specifically identify *T. borchii* in ectomycorrhiza (Bertini *et al.*, 1998).

The PCR/RFLP polymorphism of the ITS region is generally regarded as appropriate to differentiate at the species level (Egger, 1995; Erland *et al.*, 1994; Farmer and Sylvia 1998; Gardes *et al.*, 1991a,b; Gardes and Bruns 1996; Henrion *et al.*, 1992; Melo *et al.*, 1996; Nylund *et al.*, 1995, Pritsch *et al.*, 1997). ITS-RFLP analysis could differentiate among a diverse group of EM fungal species except two *Rhizopogon reaii* and *R. fuscorubens*, that form ectomycorrhiza with Southern pines (Farmer and Sylvia 1998). Pritsch *et al.* (1997) could identify mycorrhizas of eight species of EM fungi by comparing PCR-ITS/RFLP profiles obtained from eight morphophytes and sporocarps. Further evidence to show that ITS region is highly variable and suitable for identification of EM species was provided by Karen *et al.* (1997). With two endonucleases, 34 of the 44 species could be distinguished from each other. The only species that could not be separated from each other were in the genus *Cortinarius*. ITS-5.8S probe for identification of ectomycorrhizae of *Eucalyptus globulus-Pisolithus tinctorius* was developed (Carnero Diaz *et al.*, 1997).

In addition to species-specific markers, primers and probes, taxon specific primers and probes have also been developed. Gardes and Bruns (1996) developed a primer pair for the ITS region in the nuclear ribosomal repeat (ITSI-F/ITS4-B) which could preferentially allow amplification of basidiomycete component in ectomycorrhiza. Such primers are useful to study the structure of ectomycorrhizal communities. Oligonucliotide probes from the mitochondrial large subunit have also been developed which selectively hybridized to the DNA of suilloid fungi (Bruns and Gardes, 1993).

Development of taxon-specific and species-specific DNA fingerprints, probes and primers have made it convenient to understand the ecology of ectomycorrhizal fungi. These markers can be used to track the fate of inoculated fungi in the ecosystem to understand the role of mycorrhizas during successional processes as well as under varying environmental conditions; monitor the development of EM fungi during its entire life cycle, to study the survival or colonization of EM fungi over a period of time and to identify unknown symbiont (Gardes *et al.*, 1991b; Danell, 1994; Pritisch *et al.*, 1997; Farmer and Sylvia, 1998; Bertini *et al.*, 1998). These studies rely on the construction of data base or catalogue of fingerprints generated through PCR-RFLP of ITS, IGSII, 5 8S rDNA, 25 S rDNA, 18 S rDNA, mitochondrial ribosomal DNA, PCR-RAPD and other related techniques. Gardes *et al.* (1991b) were able to amplify DNA using ITS and mitochondrial Lr DNA primers from mycorrhizae from jack pine (*Pinus banksiana*) and could distinguish between the introduced *Laccaria bicolor* from the contaminant *Telephora terrestris*. Using ITS-RFLP, it was shown that *Cantharellus cibarius* successfully forms ectomycorrhiza with *Picea abies* and survived for atleast 10 months and continued to grow and colonize new roots. Moreover, the presence of alien mycorrhizae could be clearly distinguished from the *C. cibarius* (Danell, 1994). An unknown isolate S370, earlier labelled as *Pisolithus arhizus* was found to match the ITS-RFLP pattern of *Rhizopogon* and was thus classified as *Rhizopogon* (Farmer and Sylvia, 1998). The ITS region of a mycosymbiont isolated from *Pisonia grandis* was sequenced and a comparison made with sequences available in the data base. Comparisons of nucleotide sequence of ITS region of a mycosymbiont isolated from *Pisonia grandis* with the available sequences in the data base revealed high degree of homology (up to 87% over 637 bases) with a number of Thelephoraceae isolates, strongly implying that it belongs in that taxon (Chambers *et al.*, 1998).

In general a combination of mating tests, morphological data and molecular markers is preferential in taxonomic studies. Difficulty in obtaining culture due to poor germination of spores calls for more reliance on morphological and molecular markers (Liu, 1997; Martin *et al.*, 1998). Combination of morphological and molecular analysis conformed the reclassification of six species of *Rhizopogon* proposed by Smith & Zeller, as well as *R. reticulatus* as a single species, *R. villosulus* (Martin *et al.*, 1998). Phylogenetic relationships in *Dermocybe* and related *Cortinarius* taxa was derived based on sequence analysis of entire ribosomal DNA ITSI, ITS2, 5.8 S gene and partial 18 S and 25 S genes. The study revealed that *Dermocybe sensu lato* is polyphyletic and within the genus *Cortinarius* certain of the genera actually represent coherent genera and thus reorganisation of some taxa and taxonomic group have been proposed (Liu *et al.*, 1997). Similarly phylogenetic and taxonomic relationship were derived for species of *Suillus sensu lato* following sequencing of both ITS region (1 and 2) including 5.8 S rDNA of 47 isolates belonging to 38 recognized species of *Suillus sensu lato*. Alignment of sequences followed by parsimonious analysis and neighbour joining suggested that both genera *Boletinus* and *Fuscoboletinus* should be collapsed into *Suillus* and to transfer several species to *Suillus* (Kretzer *et al.*, 1996).

Ectomycorrhizal fungi exhibit substantial intraspecific variability for a number of traits including pure culture and physiological characteristics (Cao and Crawford, 1993; Cline *et al.*, 1987; Coleman *et al.*, 1989; Ho, 1987; Hung and Trappe, 1983; Molina, 1979); basidiome and basidiospore characteristics (Burgess *et al.*, 1995; Pritsch *et al.*, 1997) and mycorrhiza forming ability (Burgess, *et al.*, 1994; Marx 1981). They also exhibit variation in terms of the host species and geographic distribution. These variability suggests taxonomically diverse nature of several species of EM fungi. Testing somatic incompatibility of *in vitro* cultured samples of the population (Dahlberg, 1995) has been helpful in studying the population structure of atleast two EM fungal genera, *Suillus* and *Laccaria* (Baar *et al.*, 1994; Dahlberg and Stenlid, 1994; Sen 1990a); however it cannot be extended for majority of EM fungi as they are not easily cultured *in vitro*. Moreover compatibility does not necessarily imply that two isolates are genetically identical (Rizzo *et al.*, 1995; Rodriguez *et al.*, 1995; Sen 1990a). In recent years molecular markers have been used to determine genetic diversity among individuals (Gardes *et al.*, 1991 a,b; Hughes *et al.*, 1998; Junghans *et al.*, 1998; Karen *et al.*, 1997; Kretzer *et al.*, 1996; Lo Buglio *et al.*, 1991; Pritsch *et al.*, 1997; Timonen *et al.*, 1997) and warrants taxonomic revision for many EM fungi like *Pisolithus tinctorius* (Burgess *et. al.*, 1995; Junghans *et al.*, 1998; Anderson, et al., 1998); *Suillus pungens* (Bonello *et al.*, 1998); *Suillus sensu lato* (Kretzer *et al.*, 1996); *Cenococcum geophilum* (Lo Buglio *et al.*, 1991); *Paxillus rubicundulus* (Pritsch *et al.*, 1997); *Hebeloma cylindrosporum* (Gryta *et al.*, 1997) and many more. LoBuglio *et al.* (1991) using restriction fragment length polymorphism (RFLP) in ribosomal DNA assessed the degree of variation among 71 *Cenococcum geophilum* isolates of both geographically distinct and similar origin and concluded that *C. geophilum* are extremely heterogeneous species as a fungal complex representing a broader taxonomic rank than presently considered. Further evidence for the complexity of *C. geophilum* species was provided by Farmer and Sylvia (1998) following PCR amplification of the internal transcribed spacer (ITS) of rRNA genes and subsequent RFLP analysis. *Pisolithus tinctorius*, one of the most widespread EM fungi, exhibits ample geographic distribution and morphological diversity and hence the taxonomy of this genus is unclear. While combined morphological and polypeptide analyses (SDS-PAGE) showed that considerable variation exists within Australian *Pisolithus* isolates (Burgess *et al.*, 1995); RFLP analysis of the ITS region and ITS sequence analysis revealed large variations between isolates collected from Western Australia alone (Anderson *et al.*, 1998). ITS-RFLP and ITS sequence analyses indicated that two isolates (LJ 07 and WM 01) were considerably different from all other isolates and may represent a separate *Pisolithus* species. The morphology of basidiospores also suggest that they represent species different from *P. tinctorius* and *P. microcarpus* (Anderson *et al.*, 1998). Two genetically distinct groups, group I consisting of isolates collected from Brazil and group II those collected in Northern hemisphere, within *P. tinctorius* were identified following PCR amplification with random primers (Junghans *et al.*, 1998). Four unique genotypes were observed among 12 geographically diverse isolates of *Pisolithus arhizus* following ITS-RFLP (Farmer and Sylvia, 1998).

Sequencing of ITS region of the nuclear rDNA revealed that isolates of *Suillus granulatus* derived from either North America or Europe and Asia are polyphyletic and seem to represent atleast two different species (Kretzer *et al.*, 1996). Most examples of intraspecific variation in the ITS region of the rDNA of some fungi come from comparisons made over considerably longer distances (between continents; Feibelman *et al.*, 1994; Gardes *et al.*, 1991); but some also on a more regional scale (*Cortinarius obtusus* and *C. camphoratus*; Karen *et al.*, 1997). *Hydnum rufescens* and *H. repandum* isolates collected from various regions of Slovenia, Italy and Bavaria showed polymorphism of restriction digest of ITS region (Agerer *et al.*, 1996). Gardes *et al.* (1991a) using Mbo 1 restriction endonuclease on the amplified ITS region of fungal DNA observed intraspecific variation in three species of *Laccaria* (*Laccaria bicolor*, *L. laccata* and *L. amethystina*) that originated from North America and Sweden. Two RFLP types were found for each species and the intraspecific variation was both within (North America) and between continents. Similar trend was obtained with RFLP on mitochondrial DNA (Gardes *et al.*, 1991a). Of these species, Karen *et al.* (1997) investigated *L. laccata* and found all three vouchers (from Sweden) to be identical, but different with respect to restriction products of Mbo 1 from the two vouchers from Sweden investigated by Gardes *et al.* (1991a). RAPD analyses also recognised different genets of *L. bicolor* in the field (de la Bastide *et al.*, 1994). However no differences were revealed by RAPD markers between the two sub population of *L. bicolor* created on the basis of mating competence (Raffle *et al.*, 1995).

Intergenic spacer region (IGSII) of the nuclear ribosomal repeat is highly conserved. These non-coding regions are highly polymorphic and provide a useful tool for taxonomic and phylogenetic studies. Amplification of IGSII of the rDNA followed by digestion with restriction endonucleases showed intraspecific variation among geographically diverse isolates of *Laccaria proxima* (Albee *et al.*, 1996) and *Tuber albidium* (Henrion *et al.*, 1994). A different approach was followed by Marmeisse *et al.* (1992) to exhibit variations among strains of *Hebeloma cylindrosporum* collected within a restricted geographical area of about 100 km in S.W. France. They randomly cloned the genomic sequences of *H. cylindrosporum* and identified one clone which could hybridize only with *H. cylindrosporum* and identify a unique RFLP pattern in all ten isolates tested. The high degree of polymorphism present in the randomly cloned sequences reflect that the population is strongly outbreeding and that variations within the genome are rapidly combined (Marmeisse *et al.*, 1992).

Molecular approaches have also been helpful in recognising low levels of genetic diversity among EM fungi. Of the 44 species of EM fungi collected from the Fennoscandia (Denmark, Finland, Norway and Sweden) region, ITS-RFLP revealed only seven to be polymorphic indicating low level of intraspecific variability in majority of the species (Karen *et al.*, 1997). Low levels of genetic diversity have been reported for *Tuber magnatum* (Amicucci *et al.*, 1997; Potenza *et al.*, 1994) *T. melanosporum* (Henrion *et al.*, 1994) and in seven of the eight species identified from black alder (Pritsch *et al.*, 1997).

Evolutionary relationships have been derived using molecular techniques specially those based on PCR amplification of conserved ribosomal DNA genes. Three disjunct populations of *Pleurtopsis longinqua*, sexually compatible with each other, showed only 4 base difference in the amplified nuclear ribosomal ITS 1-5.8S-ITS 2 region, between isolates from different geographic area (Hughes *et al.*, 1998). The authors concluded that the origin of this disjunct distribution is recent and there is a recent gene flow between the three populations. They have further examined ribosomal ITS sequence differences for other disjunct fungal collections reported by different workers and found the data suggestive of patterns of old separation intermingled with possibly recent dispersals (Hughes *et al.*, 1998).

Nucleotide sequence of ribosomal DNA 5.8S and ITS were used to investigate the phylogenetic relationships among species of *Dermocybe* and selected taxa from subgenera of *Cortinarius*. The study reveal that *Dermocybe sensu lato* is polyphyletic and within the genus *Dermocybe sensu lato*, *Cortinarius* certain of the subgenera may actually represent coherent genera and hence reorganization of some taxa and taxonomic groups was suggested (Liu *et al.*, 1997).

3.   ARBUSCULAR MYCORRHIZA

For long taxonomic diagnosis of the Glomales relied entirely on the morphological characteristics of spores. In species with low degree of morphological divergence isoenzyme analyses has been used for classification (Rosendahl, 1989; Rosendalh and Hepper, 1987). Immunological methods have proved useful for detecting fungi in soil, roots and other plant materials (Burge *et al.*, 1994; Frankland *et al.*, 1981; Prestley and Devey, 1993). Monoclonal, antibodies specific for species of AMF (Hahn *et al.*, 1993; Wright *et al.*, 1987) and polyclonal antisera specific at the genus level (Aldwell *et al.*, 1983; 1985; Kough and Linderman, 1986) have been developed for detecting AMF in roots. Species - specific polyclonal antisera for *Scutellospora heterogama* was developed and successfully used to track the AMF within the plant roots (Thingstrup *et al.*, 1995). There was no cross reaction with uncolonized roots. Spore specific antigens were detected which could be useful for studying the physiology and gene expression in different phases of the fungal life-cycle.

Most of the ecological studies of AMF have been based on taxonomic data which is mostly morphological. However, the relationship between morphological divergence and genetic and functional diversity of AM fungi has not been established. Its observed that AMF are ubiquitous, found in diverse environments, and show low host specificity. The extent of genetic diversity in these organisms is not known and what genetic characteristics could be used to identify the morphological, functional and ecological differences between AM fungi also remain largely unknown. Information is slowly accumulating on the genetic makeup of AMF with the use of molecular techniques. These techniques are usually PCR-based for studying the polymorphism, quantification and identification.

The first DNA sequences from an arbuscular endomycorrhizal fungi were reported by Simon *et al.* (1992). They were obtained by directly sequencing overlapping amplified fragments of the nuclear genes coding for the small subunit rDNA. These sequences were used to develop a PCR primer (VANSI) that enables the specific amplification of portion of the vesicular-arbuscular endomycorrhizal fungus small subunit rRNA directly from a mixture of plant and fungal tissues. The specificity of this primer for arbuscular endomycorrhizal fungi was demonstrated by testing it on number of organisms and by sequencing the fragment amplified from colonized leek (*Allium porum*) roots. Using another approach, Abbas *et al.,.* (1996) did random amplification of polymorphic DNA to generate DNA fragments that were unique to isolates of two arbuscular mycorrhizal fungi, *Glomus mosseae* and *Gigaspora margarita*. Sequence analysis of these fragments allowed generation of primer pairs and subsequent specific identification of these genomes even in the presence of competing genomic DNA's. Their results confirm that the approach may be applicable to development of isolate specific primers for other fungal isolates as well and help to precisely study the interrelationships of these fungi to host plants in the soil.

SSCP (single strand conformational polymorphism) -A methodology that could be used to characterize endomycorrhizal fungi without relying on prior availability of sequences information was developed by Simon *et al.* (1993). SSCP analysis has been reported to be capable of detecting single-base substitution between homologous fragments of a few hundred base pairs. Briefly, differences in homologous fragments are detected by running them on a high-resolution polyacrylamide gel under nondenaturing conditions since the samples are denatured immediately before the gel is loaded, the two strands of each fragment migrate separately and their rate of migration is affected by the confirmation that they are allowed to adopt in the gel. SSCP analysis of the fragments allowed the distinction between *Acaulospora spinosa*, *Acaulospora rugosa* and *Entrophospora* spp. but not between *Entrophospora columbiana* and another *Entrophospora* spp.

Simon *et al.* (1993) designed two new primers, VANS22 and VANS32, other than the four family-specific primers viz.. VAACAU, VAGIGA, VAGLO and VALTEC to amplify a variable portion of the 18S gene of any endomycorrhizal fungus. Even though the specificity of these primers is not restricted to endomycorrhizal fungi, adequate specificity can be achieved by first amplifying a larger fragment by using VANS1-VANS22 and then using VANS22-VANS32. By using this approach,they characterized an endomycorrhizal fungus present in leek (*Allium porum*) roots. This methodology might be useful to rapidly characterize field collected samples, allowing a preliminary identification of the endophytes, and to help single out those that seem novel and worth further analysis.

Bonito *et al.* (1995) attempted to obtain amplification product directly from roots without DNA extraction and purification. PCR was used with the primer pair VANS1-NS2 to detect the arbuscular mycorrhizal fungus *Glomus intraradices* on roots of lettuce, zinnia, leek, pepper and endive plants. An elegant technique to identify the fungal symbionts in roots (taken directly from natural communities) was developed by Clapp *et al.,*1995. It involves selective enrichment of amplified DNA (SEAD) based on the principle of subtractive hybridization to remove interfering plant-derived DNA after amplification with PCR. The combination of SEAD with PCR is robust, requiring no DNA purification procedures other than those usual during initial DNA extraction. The technique demonstrated that natural mycorrhizas might involve a complex community of fungi and not just a pairwise interaction between a plant and a single fungal partner.

Amongst the Glomales, *Gigaspora* represents a comparatively young genus (Simon *et al.,* 1993). Species discrimination within this genus is difficult. Isozyme and SSU analyses was carried out with 28 isolates of *Gigaspora* (Bago *et al.,* 1998). The primers NS-7, NS71 and SSU 1492 were used to amplify and directly sequence the PCR fragments from the isolates. A 6 nucleotide long sequence signature that could be used to delineate three groups within this genus was found on the basis of which three *Gigaspora* groups were identified.

Little information currently exists on species diversity in communities of arbuscular mycorrhizal fungi (AMF) mainly owing to difficulties in identification of field extracted spores on the basis of morphology. Molecular techniques have now been developed to assess the genetic diversity in natural AMF communities. Sanders *et al.* (1996) used polymerase chain reaction followed by restriction fragment length polymorphism analysis (PCR-RFLP) to amplify and characterize the ITS region known to vary between species from single AMF spores. It was observed that individual spores of *G. mosseae* contain at least two different ITS sequences. Whether the different sequences are the result of different ITS repeats in the same nucleus or whether different families of nucleus exist with different ITS sequences within single AMF spores is not known. The high diversity of *Glomus* (10 genetically different spores in a sample size of 10 spores) based on the results of the PCR-RFLP indicates that diversity of AMF in natural communities might be much greater than expected.

Dodd *et al.* (1996) using a multidisciplinary approach studied the inter-and intraspecific variation within the morphologically similar arbuscular mycorrhizal fungi *Glomus mosseae* and *Glomus coronatum.* Their observations were that only spore colour could morphologically discriminate the two groups. Isozyme analysis and DNA polymorphisms also provided evidence for separation of the two species complexes. DNA fingerprints obtained with different PCR approaches demonstrated that the isolate *Glomus* sp. (BEG 49) is different from other *Glomus* sp. Based on the results the two *Glomus* isolates were considered as belonging to two genetically different groups. In an attempt to reveal DNA polymorphism in different species and isolates of individual species in AM fungi, Wyss and Bonfante (1993) utilized short (9-13 mer) oligonucleotides of arbitrary sequences for RAPD-PCR.

5. CONCLUSION

The ecology of mycorrhizal fungi has taken a leap forward with the advent of molecular techniques in general, and PCR based techniques, in particular. The development of specific primers, species and taxon - specific probes has the potential to allow precise identification and quantification of mycorrhizal fungi in the symbiotic state. Research should also focus on the potential use of PCR technology to study the selective expression of specific gaves during fungal spore development. The achievement of this future prospect depends on the ability to prepose PCR - based CDNA libraries from small amounts of fungal material. Root organ cultures, PCR methods and gene cloning technologies are powerful tools that are to be employed to study the diversity,distribution and symbiosis of ecto- and endo-mycorrhiza in natural ecosystem.

REFERENCES

Abbas, J., Hetrick, B.A.D. and Jurgenson, J.E. 1996, Isolate specific detection of mycorrhizal fungi using genome specific primer pairs, *Mycologia* 88 : 939-946.

Abbott, L.K. 1982, Comparative anatomy of vesicular arbuscular mycorrhizas formed on subterranean clover, *Aust. J. Bot.* 30: 485-499.

Agerer, R. 1987-1993, Colour Atlas of Ectomycorrhizae Ist-7th edn., Schwabisch Bmund, Einhorn-Verlag, Germany.

Agerer, R., Xraigher, H. and Javornik, B. 1996, Identification of ectomycorrhizas of *Hydnum rufescens* on Norway spruce and the variability of the ITS region of *H. rufescens* and *H. repandum* (Basidiomycetes), *Nova Hedwigia* 63 : 183-194.

Albee, S.R., Mueller, G.M. and Kropp, B.R. 1996, Polymorphisms in the large intergenic spacer of the nuclear ribosomal repeat identify *Laccaria proxima* strains, *Mycologia* 88 : 970-976.

Aldwell, F.E.B., Hall, I.R. and Smith, J.M.B. 1983, Enzyme linked immunosorbent assay (ELISA) to identify endomycorrhizal fungi, *Soil Biol. Biochem.* 15 : 377-378.

Aldwell, F.E.B., Hall, I.R. and Smith, J.M.B. 1985, Enzyme-linked immunosorbent assay as an aid to taxonomy of the Endogonaceae, *Trans Br. Mycol. Soc.* 84 : 399-402.

Amicucci, A., Ressi, I., Potenza, L., Zambenelli, A., Agostini, D., Palma, F. and Stocchi, V. 1996, Identification of ectomycorrhizae from *Tuber* species by RFLP analysis of the ITS region, *Biotechnol. Lett.* 18 : 821-826.

Anderson, I.C., Chambers, S.M. and Cairney, J.W. 1998, Molecular determination of genetic variation in *Pisolithus* isolates from a defined region in New South Wales, Australia, *New Phytol.* 138 : 151-162.

Armstrong, I.L., Fowles, N.L. and Rygiewicz, P.T. 1989, Restriction fragment length polymorphisms distinguish ectomycorrhizal fungi, *Plant Soil* 116 : 1-7.

Arnolds, E. 1981, Ecology and coenology of macrofungi in grasslands and moist heathlands in Drenthe, The Netherlands, Part 1, Introduction and synecology, *Bibliotheca Mycologia* 83 : 1-407.

Arnolds, E. 1992, The analysis and classification of fungal communities with special references to macrofungi, in : *Fungi in Vegetation Science*, W. Winter-hoff, ed., Kluwer Academic Publishsers, Dordrecht, The Netherlands, pp. 7-48.

Baar, J., Ozinga, W.A. and Kuyper, T.W. 1994, Spatial distribution of *Laccaria bicolor* genets reflected by sporocarps after removal of litter and humus layers in a *Pinus sylvestris* forest, *Mycol. Res.* 98 : 726-728.

Bago, B., Bentivenga, S.P., Brenac, V., Dodd, J.C., Piche, Y. and Simon, L. 1998, Molecular analysis of *Gigaspora* (Glomales, Gigasporaccae), *New Phytol.* 139 : 581-588.

Bertini, L., Agostini, D., Potenza, L., Rossi, I., Zeppa, S., Zambonelli, A. and Stocchi, V. 1998, Molecular markers for the identification of the ectomycorrhizal fungus *Tuber borchii*, *New Phytol.* 139 : 565-570.

Bonello, P., Bruns, T. and Gardes, M. 1998, Genetic structure of a natural population of the ectomycorrhizal fungus *Suillus pungens*, *New Phytol.* 138 : 533-542.

Bonito, R.D., Elliott, M.L. and Des Jardin, E.A. 1995, Detection of an arbuscular mycorrhizal fungus in roots of different plant species with the PCR, *Appl. Environ. Microbiol.* 61 : 2809-2810.

Bruns, T.D. and Gardes, M. 1993, Molecular tools for the identification of ectomycorrhizal fungi - taxon-specific oligonucleotide probes for suilloid fungi, *Mol. Ecol.* 2 : 233 - 242.

Burge, M.N., Msuya, J.C., Cameron, M. and Stimson, W.H. 1994, A monoclonal antibody for the detection of *Serpula lacrymans*, *Mycol. Res.* 98 : 356 - 362.

Burgess, T., Dell, B. and Malajczuk, N. 1994, Variation in mycorrhizal development and growth stimulation by 20 *Pisolithus* isolates inoculated on to *Eucalyptus grandis*, W. Hill ex Maiden, *New Phytol.* 127 : 731-439.

Burgess, T., Malajczuk, N. and Dell, B. 1995, Variation in *Pisolithus* based on basidiome and basidiospore morphology, culture characteristics and analysis of polypeptides using ID SDS-PAGE, *Mycol. Res.* 99 : 1-13.

Carnero-Diaz, E., Tagu, D. and Martin, F. 1997, Ribosomal DNA internal transcribed spacers to estimate the proportion of *Pisolithus tinctorius* and *Eucalyptus globulus* RNAs in ectomycorrhiza, *Appl. Environ. Microbiol.* 63 : 840-843.

Cao, W. and Crawford, D.I. 1993, Carbon nutrition and hydrolytic and cellulolytic activities in the ectomycorrhizal fungus *Pisolithus tinctorius*, *Can. J. Microl.* 39 : 529-535.

Chambers, S.M., Sharples, J.M. and Cairney, J.W.G. 1998, Towards a molecular identification of the *Pisonia* mycobiont, *Mycorrhiza* 7 :319-321.

Clapp, J.P., Young, J.P.W., Merryweather, J.W. and Fitter, A.M. 1995, Diversity of fungal symbionts in arbuscular mycorrhizas from a natural community, *New Phytol.* 130 : 259-265.

Cline, M.L., France, R.C. and Reid, C.P.P. 1987, Intraspecific and interspecific growth variation of ectomycorrhizal fungi at different temperatures, *Can. J. Bot.* 67 : 29-39.

Coleman, M.D., Bledsoe, C.S. and Lopushinsky, W. 1989, Pure culture response of ectomycorrhizal fungi to imposed water stress, *Can. J. Bot.* 67 : 29-39.

Dahlberg, A. 1995, Somatic incompatibility in ectomycorrhizas, in : *Mycorrhiza : Structure, Function, Molecular Biology and Biotechnology*, A. Varma and B. Hock, eds., Springer- Verlag, Berlin, pp. 115-136.

Dahlberg, A. and Stenlid, J. 1990, Population structure and dynamics in *Suillus bevinus* as indicated by spatial distribution of fungal clones, *New Phytol.* 115 : 487-493.

Dahlberg, A. and Stenlid, J. 1994, Size, distribution and biomass of genets in populations of *Suillus bovinus* revealed by somatic incompatibility, *New Phytol.* 128 : 225-234.

Danell, E. 1994, Formation and growth of the ectomycorrhiza of *Cantharellus cibarius*, *Mycorrhiza* 5 : 89-97.

Danielson, R. 1984, Ectomycorrhizal associations in jack pine stands in northeastern Alberta, *Can. J. Bot.* 62 : 932-939.

de la Bastide, P.Y., Kropp, B.R. and Piche, Y. 1994, Spatial distribution and temporal persistence of discrete genotypes of the ectomycorrhizal fungus *Laccaria bicolor* (Maire), Orton, *New Phytol.* 127 : 547-556.

De Long, E.F., Wickmann, G.S. and Pace, N.R. 1989, Phylogenetic stains : ribosomal RNA-based probes for the identification of single cells, *Science* 243 : 1360-1363.

Dodd, J.C., Rosendahl, S., Giovannetti, M., Broome, A., Lanfranco, L. and Walker, C. 1996, Inter and intraspecific variation within the morphologically similar arbuscular mycorrhizal fungi *Glomus mosseae* and *Glomus coronatum*, *New Phytol.* 133 : 113-122.

Doudrick, R.L., Raffle, V.L., Nelson, C.D. and Furnier, G.R. 1995, Genetic analysis of homokaryons from a basidiome of *Laccaria bicolor* using random amplified polymorphic DNA (RAPD) markers, *Mycol. Res.* 99 : 1361-1366.

Dupre, C., Chevalier, G., Palenzons, M., Ferrar, A.M., Nascetti, Mattiucci, S., d'Amelio, S., La Rosa, G. and Biocca, E. 1992, Differenziaz genetica dfi ascocarpi, miceli e micorrize di differenti specie di *Tuber*. in: *del Congresso*, G. Pacioni ed. Internazionale sul Tartufo. L.'Aquilu, 154-159.

Egger, K. 1995, Molecular analysis of ectomycorrhizal fungal communities, *Can. J. Bot.* 73 : 1415-1422.

Erland, S., Henrion, B., Martin, F., Glover L.A. and Alexander, I.I. 1994, Identification of the ectomycorrhizal basidiomycete *Tylospora fibrillosa* Donk by RFLP analysis of the PCR-amplified ITS and IGS regions of ribosomal DNA, *New Phytol.* 126 : 525-532.

Farmer, D.J. and Sylvia, D.M. 1998, Variation in the ribosomal DNA internal transcribed spacers of a diverse collection of ectomycorrhizal fungi, *Mycol. Res.* 102 : 859-865.

Feibelman, T., Bayman, P. and Cibula, W.G. 1994, Length variation in the internal transcribed spacer of ribosomal DNA in chanterelles, *Mycol. Res.* 98 : 614-618.

Frankland, J.C., Bailey, A.D., Gray, T.R.G. and Holland, A.A. 1981, Development of an immunological technique for estimating mycelial biomass of *Mycena galopus* in leaf litter, *Soil Biol. Biochem.* 13 : 8887-8892.

Fries, N. 1987, The third benefactors lecture : ecological and evolutionary aspects of spore germination in the higher basidiomycetes, *Trans Br. Mycol. Soc.* 88: 1-7.

Fries, N. and Neumann, W. 1990, Sexual incompatibility and field distribution of the ectomycorrhizal fungus *Suillus luteus* (Boletaceae), *New Phytol.* 107 : 735-739.

Gardes, M. and Bruns, T.D. 1996, ITS-RFLP matching for identification of fungi, in:. *Methods in Molecular Ecology*, J. Clapp, ed, Humana Press, Inc., Totowa, NJ, pp. 177-186.

Gardes, M., Fortin, J.A., Mueller, G.M. and Kropp, B.R. 1990, Restriction fragment length polymorphisms in the nuclear ribosomal DNA of four *Laccaria* spp. : *L. bicolor*, *L. laccata*, *L. proxima* and *L. amethystina*, *Phytopathol.* 80 : 1312-1317.

Gardes, M., Mueller, G.M., Fortin, J.A. and Kropp, B.R. 1991a. Mitochondrial DNA polymorphisms in *Laccaria bicolor*, *L. laccata*, *L. proxima* and *L. amethystina*, *Mycol. Res.* 95 : 206-216.

Gardes, M., White, T.J., Fortin, J.A., Bruns, T.D. and Taylor, J.W. 1991b. Identification of indigenous and introduced symbiotic fungi in ectomycorrhizae by amplification of nuclear and mitochondrial ribosomal DNA, *Can. J. Bot.* 69 : 180-190.

Gerdemann, J. and Trappe, J.M. 1974, The Endogonaceae of the pacific Northwest, Mycological Memoir No.5. Mycological Society of America, New York Botanical Garden NY.

Gronbach, E. 1988, Charakterisierung und Identifizierung von Ektomykorrhizen-in einem Fichtenbestand mit unter suchungen zur Merkmalsvariabilitat in Sauer beregneten Flachen, *Bibliotheca Mycologia* 125 : 1-217.

Gryta, H., Debaud, J.C., Effose, A., Gay, G. and Marmeisse, R. 1997, Fine-scale structure of populations of the ectomycorrhizal fungus *Hebeloma cylindrosporum* in coastal sand dune forest ecosystems, *Mol. Ecol.* 6 : 353-364.

Hahn, A., Bonfante, P., Horn, K., Pausch, F. and Hock, B. 1993, Production of monoclonal antibodies against surface antigens of spores from arbuscular mycorrhizal fungi by an improved immunization and secreening procedure, *Mycorrhiza* 4 : 69-78.

Hall, I.R. 1984. Taxonomy of VA mycorrhizal fungi, in : *VA Mycorrhiza*, C.L. Powell and D.J. Bagyaraj, eds., CRC Press, Boca Raton, Florida pp.57-94

Hall, I.R. and Fish, B.J. 1979, A key to the Endogonaceae, *Trans. Br. Mycol. Soc.* 73 : 261-270.

Harley, J.L. and Smith, S.E. 1983, *Mycorrhizal Symbiosis*, Academic Press, London.

Henrion, B., Chevalier, G. and Martin, F. 1994, Typing truffle species by PCR amplification of the ribosomal DNA spacers, *Mycol. Res.* 98 : 37-43.

Henrion, B., Le Tacon, F. and Martin, F. 1992, Rapid identification of genetic variation of ectomycorrhizal fungi by amplification of ribosomal RNA genes, *New Phytol.* 122 : 289-298.

Ho, I. 1987, Comparison of eight *Pisolithus tinctorius* isolates for growth rate, enzyme activity, and phytobormone production, *Can. J. For Res.* 17 : 31-35.

Hughes, K.W., Toyohara, T.L. and Peterson, R.H. 1998, DNA sequence and RFLP analysis of *Pleurotopsis longinqua* from three disjunct populations, *Mycologia* 90 : 595-600.

Hung,L.L. and Trappe, J.M. 1983, Growth variation between and within species of ectomycorrhizal fungi in response to pH *in vitro, Mycologia* 75 : 234 - 241.

Ingleby, K., Mason, P.A., Last, F.T. and Fleming, L.V. 1990, Identification of ectomycorrhizas, ITE research publication No. 5, Institute of Terrestrial Ecology, Edinburgh, Scotland.

Jacobsen, K.M., Milles, Jr. O.K. and Turner, B.J. 1993, Randomly amplified polymorphic DNA markers are superior to somatic incompatibility tests for discriminating genotypes in natural populations of the ectomycorrhizal fungus *Suillus granulatus, Proc. Nat. Acad. Sci.* U.S.A. 90 : 9159-9163.

Junghans, D.T., Gomes, E.A., Guimaraes, W.V., Barros, E.G. and Araujo, E.F. 1998, Genetic diversity of the ectomycorrhizal fungus *Pisolithus tinctorius* based on RAPD-RFLP analysis, *Mycorrhiza* 7 : 243-248.

Karen, O., Hogberg, N., Dahlberg, A., Jonsson, L. and Nylund, J.E. 1997, Inter and intraspecific variation in the ITS region of rDNA of ectomycorrhizal fungi in Fennoscandia as detected by endonuclease analysis, *New Phytol.* 136 : 313-325.

Kough, J. 1985, Relationship of extra-matrical hyphae of vesicular arbuscular mycorrhizal fungi to plant growth response, Ph. D. Thesis, Oregon State University, USA.

Kough, J.L. and Linderman, R.G. 1986, Monitoring extra-matrical hyphae of a vesicular-arbuscular mycorrhizal fungus with an immunofluorescence assay and the soil aggregation technique, *Soil Biol. Biochem.* 18 : 309-313.

Kretzer, A., Li, Y., Szaro, T. and Bruns, T.D. 1996, Internal transcribed spacer sequences from 38 recognized species of *Suillus sesnsu lato* : phylogenetic and taxonomic implications, *Mycologia* 88 : 776-785.

Lanfranco, L., Wyss, P., Marzachi, C. and Bonfante, P. 1993, DNA probes for identification of ectomycorrhizal fungus *Tuber magnatum* Pico, *FEMS Microbiol. Lett.* 114 : 245-252.

Liu, Y.J., Rogers, S.O. and Ammirati, J.F. 1997. Phylogenetic relationships in *Dermocybe* and related *Cortinarius* taxa basesd on nuclear ribosomal DNA internal transcribed spacers, *Can. J. Bot.* 75 :519-532.

Lo Buglio, K.F., Scott, O.R. and Wang C.J.K. 1991, Variation in ribosomal DNA among isolates of the mycorrhizal fungus *Cenococcum geophilum, Can. J. Bot.* 69 : 2331-2343.

Majer, D., Mithen, R., Lewis, B.G., Vos, P. and Oliver, R.P. 1996, The use of AFLP finger printing for the detection of genetic variation of fungi, *Mycol. Res.* 100 : 1107-1111.

Marmeisse, R., Debaud, J.C. and Casselton, L.A. 1992, DNA probes for species and strain identification in the ectomycorrhizal fungus *Hebeloma, Mycol. Res.* 96 : 161-165.

Martin, M.P., Hogberg, N. and Nylund, J.E. 1998, Molecular analysis confirms morphological reclassification of *Rhizopogon, Mycol. Res.* 102 : 855-858.

Marx, D.H. 1981, Variability in ectomycorrhizal development and growth among isolates of *Pisolithus tinctorius* as affected by source, age, and reisolation, *Can. J. For. Res.* 11 : 168-174.

Melo, A., Nosenzo, C., Meotto, F. and Bonfante, P. 1996, Rapid typing of truffle mycorrhizal roots by PCR amplification of the ribosomal DNA spacers, *Mycorrhiza* 6 : 416-421.

Molina, R. 1979, Pure culture synthesis and host specificity of red alder mycorrhizae, *Can. J. Bot.* 57 : 1223-1225.

Morton, J.B. 1985, Underestimation of most probable numbers of vesiscular arbuscular endophytes because of non-staining of mycorrhizae, *Soil Biol. Biochem.* 17 : 383-384.

Morton, J., Bentivenga, S. and Bever, J. 1995, Discovery, measurement and interpretation of diversity in arbuscular endomycorrhizal fungi (Glomales, Zygomycetes), *Can. J. Bot.* 73 : S 25-S 32.

Nylund, J.E., Dahlberg, A., Hogberg, N., Karen, O., Grip, K. and Jonsson, L. 1995, Methods for studying species of mycorrhizal fungal communities in ecological studies and environmental monitoring, in : *Biotechnology of Ectomycorrhizae*, V. Stocchi, ed., Plenum Press, New York, pp.229-239.

Potenza, L., Amicucci, A., Rossi, I., Palma, F., De Bellis, R., Cardoni, P. and Stocchi, V. 1994, Identification of *Tuber magnatum* Pico DNA markers by RAPD analysis *Biotechnol. Tech.* 8 : 293-298.

Priestley, R.A. and Dewey, F.M. 1993, Development of a monoclonal antibody immunoassay for the eyespot pathogen *Pseudocercosporella herpotrichoides, Plant Pathol.* 42 : 403-412.

Pritsch, K., Boyle, H., Munch, J.C. and Buscot, F. 1997, Characterization and identification of black alder ectomycorrhizas by PCR/RFLP analyses of the rDNA internal transcribed spacer (ITS), *New Phytol.* 137 : 357-369.

Raffle, V.L., Anderson, N.A., Frunier, G.R. and Doudrick, R.L. 1995, Variation in mating competence and random amplified polymorphic DNA in *Laccaria bicolor* (Agaricales) associated with three tree host species, *Can. J. Bot.* 73 : 884-890.

Rizzo, D.M., Blanchette, R.A. and May, G. 1995, Distribution of *Armillaria ostoyae* genets in a *Pinus resinosa-Pinus banksiana* forest, *Can. J. Bot.* 73 : 776-783.

Rodriguez, K.F., Petrini, O. and Leuchtmann, A. 1995, Variability among isolates of Xylaria cubensis as determined by isozyme analysis and somatic incompatibility tests, *Mycologia.* 87 : 592-596.

Rogers, S.O., Rehner, S., Bledsoe, K., Mueller, G.J. and Ammirati, J.F. 1989, Extraction of DNA from Basidiomycetes for ribosomal DNA hybridization, *Can. J. Bot.* 67 : 1235-1243.

Rosendahl, S. 1989, Comparisons of spore-cluster forming *Glomus* species (Endogonaceae) basesd on morphological characteristics and isoenzyme banding patterns, *Opera Bot.* 100 : 215-223.

Rosendahl, S. and Hepper, C.M. 1987, Comparative studies of medium endophytes forming VA mycorrhizae, in : *Mycorrhizae in the Next Decade, Practical Applications and Research Priorities,* D.M. Sylvia, L.L. Hung and J.H. Graham, eds., IFAS, Gainesville, Florida, p. 319.

Sanders, I.A., Alt, M., Groppe, K., Boller, T. and Wiemken, A. 1995, Identification of ribosomal DNA polymorphisms among and within spores of the Glomales : application to studies on the geneitc diveresity of arbuscular mycorrhizal fungal communities, *New Phytol.* 130 : 419-427.

Sanders, I.A., Clapp, J.P. and Wiemken, A. 1996, The genetic diversity of arbuscular mycorrhizal fungi in natural ecosystems - a key to understanding the ecology and functioning of the mycorrhizal symbiosis, *New Phytol.* 133 : 123-134.

Sanders, I.R., Ravolanirina, F., Gianinazzi-Pearson, V., Gianinazzi, S. and Lemoine, M.C. 1992, Detection of specific antigens in the vesicular arbuscular mycorrhizal fungi *Gigaspora margarita* and *Acaulospora laevis* using polyclonal antibodies to soluble spore fractions, *Mycol. Res.* 96 : 477-480.

Sastry, J.G., Ramakrishna, W., Sivaramakrishnan, S., Thakur, R.P., Gupta, V.S. and Ranjekar,P.K. 1995, DNA finger printing detects genetic variability in the pearl millet downy mildew pathogen (*Sclerospora graminicola*), *Theor. Appl. Genet.* 91 : 856-861.

Sen, R. 1990a, Intraspecific variation in two species of *Suillus* from Scots pine (*Pinus sylvestris* L.) forests baseds on somatic incompatibility and isozyme analyses, *New Phytol.* 114 : 607-616.

Sen, R. 1990b, Isozymic identification of individual ectomycorrhizas synthesized between Scots pine (*Pinus sylvestris* L.) and isolates of two species of *Suillus*, *New Phytol.* 114 : 617-626.

Simon, L., Lalonde, M. and Bruns, T.D. 1992, Specific amplification of 18S fungal ribosomal genes from vesicular-arbuscular mycorrhizal fungi colonizing roots, *Appl. Environ. Microbiol.* 58 : 291-295.

Simon, L., Levesque, R. and Lalonde, M. 1993, Identification of endomycorrhizal fungi colonizing roots by fluorescent single-strand conformation polymorphism - polymerase chain reaction, *Appl. Environ. Microbiol.* 59 : 4211-4215.

Thingstrup, I., Rozycka, M., Jeffries, P., Rosendahl, S. and Dodd, J.C. 1995, Detection of the arbuscular mycorrhizal fungus *Scutellospora heterogama* within roots using polyclonal antisesra, *Mycol. Res.* 99 : 1225-1232.

Timonen, S., Tammi, H. and Sen, R. 1997, Characterization of the host genotype and fungal diversity in Scots pine ectomycorrhiza from natural humus microcosms using isozyme and PCR-RFLP analyses, *New Phytol.* 135 : 313-323.

Tommerup, I.C. 1992, Genetics of eucalypt ectomycorrhizal fungi, in : *Proc. Intern. Symp. on Recent Topics in Genetics, Physiology and Technology of Basidiomycetes,* M. Miyaji, A. Suzuki and K. Nishimura, eds., Chiba University, Chiba, Japan, pp. 74-79.

Trappe, J.M. 1982, Synoptic keys to the genera and species of zygomycetous mycorrhizal fungi, *Phytopathol.* 72 : 1102-1108.

Vogt, K.A., Bloofield, J., Ammirati, J.F. and Ammirati, S.R. 1992, Sporocarp production by Basidiomycetes with emphasis on forest ecosystems, in : *The Fungal Community : Its organization and Role in the Ecosystem,* G.C. Carroll and D.T. Wicklow, eds. New York. Marcel Dekker, 563-581.

Wright, S.F. and Morton, J.B.1989, Detection of vesicular arbuscular mycorrhizal fungus colonization of roots by using a dot-immunoblot assay, *Appl. Environ. Microbiol.* 55 : 761-763.

Wright, S.F., Morton, J.B. and Sworobuk, J.E. 1987, Identification of vesicular-arbuscular mycorrhizal fungus by using monoclonal antibodies in an enzyme-linked immunosorbent assay, *Appl. Environ. Microbiol.* 53 : 2222-2225.

Wyss, P. and Bonfante, P. 1993, Amplification of genomic DNA of arbuscular-mycorrhizal (AM) fungi by PCR using short arbitrary primers, *Mycol. Res.* 97 : 1351-1357.

Zeze, A., Sulistyowati, E., Ophel-Keller, K., Barker, S. and Smith, S. 1997, Inter sporal genetic variation of *Gigaspora margarita*, a vesicular arbuscular mycorrhizal fungus, revealed by M13 minisatellite-primed PCR, *Appl. Environ. Microbiol.* 63 : 676-678.

Zhu, H., Higginbotham, K.O., Dancik, B.P. and Navratil, S. 1988, Intraspecific genetic variability of isozymes in the ectomycorrhizal fungus *Suillus tomentosus, Can. J. Bot.* 66 : 588-594.

THE GROWTH OF VAM FUNGI UNDER STRESS CONDITIONS

Rajni Gupta and K.G. Mukerji

Applied Mycology Laboratory
Department of Botany
University of Delhi
Delhi-110007, INDIA

1. INTRODUCTION

Since the beginning of this century the number of species found in inventories of sporocarps of ectomycorrhizal fungi has decreased drastically in central and western Europe (Arnolds, 1991). Air pollution in particular, but also forestry practices, have been suggested to be the most important factors, that may explain this decrease. Vesicular arbuscular mycorrhizae (VAM/AM), the most widespread symbioses on earth (Harley and Smith, 1983) are receiving attention because of the increasing range of their application in practical fields as diverse as sustainable agriculture, reforestation programmes and ecosystem management (Bethlenfalvay and Schuepp, 1994). The special relationships established between plant and fungus in this association implies a high degree of structural, physiological and biochemical integration from which both partners benefit (Azcon-Aguilar and Bango, 1994).

Vesicular arbuscular mycorrhizal fungi are present in most soils and are generally not considered to be host specific (Bowen, 1987). However, population levels and species composition are highly variable and are influenced by plant characteristics and a number of environmental factors such as temperature, soil pH, soil moisture, phosphorus and nitorgen levels, heavy metal concentrations (Daniels and Trappe, 1980), the presence of other microorganisms, application of fertilizers, pesticides and soil salinity, etc. (Barea and Azcon Aguilar, 1983). Species and strains of VA mycorrhizal fungi differ in thier ranges of tolerance of physical and chemical properties of soil (Abbott and Robson, 1991) therefore differ also in their effectiveness in increasing plant growth in particular soils.

VA mycorrhizas have been shown to decrease yield losses on plants in saline soils (Hirrel and Gerdemann, 1980; Poss *et al.*, 1985). This may be due to increased uptake of phosphorus leading to increased growth and subsequent dilution of toxic ion effects, to an ameliorative effect of mycorrhizas on water stress of plants or to some combination of these effects. AM fungus derives carbohydrates from the host which on the other hand improves its mineral nutrition by means of a more effective fungal mediated nutrient uptake through the external mycelial network (Harley and Smith, 1983; Srivastava *et al.*, 1996).

In spite of importance in biogeochemical soil nutrient cycling in general and in nutrient uptake by mycorrhizal plants in particular, the study of the extra radical mycelium of the AM fungi has received very little attention (Dodd, 1994; Jeffries and Barea, 1994). Colonization of plant roots by VAM fungi may increase nutrient acquisition from the soil, especially nutrients that are less mobile such as soil phosphate (P) (Allen, 1991; Nye and Tinker, 1977). When plants are well supplied with nutrients such as P, the colonization frequency and degree of VAM colonization of roots tend to decrease (Menge *et al.*, 1978). This regulation of colonization by host plants may be cost effective, since the plant can acquire adequate quantities of P without expending photosynthate on mycorrhizae (Bloom *et al.*, 1985).

2. EFFECT OF ALKALINITY

In the present article studies relating to responses of different adverse environmental conditions on mycorrhizal host plant have been reviewed. Vesicular arbuscular mycorrhization was observed on plants growing in an alkaline usar land ecosystem that supported only scanty vegetation. Janardhan *et al.* (1994) studied six plant species, four belonging to the pioneer vegetation and two introduced in the usar land. VAM colonization was observed in all the six plant species and the maximum colonization was in introduced bamboos (81.2%). Rhizosphere soil of the plants harboured a significant population of VAM fungi belonging to the genera, *Glomus, Acaulospora* and *Entrophospora*. Native VAM fungi of usar land ecosystem appear to be tolerant to soil alkalinity. The results suggest potential use of the native VAM fungi for the revegetation and reclamation of alkaline usar lands.

Usar is a term collectively used for all kinds of alkaline and saline lands in northern parts of India (Mukerji, 1966). It is estimated that presently about 7 million hectares of usar land is unfit for agriculture in India. A major part of this type of waste land is spread in the plains of Uttar Pradesh (1.29 million hectares). The soil in this area is highly alkaline and supports only marginal and scanty vegetation. VAM fungi are the most common of all the mycorrhizae and are widely recognized as components of all terrestrial ecosystems. VA mycorrhizal fungi are ubiqutous and occur in nearly all natural soils from arctic to tropics in a wide range of niches (Daniels, 1984).

3. EFFECT OF SALINITY

VA mycorrhizal fungi have been shown to occur naturally in saline environments by several workers (Ho, 1987; Khan, 1974) despite the comparatively low mycorrhizal affinity of many halophytic plants (Brundrett, 1991). Chenopods were not generally considered to form mycorrhizal association, however, there is increasing evidence that some chenopods form mycorrhizas (Pond *et al.*, 1984). Plant material and rhizosphere soils from halophytes, xerophytes and hydrophytes growing in a range of environments and soil types in Pakistan were surveyed for the presence of VA mycorrhiza (Khan, 1974). Of the 21 halophytes included in the survey, 11 contained VA mycorrhizal infection and spores of VA mycorrhizal fungi in their rhizosphere. The plants were members of the families Papaveraceae, Sapindaceae, Salvadoraceae, Apocynaceae and Gramineae. The relationship between soil salinity and the presence of mycorrhizas on halophytes has been integrated by several workers (Allen and Cunningham, 1983). Mycorrhization did not occur in plants where soil salt concentration exceeded 3.3 mg/gm. In well drained playas, predominantly vegetated with *Distichlis spicata* both percentage infection and spore number were negatively correlated with sodium content of the soil over a range of 153-1160 mg/gm (Kim and Weber, 1985). A wide ranging survey of saline soils in California and Nevada found mycorrhizal plant roots and VA mycorrhizal fungi in a range of 185 dsm^{-1} (Pond *et al.*, 1984).

Soil salinity may influence the growth and activity of VA mycorrhizal fungi *via* several mechanisms, either actively and interactively. Germination of spores of a VA mycorrhizal fungi can be described as consisting of various phases: hydration, germ tube emergence and growth of hyphae (Tommerup, 1984). Firstly, water enters the spore, the components of which become hydrated. After hydration of some or all organelles and macromolecules, ribonucleic acid enzymes become active - leading to increased metabolic activity. Two to 10 days after the spore is activated, a germ tube appears, and is followed by hyphal growth (Tommerup, 1984). Delay or prevention of all or any of the phases of spore germination by dissolved salts in the soil solution would delay or prevent growth of hyphae and therefore colonization of plant roots and establishment of the symbiosis. Very little published information is available about the effects of salinity on the germination of spores of VA mycorrhizal fungi. However, the available data indicate inhibition of spore germination by increasing concentrations of NaCl (Estaun, 1991; Hirrel, 1981).

4. MYCORRHIZAL FUNGI IN ARID AND SEMI ARID SOILS

In arid and semi arid regions many soils have ochric rather than mollic epipedons, because surface soils are massive and hard or very hard when dry. These soils are knwon to form surface crusts that reduce water infiltration rates. Soils are diverse, encompassing highly infertile acid and alkaline sands, clay of volcanic origin and soils developed from limestones. Erosion losses result in detrimental effects on the chemical, physical and microbial properties of soil. Since eroded soils have reduced numbers of mycorrhizal propagules, inoculation with VAM fungi has been proposed to rehabilitate these areas (Sarwar and Mukerji, 1998).

Soil organic matter plays a major role in terrestrial ecosystem development and functioning. In both undisturbed and cultivable systems, potential productivity is directly related to soil organic matter concentration and turnover (Reeves and Redente, 1991). Soil organic matter is the chief binding agent for soil aggregation that, in turn, controls air and water relations for root growth and provides resistance to wind and water erosion.

A survey of literature on the occurrence of mycorrhizas in desert, arid and semi arid regions reveals that most of the plant taxa probably form associations with mycorrhizal fungi. Non-mycorrhizal taxa include the cruciferae and zygophyllaceae. Although cactaceae, chenopodiaceae, cyperaceae, amaranthaceae and juncaceae typically are thought to be non mycorrhizal, most of the species were found to be infected under natural stressed range land ecosystems (Neeraj et al., 1991). VAM/AM fungi modify the plant root systems. These fungi play a critical role in nutrient cycling in the ecosystem. Since the external mycelium extends several centimeters from the root surface, it can bypass the depletion zone surrounding the root and exploit soil microhabitats beyond the nutrient depleted area where rootlets or root hairs can not thrive (O'Keefe and Sylvia, 1992). It is evident that AM fungi have a greater exploring ability than the root and overcome limitations on the acquisition of ions that diffuse to the root environments and move slowly in the soil solution of the rhizosphere. Several species of cacti grow and flourish best in arid and semi arid environments with root systems heavily mycorrhized (Mathew et al., 1991).

5. EFFECT OF NUTRIENT UPTAKE

The possibility of nutrient transport among plants in patchy environments has also been hypothesized. In this scenario, patches with high nutrients as under shrub would be occupied by plants, that are interconnected by mycorrhizal mycelia to plants existing in low nutrient patches. Mycorrhizal fungi, by interconnecting several individuals, equalize resource allocations and allow less competitive species to co-exist (Sylvia et al., 1993). Mycorrhizas regulate the composition and functionality of plant communities by regulating resource allocation and growth characteristics of interacting plants. This regulation can take the form of transporting nutrients preferentially to one individual over another or of redistributing nutrients throughout the community. Mycorrhizal fungi, by growing into soil matrix, can again access to bulk soil P beyond the depletion zones created by the plant roots. The soil P concentrations under shruby trees increased with time in successional shrub deserts, but declined in the associated inter space regions occupied by arbuscular mycorrhizas. It seems that mycorrhizas may be involved in the development of these "islands of fertility" that characterise arid regions (Newman *et al.*, 1992).

6. EFFECT OF PHOSPHORUS

Colonization of plant roots by vesicular - arbuscular mycorrhizal fungi may increase nutrient acquisition from the soil, especially nutrients that are less mobile such as soil phosphate (Allen, 1991; Nye and Tinker, 1977). When plants are well supplied with nutrients such as P, the colonization frequency and degree of VAM colonization of roots tend to decrease (Braunberger *et al.*, 1991; Menge *et al.*, 1978). Since plant can acquire adequate quantities of P without expending photosynthate on mycorrhizae (Bloom *et al.*, 1985; Koide and Elliot, 1989). Nutrients are rarely homogeneously distributed with in the rooting zone of individual plants (Charley and West, 1975; Jackson and Caldwell, 1993). Microsites on nutrient enrichment are assumed to be important to the nutrient status of plants (Chapin, 1980) and VAM hyphae can be important in the exploitation of soil heterogenecity (St. John's *et al.*, 1983). Development of VAM in roots experiencing localized areas of nutrient enriched soils may be reduced if the nutrient demand of the whole plant is decreased or if roots in these microsites have greater nutrient concentrations (Koide and Li, 1990). Jackson *et al.* (1990) showed that roots of these species can increase their physiological uptake capacity for N and P in response to local nutrient rich patches. Since these species are VAM symbionts and VAM can contribute significantly to nutrient exploitation. Duke *et al.* (1994) noted that arbuscule frequency was significantly reduced for *Agropyron desertorum* roots in enriched soil patches. Roots from nutrient enriched patches had an average arbuscule colonization frequency of *Agropyron* roots in enriched patches was lower for the 3 week time course. There were no changes in vesicle frequency and the overall VAM colonization decreased slightly in enriched patches. There were no significant treatment differences in any roots (p >0.11) in each case. Although the relative pattern of lower arbuscule frequencies of nutrient patches was similar to that of *Agropyron*, the data are less conclusive for *Artemisia* because the dark colour of its roots made it difficult to accurately identify branches. The process whereby plants reduce or regulate VAM colonization is not well understood (Koide and Schreiner, 1992). Braunberger *et al.* (1991) found that arbuscule development in maize roots was reduced by a general application of phosphate fertilizer and they concluded that increased shoot P concentrations had a significant influence on reducing arbuscule development. Root proliferation is a primary strategy by which plants can more effectively exploit nutrient patches. Root proliferation was likely to be occurring during the 3-week experimental period (Jackson and Caldwell, 1989). However, it is not apparent that root growth was more rapid than fungal development, since neither *Agropyron desertorum* or *Artemisia tridentata* showed reduced vesicle formation or overall colonization. *Agropyron desertorum* and *Artemisia tridentata* both exhibited rapid physiological responses and increased nutrient procurement from nutrient enriched patches in the soil (Caldwell *et al.*, 1991). Since mycorrhizal colonization does not increase and arbuscule frequency can decrease in enriched soil patches.

Mycorrhizal fungi can also increase the uptake of mineralized P by occupying the microsites of active decomposition and possibility by being involved in the degradation of litter. Herrera *et al.* (1978) found mycorrhizal hyphae on a decomposing leaf and rapid transport of radiolabelled P from the leaf to a host plant *via* mycorrhizal hyphae. Allen and Mac Mohan (1985) found high spatial correlation between decomposer activity and labile organic matter with AM mycorrhizal hyphae. It was observed that plants with a greater rooting density and greater mycorrhizal fungal density gained more labelled P from the interspace than neighbouring plants. This could result in the redistribution of

P in the soil, creating habitats where resources are distributed in discrete patches. Three hypotheses have been put forward in which mycorrhizal activity can contribute soil P to the host plant: (i) The interaction of mycorrhizal fungi and P-solubilizing bacteria, (ii) the production of phosphatases by the mycorrhizal fungus and (iii) the production of organic acids by mycorrhizal hyphae which mineralize P.

The role of mycorrhizas in N cycling has become a central focus in ectomycorrhizal research. Mycorrhizas can affect N cycling *via* several means. As they are a large biomass component, both the fungus and fine roots will immobilize substantial quantities of N in promoting their own growth. Mycorrhizal hyphae have the capacity to extract N and transport it from soil to plant because of the enhanced absorptive surface area. In soils where ectomycorrhizas generally occur, low pH and temperature do not favour mineralization and there is a consequent build up of organic matter and lack of labile nutrients. Ectomycorrhizal mycelia are closely associated with the organic decomposition horizon in such soils (Read, 1991) and might benefit their hosts by accessing this otherwise unavailable source of phosphorus (P) *via* the production of external phosphatase enzymes (Gianinazzi-Pearson and Gianinazzi, 1989).

Mycorrhizal hyphae have the capacity to extract N and transport it from soil to plant because of the enhanced absorptive surface area. In addition, mycorrhizal fungi contain enzymes that break down organic nitrogen and contain N-reductase for altering forms of N in soil. As mycorrhizae can increase the water flow through plants, there is a potential for increased nitrate migration to the roots for uptake. AM fungi readily transport NH_4 from soil to plant. This may be important where N is distributed in discrete patches. Mycorrhizal associations can also enhance N gain in ecosystems by increasing N fixing associations. Increases in the nodulation status and rates of N-fixation with mycorrhizas have been reported in legumes, actinorrhizal associations and in free living associations (*Azospirillum, Azotobacter*) (Barea *et al.*, 1992). One means whereby mycorrhiza enhanced N fixation may increase the N incorporated into the substrate through the hyphal connections between plants. Several investigators have demonstrated that N fixed in association with one plant can be transported to an adjacent, non fixing plant *via* the mycorrhizal fungal mycelium (Frey and Schuepp, 1993). Wallander (1995) proposed that in a situation with increased availability of N, the fungus allocates more carbon to energy demanding nitrogen assimilation than to vegetative growth.

7. EFFECT OF SOIL ACIDITY

Soil acidity is an important factor regulating spore germination and may in part, explain the distribution of VAM fungi in different soils (Mosse *et al.*, 1981; Porter *et al.*, 1987). Green *et al.* (1976) observed that neutral to alkaline pH's favoured the germination of *Glomus mosseae*, while spores of *Gigaspora coralloidea* and *Gigaspora heterogama* germinated best at pH 5 and 6 respectively. The best pH range for *Glomus epigaeus* germination appears to be 6 to 8, with the percent germination declining significantly as pH decreased much below 7 (Daniels and Trappe, 1980). Siqueira *et al.* (1984) investigated the influence of liming on the germination of spores of *G. mossae* and *Gigaspora margarita* and noted a 58 to 320% increase in germination and germ tube growth of spores of *G. mossae* as pH increased from 5.5 to 6.4. Hepper (1984) determined the germination of *Acaulospora laevis* in soils having different pH and concluded that the optimum range for germination was 4 to 5. The solubility of MnO_2 increases with decrease in soil pH and this can have an inhibitory effect on spore germination. Hepper and Smith (1976) observed that spore germination was inhibited by Mn at a concentration as low as 136 mg l^{-1}. Subsequently, Hepper (1979) demonstrated that spores exposed to 1.4 mg of manganese l^{-1} for 28 days germinated normally when transferred to water agar.

AM fungi have been reported in roots of plants growing in soils having pH 9.2 and in bituminous mine soils at pH 2.7 (Siqueria *et al.*, 1984). As with spore germination, some AM fungi colonize roots across with pH ranges, while others have specific pH preferences for maximal root colonization. Sparling and Tinker (1978) monitored root colonization in grassland sites pH 4.9 to 6.2 and noted little or no variation in the extent to which the natural vegetation was colonized by AM fungi. The proportion of root length of subterranean clover colonized by *Glomus fasciculatum* remained unchanged if pH was elevated from 5.3 to 7.0. Some endophytes appear to exhibit consistent preferences for low pH (e.g. *Acaulospora* sp.) or near neutral pH (e.g., *G. mosseae*) for maximal colonization of roots (Mosse, 1972; Siqueira *et al.*, 1984). Davis *et al.* (1983) grew sweetgum at pH 5.1 or 5.9 in the presence of *Glomus mosseae* or *G. fasciculatum* the earliest and greatest response of sweetgum was observed in the soils inoculated with *G. fasciculatum*. Skipper and Smith (1979) compared *Gigaspora giagantea* and *Glomus mosseae* in an unlimited pH 5.1 and in a limed pH 6.2 soil and noted that the growth response of soybean in the unlimed soil was greater if infected by *G. gigantea* than *G. mosseae*. Abbot and Robson (1985) also demonstrated the preference of *G. mosseae* for pH in the neutral to slightly alkaline range. Decreased soil pH means increased concentration of hydrogen ions, it also means a host of other things, including increased concentration of aluminium, manganese or both, and decreased concentrations of Ca, Mo, Mg, P or all of these depending on soil chemical and physical properties.

8.   EFFECT OF SOIL pH

Influence of soil pH on the VAM symbiosis, there is not sufficient basis for identifying which component of soil acid was responsible for the observed effects (Robson and Abbot, 1989). In the investigations undertaken by Wang *et al.* (1985) a significant decrease in fractional infection was obtained by increasing the Al or Mn contents of an acid sand inoculated with *G. caledonicum*. While the data suggest that Al and Mn toxicity do not help one to appraise that Al was responsible for the inhibition of spore germination and germ tube elongation of *G. mosseae* at pH 5.5 is equivocal, since the investigations extracted Al from soil using a solution that was far more acid than the ambient solution. Moreover the concentration of Al at pH 5.5 is negligible. In Al rich acid soils, the effect of Al toxicity on the VAM symbiosis may be delineated from the toxic effects of H^+ by monitoring VAM activity prior to and after fresh organic matter amendment. Some host species can alter the pH of their rhizosphere to a greater extent than others, and pH measured in bulk soil solution may not accurately reflect pH in the root zone. Host species also vary in their requirements for VAM fungi (Habte and Manjunath, 1991) and the absence of response to VAM infection in acid soils may not necessarily reflect the adverse effects of pH.

VAM endophytes seem to have preferences for particular soils that may not necessarily be related to soil pH. In general, they seem to be more active in soils from which they are isolated. The tendency of VAM endophytes to be better adapted in one soil than another may reflect the degree to which soils differ in their ability to support optimal VAM effectiveness. However, a much wider pool of VAM genotypes and a more rational basis for isolating and evaluating endophytes is needed.

9. EFFECT OF WATER STRESS

Mycorrhizal fungi are believed to improve water relation of host plant by increasing root hydraulic conductivity,increasing transpiration rate and lowering stomatal resistance (Levy and Krikum, 1980; Sanders and Tinker, 1973) by altering the balance of plant hormones (Allen *et al.*, 1982). Studies on the response of mycorrhizal corn, sudan grass, big blue stem under drought stress, have indicated the benefit derived by each plant species under adequate water condition (Hetrick *et al.*, 1987). He also noted that ability of VAM to benefit plant growth under drought stress was apparently plant mediated and possibly related to the dependency of the plant on the mycorrhizal fungus. Under adequately watered condition, inoculated corn and sudan grass were respectively 1.23 and 1.13 times larger than non inoculated plant, while inoculated big blue stem was

6.56 fold larger than non-inoculated control plants.

During periods of soil water deficit nutritional status of crops is reportedly improved following the enhancement of drought resistance (Fitter, 1988). Spore germination, spore production and VAM root colonization are reportedly maximum at optimal soil moisture supporting plant growth (Redhead *et al.*, 1975). Yocom *et al.* (1987) noted the differences in intensity of VAM infection between the four species of *Glomus*. However, *G. deserticola* registered for higer root infection at high water level and low water level.

REFERENCES

Abbot, L.K. and Robson, A.D. 1991, Factors influencing the occurrence of vesicular arbuscular mycorrhizas, *Agric. Ecosyst. Environ.* 35: 121-128.

Abbot, L.K. and Robson, A.D. 1985, The effect of soil pH on the formation of VA mycorrhizas by two species of *Glomus*, *Aust. J. Soil. Res.* 23: 253-261.

Allen, E.B. and Cunningham, G.L. 1983, Vesicular-arbuscular mycorrhizae on *Distichlis spicata* under three salinity levels, *New Phytol.* 93: 227-236.

Allen, M.E., Moore, T.S. Jr. and Christensen, M. 1982, Phytochromone changes in *Bouteloua gracilis* infected by vesicular arbuscular mycorrhizae, Cytokinin increases in the host plant, *Can. J. Bot.*, 58: 371-374.

Allen, M.F. 1991, *The Ecology of Mycorrhizae*, Cambridge University Press, Cambridge.

Allen, M.F. and Mac Mohan, J.A. 1985, Importance of disturbance of cold desert fungi: comparative microscale dispersion patterns, *Pedologia* 28: 215-224.

Arnolds, E. 1991, Decline of ectomycorrhizal fungi in Europe, *Agric. Ecosyst. Environ.* 35: 209-244.

Azcon, Aguilar, C. and Bago, B. 1994, Physiological characteristics of the host plant promoting an undisturbed functioning of the mycorrhizal symbiosis, in: *Impact of Arbuscular Mycorrhizas on Sustainable Agriculture and Natural Ecosystems*, S. Gianinazzi. and H. Schwepp, eds., Basal, Birkauser, Verlag, pp. 47-60.

Barea, J.M. and Azcon-Aguilar, C. 1983, Mycorrhizas and their significance in nodulating nitrogen fixing plants, *Adv. Agron.* 30: 1-54.

Barea, J.M., Azcon, R. and Azcon Aguilar, C. 1992, Vesicular arbuscular mycorrhizal fungi in nitrogen fixing systems, in: *Methods in Microbiology* Vol. 24, J.R. Norris, D.J. Read and A.K. Verma, eds., Academic Press, London, pp. 391-416.

Bethlenfalvay, G.J. and Schwepp, H. 1994, Arbuscular mycorrhizas and agrosystem stability, in: *Impact of Arbuscular Mycorrhizas on Sustainable Agriculture and Natural Ecosystems*, S. Gianinazzi, H. Schwepp, eds., Basel, Birkhauser Verlag, pp. 117-131.

Bloom, A.J., Chaplin, F.S. and Mooney, H.A. 1985, Resource limitation in plants - an economic analogy, *Annu. Rev. Ecol. Syst.* 16: 363-392.

Bowen, G. 1987, The biology and physiology of infection and its development, in: *Ecophysiology of VA Mycorrhizal Plants*, G.R. Safir, ed., CRC Press, Boca Raton, Flo., pp. 27-57.

Braunberger, P.G., Miller, M.H. and Peterson, R.L. 1991, Effect of phosphorus nutrition on morphological characteristics of vesicular arbuscular mycorrhizal colonization of maize, *New Phytol.* 119: 107-113.

Brundrett, M.C. 1991, Mycorrhizas in natural ecosystems, *Adv. Ecol. Res.* 21: 171-313.

Caldwell, M.M., Manwaring, J.H. and Jackson, R.B. 1991, Exploitation of phosphate from fertile soil microsites by three great Basin Perennials when in competition, *Funct. Ecol.* 5 : 757-764.

Chapin, F.S. 1980, The mineral nutrition of wild plants, *Ann. Rev. Ecol. Syst.* 11: 233-260.

Charley, J.L. and West, N.E. 1975, Plant induced soil chemical patterns in some shrub dominated semi-desert ecosystems of Utah, *J. Ecol.* 63: 945-964.

Daniels, B.A. and Trappe, J.M. 1980, Factors affecting the vesicular arbuscular mycorrhizal fungus *Glomus epigeous, Mycologia* 72: 457-471.

Davis, E.A., Young, J.L. and Linderman, R.G. 1983, Soil lime level (pH) and VA mycorrhizae effects on growth response of sweetgum seedlings, *Soil Sci. Soc. Amer.* 47: 251-256.

Dodd, J.C. 1994, Approaches to the study of the extra-radical mycelium of arbuscular mycorrhizal fungi, in: *Impact of Arbuscular Mycorrhizas on sustainable Agriculture and Natural Ecosystem*, S. Gianinazzi and H. Schuepp, eds., Basel, Birkhauser Verlag, pp. 147-166.

Duke, S.E., Jackson, R.B. and Caldwell, M.M. 1994, Local reduction of mycorrhizal arbuscule frequency in enriched soil microsites, *Can. J. Bot.* 72: 998-1001.

Estaun, M.V. 1991, Effect of NaCl and Mannitol on the germination of two isolates of the vesicular arbuscular mycorrhizal fungus *Glomus mosseae*, (Abstract), 3rd European Symposium on Mycorrhizas, University of Sheffield, Sheffield, U.K.

Fitter, A.H. 1986, Effect of benomyl on leaf phosphorus concentration in alpine grasslands, A test of mycorrhizal benefit, *New Phytol.* 103: 767-776.

Frey, B. and Schuepp, H. 1993, Transfer of symbiotically fixed nitrogen from berseem (*Trifolium alexandrinum*) to maize via vesicular arbuscular mycorrhizal hyphae, *New Phytol.* 122: 447-454.

Gianinazzi, Pearson, V., Branzanti, B. and Gianinazzi, S. 1989, *In vitro* enhancement of spore germination and early growth of vesicular arbuscular mycorrhizal fungal by root exudates and plant flavonoids, *Symbiosis* 7: 243-255.

Gianinazzi-Pearson, V. and Gianinazzi, S. 1989, Phosphorus metabolism in mycorrhizas, in: *Nitrogen, Phosphorus and Sulphur Utilisation by Fungi*, L. Boddy, R. Marchant and D.J. Read, eds., Cambridge University Press, Cambridge, U.K., pp. 227-242.

Green, N.E., Graham, S.O. and Schenck, N.C. 1976, The influence of pH on germination of vesicular arbuscular mycorrhizal spores, *Mycologia* 68: 929-934.

Habte, M. and Manju Nath, A. 1991, Categories of VAM dependency of host species, *Mycorrhiza* 1: 3-12.

Harley, J.L. and Smith, S.E. 1983, *Mycorrhizal Symbiosis*, Academic Press, London.

Hepper, C.M. 1979, Germination and growth of *Glomus caledonius* spores: the effects of inhibitors and nutrients, *Soil Biol. Biochem.* 11: 269-277.

Hepper, C.M. 1984, Regulation of spore germination of the VAM fungus *Acaulospora laevis* by soil pH, *Trans. Br. Mycol. Soc.* 83: 154-156.

Hepper, C.M. and Smith, G.A. 1976, Observations on the germination of *Endogone* spores, *Trans. Br. Mycol. Soc.* 66: 189-194.

Herreara, R., Merida, T., Stark, N. and Jordan, C.F. 1978, Direct phosphorus transfer from leaf litter to roots, *Naturwissenschaften* 65: 208-209.

Hetrick, B.A.D., Kitt, D.G. and Wilson, G.T. 1987, Effect of drought stress on growth response in corn, sudan grass and big blue stem to *Glomus etunicatum*, *New Phytol.* 105: 403-410.

Hirrel, M.C. 1981, The effect of sodium and chloride salts on the germination of *Gigaspora margarita*, *Mycologia* 73: 610-617.

Hirrel, M.C. and Gerdemann, J.W. 1980, Improved growth of onion and bell pepper in saline soils by two vesicular arbuscular mycorrhizal fungi, *Soil Sc. Soc. Am. J.* 44: 654-655.

Ho, I. 1987, Vesicular arbuscular mycorrhizae of halophytic grasses in the Aluvard desert of Oregon, *Northwest Sci.* 61: 148-151.

Jackson, R.B. and Caldwell, M.M. 1989, The timing and degree of root proliferation in fertile soil microsites for three cold desert perennials, *Oecologia* 81: 149-154.

Jackson, R.B. and Caldwell, M.M. 1993, The scale of nutrient heterogenecity around industrial plants and its quantification with geostatistics, *Ecology* 74: 612-614.

Jackson, R.B., Manwaring, J.H. and Caldwell, M.M. 1990, Rapid physiological adjustment of roots to localized soil enrichment, *Nature* 344: 58-60.

Janardhan, K.K., Khaliq Abdul, Naushin Fauzia and Ramaswamy, K. 1994, Vesicular arbuscular mycorrhiza in an alkaline usar land ecosystem, *Curr. Sci.* 67(6): 465-469.

Jeffries, P. and Barea, J.M. 1994, Biogeochemical cycling and arbuscular mycorrhizas in the sustainability of plant-soil systems, in: *Impact of Arbuscular Mycorrhizas on Sustainable Agriculture and Natural Ecosystems*, S. Gianinazzi and H. Schuepp, eds., Basel, Birkhauser Verlag, pp. 101-115.

Khan, A.G. 1974, The occurrence of mycorrhizas in halophytes, hydrophytes, xerophytes and *Endogone* spores in saline soils, *J. Gen. Microbiol* 81 : 7-14.

Kim, C.K. and Weber, D.J. 1985, Distribution of VA mycorrhizal halophytes on inland salt palayas, *Plant Soil* 83: 207-214.

Koide, R. and Li, M. 1990, On host regulation of the vesicular arbuscular mycorrhizal symbiosis, *New Phytol.* 114: 59-64.

Koide, R. and Elliot, G. 1989, Cost-benefit and efficiency of the vesicular arbuscular mycorrhizal symbiosis, *Funct. Ecol.* 31: 252-255.

Koide, R.T. and Schreiner, R.P. 1992, Regulation of the vesicular arbuscular mycorrhizal symbiosis, *Ann. Rev. Plant Physiol, Plant Mol, Biol.* 43: 557-581.

Levy, Y. and Krikum, J. 1980, Effect of vesicular arbuscular mycorrhiza on *Citrus jambhiri* water relations, *New Phytol.* 85: 25-31.

Mathew, J., Shankar, A., Neeraj and Varma, A. 1991, Glomaceous fungi associated with spineless cacti, a fodder supplement in deserts, *Trans. Jpn. Mycol. Soc. Jpn.* 32: 225-233.

Menge, J.A., Steirle, D., Bagyaraj, D.J., Johnson, A. and Leonard, R.T. 1978, Phosphorus concentration in plants responsible for inhibition of mycorrhizal infection, *New Phytol.* 80: 575-578.

Mosse, B. 1972, The influence of soil type and *Endogone* strain on the growth of mycorrhizal plants in phosphate deficient soils, *Rev. Ecol. Biol. Soil.* 3: 529-537.

Mosse, B., Stribley, D.P. and Letacon, F. 1981, Ecology of mycorrhizae, *Adv. Microb. Ecol.* 5: 137-210.

Mukerji, K.G. 1966, Ecological studies on the microorganic population of usar soils, *Mycopath. Mycol. Appl.* 29: 330-349.

Neeraj, Shankar, A., Mathew, J. and Varma, A. 1991, Occurrence of VA mycorrhizae within Indian semiarid soils, *Biol. Fertil. Soils* 11: 140-144.

Newman, E.I., Eason, W.R., Eissenstat, D.M. and Ramos, M.I.R.F. 1992, Interactions between plants: the role of mycorrhizae, *Mycorrhiza* 1: 47-53.

Nye, P.H. and Tinker, P.B. 1977, Solute movement in the soil - root system, University of California Press, Berkeley, California.

O'Keefe, D.M. and Sylvia, D.M. 1992, The chronology and mechanisms of mycorrhizal plant growth responsible for sweet potato, *New Phytol.* 122: 651-659.

Pond, E.C., Menge, J.A. and Jarrell, W.M. 1984, Improved growth of tomato in salinized soil by VAM collected from saline soils, *Mycologia* 76: 74-84.

Porter, W.M., Robson, A.D. and Abbott, L.K. 1987, Field survey of the distribution of VAM fungi in relation to soil pH, *J. Appl. Ecol.* 24: 659-662.

Poss, I.A., Pond, E., Menge, J.A. and Jarrell, W.M. 1985, Effect of salinity on mycorrhizal onion and tomato in soil with and without additional phosphate, *Plant Soil* 88: 307-319.

Read, D.J. 1991, Mycorrhizas in ecosystems - nature's response to the "law of the minimum", in: *Frontiers in Mycology, Proceedings of the Fourth International Mycological Congress*, D.L. Hawksworth, ed., CAB International, Wallingford, pp. 101-130.

Readhead, J.E. 1975, In: *Endomycorrhizas*, Academic Press, New York, pp. 447.

Reeves, F.B. and Redente, E.E. 1991, Importance of mutualism in succession, in: *Semiarid Lands and Deserts: Soil Resource and Reclamation*, J. Skujins, ed., Marcel Dekker, New York, pp. 423-442.

Sanders, F.E. and Tinker, P.B. 1973, Phosphate flow into mycorrhizal roots, *Pesticide Sci.* 4: 385-395.

Sarwar, N. and Mukerji, K.G., 1998, Mycorrhiza in forestation in arid, semiarid and wastelands, in: "*Ecological Techniques and Approaches to Vulnerable Environment*", R.B. Singh ed., Oxford and I.B.H. Publ. Co. Ltd., New Delhi, pp. 263-271.

Siqueira, J.O., Hubbel, D.H. and Mahmud, A.W. 1984, Effect of liming on spore germination, germ tube growth and root colonization by VAM fungi, *Plant Soil* 76: 115-124.

Skipper, H.D. and Smith, G.W. 1979, Influence of soil pH on the soybean-endomycorrhiza symbiosis, *Plant Soil* 53: 559-563.

Sparling, G.P. and Tinker, P.B. 1978, Mycorrhizal infections in Pennine grassland I, levels of infections in the field, *J. Appl. Ecol.* 15: 943-950.

Srivastava, D., Kapoor, R., Srivastava, S.K. and Mukerji, K.G. 1996, Vesicular arbuscular mycorrhiza - an overview, in: *Concepts in Mycorrhizal Research*, K.G. Mukerji, ed., Kuluwer Academic Publihsers, Netherland, pp. 1-39.

St John, T.V., Coleman, D.C. and Reid, C.P.P. 1983, Association of vesicular-arbuscular mycorrhizal hyphae with soil organic particles, *Ecology* 64: 957-959.

Tommerup, I.C. 1984, Effect of soil water potential on spore germination by VAM fungi, *Trans. Br. Mycol. Soc.* 83: 193-202.

Wallender, H. 1995, A new hypothesis to explain allocation of dry matter between mycorrhizal fungi and pine seedlings in relation to nutrient supply, *Plant Soil* 168-169: 243-248.

Wang, G.M., Stribley, D.P., Tinker, P.B. and Walker, C. 1985, Soil pH and vesicular arbuscular mycorrhizas, in: *Ecological Interaction in Soil, Plants and Animals*, A.H. Fitter, ed., Blackwell, Oxford, pp. 219-224.

Yocom, D.H., Boosalis, M.G. and Flessuer, T.R. 1987, In: *Mycorrhizae in the Next Decade*, D.H. Sylvia, L.L. Hung and J.H. Graham, eds., 6th NACOM, Oregon, USA, pp. 41.

MYCORRHIZAL PLANTS IN RESPONSE TO ADVERSE ENVIRONMENTAL CONDITIONS

R.K. Gupta[1] and Pradeep Kumar[2]

Department of Environmental Science[1] and Plant Pathology[2]
G.B. Pant University of Agriculture and Technology
Pantnagar - 263145, U.P., INDIA

1. INTRODUCTION

Plant roots provide an ecological niche for many of the microorganisms that abound in soil. In natural ecosystem mycorrhizal fungi occur in nearly all soils on the earth and forms a symbiotic relationship with root of most terrestrial plants (Bagyaraj, 1991). The two major types of mycorrhizae are the ectomycorrhizae (EM) and the endomycorrhizae (Vesicular-arbuscular mycorrhizae, VAM), so named and distinguished by the anatomical nature of the symbiotic relationship. VAM fungi are known to be associated with most agricultural crops and appear over a wide range of ecological zones; they have 1000 genera of plants representing some 200 families (Bagyaraj, 1991). There are atleast 30,000 receptive hosts in the world flora (Kendrick and Berch, 1985), and about 150 species of VAM fungi(Schenck and Perez 1985). If the hosts are divided up evenly among the fungi with no overlap in host range, each fungus would have more than 2500 potential partners (Bagyaraj,1991). The beneficial effects of VAM fungi on plant growth have long been recognized. Increased growth and yield of many plants owing to infection with VAM fungi has been attributed to enhance nutrition status, especially of phosphorous (Rhodes and Gerdemann, 1978; Safir, 1980; Safir and Nelson, 1981), and to improve water relations (Hardie and Layton, 1981; Levy and Krikun, 1980; Safir *et al.*, 1972; Sievertding, 1979).

Response of plants to adverse environmental conditions has been extensively studied, and this has been excellently reviewed (Levitt, 1972). Crop growth and yield are controlled by environmental factors (light, CO_2 supply, temperature, watersupply, nutrients, etc.) interacting with the genetically determined physiological and biochemical system (Lawlor, 1979). The greatest effects of stress occur during leaf development,when new cells and their constituent organelles, walls, etc. are formed (Lawlor and Leach,1985).

A number of review papers have been published regarding various physio-biochemical aspects in relation to water deficit stress in homoiohydrous plants (Hsiao, 1973; Iljin, 1957) and more recently in poikilohydrous plants (Bewley, 1929; Bewley and Kochko, 1982; Dhindsa and Matowe, 1981; Farrar, 1976). Recently, Gupta (1991) extensively reviewed the studies on drought response in fungi and mycorrhizal plants. However, investigations on mycorrhizal plants in response to different adverse environmental conditions have not been reviewed previously. In the present paper such studies are, therefore, reviewed extensively.

Mycorrhizal Biology, edited by Mukerji *et al.*
Kluwer Academic/Plenum Publishers, 2000

2.   RESPONSE TO WATER DEFICIT IN MYCORRHIZAL PLANTS

2.1.   Water Relations

Mycorrhizal associations significantly alter the water relationships of the host plants (Dixon et al., 1983; Osonuki et al., 1991; Parke et al., 1983). Safir et al. (1971, 1972) were the first to demonstrate that mycorrhizal soybean plants had lower resistance to water transport than non-mycorrhizal plants. Lavy et al. (1983) studied the role of VAM fungi on the hydraulic conductivity of root in plant during and after recovery from drought stress. The hydraulic conductivity of root was significantly decreased by previous water stress treatments and further reduced by VAM. Since water stress alone can reduce hydrualic conductivity (Ramos and Kaufmann, 1979), water could have been the determining factor for decrease of hydranulic conductivity. It is interesting to note that root hydraulic conductivity responded to mycorrhizae and water stress in ways parallel to shoot : root ratios. Plants with highest conductivity also had the highest shoot : root ratios. The longer root length of mycorrhizal plants could have compensated for decreases in root hydraulic conductivity leading to higher transpiration rate. Graham et al. (1981) found that whole plant transpiration, leaf water potential, and root hydraulic conductivity were similar in VAM and non-VAM plants, under well-watered conditions. During drought stress and recovery periods, VAM plants also had very comparable whole plant transpiration rates and leaf water potential to non-VAM plants, but mycorrhizal root hydraulic conductivity of *Carrizo citrange* and sour orange was 66 and 49%, respectively. These data do not support the hypothesis that mycorrhizae significantly enhance water relations of citrus, under the drought stress conditions.

A large number of reports are available regarding influence of VAM on transpiration and its associated parameters in host plants, in well-watered and water stress conditions. VA-mycorrhizal infection increased transpiration rates in well-watered plants (Allen et al., 1981; Levy and Krikun, 1980). Under drought stress conditions, several studies indicate that stomatal conductivity or transpiration rate of mycorrhizal plants was increased compared to similar sized non-mycorrhizal plants (Allen and Boosalis, 1983; Allen et al., 1981; Hardie and Leyton, 1981; Levy and Krikun, 1980). Levy and Krikun (1980) found higher transpiration fluxes and lower diffusion resistance in infected rough lemon seedlings before and after water stress, but the differences decreased rapidly with the increasing water stress. Whittingham (1980) reported greater stomatal resistance in mycorrhizal than in non-mycorrhizal *Plantago lanceolata* under water stress. It appears that mycorrhizal infection can indeed influence stomatal opening by inducing greater water stress and lowering leaf water potential (Hardie and Leyton, 1981). Lavy et al.(1983) found that mycorrhizal infection increased significantly the mean daily transpiration rate of well-watered as well as previously stressed-plants over that of non-inoculated high phosphorous control plants. Within the mycorrhizal plants, there were significant differences between the transpiration rate of well-watered and previously drought-stressed seedlings. Stomatal conductance were higher in mycorrhizal than in non-mycorrhizal plants in both wet and dry treatments (Hardie and Leyton, 1981). Stomatal closure in mycorrhizal plants occurred at lower leaf water potentials and after greater desiccation than in non-mycorrhizal plants, but some leaves of *Glomus mosseae* infected plants showed no stomatal response to drought and continue to transpire at water potential as low as -4.1 MPa (Allen and Boosalis, 1983). Bethlenfalvay et al. (1987) investigated the influence of VA-mycorrhizae on soybean plants, both under water deficit stress and non stress conditions. In stressed plants, transpiration and leaf conductance in VAM plants were significantly greater than in non-VAM P-fed plants during the first part of the stress cycle (Fig. 1). Transpiration rates and leaf conductance declined linearly with the time for VAM and P-fed treatments and converged towards the end of the stress cycle. This decline occurred even during the early part of the drought cycle where moisture sensors were not sensitive enough to detect a change in soil water potential (Fig. 2). The relationship between transpiration and leaf conductance was essentially linear towards the end of the stress cycle. However, during the first day of the cycle, while soil water potential was still high (Fig. 2), large changes in leaf conductance were

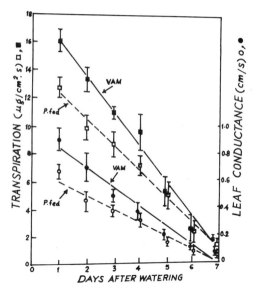

Fig. 1 : Change in transpiration rates and leaf conductance in VAM and Non-VAM, P-fed plants during the final stress cycle. Correlation coefficient were r = -0.99 (transpiration) and r = -0.98 (leaf conductance) for both VAM and non-VAM plants. (Source : Bethlenfalvay *et al.*, 1987).

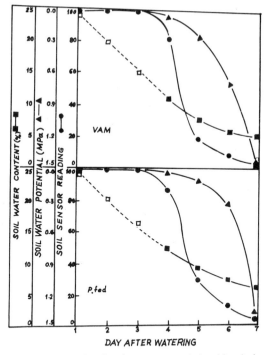

Fig. 2 : Soil water content, water potential and moisture sensor relationships during the final stress cycle of VAM and non-VAM, P-fed soybean plants. Soil sensor data points reflect the average of six replications, and are related to soil water data by celibration in a pressure plate apparatus. The dotted portion of the soil water content curve indicates an approximate position of the data point for days 2 and 3 of the cycle. Soil water content values were not measured, but calculated from a celibration curve based on soil sensor readings and the saturation value on day 1 (24%, w/w). Since soil sensors were not sensitive to change during the first 3 days, the curve was interpolated, and the symbols placed at the appropriate position. (Source : Belhlenfalvay *et al.*, 1987).

accomplished by relatively small change in transpiration. Koide (1985) also found that mycorrhizal condition enhanced transpiration and leaf conductance during an imprtant portion of recurring drought cycles. Unlike other workers, Nelson and Safir (1982) found that there was no significant differences in plant water relations (transpiration and leaf water potential) under either well-watered and drought-stressed conditions between mycorrhizal and non-mycorrhizal plants.

Some work has been done regarding the response of mycorrhizal infection on leaf/soil water potential of host plants, under well-watered and water-stressed condtions. Allen (1982) reported no change in leaf water potential due to mycorrhizal infection under well-watered conditions. No difference in leaf water potential was reported between mycorrhizal and non-mycorrhizal plants under either water deficit or well-watered conditions (Allen *et al.*, 1981; Dixon *et al.*, 1980; Graham *et al.*, 1981; Nelsen and Safir, 1982; Ramakrishnan *et al.*, 1988) Parke *et al.* (1983) in a growth chamber study, demonstraed that ectomycorrhizal Doughlas-fir [*Pseudotsuga menziessii* (Mirb.) Franco] seedlings had lower needle water potential and greater transpiration rates than non-inoculated seedlings during short-term soil water stress. *Pisolithus tinctorius* - inoculated seedlings of *Eucalyptus camaldulensis* maintained more negative leaf water potential during pre-drought than plants colonized by *Thelephora terrestris.* Recently, Huang *et al.* (1985) studied diurnal changes of xylem pressure potential and soil water potential for mycorrhizal and non-mycorrhizal *Leucaena leucocephala* (Lam.) de Witt. Soil water potential remained high (-0.3 MPa) in the non-mycorrhizal pots during the entire day compared to a comparatively large decrease in mycorrhizal pots (-0.3 to -2.2 MPa). Xylem pressure potential also decreased markedly for both mycorrhizal and non-mycorrhizal plants during the morning. By late afternoon, a maximum xylem pressure potential of nearly -2.0 MPa, occurred for both treatments, but the difference between the xylem pressure potential and the soil water potential was less for mycorrhizal plants (-0.1 vs. - 1.6 MPa). Lavy *et al.* (1983) reported that leaf water potential was nearly the same in mycorrhizal and non-mycorrhizal plants under well-watered conditions, but it was greater in mycorrhizal plants under water deficit stress. Hardie and Leyton (1981) found that water potential of soil, and hence leaf, at which mycorrhizal plants wilted, was significantly lower than in non-mycorrhizal plants. According to them, there are two, not mutually incompatible, explanations for this; (i) The soil water potentials quoted are bulk values, if, because of their smaller root systems, exploitation of the soil moisture of the non-mycorrhizal plants was not so effective; there may have been drier regions in the vicinity of the roots and wetter regions elsewhere. In other words, water potential of soil around the roots may be lower than the bulk value for the non-mycorrhizal plants, but much nearer to it in the case of the mycorrhizal plants of their more extensive root system, including fungal hyphae (ii) In some way, mycorrhizal infection enables the leaf to develop a lower water potential, thus drying out the soil more efficiently. It is well established that when subjected to water stress, leaf turgor can be maintained by decreasing the osmotic potential of the vacuolar sap by an increase in inorganic/or organic ions, sugars, etc. In case of inorganic ions, this is much more likely to be achieved by an increased uptake of, say, K rather than of P. There are indeed cases where endomycorrhizal infection has led to increased concentrations of K (Mosse, 1957)), but this response is not consistent (Schultz *et al.*, 1979).

2.2. Growth Parameters

It has been stated that mycorrhiza may be relatively more important to plant growth under dry conditions than when soil moisture is plentiful (Fitter, 1985; Nelson and Safir, 1982). The water-stressed non-mycorrhizal onion plants, despite high level of P fertilization, were only 23% as large as the stressed mycorrhizal onions grown at low P level (Rama and Kaufman, 1979). Under both nonstress and water stress conditions, total root length and total leaf area were markedly greater in mycorrhizal than in non-mycorrhizal plants; however, mycorrhizal infection caused slight decreases in leaf phosphorous and shoot-root ratio (Levy *et al.*, 1983) with *G. fasciculatum* had dry weight similar to non-mycorrhizal wheat plants; infection with *G. mosseae* significantly reduced growth of wheat under both wet and dry conditions. Dry weights of both non-mycorrhizal and *G. fasciculatum*-infected plants were significantly reduced (27 and 33%, respectively) by drought. Leaf dry weights were 34% lower in *G. mosseae*-infected plants with dry versus wet treatments, but there was no difference in root dry weights (Allen and Boosalis, 1983). VAM condition in soybean stimulated nodule development and activity under drought stress, and that this effect may result not only from better P nutrition, but also from increased CO_2 uptake due to better leaf conductance (Bethlenfalvay *et al.*, 1988). Colonization of VAM fungi confers a growth advantage on host plant under drought stress (Bethlenfalvay *et al.*, 1988). Michelsen and Rosendahl (1990) studied the effect of VAM fungi on growth and drought resistance of *Acacia nilotica* and *Leucaena leucocephala* seedlings. The drought treatment significantly reduced shoot and root dry weights of both *A. nilotica* and *L. leucocephala*. The total root length was not significantly affected in two species. Leaf area was reduced with *Glomus deserticola* by 9% due to drought, while with *G. occultum* the reduction was 70% (Ruiz-Lozano *et al.*, 1995). Under water stress conditions, VAM decreased root : shoot ratio and increased seed yield (Reichanbach and Schonback, 1995). The exact reason for increased growth in VAM-infected plants due to drought is not known. Greater growth of the drought stressed mycorrhizal plants can be attributed to improved P nutrition when compared to stressed non-mycorrhizal plants, despite the presence of lower level of soil P available to the mycorrhizal plants (Nelsen and Safir, 1982). Fitter (1985) stated that mycorrhizal phosphorus supplies are likely to be much more advantageous for plant growth under arid than in humid condition, as the diffusion coefficient of phosphate in soil is linearly related to soil moisture content. Phosphorus concentrations in the roots and leaves of non-mycorrhizal plants are significantly higher at all stress levels than in mycorrhizal plants, suggesting that P uptake was not a decisive factor in the growth enhancement of mycorrhizal plants at high drought stress. It is quite likely that growth enhancement in VAM-plants under drought condition might be related to phytohormones. Edriss *et al.* (1984) reported that cytokinin levels in mycorrhizal sour orange are greater than in comparable P-amended non-VAM plants. This indicates a potential for VAM fungi to alter hormones production, especially in root tissues which they occupy. Further studies are required to investigate various growth promoting and growth inhibiting phytohormones in VAM and non-VAM plants under drought conditions.

2.3. Root Colonization

Very little work has been done on the response of drought on root colonization of plants by VAM fungi. Hetrick *et al.* (1984) studied root colonization of corn plants by the VAM fungus *Glomus mosseae* as influenced by phosphorus level and moisture regime. Root colonization decreased significantly with increasing phosphorus concentration in both droughted and adequately watered plants; it was also depressed in adequately watered plants at the lower phosphorus levels as compared with droughted plants at these P levels (Table 1). Seedlings colonized with various dikaryotic strains of *Pisolithus* species were more sensitive to water stress and showed less mycorrhizal formation under water stress (Lamhamedi *et al.*, 1992). Higher efficiency in water and P uptake was apparently related to the greater development of extraradical VAM fungal mycelium in the stressed plants compared to those grown at field capacity (Bethlenfalray *et al.*, 1988). Such changes in fungal development related to soil water status were found by others also (Reid and Bowen, 1979; Sieverding, 1984). These reports note a decline in fungal development at high and low level of soil water content. However, Bethlenfalvay *et al.* (1988) reported decline in extraradical mycelium development at a high, but not at low level of soil water content.

Table 1 : Percentage root colonization of corn plants by the VAM fungus *Glomus mosseae* as influenced by phosphorus level and moisture regime

Phosphorus amendment (ppm)	Root colonization (%)*	
	Adequately watered, inoculated	Droughted, inoculated
0	12.9 bc	26.82 a
7.5	12.76 bc	19.28 b
15.0	9.08 cd	12.70 bc
22.5	8.18 cd	7.84 cd
30.0	7.80 cd	4.30 de

* Means followed by the same letters are not significantly different (P=0.05) as detrmined by Duncan's multiple-range test. a.b,c,d indicates that treatments having same superstate do not differ significantly. Non-inoculated control plants were not colonized. (Source : Hetrick *et al.*, 1984)

2.4. Photosynthesis and Photorespiration

Some reports are available regarding response of VAM fungi to photosynthesis and photorespiration in plants, under different water deficit conditons. Despite more negative leaf water potential, seedling with *Pisolithus tinctorius* ectomycorrhizae maintained a high rate of photosynthesis than plants with *Pisolithus terrestris* during short term drought (Dixon and Hiot-Hiot, 1992). Similar results have been reported previously for ectomycorrhizal *Pseudotsuga* and *Quercus* seedlings (Dixon *et al.*, 1983; Parke *et al.*, 1983). CO_2 assimilation in plants colonized by *G. etunicatum, G. mosseae* and *G. occultum* was highly sensitive to even mild stress (-0.06 MPa), while *G. fasciculatum* - infected plants showed a decrease only under severe stress (-0.17 MPa) (Ruiz - Lozano *et al.*, 1995). VAM plants tended to have high net photosynthetic flux and stomatal conductance than non-VAM plants under drought stress (Davies *et al.*, 1993). Dosskey *et al.* (1991) found that *Rhizopogon vinicolor* FS L 788-5 enhanced both net photosynthesis rate and stomatal conductance compared to non-mycorrhizal control in the soil water potential range -0.5 to -5.0 MPa, despite -0.2 to -0.3 MPa lower leaf water potential. Phtosynthesis and chlorophyll concentration in water-stressed wheat plants were generally increased by inoculation with *Azospirillum brasilense* or VAM, and highest in dual inoculated plants (Panwar, 1992). Recently Ramakrishan *et al.* (1988) studied the influence of introduced VAM fungus *Glomus caledonium* and that of indigenous population on photosynthesis, photorespiration, and chlorophyll content in osmotically stressed maize grown in sterile/unsterile soil, with or without P additions. There was no significant increase in photosynthesis under different treatments and soil conditions in maize plants at -0.3 MPa water potential. However, at severe stress (-1.2 MPa) there was significant increase in photosynthetic rate in VAM (10.4%) and P-amended plants (23.9%) in sterile soil. In sterile soil, some increase in chlorophyll -a and -b content was found in VAM, IVAM +GC, and P-amended plants over the non-VAM at zero osmotic potential. In non-stressed plants, evolution of $^{14}CO_2$ by decarboxylation of ^{14}C glycine was considerably greater in VAM and IVAM + GC treatments than in non-VAM plants. At a mild water stress (-0.6 and -0.8 MPa) there was no significant change in photorespiration in sterile soil; at severe stress (-1.2 MPa), it was considerably more in VAM, IVAM +GC, and P-amended soil than in non-VAM plants.

The above studies clearly shows that little work has been done on photosynthesis and photorespiration in repsonse to drought in mycorrhizal plants. Further work is needed to investigate the effect of mycorrhizal fungi on various metabolites of photosynthetic and photorespiratory pathways, under drought condition. In order to explain changes in photosynthesis and photorespiration, further studies are required to investigate various photoshythetic and photorespiratory enzymes in VAM and non-VAM plants under drought conditions. Such work has been reported extensively in non-mycorrhizal plants (Lawlor, 1979; Lawlor and Leach, 1985; Kumar and Gupta, 1986a,b).

2.5.　Metabolic Changes

A number of studies are available regarding the influence of mycorrhizae on various metabolic changes in plants, under different drought conditions. Subramanian and Charest (1995) studied the influence of *Glomus intraradices* on the metabolic changes in tropical maize (*Zea mays*) under drought. Chlorophyll content was not altered either by water stress or in the presence of mycorrhizae. Mycorrhizal plants (M+) had higher level of total reducing sugars than non-mycorrhizal plants (M-) at the end of three weeks of drought cycle . An increase in total soluble portein content was observed with drought stress in (M+) plants. Most of the amino acids showed a linear increase during the period of water stress in M+ and M- plants. Photosynthesis, chlorophyll concentration, nitrate reductase activity and glutamine synthetase activity in water-stressed wheat plants generally increased by inoculation with *Azospirillum brasilense* or VAM, and highest in dual inoculated plants (Panwar, 1992). Panwar (1993) studied the response of VAM fungus *Glomus fasciculatum* and *Azospirillum brasilense* on wheat plants under different stress conditions. The relative total chlorophyll content and nitrate reductase activity increased, and leaf ion leakage decreased with the combined inoculation compared with the single inoculation, resulting in higher biomass production and grain yield under water stress conditions. Pai *et al.* (1994) studied the role of VAM colonization on calcium uptake under different levels of moisture stress. Mycorrhizal plants showed high ^{45}Ca activity in all part of plants compared to non-mycorrhizal plants, at all the levels of moisture stress; the ^{45}Ca activity in stress has been postulated as a major factor for improved drought the roots of mycorrhizal plants increased, while the activity in the leaf showed marked reduction. Tobar *et al.* (1994) investigated the significance of external mycelium of arbuscular mycorrhiza for uptake and transport N from ^{15}N-labelled nitrate in benefiting plant nutrition, under well-irrigated and water-stressed conditions. The ^{15}N enrichment in plant tissues was the same for both mycorrhizal and non-mycorrhizal plants under optimum water supply conditions. However, under water conditions,where the mass flow and diffusion of NO_3^- ions to the root can be affected, the ^{15}N enrichment was 4 times higher in mycorrhizal than in non-mycorrhizal plants.

2.6.　Drought Resistance

A number of reports are available regarding the influence of mycorrhizal fungi on drought resistance in plants. VAM fungi can improve survival of host plants in diversified habitats (Gupta, 1991). Although VAM fungi benefit plant growth, mainly by producing supplemental nutrients required for optimum plant growth, these fungi have also been shown to improve the drought tolerance of colonized plants (Hetrick and Bloom, 1984). The fungal symbiont *Rhizopogon vinicolor* was reported to lessen the severity of drought effects on Douglas-fir seedlings better than *Pisolithus tinctorius* and *Laccaria laccata* (Parke *et al.,* 1983). The *Rhizopogon* symbiont enhanced drought tolerance in Douglas-fir only after preconditioning of inoculated seedlings to cyclic water stress (Dixon and Hiol-Hiol, 1992). In the uninoculated and unfertilized treatment the seedlings of *Acacia nilotica* were more drought tolerant than *Leucaena leucocephala* (Michelsen and Rosendahl, 1990).

It will be interesting to know why mycorrhizal plants yield more drought resistance than non-mycorrhizal plants. However, little information is available to explain the mechanism of drought resistance in mycorrhizal plants. According to Levitt (1972), plant response to environmental stresses may be considered in terms of 'avoidance' or 'tolerance'. Evidences for both types of resistance exists in the literature. The ability of VAM fungi to maintain adequate P nutrition in plants under drought stress has been postulated as a major factor for improved drought tolerance (Bethlenfalvay *et al.,* 1988). In contrast, many workers presented the evidence for water as an important factor for improved drought resistance in plants. The increased survival of mycorrhizal coffee and tea seedlings after drought stress could be the result of improved water relations (Sieverding and Toro, 1987).

Hardie (1985) also showed water as an important factor for better drought survival by mycorrhizal plants. Ectomycorrhizae reduce seedling water stress by several mechanisms including: (i) decreased resistance to water flow from bulk soil to roots (Schulze, 1986), (ii) increased root fungus absorptive surface area (Dixon et al., 1983; Parke et al., 1983) or (iii) greater potential for increased fungal hyphae contact with soil compared to root hairs (Brownlee et al., 1985; Reid, 1978). Mycorrhizal plants showed greater drought avoidance by maintaining higher leaf water potential and transpiration rate, and lower stomatal resistance (Subramanian et al., 1995). The VAM seedlings maintained slightly greater leaf water potential, leaf stomatal conductance and photosynthesis relating to non-VAM seedlings at the peak of drought treatment. This suggests that VAM fungi help *L. leucocephala* to avoid drought stress. VAM plants maintained higher turgor than non-VAM plants, during and following drought stress, indicate superior drought avoidance of these plants (Davies et al., 1993). Mycorrhizal *Acacia albida* avoid drought stress by producing long tap roots (Osonuki et al., 1991). During peak drought stress, plants with the combination of VAM-drought treatments had the greatest drought resistance, as indicated by the highest leaf water potential, turgor and relative water content. Nutrition and plant size were not associated with this drought resistance.

Bethlenfalvay et al. (1988) reported that proline level was much higher under water stress condition in leaves of soybean plants. Its level in the roots of wilted VAM plants, however, were not consistently higher than those of non-VAM plants. Thus, the implication of increased proline production in the leaf and transported to the root as a mechanism facilitating in uptake of 'unavailable' water by VAM roots remains inconclusive. However, if proline accumulation provides a measure of the extent of stress experienced by plants (Paleg and Aspinall, 1981), the low levels of proline in the leaves of stressed VAM plants relative to non-VAM plants indicate that VAM plants were suffering less stress (Table-2). In contrast, at all the water stress treatments the accumulation of proline was generally greater in P-amended plants than in plants given VAM, IVAM (indigenous inoculum) + *G. caledonium,* and non-VAM treatments (Ramakrishan et al., 1988).

Table 2 : Influence of water deficit stresses on free proline content in soybean plants colonized by the VAM fungus *Glomus mosseae* or fertilized with KH_2PO_4 (P-fed)

| Plant part | P nutrition | Proline content (mg g^{-1}) | | | | | |
| | | Turgid | | | Wilted | | |
		-0.05 MPa	-0.30 MPa	-1.00 MPa	-0.05 MPa	-0.30 MPa	-1.00 MPa
Leaf	VAM	0.14*a	0.34*b	1.07*c	3.27*a	2.65*a	8.51*b
	P-fed	0.10a	1.14b	2.01c	1.50a	5.16b	18.21c
Root	VAM	-	-	-	1.63*a	1.48*a	2.38b
	P-fed	-	-	-	0.71a	1.49b	2.26c

Number are the means of 6 replications. Comparisons between P-nutrition methods (VAM Vs P-feel) were evaluated for significance by students t-Test (NS, P>0.05*; P<0.05**; P<0.01; P<0.001***). The effects of stress were evaluated by Duncan's multiple range test and were not significatnly different (P>0.05) where numbers are followed by the same letter. a,b,c indicate that treatments (Source : Bethlenfalvay et al., 1988) having same superstate do not effort significants. (-) Data for turgial root and available.

The above review of literature clearly shows that insufficient work has been done to explain the induction of drought hardness in mycorrhizal plants. It is known that a number of physio-biochemical parameters are associated with drought tolerance in non-mycorrhizal plants. At a specific water stress treatment, there was more increase in membrane permeability in drought sensitive plant tissues (Gupta, 1977; 1981; 1982; 1983; 1984). Drought resistant plants revealed greater aminats of phospholipids (Gupta, 1977 a,b). Photosynthesis and respiration has been correlated with the drought tolerance in moss plants (Gupta, 1977a, b). Chlorophyll stability index can be correlated with drought tolerance in lower (Gupta, 1978) and higher plants (Kaloyereas, 1958). Levitt (1972) considered that the denaturation of proteins, and cell death following freezing and desiccation are caused primarily by formation of disulphide bonds (-S-S-) between neighbouring protein molecules, and he suggested that stress resistance (freezing, drought, high temperature) is related to a low incidence and reactivity of thiol (-SH) groups. Betaine is a metabolic end product which accumulates in stressed tissues, and may have a protective role (Hanson and Hitz, 1992). Accumulated organic molecules, which are "compatible" with metabolism, and accumulated ions decrease osmotic potential, providing "osmotic adjustment" and increasing turgor, thus enabling growth and metabolism to be maintained under stress (Lawlor and Leach, 1985). Further studies are needed to investigate whether these physio-biochemical parameters can be correlated with the increase of drought tolerance due to mycorrhizal infection.

3.   RESPONSE TO TEMPERATURE IN MYCORRHIZAL PLANTS

3.1.   Plant Growth and Seed Germination

Some work has been done on the responses of VAM-fungi and temperature on plant growth and seed germination. Onion (*Allium cepa* L.) growth was stimulated with *Glomus mosseae* at 14°C (Hayman, 1974). Maximum shoot growth of mycorrhizal cotton (*Gossypium hirsutum* L.) was obtained at 30°C when plants were inoculated with *Gigaspora calospora* Becker and Hall and *Glomus intraradices* Schenck and Smith, but at 36°C when plants were inoculated with *Glomus ambisporum* Smith and Schenck (Smith and Roncadori, 1986). Infected root length and number of vesicles increased at temperature (and soil oxygen level) increased from 20 to 30° C in *Eupatorium odoratum* L. inoculated with *Glomus macrocarpum* (Saif, 1983). Raju *et al.* (1990) investigated the effects of various *Glomus* species on growth of sorghum grown at 20, 25and 30°C. Sorghum plants colonized by *G. fasciculatum* had higher shoot dry matter yields, and longer roots at 20 and 25°C, and root dry matter yields 20°C than non-mycorrhizal plants grown under the same conditions. Plants colonized by *G. intraradices* and non-mycorrhizal plants had similar shoot and root dry matter yields and root length at both 20 and 25°C. The reason why sorghum plants did not respond favourably to *G. intraradices* is not known. Saito and Kato (1994) examined the effect of low temperature (15°C 14h/13°C 10h) on *Glycine-Bradyrhizobium-Glomus* species symbiosis in pot experiments. Nodulation (weight and number) was slightly reduced by this treatment, but the proportion of large nodules was increased. The root infected by the VAM fungus was less affected at low temperature treatment. Leaf area growth was significantly greater in mycorrhizal (*Glomus etunicatum*) seedlings of *Fraxinus pennsylvanica* than in uninfected seedlings at various temperature from 7.5 to 20°C. Relative leaf area growth rate was greater in mycorrhizal seedlings at 7.5 and 11.5°C, similar in all treatments at 15.5°C, and was greater in non-mycorrhizal seedlings at 20°C (Anderson *et al.*, 1987).

Rasmussen and Anderson (1990), investigated temperature sensitivity of *in vitro* germination and seedling development of *Dactylorrhiza majalis* (Orchidaceae) with or without a mycorrhizal fungus. There was a marked dacline in germination percentage at temperatures >23-25 °C. Seeds that germinate at higher temperatures exhibited only slight or no mycorrhizal development and developed few or no rhizoids compared with seedlings raised at optimal or lower temperatures. When grown at 26°C, seedling had small starch reserves than those grown at lower temperatures and increased in length as much as kept at 13°C. At 23-24.5°C seed germination in the presence of *Rhizoctonia* was about double than in the absence of fungus, and seedling length increased at 45% per week in the presence of fungus compared with 30% in its absence. Seedling growth responded rapidly to mycorrhizal inocoulation when room temperatures were 25 or 35°C; response to inoculation at 15°C was delayed, but seedlings at this temperature finally attained the size of those grown at 25 and 35°C (Borges and Chaney, 1989).

3.2. Root Colonization

A number of reports are available regarding the response of different temperatures on VAM root colonization. Grey (1991) investigated the effect of three soil temperatures on the growth of spring barley (*Hordeum vulgare* L.) and on their root colonization by VAM fungi from agricultural soils in Montana (USA) or Syria at different inoculum concentrations. Mycorrhizae developed at a range of temperatures from 11 to 26°C, but a greater proportion of colonized roots occurred at warm temperature. VAM fungi from Montana, primarily *Glomus macrocarpum*, were more tolerant to cool soils at 11 to 14°C, whereas, VAM fungi from Syria, primarily *Glomus hoi*, were more tolerant of warm soils at 24 to 26°C. Differences in optimum temperatures for development of mycorrhizae among *Glomus* species have also been reported in wheat (Hetrick and Bloom, 1984). Optimum infection of winter wheat with *G. epigaeum* occurred at 25°C, but not at 10°C. The wide range of temperatures indicates a variability for temperature tolerance among different *Glomus* species. Schenck and Schroder (1994) observed maximum arbuscule development in soybean [*Glycine max* (L.) Merr.] roots at 30°C with *Endogone gigantea* Nicol. and Gerd., while maximum mycelial development on root surface occurred at 28 to 34°C, and maximum sporulation and vesicles development at 35°C. Soybean growth responses to various combinations of soil temperatures and VAM fungal species were also variable (Schenck and Smith, 1982). *Glomus macrocarpum* inoculation resulted in higher percentage root colonization by VAM fungi and higher colonized root lengths than either *G. fasciculatum* or *G. intraradices* inoculations. The highest percentage of root colonization and colonized root lengths occurred at 25°C for *G. fasciculatum* and *G. intraradices*, and at 30°C for *G. macrocarpum* (Raju *et al.*, 1990).

Bentivenga and Hetrick (1992) conducted growth chamber experiment to examine the effect of temperature on mycorrhizal dependence of cool and warm-season grasses. For both types of grasses, the dependence on mycorrhizal symbiosis was greatest at the temperature that favoured plant growth. Mycorrhizal colonization was some what higher at 29°C than at 18°C in various host species (Table-3). Vogelzang *et al.* (1993) investigated the effects of temperatures between 30 and 38°C on the root length of mung bean and the early phases of infection by *Glomus versiforme*. Fungal colonization, and hence mycorrhizal growth response, were more sensitive than root length. Lower infection at higher temperature was apparently not related to the amount of root available (i.e. reduced root length), but to factors directly affecting the ability of the fungus to colonize and become established within the plant root.

3.3. Mineral Element Uptake

Root colonization with VAM fungi normally enhances uptake of many mineral nutrients (Cooper, 1984; Janos, 1987; Smith and Roncadori, 1986; Stribley, 1987); this is because of higher biomass production by mycorrhizal plants. Raju *et al.* (1990) studied the effects of VA-mycorrhizal fungi on mineral uptake of sorghum plants at different temperatures. Sorghum plants colonized by *G. macrocarpum* had considerably higher shoot mineral nutrient content (and concentrations) than plants colonized by *G. fasciculatum* and *G. intraradices* or non-mycorrhizal plants over the 20 to 30°C temperature range (Table-4). This probably occurred because *G. macrocarpum* colonized plants had a greater absorption surface area offered by the extensive fungal hyphae and enhance root growth. *G. fasciculatum* enhanced shoot growth at 20 and 25°C, and mineral uptake only at 20°C, *G. intraradices* depressed shoot growth and mineral uptake at 30°C, and *G. macrocarpum* enhanced shoot P, K and Zn at all temperatures, and Fe at 25 and 30°C. Enhancement of these nutrients might have been due to the fungal hyphae enhancement of availability or transport (absorption and/or translocation).

Table 3 : Mycorrhizal dependency and colonization of cool-season (C_3) and warm-season (C_4) grasses grown at two temperatures

Species and temp.	Dry wt.* (g/plant)		Mycorrhizal dependency (%)	Colonization (%)
	+myc	-myc		
Cool-season C_3				
Agropyron smithii				
18°C	6.61a	4.80b	27.3	10.0b
29°C	1.33c	1.27c	-	19.3a
Bromus inermis				
18°C	10.63a	7.24b	31.9	11.0a
29°C	2.04c	2.64c	-	10.4a
Poa pratensis				
18°C	5.16a	6.21a	-	8.1b
29°C	2.30b	3.10b	-	17.2a
Warm-season C_4				
Andropogon gerardii				
18°C	0.13b	0.04c	67.2	13.2b
29°C	3.38a	0.05c	98.5	30.2a
Schizachyrium scoparium				
18°C	0.09b	0.03c	66.7	14.3b
29°C	4.94a	0.03c	99.4	51.2a
Sorghastrum nutans				
18°C	0.12b	0.5c	58.3	15.9b
29°C	5.51a	0.04c	99.2	44.7a

Note : Within a species, means with the same letter are not significantly different (p = 0.05) according to the least significant difference procedure.

*+myc, inoculated with 400 spores of *Glomus etunicatum,* -myc, uninoculated control.

(Source : Bentivenga and Hetrick, 1992).

3.4. Metabolic Changes

Very little work has been done on response of different temperatures on mycorrhizal plants. The combined effect of arbuscular mycorrhizae and low temperature on 7 weeks old seedlings of spring and winter wheat cultivars (Glenlea and AC Ron, respectively) was determined for several physiological parameters including biomass, chlorophyll, proteins and sugar contents (Paradis *et al.*, 1995). The dry biomass was higher in Glenlea than AC Ron at week five and did not significantly change at week eight in both cultivars with either the mycorrhizal or the cold treatment. The chlorophyll content was higher in mycorrhizal (M+) than in non-mycorrhizal (M-) Glenlea at 5°C, but was unaltered in AC Ron or in either cultivar at 25°C. The reducing and total sugar contents were higher in M+ AC Ron then M+ Glenlea. The protein content was greater in Glenlea than in AC Ron at 25°C, but remained constant regardless of the mycorrhizal or cold treatment. It is suggested that the spring wheat benefits more than winter wheat from mycorrhizal association after a short-term exposure to low temperature. Borges and Chaney (1989) investigated growth parameters and carbohydrate contents for seedlings inoculated with *Glomus macrocarpum* or *G. fasciculatum* and for uninoculated controls that grown in a greenhouse for 16 weeks at root temperature of 15, 25 or 35°C. Sugar content increased as root temperature increased whereas starch content was inversely related. Both soluble sugar and starch content were less in mycorrhizal than in non-mycorrhizal seedlings.

Table 4 : Phosphorus, N,S,K,Ca,Mg, Mn, Fa, Zn and Cu contents in shoots of sorghum plants grown with varied VAMF species at 20, 25, and 30°C

Mineral nutrient	Temperature (°C)	VAM fungal species				
		Nonmycorrhizal plants	*Glomus fasciculatum*	*Glomus intraradices*	*Glomus macrocarpum*	LSD P<0.05
P(mg Plant⁻¹)	20	0.23	0.44(1.9)[a]	0.24(1.0)	1.69(7.3)	0.20
	25	0.29	0.74(2.6)	0.30(1.0)	5.86(20.2)	0.53
	30	0.36	0.51(1.4)	0.17(0.5)	5.33(14.8)	0.19
N(mg Plant⁻¹)	20	9.8	15.7(1.6)	11.3(1.2)	28.8(2.9)	4.4
	25	17.8	39.6(2.2)	14.5(0.8)	122.0(6.9)	11.7
	30	20.6	27.6(1.3)	10.9(0.5)	114.3(5.5)	5.5
S(mg Plant⁻¹)	20	0.33	0.59(1.8)	0.36(1.1)	1.52(4.6)	0.19
	25	0.51	1.18(2.3)	0.45(0.9)	6.58(12.9)	0.65
	30	0.62	0.81(1.3)	0.35(0.6)	5.71(9.2)	0.28
K(mg Plant⁻¹)	20	2.7	7.4(2.7)	3.8(1.4)	29.7(11.0)	3.1
	25	3.6	10.8(3.0)	5.4(1.5)	156.0(43.3)	11.6
	30	6.5	8.2(1.3)	2.5(0.4)	124.5(19.2)	6.8
Ca(mg Plant⁻¹)	20	3.2	5.7(1.8)	3.5(1.1)	9.1(2.8)	1.4
	25	4.1	9.7(2.4)	4.1(1.0)	38.5(9.4)	4.5
	30	5.2	6.6(1.3)	2.8(0.5)	37.5(7.2)	1.8
Mg(mg Plant⁻¹)	20	1.3	2.2(1.7)	1.4(1.1)	3.6(2.8)	0.7
	25	1.7	4.1(2.4)	1.8(1.1)	19.7(11.6)	3.0
	30	2.2	2.9(1.3)	1.0(0.5)	19.9(9.0)	1.7
Mn(μg Plant⁻¹)	20	19	33(1.7)	22(1.2)	66(3.5)	9
	25	40	106(2.6)	39(1.0)	403(10.1)	50
	30	59	75(1.3)	32(0.5)	470(8.0)	25
Fe(μg Plant⁻¹)	20	29	64(2.2)	39(1.3)	117(4.0)	22
	25	38	108(2.8)	63(1.7)	1045(27.5)	353
	30	56	89(1.6)	43(0.8)	1484(26.5)	333
Zn(μg Plant⁻¹)	20	375	810(2.2)	601(1.6)	1904(5.1)	446
	25	554	1070(1.9)	456(0.8)	10610(19.2)	2469
	30	658	971(1.5)	353(0.5)	12060(18.3)	1340
Cu(μg Plant⁻¹)	20	3.0	4.9(1.6)	3.4(1.1)	11.5(3.8)	2.7
	25	6.3	12.0(1.9)	7.2(1.1)	65.9(10.5)	8.6
	30	72	100(1.4)	57(0.8)	381(5.3)	96

[a] Numbers in ( ) refer to the relative changes due to the VAMF species compared to non-mycorrhizal plants.
(Source : Raju *et al.*, 1990).

Thus, it is generally observed that mycorrhizal plants revealed higher matabolites than non-mycorrhizal plants, at various temperatures, consequently the growth and yield may be increased in mycorrhizal plants.

4. RESPONSE TO OTHER ENVIRONMENTAL STRESSES

4.1. Air Pollution

Large areas of forests especially comprising of Norway spruce (*Picea abies*) in Central European countries, have been affected due to various factors including air pollution resulting in forest decline. Forest decline due to air pollution is generally associated with soil degradation, nutrient deficiencies, root decline, and mycorrhizal disfunction (Majstrik, 1990). In Germany, among different categories of mycorrhiza, it was found that trees with obligate ectomycorrhiza to which forest trees belong, react sensitively to pollutants and are usually first to be damaged. Plant species with no mycorrhiza or with only endomycorrhiza or with endomycorrhiza and facultative ectomycorrhiza were found resistant to pollution damage (Heyser et al., 1988). VAM infection of the grass was highest in soil taken from stands with full grass cover. VAM infection of the grass was highest in the plot with severest pollution damage and lowest in least damaged plot (Vosatka et al., 1991). Sulphur dioxide treatments resulted in significant reduction in the proportion of ectomycorrhizal root-tips and increases in the number of necrotic spots (Blaschke, 1989; Wollmer and Kottke, 1990). Low concentration (0.05 - 0.07 μl/l of SO_2) affected the ability of VAM fungi to colonize and also to proliferate within roots (Clapperton et al., 1990; Clapperton and Reid, 1991; 1992). Mycorrhizal formation with Laccaria proxima was not affected while that of Paxillus involutus decreased significantly from 120 μg/m^3 SO_2 onwards (Termorshnizen et al., 1990). Ozone is another gaseous pollutant which affect mycorrhizal development in trees if it exceeds its normal percentage in the air (Singh, 1996). Low percentage of ectomycorrhizae with well-developed mantle in both neutral (pH 6.9) and acid soil (pH 4.4) when exposed to 100 μg/m^3 ozone (Blaschke, 1989) were also observed. Soybean plants infected by Glomus geosporum were least sensitive to the adverse effects caused by 0.079 ppm O_3 exposure. Also pot yield of mycorrhizal plants was reduced by 25% and that of non-mycorrhizal plants by 48% (Singh, 1996). In Pinus taeda seedlings exposed to 0.50, 100, and 150 ppb O_3 for 5 hours per day, five days a week for 6-12 weeks, a linear relationship between the increase in O_3 and decreased number of ectomycorrhizae was observed (Meier et al., 1990). Bonello et al. (1993) investigated interaction of ozone, mycorrhiza and pathogens. Interaction between O_3 and Heterobasidium annosum was studied in Pinus sylvestris seedlings with Habeloma crustuliniforme mycorrhizae. O_3 exposure (200 μl/l for eight hours per day for 28 days) increased disease incidence significantly, but mycorrhizal infection completely prevented this negative effect. Presence of the pathogen on the root system was necessary for the induction of changes in the soluble and wall-bound secondary compounds of roots and needles. O_3 alone did not induce the changes while mycorrhizal infection appeared to have had a dampening effect on the induction of these compounds.

4.2. Acid Rain/pH

Mycorrhizal fungi differ in their response to rainfall acidity; generally mycorrhizal infection is more sensitive than seedling root growth due to acid rainfall (Simmons and Kelly, 1989). Mycorrhiza can withstand environmental pressure when Picea abies seedlings with established mycorrhiza were planted in soil collected from acid rain damaged stand and subjected to acid rain (Kattner, 1990). Soybean plants were exposed twice a week for 13 weeks to simulated rain of pH 5.5, 4.0, 3.2 and 2.8 (Brewer and Heagle, 1983). Neither rain activity nor Glomus geosporum significantly affected foliar nutrient ion content, vegetative growth or yield. However, Glomus geosporum produced 39 fewer chlamydospores per gram root when plants received simulated rain of pH 2.8 than of pH 5.5. Pinus taeda were exposed 37 times over 16 weeks to simulated acid rain adjusted to pH 5.6, 4.0, and 2.4. Mean percentage of short roots were greatest (62.2) for seedlings exposed to rain of pH 2.4. Values of root/shoot ratio were negatively correlated with mycorrhizae. This suggests that the rain of intermediate acidity (pH 4.0 and 3.2) inhibited ectomycorrhiza formation or increased soil acidity or other factors induced by rain at pH 2.4 enhanced ectomycorrhiza formation (Shafler et al., 1985). Khanaqa (1987) investigated the influence of combination of soil pH and temperature on Burley tobacco without or with VAM. Burley tobacco, fertilized with monocalcium phosphate (MCP) or hydrooxylapatite (HA), was grown at pH 5,6 or 7 and soil temperature of 20, 25, 30 or 35°C. HA plants inoculated with Glomus mosseae and non-inoculated MCP plants respond similarly to change in pH or temperature. The temperature optimum tended to shift from 30°C at pH 5.0 to 25°C at higher pH.

4.3.   Cu, Al and Coal Mines

Bisen *et al.* (1996) investigated VAM colonization in tree species planted in Cu, Al and coal mines of Madhya Pradesh, India with special reference to *Glomus mosseae*. Most of the tree species planted in mine dumps demonstrated colonization of VAM fungi. Maximum percentage of infection was observed in *Dalbergia sisoo* (73%) and *Eucalyptus* species (68%) planted at copper and coal mining sites, respectively. Whereas, the maximum percentage infection was observed in *Eucalyptus* species planted at both Cu (55%) and Al (40%) mine dumps.

Further studies are required to investigate various physio-biochemical changes in mycorrhizal and non-mycorrhizal plants, grown in Cu, Al and coal mines.

5.   CONCLUSIONS

Mycorrhizal associations significantly alter the water relationships of host plant, under both well-watered and drought conditions. Generally, stomatal conductivity and transpiration rate was significantly higher in water-stressed mycorrhizal than in non-mycorrhizal plants. Leaf water potential was same in mycorrhizal and non-mycorrhizal plants, under well-watered and water deficit conditions, but in some cases it was significantly greater in mycorrhizal plants. Under both water-stressed and well-watered conditions, various growth contributing parameters were markedly greater in mycorrhizal than in non-mycorrhizal plants. Root colonization by VAM fungi was depressed in adequately watered plants at the low phosphorous levels as compared with droughted plants at these P levels. Generally, photosynthesis, photorespiration, and photosynthetic pigments were greater in water stressed mycorrhizal than in non-mycorrhizal plants. Other metabolites like total reducing sugars, proteins, amino acids, nitrate reductase and glutamine synthatase activity were also markedly greater in water-stressed mycorrhizal plants. A number of reports show that mycorrhizal plants revealed greater resistance to drought. Insufficient information is available to explain machanism of drought resistance in mycorrhizal plants. The available evidences clearly show that resistance in mycorrhizal plants could be due to avoidance or tolerance. Mycorrhizal plants revealed greater drought avoidance by maintaining higher leaf water potential, transpiration rate and longer tap roots. Various growth contributing parameters were increased due to VAM infection, at different temperatures. Root colonization with VAM fungi normally enhanced uptake of many minerals, at various temperatures. Various biochemical parameters namely, chlorophyll content, reducing sugars and soluble proteins were also higher in mycorrhizal plants, at specific temperatures. Plant species with no mycorrhiza or with only endomycorrhiza or with endomycorrhiza and facultative ectomycorrhiza were found resistant to air pollution. Mycorrhizal fungi differ in their response to rainfall acidity. Generally, mycorrhizal infection is more sensitive than seedling root growth due to acid rainfall. Most of the tree species, planted in Cu, Al and coal mines, demonstrated colonization of VAM fungi.

REFERENCES

Allen, M.F. 1982, Influence of vesicular-arbuscular mycorrhizae on water movement through *Bouteloua gracilis* (H.K.B.) Lag Ex Staud, *New Phytol.* 91 : 191-196.
Allen, M.F. and Boosalis, M.G. 1983, Effect of two species of VA mycorrhizal fungi on drought tolerance of winter wheat, *New Phytol.* 93 : 67-76.
Allen, M.F., Smith, W.K., Moore, T.S. Jr., and Christensen, M. 1981, Comparative water relations and photosynthesis of mycorrhizal and non-mycorrhizal *Bouteloua gracilis* (H.B.K.) Lag Ex Steud, *New Phytol.* 88 : 683-693.

Andersen, C.P., Sucoff, E.I. and Dixon, R.K. 1987, The influence of low soil temperature on the growth of vesicular-arbuscular mycorrhizal *Fraxinus pennsylvanica, Can. J. For. Res.* 17 : 951-956.

Bagyaraj, D.J. 1991, In : *Hand Book of Applied Mycology*, Vol.I, Soil and Plants, D.K. Arora, B. Rai, K.G. Mukerji and G.R. Knudson, eds., Marcel Dekker Inc., New York, USA, pp.3-34.

Bentivenga, S.P. and Hetrick, A.B.D. 1992, Seasonal and temperature effects on mycorrhizal activity and dependence of cool and warm-season tall grass prairie grasses, *Can. J. Bot.* 70 : 1596-1602.

Bethlenfalvay, G.J., Brown, M.S., Ames, R.N. and Thomas, R.S. 1988, Effects of drought on host and endophyte development in mycorrhizal soybeans in relation to water use and phosphate uptake, *Physiol. Plant.* 72 : 565-571.

Bethlenfalvay, G.J., Brown, M.S., Mishra, K.L. and Stafford, A.E. 1987, *Glycine-Glomus-Rhizobium* symbiosis, V, Effect of mycorrhiza on nodule activity and transpiration in soybeans under drought stress, *Plant Physiol.* 85 : 115-119.

Bethlenfalvay, G.J., Brown, M.S., Mishra, K.L. and Stafford, E. 1987, *Glycine-Glomus-Rhizobium* symbiosis, *Plant Physiol.* 85 : 115-119.

Bewley, J.D. 1979, Physiological aspects of desiccation tolerance, *Annu. Rev. Plant Physiol.* 30 : 195-238.

Bewley, J.D. and Krochko, J.E. 1982, In : *Physiological Plant Ecology, Encyclopedia of Plant Physiology*, O.L. Lang, P.S. Novel, C.B. Osmond and H. Ziegler, eds., Springer-Verlag, Heidelberg, pp.325-370.

Bisen, P.S., Gour, R.K., Jain, R.K., Dev, A. and Sengupta, L.K. 1996, VAM colonization in tree species planted in Cu, Al and coal mines of Madhya Pradesh with special reference to *Glomus mosseae, Mycorrhiza News* 8 : 9-11.

Blaschke, H. 1989, Biology of fine roots of Norway spruce (*Picea abies*) in the open and after controlled fumigation with forest decline, *Environ. Pollution* 73 : 263-270.

Bonello, P., Heller, W. and Sandermann, H. Jr. 1993, Ozone effects on root-diseases susceptibility and defence responses in mycorrhizal and non-mycorrhizal seedlings of scots pine (*Pinus sylvastris* L.), *New Phytol.* 124 : 653-663.

Borges, R.G. and Chaney, W.R. 1989, Root temperature affects and mycorrhizal efficacy in *Fraxinus pennsylvanica* Marsh, *New Phytol.* 112 : 411-417.

Brewer, P.F. and Heagle, A.S. 1983, Interactions between *Glomus geosporum* and exposure of soybeans to ozone or simulated acid rain in the field, *Phytopath.* 73 : 1035-1040.

Brownlee, C., Duddridge, J.A., Malibari, A. and Read, D.J. 1985, The structure and function of mycelial systems of ectomycorrhizal roots with special reference to their role in forming inter-plant connection and providing pathways for assimilate and water transport, *Plant Soil* 71 : 433-443.

Clapperton, M.J. and Reid, D.M. 1991, Exposure of *Phleum pratense* to SO_2 and its subsequent effect on VAM root infectivity, *The Rhizosphere and Plant Growth* 14 : 369.

Clapperton, M.J. and Reid, D.M. 1992, Effects of low concentration sulphur dioxide fumigation and vesicular-arbuscular mycorrhizas on ^{14}C-partitioning in *Phleum pratense* L., *New Phytol.* 120 : 381-387.

Clapperton, M.J., Reid, D.M. and Parkinson, D. 1990, Effects of sulphur-dioxide fumigation on *Phleum pratense* and vasicular-arbuscular mycorrhizal fungi, *New Phytol.* 115 : 465-469.

Cooper, K.M. 1984, In : *VA mycorrhiza*, C.L. Powell and D.J. Bagyaraj, eds., CRC Press, Boca Raton, FL. pp.155-1286.

Davies, F.T., Potter, J.R. and Linderman, R.G. 1993, Drought resistance of mycorrhizal pepper plants independent of leaf P concentration response in gas exchange and water relations, *Physiol. Plant.* 87 : 45-53.

Dhindsa, R.S. and Matowe, W. 1981, Drought tolerance in two mosses : Correlated with enzymatic defence against lipid peroxidation, *J. Explo. Bot.* 32 : 79-91.

Dixon, R.K. and Hiol-Hiol, F. 1992, Gas exchange and photosynthesis of *Eucalyptus camaldulensis* seedlings inoculated with different ectomycorrhizal symbionts, *Plant Soil* 147 : 143-149.

Dixon, R.K., Pallardy, S.G., Garrett, H.E., Cux, G.S. and Sander, I.L. 1983, Comparative water relations of container grown and bare-root ectomycorrhizal and non-mycorrhizal *Quercus velutina* seedlings, *Can. J. For. Res.* 61 : 1559-1565.

Dixon, R.K., Wright, G.M., Behrns, G.T., Teskey, R.O. and Hinckley, T.M. 1980, Water deficits and root growth of ectomycorrhizal white oak seedlings, *Can. J. For. Res.* 10 : 547-548.

Dosskey, M.G., Boersma, L. and Linderman, R.G. 1991, Role for the photosynthate demand of ectomycorrhizas in the response of Douglas fir seedlings to drying soil, *New Phytol.* 117 : 327-334.

Edriss, M.H., Davis, R.M. and Burger, D.W. 1984, Influence of mycorrhizal fungi on cytokinin production in sour orange, *Jour. Amerc. Soc. Hort. Sci.* 109 : 587-590.

Farrar, J.F. 1976, In : *Lichenology : Progress and Problems*, D.H. Brown, D.L. Hawksworth and R.H. Bailey, eds., Academic Press, New York, pp.385-406.

Fitter, A.H. 1985, Functioning of vesicular-arbuscular mycorrhiza on growth and water relations of red clover, *New Phytol.* 89 : 599-608.

Graham, J.H., Leonard, R.T. and Menge, J.A. 1981, Membrane-mediated decrease in root exudation responsible for phosphorus inhibition of vesicular-arbuscular mycorrhiza formation, *Plant Physiol.* 68 : 548-552.

Grey, W.E. 1991, Influence of temperature on colonization of spring bareleys by vesicular-arbuscular mycorrhizal fungi, *Plant Soil*, 137 : 181-190.

Gupta, R.K. 1977, Morphological and physiological studies in bryophytes : A comparative study of cell size, soluble sugars, proteins and phospholipids in resistant and non-resistant liverwort, *Ind. J. Exp. Biol.* 15 :695-697.

Gupta, R.K. 1977a, The study of photosynthesis and leakage of solutes in relation to the desiccation effects in bryophytes, *Can. J. Bot.* 55 : 1186-1194.

Gupta, R.K. 1977b, An artefact in studies of the responses of respiration of bryophytes to desiccation, *Can. J. Bot.* 55 : 1195-1200.

Gupta, R.K. 1978, The physiology of desiccation resistance in bryophytes : Effect of desiccation on water status and chlorophyll-a and -b in bryophytes, *Ind. J. Exp. Biol.* 16 : 354-356.

Gupta, R.K. 1981, Effects of various rates of desiccation on photosynthesis and photosynthates leakage in young and aged tissues of the moss *Fissidens adianthoides* (Hedw), *Photosynthetica* 15 : 347-350.

Gupta, R.K. 1982, Drought tolerance in poikilohydrous plants, *Ind. Rev. Life Sci.* 2 : 189-212.

Gupta, R.K. 1983, Drought injury and membrane integrity in young and aged tissues of a moss *Neckera crispa* (Heds.), *Trop. Plant Sci. Res.* 1 : 259-261.

Gupta, R.K. 1984, Leakage phenomenon and plant cellular membranes under drought, *Ind. Rev. Life Sci.* 4 : 69-86.

Gupta, R.K. 1991, In : *Hand Book of Applied Mycology*, Vol. I, Soil and Plants; D.K. Arora, B. Rai, K.G. Mukerji and G.R. Knudson, eds., Marcel Dekker Inc., New York, USA, pp.55-75.

Hanson, A.D. and Hitz, W.D. 1992, Metabolic responses of mesophytes to plant water deficits, *Ann. Rev. Plant Physiol.* 33, 163-203.

Hardia, K. and Leyton, L. 1981, The influence of vesicular-arbuscular mycorrhiza on growth and water relations of red clover, *New Phytol.* 89 : 599-608.

Hardie, K. 1985, The effect of removal of extraradical hyphae on water uptake by vesicular-arbuscular mycorrhizal plants, *New Phytol.* 101 : 677-684.

Hayman, S. 1974, Plant growth responses to vesicular-arbuscular mycorrhiza, IV, Effect of light and temperature, *New Phytol.* 73 : 71-80.

Hetrick, B.A. and Bloom, J. 1984, The influence of temperature on colonization of winter wheat by vesicular-arbuscular mycorrhizal fungi, *Mycologia* 76 : 953-956.

Hetrick, B.A.D., Hetrick, J.A. and Bloom, J. 1984, Interaction of mycorrhizal infection, phosphorus level, and moisture stress in growth of field corn, *Can. J. Bot.* 62 : 2267-2271.

Heyser, W., Ike, J. and Meyer, F.H. 1988, Tree decline and mycotrophy, *Allgem. Forstzeitschrift.* 43 : 1174-1175.

Hsiao, T.C. 1973, Plant responses to water stress, *Annu. Rev. Pl. Physiol.* 24 : 519-570.

Huang, R.S., Smith, W.K. and Yost, R.S. 1985, Influence of vesicular-arbuscular mycorrhiza on growth, water relations and leaf orientattion in *Leucaena leucocephala* (Lam.) de Wit., *New Phytol.* 99 : 229-243.

Iljin, W.S. 1957, Drought resistance in plants and physiological processes, *Annu. Rev. Plant Physiol.* 8 : 257-274.

Janos, D.P. 1987, In : *Ecophysiology of VA mycorrhizal Plants,* G.R. Safir, ed., CRC Press, Boca Raton, FL. pp.107-134.

Kaloyereas, S.A. 1958, A new method of determining drought resistance, *Plant Physiol.* 33 : 232-234.

Kattner, D. 1990, Development of very fine roots of young Norway spruce (*Picea abies* Karst.) exposed to acid rain-results of a simulation experiment in the green-house, *Europ. J. For. Patho.* 20 : 247-250.

Kendrick, B. and Berch, S. 1985, In : *Comprehensive Biotechnology,* Vol. IV, C.W. Robinson, ed., Pergamon Press, Oxford, pp.109-150.

Khanaqa, A. 1987, Influence of soil pH and temperature on Burley tobacco without and with VA, *Angew. Bot.* 61 :337-345.

Koide, R. 1985, The effect of VA mycorrhizal infection and phosphotus status on sunflower hydraulic and stomatal properties, *J. Exp. Bot.* 36 : 1087-1098.

Kumar, S. and Gupta, R.K. 1986a, Influence of different leaf water potentials on photosynthetic carbon matabolism in sorghum, *Photosynthetica* 20 : 391-396.

Kumar, S. and Gupta, R.K. 1986b, Photorespiratory - CO_2 evolution and ^{14}C-glycine metabolism under different leaf water potentials in sorghum, *Photosynthetica* 20 : 401-404.

Lamhamedi, M.S., Bernier, P.Y. and Fortin, J.A. 1992, Growth nutrition and response to water stress of *Pinus pinaster* inoculated with ten dikaryotic strains of *Pisolithus* sp., *Tree Physiol.* 10 : 153-167.

Lawlor, D.W. 1979. In : *Stress Physiology in Crop Plants*, H. Mussell and R.C. Staples, eds., John Wiley and sons, New York, pp. 303-306.

Lawlor, D.W. and Leach, J.E. 1985, In : *Control of Leaf Growth*, N.R. Baker, W.J. Davies and C.K. Ong, eds., Cambridge University Press, Great Britain, pp.267-294.

Levitt, J. 1972, Responses of plants to environmental stresses, Academic Press, New York.

Levy, I. and Krikun, J. 1980, Effect of vesicular-arbuscular mycorrhiza on *Citrus jambhiri* water relations, *New Phytol.* 85 : 25-32.

Levy, Y., Syvertsen, J.P. and Memec, S. 1983, Effect of drought stress and vesicular-arbuscular mycorrhiza on citrus transpiration and hydraulic conductivity of roots, *New Phytol.* 93 : 61-66.

Majstrik, V. 1990, Ectomycorrhizas and forest decline, *Agricul. Ecosyst. Environ.* 28 : 325-337.

Meier, S., Grand, L.F., Schoeneberger, M.M., Reinert, R.A. and Bruck, R.I. 1990, Growth, ectomycorrhizae and nonstructural carbohydrates of loblolly pine seedlings exposed to ozone and soil water deficit, *Environ. Pollution* 64 : 11-27.

Michelsen, A. and Rosendahl, S. 1990, The effect of VA mycorrhizal fungi, phosphorus and drought stress on the growth of *Acacia nilotica* and *Leucaena Leucocephala* seedlings, *Plant Soil.* 124 : 7-13.

Mosse, B. 1957, Growth and chemical composition of mycorrhizal and non-mycorrhizal apples, *Nature* 179 : 922-924.

Nelsen, C.E. and Safir, G.R. 1982, The water relations of well-watered mycorrhizal and non-mycorrhizal onion plants, *J. Amer, Soc. Hort. Sci.* 107 : 271-274.

Nelson, C.E. and Safir, G.R. 1982, Increased drought tolerance of mycorrhizal onion plants caused by improved phosphorus nutrition, *Planta* 154 : 407-413.

Osonuki, O., Mulongoy, K., Awotoye, O.O., Atayese, M.O. and Okail, D.U.U. 1991, Effects of ectomycorrhizal and vesicular-arbuscular mycorrhizal fungi on drought tolerance of four leguminous woody seedlings, *Plant Soil*, 136-131-143.

Pai, G., Bagyaraj, D.J., Ravindra, T.P. and Prasad, T.G. 1994, Calcium uptake by cowpea as influenced by mycorrhizal colonization and water stress, *Curr. Sci.* 66 : 444-445.

Paleg, L.G. and Aspinall, D. 1981, In : *The Physiology and Biochemistry of Drought Resistance in Plants*, L.G. Paleg and D. Aspinall, eds., Academic Press, Sydney, pp.205-241.

Panwar, J.D.S. 1992, Effect of VAM and *Azospirillum* inoculation on metabolic changs and grain yield of wheat under moisture stress condition, *Ind. J. Pl. Physiol.* 35 : 157-161.

Panwar, J.D.S. 1993, Response of VAM and *Azospirillum* inoculation to water status and grain yield in wheat under water stress condition, *Ind. J. Pl. Physiol.* 36 : 41-43.

Paradis, R., Dalpe, Y. and Charest, C. 1995, The combined effect of arbuscular mycorrhizas and short-term cold exposure on wheat, *New Phytol.* 129 : 637-642.

Parke, J.L., Linderman, R.G. and Black, C.H. 1983, The role of ectomycorrhizas in drought tolerance of Douglas-fir-seedlings, *New Phytol.* 95 : 83-95.

Raju, P.S., Clark, R.B., Ellis, J.R. and Maranville, J.W. 1990, Effects of species of VA mycorrhizal fungi on growth and mineral uptake of sorghum at different temperatures, *Plant Soil* 121 : 165-170.

Ramakrishnan, R., Johri, B.N. and Gupta, R.K. 1988, Influence of VAM-fungus *Glomus calendonius* on free proline accumulation in water-stressed maize, *Curr. Sci.* 57 : 1082-1084.

Ramakrishnan, R., Johri, B.N., and Gupta, R.K. 1988, Effect of vesicular-arbuscular mycorrhizal fungus on photosynthesis and photorespiration in water-stressed maize, *Photosynthetica* 22 : 443-447.

Ramos, C. and Kaufmann, M.R. 1979, Hydraulic resistance of rought lemon roots, *Physiol. Plantarum* 45 : 311-314.

Rasmussen, H. and Anderson, T.F. 1990, Temperature sensitivity of *in vitro* germination and seedling development of *Dactylorhiza majalis* (Orchidaceae) with and without a mycorrhizal fungus, *Pl. Cell Environ.* 13 : 171-177.

Reichanbach, H.G.V. and Schonback, F. 1995, Influence of VA mycorrhiza on drought tolerance of flax (*Linum usitatissimum* L.), I, Influence of VAM on growth and morphology of flax and on physical parameters of the soil, *Angew. Botanik* 49 : 49-54.

Reid, C.P.P. 1978, In : *Root Physiology and Symbiosis*, A. Rildacher and J. Gognoire-Michaud, eds., IUFRO Symp. Proc. Nancy, France, pp.392-408.

Reid, C.P.P. and Bowen, G.D. 1979. In : *The Soil-Root Interface*, J.L. Harlay and R.S. Russell, eds., Academic Press, London, pp.211-219.

Rhodes, L.H. and Gerdemann, J.W. 1978, Influence of phosphorus nutrition on sulfur uptake by vesicular-arbuscular mycorrhizae of onion, *Soil Bio. Biochem.* 10 : 361-364.

Ruiz-Lozano, J.M., Gomez, M. and Azcon, R. 1995, Influence of different *Glomus* species on the time-course of physiological plant responses of lettuce to progressive drought stress periods, *Plant Science Limerick* 110 : 37-44.

Safir, G.R. 1980, In : *The Biology of Crop Productivity*, P.S. Carlson, ed., Academic Press, New York, pp.231-252.

Safir, G.R. and Nelsen, C.E. 1981, In : *Role of Mycorrhizal Associations in Crop Production*, R. Myers, ed., New Jersey Agr. Expt. Stn. pp.25-31.

Safir, G.R., Boyer, J.S. and Gerdemann, J.W. 1971, Mycorrhizal enhancement of water transport in soybean, *Science* 172 : 581-583.

Safir, G.R., Boyer, J.S. and Gerdemann, J.W. 1972, Nutrient status and mycorrhizal enhancement of water transport in soybean, *Plant Physiol.* 49 : 700-703.

Saif, S.R. 1983, Soil temperature, soil oxygen and growth of mycorrhizal and non-mycorrhizal plants *Eupatorium odoratum* L. development of *Glomus macrocarpum*, *Angew. Botanik* 57 : 143-155.

Saito, M. and Kato, T. 1994, Effects of low temperature and shade on relationships between nodulation, vesicular-arbuscular-mycorrhizal infection and shoot growth of soybeans, *Biol. Fert. Soil* 17 : 206-211.

Schenck, N.C. and Perez, Y. 1987, *Manual of the Identification of VA Mycorrhizal Fungi*, University of Florida, Gainesville, FL, pp.1-245.

Schenck, N.C. and Schroder, V.N. 1994, Temperature response of *Endogone* mycorrhiza on soybean roots, *Mycologia* 66 : 600-605.

Schenck, N.C. and Smith, G.S. 1982, Responses of six species of vesicular-arbuscular mycorrhizal fungi and their effects on soybean at four soil temperatures, *New Phytol.* 92 : 193-201.

Schultz, R.C., Kormanik, P.P., Bryan, W.C. and Brister, G.H. 1979, Vesicular-arbuscular mycorrhizae influence growth but not mineral concentrations in seedlings of eight sweetgum families, *Can. J. For. Res.* 9 : 218-223.

Schulze, E.D. 1986, Whole plant responses to drought, *Aust. J. Plant Physiol.* 13 : 127-141.

Shafler, S.R., Grand, L.F., Bruck, R.I. and Heagle, A.S. 1985, Formation of ectomycorrhizae on *Pinus taeda* seedlings exposed to simulated acid rain, *Can. J. For. Res.* 15 : 66-67.

Sieverding, E. 1979, Einfluss der Bodenfeucht auf die Effektivitat der VA-mykorrhiza, *Angew. Botanik* 53 : 91-98.

Sieverding, E. 1984, Influence of soil water regimes on VA mycorrhiza, III, Comparison of three mycorrhizal fungi and their influence on transpiration, *Z. Acker. Pflanzenbau.* 153 : 52-61.

Sieverding, E. and Toro, S.T. 1987, In : *Proceedings of the 7th North American Conference on Mycorrhizae*, D.M. Sylvia, L.L. Hung and J.H. Graham, eds., IFAS, Gainesville, FL, pp.58.

Simmons, G.L. and Kelly, J.M. 1989, Influence of O_3, rainfall acidity and soil Mg status on growth and ectomycorrhizal colonization of loblloly pine roots, *Water, Air and Soil Pollution* 14 : 150-171.

Singh, S. 1996, Effect of atmospheric pollution on mycorrhiza, *Mycorrhiza News* 8 : 1-7.

Smith, G.S. and Roncadori, R.W. 1986, Responses of three vesicular-arbuscular mycorrhizal fungi at four soil temperatures and their effects on cotton growth, *New Phytol.* 104 : 89-95.

Stribley, D.P. 1987, In : *Ecophysiology of VA Mycorrhizal Plants*, G.R. Safir, ed., CRC Press, Boca Raton, FL. pp.59-70.

Subramanian, K.S. and Charest, C. 1995, Influence of arbuscular mycorrhizae on the metabolism of maize under drought stress, *Mycorrhiza* 5 : 273-278.

Subramanian, K.S., Charest, C., Dwyer, L.M. and Hamilton, R.I. 1995, Arbuscular mycorrhizas and water relations in maize under drought stress at tasselling, *New Phytol.* 129 : 643-650.

Termorshnizen, A.J., Eerden, L.J. van Der, and Dueck, T.A. 1990, The effects of SO_2 pollution on mycorrhizal and non-mycorrhizal seedlings of *Pinus sylvestris*, *Agricul. Ecosys. Environ.* 28 : 513-518.

Tobar, R., Azcon, R. and Barea, J.M. 1994, Improved nitrogen uptake and transport from ^{15}N-labelled nitrate by external hyphae of arbuscular mycorrhiza under water-stressed conditions, *New Phytol.* 126 : 119-122.

Vogelzang, B., Parsons, H. and Smith, S. 1993, Separate effects of high temperature on root growth of *Vigna radiata* L. and colonization by the vesicular-arbuscular mycorrhizal fungus *Glomus versiforme*, *Soil Biol. Biochem.* 25 : 1127-1129.

Vosatka, M., Cudlin, P. and Mezstrik, V. 1991, VAM populations in relation to grass invasion associated with forest decline, *Environ. Pollution* 73 : 263-270.

Whittingham, J. 1980, The biology of mycorrhizae with special reference to the ecology of semi-natural limestone grasslands, Ph.D. thesis, University of Sheffield, U.K.

Wollmer, H. and Kottke, I. 1990, Fine root studies *in situ* and in the laboratory, *Environ. Pollution* 68 : 383-407.

THE GLOBAL CARBON CYCLE AND GLOBAL CHANGE : RESPONSES AND FEEDBACKS FROM THE MYCORRHIZOSPHERE

Robert K. Dixon

US Country Studies Program
1000 Independence Avenue, SW
Washington, DC 20585
USA

1.   INTRODUCTION

The accumulation of greenhouse gases (GHG) in the atmosphere due to deforestation, fossil fuel combustion and other anthropogenic activities may have begun to change the global climate system (Dickinson, 1989; Watson et al., 1995). According to most general circulation models e.g. Geophysical Fluid Dynamics Laboratory (GFDL) and Godard Institute for Space Studies (GISS), a continued build-up of greenhouse gases (e.g., CO_2, CH_4, N_2O) is likely to change mean atmospheric temperature (Hansen et al., 1988; Manabe and Wetherald, 1987; Watson et al., 1995). The amount and distribution of precipitation is also predicted to shift over the next century (Schnieder, 1989). Changes of this magnitude in the Earth's climate system will have significant consequences for the terrestrial biosphere, including above- and below-ground components (Shukla et al., 1990) (Fig.1).

The responses and feedbacks of terrestrial biomes, including below-ground systems, to projected global change are expected to be profound (Lal et al., 1995; Smith and Tirpak, 1989). The areal distribution of deserts and grasslands may increase, while that of boreal forest and tundra ecosystems will decrease due to global change expected in the next century (Dixon et al., 1994a; Smith et al., 1993). Regional effects on vegetation distribution and composition will also be dramatic (Smith et al., 1993; Watson et al., 1995). The decline and redistribution of vegetation types could significantly alter existing terrestrial carbon (C) pools and result in a large flux of CO_2 into or out of the atmosphere (Prentice and Fung, 1990; Tans et al., 1990).

The flux of other GHG (e.g. H_2O, CH_4 and N_2O) between the terrestrial biosphere and the atmosphere is also significant and may change in response to global change (Dixon and Wisniewski, 1995; Trabalka and Reichle, 1986). The global contribution of below-ground processes to the net flux of greenhouse gases is approximately 30% for CO_2, 70% for CH_4, and 90% for N_2O (Burke et al., 1989; Lal et al., 1995). The mechanisms of GHG emissions from soils are, in general, poorly understood and preliminary point estimates are difficult to adapt to a global scale (Bouwman, 1989; Oechel et al., 1993). For example, methanogenesis from man-made and natural wetlands is a complex biochemical process. An assessment of CH_4 emissions is complicated by the wide-scale

Mycorrhizal Biology, edited by Mukerji et al.
Kluwer Academic/Plenum Publishers, 2000

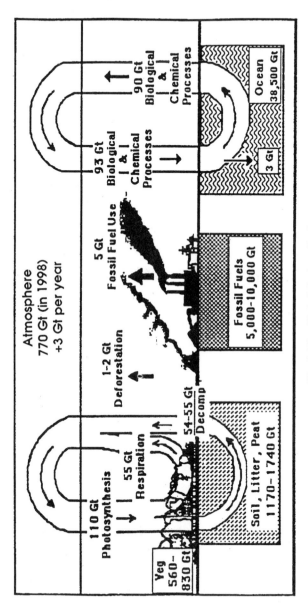

Fig.1. The Global Carbon Cycle

geographic distribution and highly diverse methods of managing wetlands. Trends in GHG flux from representative ecosystems due to changes in land-management or climate changes are presented in Table 1. Preliminary estimates reveal that global change in northern latitudes will significantly increase CH_4 and N_2O emissions (Mathews and Fung, 1987; Turner *et al.*, 1990). Examination of trends in CH_4 and N_2O concentrations in relation to records over the past 10,000-20,000 years suggests a positive feedback of GHGs to the global climate system (Khalil and Rasmussen, 1989).

Table 1 . Projection of greenhouse gas flux relative to current and modified land-use management options (after Bouwman, 1989; Dixon *et al.*, 1993)

	Greenhouse gas component				
	CO_2	CH_4	H_2O	N_2O	Albedo
Current land-use options					
Deforestation	+	+	-	+	+/-
Increase wetlands	0	+	0	+/-	0
Warm permafrost	+	+	0	+	0
N-fertilization	0	0	0	+	0
Alternative options					
Forestation	-	-	+	-	+
Conservation agriculture	-	-	-	-	+
Wetland management					
Inorganic fertilizers	0	-	0	+	0
C addition	+	+	0	-	0
Water management	+	-	0	+	0
Conservative N-fertilization	0	0	0	-	0
Slow desertification	-	0	+	0	+

+ indicates an increase in gas flux or forcing of climate change; - indicates a decrease in flux or property that diminishes climate change; and 0 indicates negligible.

Our knowledge of the role of below-ground systems in feedbacks and responses to global change is limited (Bouwman, 1989; Dixon and Turner, 1991). Terrestrial ecosystem structure and function is highly dependent on the relationship between plants and soil systems, especially the rhizosphere (Curl and Truelove, 1986; Mukerji, 1996). The projected changes in the distribution and condition of the Earth's vegetaton, induced by global change, will be tightly linked to below-ground processes (Dixon *et al.*, 1994a,b; Lal *et al.*, 1995). Because C cycling is a central theme in global change, characterization of the size and flux of C pools from below-ground systems is of keen interest. The objectives of this chapter are three-fold: (i) characterize the role of below-ground systems in the global C cycle; (ii) identify the potential responses and feedbacks of rhizosphere systems in global change processes, and (iii) evaluate a case study of rhizosphere organism response to severe disturbance (simulated global change).

2.   ROLE OF BELOW-GROUND SYSTEMS IN THE GLOBAL C CYCLE

The terrestrial biosphere plays a major role in Earth's C cycle (Fig. 1). Estimates of the global above-ground biomass carbon pool range from 560 to 830 Pg (Dixon *et al.*, 1994a; Sundquist, 1993). Other global C pools include the atmosphere (770 Pg), and oceans (38500 Pg). Storage of C as organic matter and litter in global soils is estimated to range from 1400 to 1700 Pg, using approaches based on soil and vegetation groups (Schlesinger, 1990), life zone classes (Post *et al.*,

1982) and model projections (Smith *et al.*, 1993). Estimates of soil C on a per unit area basis are variable, especially in histosols, tundra and boreal ecosystems (Bouwman, 1989; Oechel *et al.*, 1993). Although the literature is relatively consistent regarding C content of low- and mid-latitude soils (Dixon *et al.*, 1994b), there is substantial disagreement regarding areal distribution of C among life zones. Storage of C in arid and semi-arid ecosystems is primarily as calcium carbonate (Trabalka and Reichle, 1986).

The gross annual flux of C into the terrestrial biosphere *via* photosynthesis is estimated to be 110 Pg (Dixon *et al.*, 1994a). About half is returned to the atmosphere *via* plant respiration (Dorr and Munnich, 1987). The remainder is considered net primary production, including a significant proportion as root production. In terms of the global C budget, the primary productin is largely balanced by C loss due to heterotrophic respiration associated with decomposition of litter and soil organic matter (Sundquist 1993). Estimates of litter fall range from 40 to 50 Pg (Dixon *et al.*, 1994a; Oades, 1988). Despite the growing understanding of C pools and flux, large uncertainties remain, particularly regarding below-ground processes (Dixon and Turner, 1991; Tans *et al.*, 1990). The Earth's climate system plays a prominent role in shaping soil properties, such as pH, organic matter content and nutrient cycling (Mentenmeyer *et al.*, 1981).

3. RESPONSE OF BELOW-GROUND SYSTEMS TO PROJECTED GLOBAL CHANGE

The response of below-ground systems to projected global change is likely to be profound (Dixon and Turner, 1991; Lal *et al.*, 1995). Shifts in the Earth's temperature and precipitation patterns will modify the chemical, physical and biological properties of soils and root systems which will in turn influence the distribution, condition and productivity of the vegetation (Oechel *et al.*, 1993). Once species composition begins to change, there will be additional impacts on soil properties because of species-specific influences on nutrient cycling processes (Cole *et al.*, 1993; Dixon and Wisniewski, 1995).

The below-ground community of organisms is intimately linked to plant root systems (Fogel and Hunt, 1979; Lal *et al.*, 1995). Approximately 90% of microbial populations are concentrated within the plant rhizosphere (Curl and Truelove, 1986). Resources (e.g. C metabolites) transported from plants to roots benefit associated microbes. At the same time, plants receive benefits from many rhizosphere organisms (e.g. mutualists) including: uptake of water and nutrients, atmospheric nitrogen fixation, projection of host from pathogens and toxins, production of growth regulators and chelating agents, and stabilization of soil physical properties (Curl and Truelove, 1986; Perry *et al.*, 1987).

Projected global change has the potential to disrupt plant-soil linkages and lead to long-term ecosystem instability (Perry *et al.*, 1990). The unprecedented rate and magnitude of projected global change is of critical importance. It may lead to wide-scale decline, migration and/or extinction of plants and microbes (Dixon and Turner, 1991; Watson *et al.*, 1995). In numerous examples of massive degradation (e.g. deforestation of wet tropical forests, desertification of savanna), long-term, sometimes irreversible changes in ecosystems are exhibited (Schlesinger *et al.*, 1990; Shukla *et al.*, 1990). Although the knowledge of below-ground processes following massive disturbance is limited, evidence suggests the integrity and stability of terrestrial ecosystems are based in soil systems (Bouwman, 1989; Smith *et al.*, 1993).

The ability of various plant genera to adapt or migrate in response to projected global change will depend on many below-ground factors (Perry *et al.*, 1990). Adaptation to global environment stress will be influenced by soil chemical and physical factors, root morphology and physiology, and rhizosphere associates (Bouwman, 1989; Dixon and Turner, 1991). For example, the ability of perennial plant species to tolerate or adapt to a warm and dry environment is intimately linked to root system structure and function (Dixon *et al.*, 1980). The successful long-distance migration of many plant species may depend, in part, on the availability of requisite propagules in soils (e.g. mycorrhizal fungi or rhizosphere bacteria). For example, survival of transplanted trees is impaired in the absence of appropriate mycorrhizal fungi (Perry *et al.*, 1987). Thus, dispersal of palnt species will depend on both successful co-migration of microbes and adequate site conditions.

Atmospheric CO_2 enrichment is an incontrovertible aspect of projected global change, yet the knowledge of below-ground responses and feedbacks is limited (Mooney *et al.*, 1990). Woody and herbaceous perennials and annuals show a wide range of above- and below-ground morphological and physiological responses to elevated CO_2 (Curtis, 1990; Fajer, 1989). Photosynthesis, water-use efficiency and patterns of nutrient and C allocation are all affected by ambient CO_2. Root growth and associated symbiotic relationships/organisms (e.g., mycorrhizal fungi) are often stimulated by CO_2 enrichment even in the absence of a major shoot response (Norby 1987; O'Neill *et al.*, 1987; Sinoit *et al.*, 1985). Short-term CO_2 enrichment in a mixed *Spartina-Scirpus* community revealed that root and rhizome growth of the C_3 sedge (*Scirpus*) was significantly greater than for the C_4 (*Spartina*) (Curtis *et al.*, 1990). Stimulation of root biomass and associated rhizosphere organisms may help plants to adapt to massive disturbance of ecosystems and shifts in temperature and precipitation associated with global change (Norby, 1987).

Global change effects on below-ground processes will also influence above-ground productivity *via* influences on nutrient availability. For example, high-latitude forests may become more productive at sites where water does not become limiting, mainly due to higher rates of N mineralization in the litter and soil (Apps *et al.*, 1993; Dixon and Wisniewski, 1995). At sites where water limitations become aggravated by global change, such as on sandy soils with low water-holding capacity, an interesting positive feedback develops. Lower water availability leads to plant foliage with a wider C to N ratio, which in turn results in slower rates of N mineralization and lower productivity. Simulations of grasslands by Schimel *et al.* (1990) using the CENTURY model show similar trends, with higher productivity due to greater N availability where water does not limit plant productivity. Nutrient cycling responses and feedbacks of arctic and tundra soil systems are more complex and may differ from low-and mid-latitude soil systems (Oechel *et al.*, 1993).

4. DISTRIBUTION AND OCCURRENCE OF MYCORRHIZAL ASSOCIATIONS IN RESPONSE TO DISTURBANCE

Populations of soil bacteria and fungi shift in response to changes in the above-and below-ground environment following disturbance (Pilz and Perry, 1984). Previous studies in Australia, Europe, and the USA Pacific Northwest and Rocky Mountain region, revealed that ectomycorrhizal and vesicular-arbuscular mycorrhizal (VAM) propagules generally decrease in response to site disturbance (Harvey *et al.*, 1980; Mason *et al.*, 1983; Perry *et al.*, 1990; Powell, 1980). However, the occurrence and distribution of fungal propagules varies with soil physical and chemical properties and ambient environment (Parke *et al.*, 1983). On harsh sites, such as the relatively dry and fragile soils, disturbance may jeopardize survival of mycorrhizal populations (Baer and Otta, 1981; Danielson, 1984). Soil temperature influences occurrence and development of mycorrhizal and some fungal species are probably adapted to a specific microclimate (Parke *et al.*, 1983). Similarly, soil moisture and fertility are widely known to influence distribution and growth of mycorrhizae (Perry *et al.*, 1987).

Ectomycorrhizal and VAM fungal propagules introduced to the rhizosphere of plants in nurseries or greenhouses should persist or spread following outplanting (Perry *et al.*, 1987). However, field assessments in the USA Pacific Northwest (Pilz and Perry, 1984), Europe (Fleming, 1985), and Australia (Marks and Foster, 1967) report ectomycorrhiza present on tree root systems prior to planting may be replaced by indigenous mycorrhizal fungi after transplanting. Survey of literature reveals a succession of mycorrhizal associates in mid-latitude ecosystems can occur with a single host over a period of months to years (Fleming, 1985; Mason *et al.*, 1982). Field assessments suggest both occurrence and succession of mycorrhizal fungi are influenced by disturbance and extreme environmental events. Preliminary surveys suggest the occurrence and distribution of mycorrhizal associates in the fragile soils of boreal and sub-boreal forests differs from that of mid- or low-latitude ecosystems (Danielson, 1984; Danielson *et al.*, 1984).

This study was designed to : (i) identify ectomycorrhizal and VAM fungi associated with three different terrestrial ecosystems in northern Minnesota, and, (ii) characterize the relative influence of disturbance (simulated global change impacts) on occurrence, distribution and abundance of VAM and ectomycorrhizal fungi and their propagules.

4.1. Study Sites

The study sites were located near Cloquet, Warba and Bemidji, Minnesota, USA, respectively. These sites were selected along a 200k transect to represent a range of soils, vegetation types, and local climate (Table 2). The climate in northern Minnesota is characterized as continental with extreme ambient temperatures in winter and summer months (Critchfield, 1966). Soils at each site were derived from glacial deposits (Curtis, 1959). The soil horizons of the sites were disturbed by heavy machinery before the study was initiated. Sample plots of undisturbed ecosystems (control plots) were located at least 25 meters from the harvest zone to eliminate edge effects.

Table 2. Description of site conditions at Cloquet, Warba and Bemidji, Minnesota, USA

Site	Aspect, Slope	Dominant vegatation (prior to disturbance)	Fine litter depth (cm)	Soil texture	Soil temperature[a] (°C) (after disturbance)	Soil moisture[a] (%)	Average annual precipitation (cm)
Cloquet	NNW, 20%	*Pinus resinosa*	0.7	loamy sand	14.6	21	71
Warba	SSW, 25%	*Picea, Betula Populus* sp.	1.4	loam	15.3	32	64
Bemidji	SSE, 20%	*Pinus banksiana*	0.5	loamy sand	15.9	19	51

[a]Mean values based on bi-monthly measurements, May to October.

4.1.1. Cloquet

This site is located in Carlton County, Minnesota, USA, and was disturbed by tracked vehicles approximately 6 months before the study was initiated. Prior to harvest the site was dominated by jack pine (*P. banksiana* Lamb.) with scattered individuals or stands or red pine, quaking aspen (*Populus tremuloides* Michx.), bigtooth aspen (*P. grandidentata* Michx.) and paper birch (*Betula papyrifera* Marsh.). Site debris was piled and burned in December prior to the first growing season. Vegetation recolonizing the site in the first growing season included coppice of aspen and birch, hazel (*Corylus americana* Walt. and *Corylus cornuta* Marsh.), serviceberry (*Amelanchier* sp.), pin cherry (*Prunus pennsylvanica* L.) and blueberry (*Vaccinium*). Herbaceous species, dominated by goldenrod (*Solidago gigantea* Ait), stinging nettle (*Urtica dioica* L.) and thistle [*Cirsium discolor* (Muhl) Spreng.] occurred across the site. The soil at this site is an omega loamy sand with chemical and physical characteristics (Table 2).

4.1.2. Warba

Located in Itasca County, Minnesota, USA, this site was disturbed by shearing and soil raking approximately 18 months before initiating the study. Principal tree species of this site included quaking and bigtooth aspen with scattered individuals or stands of paper birch, red maple (*Acer rubrum* L.), yellow birch (*Betula alleghaniensis* Britton), and American basswood (*Tilia americana* L.). After disturbance, vegetation recolonizing the site in the first growing season included coppice of aspen and birch, hazel, pin cherry, blueberry, sweetfern [*Comptonia peregrina* (L.) Coult.], dwarf bush honeysuckle (*Diervilla lonicera* Mill.) and trailing arbutus (*Epigaea repens* L.). The soil at the Warba site is an Itasca-Goodland silt loam (Table 2).

4.1.3. Bemidji

Jack pine (*Pinus banksiana* Lamb.) was the dominant tree species at this Beltrami County, Minnesota, USA, site prior to soil disturbance. Quaking and bigtooth aspen were minor components of this site before disturbance. The site was disturbed by tracked vehicles and scarification approximately 8 months before the initiation of the study. Woody vegetation recolonizing the site during the study included coppice of aspen, birch and hazel but herbaceous vegetation were dominant across the site. Perennial grasses including quackgrass [*Agropyron repens* (L.) Beauv.], fowl meadow grass (*Poa palustris* L.), forbs such as red sorrel (*Rumex acetosella* L.), goldenrod, vetch (*Vicia villosa* Roth.), stinging nettle and thistle recolonized the site in the first growing season. The soil at the Bemidji site is a loamy sand (Table 2).

4.2. Methods

At each study site, 3-5 transects were identified with ten randomly located sample plots established along each transect (Joslin and Henderson, 1987; Yin *et al.*, 1989). Soil, vegetation and fungi samples were collected at these plots over the course of the two-year study using methods described below. Samples were collected using procedures described below and stored at 5°C prior to processing (Yin *et al.*, 1989).

Litter depth was measured in April of the first growing season immediately after snowmelt. One-liter soil samples were collected from the top 0-15 and 15-30 cm of mineral soil for each site. Soil was sieved through 2 mm sieve for chemical analysis using standard methods (Black, 1965). The pH levels were determined in a saturated paste using a glass electrode. Cation exchange capacity (CEC) was determined by NH_4^+ saturaton. Ammonium N was extracted with KCI and analyzed by distillation. Water extractable NO_3-N and Bray II P was determined by an autoanalyzer. Exchangeable Ca, K and Mg were extracted with NH_4OAc and analyzed by an inductively-coupled plasma spectrophotometer (Yin *et al.*, 1989). Bulk density was measured in the field for the upper 15 cm. Soil temperature was measured monthly 2-3 cm below soil surface using a thermometer. Soil moisture content was measured volumetrically. Soil physical properties were determined by using standard methods (Black, 1965).

Standing fine-root biomass was sampled monthly, June-September, during the first growing season at sample plots along the 3-5 transects at each site: Cloquet, Warba, Bemidji (Yin *et al.*, 1989). This sampling scheme was chosen to investigate temporal changes in standing root biomass. Prior investigations revealed the upper 30 cm contained more than 80% of fine roots, thus sampling was limited to this depth. On each sampling date two cores were randomly collected from each sample plot with the restriction that no core should be less than 40 cm from any tree stump or previous core. Root biomass samples were collected with a 10 cm soil cover which was mechanically driven into the soil profile. Soil cores were divided into 5 cm increments for analysis. Ectomycorrhizae and VAM progagules were extracted from soil core samples using methods described below.

The sample plots were surveyed bi-weekly to collect fungal fruiting bodies of mycorrhizal fungi. Fruiting bodies were collected and stored using methods described by Molina and Palmer (1984). Ectomycorrhizal isolation procedures, from soil cores, were used to compare morphological types on standard media. Fungi forming ectomycorrhizae were excised, washed under cold running water, sterilized in 30% H_2O_2 for 15-30s. The isolation media was modified Melin-Norkrans (MMN) with a dextrose substitute and 100 mg/L streptomycin and 5 mg/L benomyl (Molina and Trappe, 1982).

A modified sucrose centrifugation technique was used to recover VAM chlamydospores from soil samples. A bioassay technique employigng sorghum (*Sorghum bicolor*) was used to determine viability of suspected VAM propagules (Daniels and Skipper, 1984). Roots of seedling bioassay and field samples were washed, blotted dry, and analyzed for presence of VAM. Roots were cleared, stained, and VAM quantified using methods described by Phillips and Hayman (1970).

Soil vegetation and fungal symbiont data from field and laboratory samples were subject to analysis of variance of t-test (P=0.05). The relationship between soil variables, root systems and mycorrhizal propagules were subject to regression analysis (Steel and Torrie, 1980).

4.3. Results

The upper soil profiles were altered by disturbance and soil pH and bulk density declined in year one (Table 3). Generally, exchangeable cations and soluble P increased on the disturbed sites relative to undisturbed control plots. Soil moisture on the disturbed plots was similar to that of sites not disturbed, perhaps due to incorporation of organic matter into surface soil layers. Soil chemical and physical properties did not significantly change in year two relative to year one.

The number and biomass of roots occurring at the 0-30 cm depth increased from June to September at all three field sites examined (Table 4). Root growth was less at the Bemidji site relative to Cloquet and Warba. The number of ectomycorrhizal roots increased at all three sites examined. However, the proportion of ectomycorrhizal roots relative to total number of roots declined from June to September.

Sclerotia of *C. geophilum* Fr. was most abundant at the Bemidji site relative to the Cloquet or Warba sites (Table 5). Moreover, the proportion of viable sclerotia was also greatest at Bemidji. The sclerotia of *C. geophilum* declined from May to June at all disturbed sites. In contrast, sclerotia occurrence was unchanged in the undisturbed forest.

The number of nonmycorrhizal and ectomycorrhizal roots was positively correlated with soil N and exchangeable cations (K, Ca, Mg) (Table 6). Root dry weight was not well correlated with soil variables. Multiple regression with soil variables did not improve the ability to predict root growth or ectomycorrhizal characteristics. The occurrence of *C. geophilum* sclerotia was significantly correlated with soil N, P, K and Mg but not pH or Ca content.

Table 3. Description of soil chemical properties at Cloquet, Warba adn Bemidji, Minnesota, USA, sites in year one

Site	Plot type	Soil pH (0-15cm)	Bulk density (0-15cm)	Nitrogen NH$_3$·NO$_3$	IIP (ppm)	Bray (ppm)	CEC (meg/100g) (ppm)	Exchangeable cations			
								NH$_4$ (ppm)	K (ppm)	Mg (ppm)	Ca (ppm)
Cloquet	undisturbed forest	5.1	0.64	0.5	3.0	10	11.4	4	31	11	15
	disturbed	5.3	0.51	0.5	1.0	18	9.8	5	42	15	16
Warba	undisturbed forest	5.9	0.78	0.3	4.0	16	5.3	6	37	19	24
	disturbed	6.1	0.70	0.3	1.0	12	4.2	8	40	21	18
Bemidji	undisturbed forest	6.4	0.60	0.1	3.0	12	11.8	3	20	13	45
	disturbed	6.2	0.58	0.1	0.5	17	11.5	5	25	14	34

Table 4. Number and biomass of roots and number of ectomycorrhizal roots occurring in plots of indigenous vegetation regenerating on harvested sites at Cloquet, Warba and Bemidji, Minnesota, USA, in June of year one and September of year two

Site Plot Type	Woody Roots (#/100g)			Woody Root Weight (mg/cm^2)			Ectomycorrhizal Roots (%)		
	Year 1 June	Year 1 Sept.	Year 2 Sept.	Year 1 June	Year 1 Sept.	Year 2 Sept.	Year 1 June	Year 1 Sept.	Year 2 Sept.
Cloquet undisturbed forest	28	23	30	0.09	0.08	0.11	63	79	61
disturbed	19	47	51	0.02	0.12	0.19	42	30	23
Warba undisturbed forest	35	38	31	0.12	0.12	0.15	48	65	51
disturbed	23	52	57	0.03	0.14	0.21	30	37	16
Bemidji undisturbed forest	24	21	29	0.08	0.09	0.16	57	53	59
disturbed	17	39	34	0.04	0.09	0.17	47	33	31

Table 5. Occurrence of *C. geophilum* sclerotia in upper soil profile (0-15 cm) of harvested and undisturbed sites at Cloquet, Warba and Bemidji, Minnesota, USA, in June of year one and September of year two

Site	Plot Type	Sclerotia/g soil			Viable Sclerotia (%)		
		June Year 1	September Year 1	September Year 2	June Year 1	September Year 1	September Year 2
Cloquet	undisturbed forest	0.10	0.10	0.10*	23*	18*	19*
	disturbed	0.10	0.10	0.05	12	8	3
Warba	undisturbed forest	0.50	0.60*	0..60	43*	49*	54*
	disturbed	0.30	0.10	0.05	27	15	18
Bemidji	undisturbed forest	1.80*	1.40*	1.40*	61	53*	68*
	disturbed	0.60	0.40	0.60	47	16	20

Within a site (columns), pairs or means (undisturbed vs. harvest) followed by an asterisk(*) are significantly different by t-test (P=0.05).

Table 6. Linear correlation coefficient (r) between soil variables and number of roots, root dry weight, number of ectomycorrhizal roots of indigenous woody vegetation, and numbe of *C. geophilum* sclerotia at Cloquet, Warba and Bemidji, Minnesota, USA, sites

Soil Variable	Number of Roots	Root Dry Weight	Number of Ectomycorrhizal Roots	Number of Sclerotia
NH$_3$-N	NS[a]	NS	0.18*	0.35**
NO$_3$-N	0.76**	NS	0.65*	0.52*
P	0.75*	NS	NS	0.54*
K	NS	NS	0.65*	0.46**
Ca	0.80**	NS	0.66*	NS
Mg	0.62**	0.58*	0.24*	0.57*
CEC	NS	0.62**	NS	NS
pH	0.43*	0.25**	0.13*	NS

[a] Significant at 0.1* or 0.01** level. NS = Significant

Ectomycorrhizal fruiting bodies were present at all three sites (Table 7). A total of 55 collections were recorded during May to September. Five genera of ectomycorrhizal fungi were dominant: *Hebeloma, Inocybe, Laccaria, Rhizopogon* and *Suillus*. The most frequently collected fungus was *Laccaria laccata* (Scop. ex Fr.) followed by *Hebeloma* and *Rhizopogon* species. Non-mycorrhizal fungi were also observed.

Ectomycorrhizal roots and inoculum were present in soils of all three sites but their relative abundance varried. Fungal symbionts such as *Rhizopogon* sp. formed coralloid ectomycorrhizae with abundant extramatrical mycelium. Other ectomycorrhizae observed were monopodial or bifurcate with less well-developed fungal mantles. Unidentified E-strain fungi were observed infrequently but were characterized by thin mantles and intracellular hyphae in roots (Wilcox, 1968).

Intact or damaged chlamydospores of VAM fungi were detected at all three sites, but the distribution was not uniform (Table 8). In May, soils of the Warba site yielded the most chlamydospores and highest proportion of roots colonized by VAM. The abundance of VAM propagules and root system colonization significantly increased at the Bemidji and Cloquet sites when sampled in September. The chlamydospores were spherical, thick-walled and the color ranged from light to brown. Chlamydospores observed in association with sorghum roots in the bioassay were identified as *Glomus fasciculatum*, *G. tenue* (Green) Hall and *G. mosseae* (Nic. and Gerd.) Gerd. and Trappe.

Table 7. Host species, ectomycorrhizal general and number of fruiting bodies observed in sample plots at Cloquet, Warba and Bemidji, Minnesota, USA, sites

Site	Host Species	Ectomycorrhizal genera	Observations (#) Year 1	Year 2
Cloquet	*Pinus banksiana*	*Suillus*	3	8
		Rhizopogon	9	13
	Pinus resinosa	*Hebeloma*	8	6
		Laccaria	10	12
Warba	*Pinus glauca*	*Hebeloma*	9	4
		Laccaria	12	-
		Suillus	1	5
Bemidji	*Pinus banksiana*	*Rhizopogon*	3	4

Table 8. *Glomus* species chlamydospores and vesicular-arbuscular mycorrhizal (VAM) colonization observed in soil at Cloquet, Warba and Bemidji, Minnesota, USA, sites in years one and two

Site	Plot Type	spores/g soil (0-15 cm)			(15-30 cm)			VAM Root System Colonization (%) (0-15 cm)		
		June (Yr 1)	September (Yr 1)	September (Yr 2)	June (Yr 1)	September (Yr 1)	September (Yr 2)	June (Yr 1)	September (Yr 1)	September (Yr 2)
Cloquet	undisturbed forest	3*a	1	2	1	2	2	71*	53	62
	disturbed	1	13	17	-	7	9	30	45	28
Warba	undisturbed forest	12*	9	4	8*	6	6	48*	44	37
	disturbed	4	32	16	1	5	4	17	58	39
Bemidji	undisturbed forest	2	4	-	1	1	5	26*	42	38
	disturbed	-	28	16	-	14	9	4	36	20

[a]Within a site, pairs of means (undisturbed vs. disturbed) followed by an astrisk (*) are significantly different by t-test (P=0.05).

4.4. Discussion

Changes in above- and below-ground site conditions significantly altered soil chemical and physical properties, the distribution of fungal propagules, and the occurrence of ectomycorrhizae and VAM at Cloquet, Warba and Bemidji. The impacts of disturbance were dynamic and varied with site conditions and fungal organisms. Prior reports revealed mycorrhizal propagules and symbioses diminished following disturbance (Harvey *et al.*, 1980; Parke *et al.*, 1983; Shaw and Sidle, 1983) and the results of this study partially confirm this conclusion. Disturbed sites subject to burning often experience a dramatic decline in fungal propagules (Harvey *et al.*, 1980).

The diversity of mycorrhizal fungi on the three Minnesota, USA, sites, and their differential response to disturbance, suggest a biologic capacity to buffer dramatic environmental changes such as global change. The relative diversity of mycorrhizae varies widely in North America. Danielson *et al.* (1984); Pilz and Perry (1984) and Shaw and Sidle (1982) reported abundant mycorrhizal diversity in Alberta, Canada, Oregon and Alaska. Mycorrhizal fungi are known to occupy different successional niches (Fleming *et al.*, 1984). Sites with tree species diversity in harsh habitats foster a wide range of rhizosphere symbionts. This phenomenon ensures rapid colonization of the site by pioneer species following disturbance and also maintains a pool of soil organism that will benefit later successional stages (Amaranthus and Perry, 1989).

The dominant ectomycorrhizae were *C. geophilum*, *Hebeloma*, *Laccaria* and *Rhizopogon* species. The disproportionately large number of mycorrhizae found at the Warba site may be associated with the diversity of plant species present prior to disturbance and favourable microsite conditions (Amaranthus and Perry, 1989). *Rhizopogon* species do not compete well on all sites, but with their characteristic rhizomorphs, may have a competitive advantage under these conditions (Castellano and Trappe, 1983). Other ectomycorrhizal fungi such as *Laccaria laccata* can withstand mechanical manipulation and may be well adapted to survival following disturbance (Molina and Palmer, 1984). The relative abundance of *Hebeloma* was not expected as earlier assessments revealed this species was not able to compete with indigenous species in the field.

Formation of ectomycorrhizae was a function of inocula potential, as well as the physical and chemical properties of the soil at all three sites. This response has been observed previously in Oregon, USA (Amaranthus and Perry, 1987). The presence of non-host specific fungi, such as E-strain or *C. geophilum*, ensure that indigenous inocula are compatible with a wide range of species. In this study, the proportion of non-mycorrhizal roots increased with a decline in the abundance of ectomycorrhizal propagules . Propagule density may not be well correlated with ectomycorrhizae formation (Korpp, 1982; Pilz and Perry, 1984). The significant influence of soil chemistry on mycorrhizal symbioses is consistent with earlier field analyses (Shaw and Sidle, 1983).

In contrast to the ectomycorrhizal fungi, the occurrence of VAM fungi increased at the Cloquet and Bemidji sites. Disturbance is reported to both increase (Parke *et al.*, 1983) and decrease (Powell, 1980) the number of VAM propagules. Vesicular-arbuscular mycorrhizal fungi appear to have a wide range of tolerance to extremes of soil pH, temperature, moisture and chemical properties (Mukerji, 1996). Vesicular-arbuscular mycorrhizal fungi can form an effective link between plants and soil systems, increasing ecosystem integrity and stability, based on critical determinants such as edaphic and climatic conditions, spatial relationships of the landscape, and host-fungi genotypes and diversity. Although disturbance may have altered the stability of VAM in this study, adequate propagules were present to colonize plant hosts invading or proliferating at all three sites. Many of the plant genera which dominated the canopy of the sites in year one are hosts for VAM (e.g., *Populus*). Although ectomycorrhizal fungi were relatively diverse on the disturbed sites, the VAM fungi were not. Parke *et al.* (1983) observed similar trends in VAM populations of disturbed sites in southern Oregon, USA.

5.   CONCLUSIONS

Mycorrhizal fungi and other microbes play important roles in plant community structure and processes (Lal *et al.*, 1995). The mycorrhizal relationships of single plants are probably both functionally and structurally significant at the ecosystem level. In the context of potential global change, mycorrhizal fungi can enhance resilience and elasticity of ecosystems, providing a capacity to mitigate or adapt to future extreme events (e.g., disturbance, droughts, floods, changing atmospheric chemistry). Mycorrhizal fungi can modulate plant competitive relationships (Mason *et al.*, 1983), offer resilience to disturbance (Parke *et al.*, 1983), enhance nutrient cycling and storage (Nilsson *et al.*, 1995), and affect plant productivity, health, reproduction (Mukerji, 1996). Diversity of mycorrhizal fungi and plant species have been associated with ecosystem stability in stressed environments (Fogel and Hunt, 1979; Schimmel *et al.*, 1990).

The likelihood of significant changes in global change has received increasing attention in recent decades (Watson *et al.*, 1995). Many of the predicted changes are latent and irreversible. Our knowledge of the responses and feedbacks of below-ground systems, especially the rhizosphere, to global change is limited. However, preliminary assessments reveal that below-ground systems will play a prominent role in vegetation response to global change (Shukla *et al.*, 1990). Moreover, world soils influence albedo and are significant sources and sinks of GHGs such as water vapor, CO_2, CH_4 and N_2O. The role of below-ground systems in the global C cycle and environmental change should be systematically assessed (Parton *et al.*, 1987). Future assessment should include an examination of below-ground systems, rhizosphere systems, and their components and processes (Lal *et al.*, 1995).

ACKNOWDGEMENTS

A. Alm and M. Sword provided valuable technical contributions to the assessments presented in this chapter.

This U.S. Country Studies Program is co-financed and co-managed by the U.S. Department of States, U.S. Environmental Protection Agency, U.S. Department of Energy, and U.S. Agency for International Development. The views expressed in this manuscript do not represent any government or intergovernmental body.

REFERENCES

Amaranthus, M.P. and Perry, D. 1987, Effect of soil transfer of ectomycorrhizal formation and the survival and growth of conifer seedlings on old, nonreforested clear-cuts, *Can.J.For.Res*. 17: 944-950.

Amaranthus, M.P. and Perry, D. 1989, Interaction effect of vegetation type and Pacific mandrone soil inoculum on survival, growth and mycorrhiza formation of Douglas-fir, *Can.J.For.Res*.19: 55-56.

Apps, M.J., Kurz, W.A., Luxmoore, R.J., Nilsson, L.O., Sedjo, R., Schmidt, R., Simpson, L.G. and Vinson, T.S. 1993, Boreal forests and tundra, *Water, Air and Soil Pollution* 70:39-54.

Baer, N.W. and Otta, J.D. 1981, Outplanting survival and growth of ponderosa pine seedlings inoculated with *Pisolithus tinctorius* in South Dakota, *For. Sci*.27:277-280.

Bhatia, N.P., Sundari, K. and Adholeya, A. 1996, Diversity and selective dominance of vesicular-arbuscular mycorrhizal fungi, in: *Concepts in Mycorrhizal Research*, K.G Mukerji, ed., Kluwer Academic Publishers, Dordrecht, Netherlands pp.133-178.

Black, C.M. 1965, *Methods of Soil Analysis*, American Society of Agronomy, Madison, WI.

Bouwman, A.F. (ed.) 1989, *Soils and the Greenhouse Effect*, John Wiley, New York.

Burke, I.C., Yonker, C.M., Parton, W.J., Cole, C.V., Flach, K. and Schimel, D.S. 1989, Texture, climate and cultivation effects on soil organic matter content in US grassland soils, *Soil Sci. Soc. Amer.* J.53:800-805.

Castellano, M.A. and Trappe, J.M. 1985, Ectomycorrhizal formation and plantation performance of Douglas-fir nursery stock inoculated with *Rhizopogon* spores, *Can J. For. Res*.15:613-617.

Critchfield, H.J. 1966, *General Climatology*, Prentice-Hall, Englewood Cliffs, NJ, USA.

Cole, C.V., Flach, K., Lee, J., Sauerbeck, D. and Stewart, B.1993, Agricultural sources and sinks of carbon, *Water, Air and Soil Pollution* 70:111-122.

Curl, R. and Truelove, B.1986, *The Rhizosphere*, Springer-Verlag, New York.

Curtis, J.T. 1959, *The Vegetation of Wisconsin*, University of Wisconsin Press, Madison, WI, 657p.

Curtis, P.S., Balduman, L.M., Drake, B.G. and Whigham, D.F.1990, The effect of elevated atmospheric CO_2 on below-ground processes in C_3 and C_4 estuarine marsh communities, *Ecology* 71:2001-2006.

Daniels, B. and Skipper, D.H. 1984, Methods for recovery and quantitative estimation of propagules from soil, in : *Methods and Principles of Mycorrhizal Research*, N.C. Schenck, ed., Am. Phytopathol. Soc., St. Paul, MN, USA.

Danielson, R.M. 1984, Ectomycorrhizal associations in jack pine stands in northeastern Alberta, *Can. J. Bot.* 62 : 932-939.

Danielson, R.M., Zak, J.C. and Parkinson, D.1984, Mycorrhizal inoculum in a peat deposit under a white spruce stand in Alberta, *Can. J.Bot.* 63 : 2557-2560.

Dickinson, R.E. 1989, Uncertainties of estimates of climate change, *Clim. Change* 15 : 5-13.

Dixon, R.K., Wright, G.M., Behrns, G.T., Teskey, R.O. and Hinckley, T.M. 1980, Water deficits and root growth of ectomycorrhizal white oak seedlings, *Can.J.For.Res.* 10:545-548.

Dixon, R.K. and Turner, D.P. 1991, The global carbon cycle and climate change:responses and feedbacks from below-ground systems, *Environ. Poll.* 73:245-262.

Dixon, R.K., Winjum, J.K. and Schroeder, P.E. 1993, Conservation and sequestration of carbon: the potential of forest and agroforest management practices, *Glob. Environ. Change* 2 : 159-173.

Dixon, R.K., Brown, S., Houghton, R.A., Solomon, A.M., Trexler, M.C. and Wisniewski, J. 1994a, Carbon pools and flux of global forest systems, *Science* 263 : 185-190.

Dixon, R.K., Winjum, J.K., Andrasko, K.J., Lee, J.J. and Schroeder, P.E. 1994b, Integrated land-use systems: assessment of promising agroforest and alternative land-use practices to enhance carbon conservation and sequestration, *Clim. Change* 27:71-92.

Dixon, R.K. and Wisniewski, J.1995, Global forest systems:an uncertain response to atmospheric pollutants and global climate change, *Water, Air and Soil Pollution* 85:101-110.

Dorr, H. and Munnich, K.O. 1987, Annual variation in soil respiration in selected areas of the temperate zone, *Tellus* 39B:114-121.

Fajer, E.D. 1989, How enriched carbon dioxide environments may alter biotic systems even in the absence of climatic changes, *Conserv. Bio.* 3:318-320.

Fleming, L.V., Deacon, J.W., Last, F.T. and Donaldson, S.J. 1984, Influence of propagating soil on the mycorrhizal succession of birch seedlings transplanted to a field site, *Trans. Br. Mycol. Soc.* 82:707-711.

Fleming, L.V. 1985, Experimental study of sequences of ectomycorrhizal fungi on birch (*Betula* sp.) seeling root systems, *Soil Biol.Biochem.* 17:591-600.

Fogel, R. and Hunt, G.1979, Fungal and arboreal biomass in a western Oregon Douglas-fir ecosystem:distribution patterns and turnover, *Can.J.For.Res.* 9 : 245-256.

Hansen, J. 1988.Global climate changes as forecast by Goddard Institute of Space Studies three-dimensional model, *J. Geophys.Res.* 93:9341-9364.

Harvey, A.E., Larsen, M.J. and Jurgensen, M.F. 1980, Partial cut harvesting and ectomycorrhizae:early effects in Douglas-fir/larch forests of western Montana, *Can.J.For.Res.* 10:436-440.

Joslin, J.D. and Henderson, G.S. 1987, Organic matter and nutrients associated with fine root turnover in a white oak stand, *For.Sci.* 33:330-346.

Khalil, M.A.K. and Rasmussen, R.A. 1989, Climate-induced feedbacks for the global cycles of methane and nitrous oxide, *Tellus*, 41B:554-559.

Korpp, B.R.1982, Formation of mycorrhizae on nonmycorrhizal western hemlock outplanted on rotten wood and mineral soil, *For.Sci.*28:706-710.

Lal, R., Kimble, J., Levine, E. and Stewart, B.A.(eds.), 1995, *Soils and Global Change*, CRC Press, Boca Raton, USA, 440p.

Manabe, S. and Wetherald, R.T. 1987, Large-scale changes of soil wetness induced by an increase in atmospheric carbon dioxide, *J.Atmos.Sci.* 44:1211-1235.

Marks, G.C. and Foster, R.1967, Succession of mycorrhizal associations on individual roots of radiata pine, *Aust. For.*31:193-201.

Mason, P.A., Last, F.T., Pleham, J. and Ingleby, K.1982, Ecology of some fungi associated with an aging stand of birches (*Betula pendula* and *B.pubescens*), *For.Ecol.Manage.* 4:19-39.

Mason, P.A., Wilson, J., Last, F.T. and Walker, C.1983, The concept of succession in relation to the spread of sheathing mycorrhizal fungi on inoculated tree seedlings growing in unsterile soils, *Plant Soil* 71:247-256.

Mathews, E. and Fung, I.1987, Methane emission from natural wetlands:Global distribution, area and environmental characteristics of sources, *Biogeochem. Cycles* 1:61-86.

Meentenmeyer, V., Box, E.O., Folkoff, M. and Gardner, J.1981, Climatic estimation of soil properties:Soil pH, litter accumulation and soil organic content, *Ecol. Soc.Amer.Bull.* 62:104.

Molina, R. and Trappe, J.M. 1982, Patterns of ectomycorrhizal host specificity and potential among Pacific Northwest conifers and fungi, *For.Sci.* 28:423-458.

Molina, R. and Palmer, J.G.1984, Isolation, maintenance and pure culture manipulation of ectomycorrhizal fungi, in:*Methods and Principles of Mycorrhizal Research*, N.C.Schenck, ed., Am. Phytopathol.Soc., St. Paul, MN, USA, pp.115-130.

Mooney, H.A., Drake, B.G., Luxmoore, R.J., Oechel, W.C. and Pitelka, L.F.1991, Predicting ecosystem responses to elevated CO_2 concentrations, *Bioscience* 41:96-104.

Mukerji, K.G. (ed). 1996, *Concepts in Mycorrhizal Research*, Kluwer Academic Publishers, Dordrecht, Netherlands, 374p.

Nilsson, L.O., Huttl, R.F. and Johansson, U.T. (eds.), 1995, *Nutrient Uptake and Cycling in Forest Ecosystems*, Kluwer Academic Publishers, Dordrecht, Netherlands, 685p.

Norby, R.J.1987, Nodulation and nitrogenase activity in nitrogen-fixing woody plants stimulated by CO_2 enrichment of the atmosphere, *Physio.Plant* 71:77-82.

Oades, J.M. 1988, The retention of organic matter in soils, *Biogeochem.* 5:35-70.

Oechel, W.C., Hastings, S.J., Vourilitis, G., Jenkins, M., Riechers, G. and Grulke, N.1993, Recent change of Arctic tundra ecosystems from a net carbon dioxide sink to a source, *Nature* 361:520-523.

O'Neill, E.G., Luxmoore, R.J. and Norby, R.J. 1987, Increases in mycorrhizal colonization and seedling growth in *Pinus echinata* and *Quercus alba* in an enriched CO_2 atmosphere, *Can.J.For.Res.* 17:878-883.

Parke, J.L., Linderman, R.G. and Trappe, J.M.1983, Effect of root zone temperature on ectomycorrhiza and vesicular-arbuscular mycorrhiza formation in disturbed and undisturbed soils of southwest Oregon, *Can.J.For.Res.*, 13:657-665.

Parton, W.J., Schimel, D.S., Cole, C.V. and Ojima, D.S. 1987, Analysis of factors controlling organic matter levels in Great Plains grasslands, *Soil Sci. Soc. Amer.J.* 51:1173-1179.

Perry, D.A., Molina, R. and Amaranthus, M.P. 1987, Mycorrhizae, mycorrhizospheres and reforestation: Current knowledge and research needs, *Can. J.For.Res.* 17:929:940.

Perry, D.A., Borchers, J.G., Borchers, S.L. and Amaranthus, M.P. 1990, Species migrations and ecosystems stability during climatic change: The below-ground connection, *Conser.Biol.* 4:266-274.

Phillips, J.M. and Hayman, D.S. 1970, Improved procedures for clearing and staining parasitic and vesicular-arbuscular mycorrhizal fungi for rapid assessment of infection, *Trans.Br.Mycol.Soc.* 55:158-161.

Post, W.M., Emanuel, W.R., Zinke, P.J. and Stangenberger, A.G. 1982, Soil carbon pools and world life zones, *Nature* 298:156-159.

Pilz, D.P. and Perry, D.A. 1984, Impact of clear-cutting and slash burning on ectomycorrhizal associations of Douglas-fir seelings, *Can.J.For.Res.* 14:94-100.

Powell, C.L. 1980, Mycorrhizal infectivity of eroded soils, *Soil Biol.Biochem.* 12:247-250.

Prentice, K.C. and Fung, I.Y.1990, Bioclimatic simulations test the sensitivity of terrestrial carbon storage to perturbed climates, *Nature* 346:48-51.

Schimel, D.S., Parton, W.J., Kittel, T.G.F., Ojima, D.S. and Cole, C.V. 1990, Grassland biogeochemistry:Links to atmospheric processes, *Clim.Change* 17:13-25.

Schlesinger, W.H. 1990, Evidence from chronosequence studies for a low carbon-storage potential of soils, *Nature* 348:232--234.

Schlesinger, W.H.1990, Biological feedbacks in global desertification, *Science*, 247:1043-1048.

Schneider, S.H.1989, The greenhouse effect:Science and policy, *Science* 243:771-781.

Shaw, G.C. and Sidle, R.C. 1983, Evaluation of planting sites to a southeast Alaska clear-cut, II, Available inoculum of the ectomycorrhizal fungus *Cenococcum geophilum*, *Can.J.For.Res.*13:9-11.

Shukla, J., Nobre, C. and Sellers, P.1990, Amazon deforestation and climate change, *Science* 247:1322-1325.

Sionit, N., Strain, B.R., Hellman, H., Riechers,G.H. and Jaeger, C.H. 1985, Long-term atmospheric CO_2 enrichment affects the growth and development of *Liquidambar styraciflua* and *Pinus taeda* seedlings, *Can.J.For.Res.* 15:468-471.

Smith, J.B. and Tirpak, D.A.1989, The potential effects of global climate change on the United States, EPA-230-05-89-050, US Environmental Protection Agency, Washington, DC, USA.

Smith, T.M., Cramer, W.P., Dixon, R.K., Leemans, R., Neilson, R.P. and Solomon, A.M. 1993, The global terrestrial carbon cycle, *Water, Air and Soil Pollution* 70:19-38.

Steel, R.G.D. and Torrie, J.H.1980, *Principles and Procedures of Statistics*, McGraw-Hill, New York, USA.

Sundquist, E.T.1993, The global carbon dioxide budget, *Science* 259:934-941.

Tans, P.P., Fung, I.Y. and Takahashi, T.1990, Observational constraints on the global atmospheric CO_2 budget, *Science* 247:1431:1438.

Trabalka, J.R. and Reichle, D.E. 1986, *The Changing Carbon Cycle:A Global Analysis*, Springer-Verlag, New York, USA.

Turner, D.P., Myrold, D.D. and Bailey, J.D. 1990, Cl'mate change and patterns of denitrification in the Wilamette basin of western Oregon, USA, in : *Soils and the Greenhouse Effect*, A.F.Bouwman, ed., John Wiley, New York, USA.

Watson, R.T., Zinyowera, M.C., Moss, R.H. and Dokken, D.J. (eds.), 1995, *Climate Change 1995*, Impacts, Adaptations and Mitigation of Climate Change:Scientific-Technical Analysis, Intergovernmental Panel on Climate Change, Cambridge University Press, 879p.

Wilcox, H.E. 1968, Morphological studies of the roots of red pine *Pinus resinosa*, II, Fungal colonization of roots and the development of mycorrhizae, *Am.J.Bot.* 55:686-700.

Yin, X., Perry, J.A. and Dixon, R.K. 1989, Fine-root dynamics and biomass distribution in *Quercus* ecosystem following harvesting, *For.Ecol.Manage.* 27:159-177.

ECTOMYCORRHIZA - AN OVERVIEW

T.N. Lakhanpal

Department of Biosciences, Himachal Pradesh University
Summer Hill, Shimla - 171005, H.P., INDIA

1.   INTRODUCTION

The term mycorrhiza, coined by Frank literally means 'fungus-root'. It refers to a symbiotic association between fungi and the feeder roots of higher plants. It is now well documented that with the exception of aquatics, some halophytes and few other plants, all form mycorrhizae in natural roots to varying degrees (Mukerji and Mandeep, 1998). Harley and Smith(1983) classified mycorrhiza into seven types e.g. Ectomycorrhiza, Ectendomycorrhiza, Endomycorrhiza (Vesicular-arbuscular mycorrhiza), Arbutoid, Monotropoid, Ericoid and Orchidaceous type. This chapter concerns primarily with the ectomycorrhiza.

The ectomycorrhizae are most common among forest and ornamental tree species in the family Pinaceae, Salicaceae, Betulaceae, as well as in some members of Rosaceae, Leguminoseae, Ericaceae and Juglandaceae. The ectomycorrhizae are characterized by the presence of 'mantle' and the 'Hartig net'. Ectomycorrhizal infection is initiated from spores, sclerotia, hyphae and/or rhizomorphs of the fungi. Spores or hyphae come into contact with and invade the actively growing roots of trees. The propagules are stimulated by root exudates to grow rapidly and cover the entire root tip with dense sheath of hyphae (Fungal-Mantle). Such distinguishing features of ectomycorrhiza, the fungi that enter into ectomycorrhizal association and the role that ectomycorrhizal fungi can play in forest biotechnology, are discussed in details in this article.

2.   MORPHO-ANATOMICAL FEATURES OF ECTOMYCORRHIZAE

The ectomycorrhizae are easy to distinguish from endomycorrhizae because in them the root system is heterorrhizic, comprised of two kinds of roots; long roots of potentially unlimited extension and short roots with restricted growth and life span. The long and short roots in most of ectomycorrhizal plants, whether angiosperms or gymnosperms, differ only in degree. The root apices grow for a period and all may become infected in some way by a mycorrhizal fungus so that the whole root system is mycorrhizal. Those apices which become fully infested by a sheath of fungus are usually short and continue to grow slowly and give rise to racemose system of branches. Those apices which are not permanently infested by a sheath or which remain uninfected, either maintain very active growth and are the leading apices of the system or especially, if of small diameter, may abort or become dormant.

Mycorrhizal Biology, edited by Mukerji et al.
Kluwer Academic/Plenum Publishers, 2000

The genus *Pinus* departs from this general form because its ultimate laterals - its short roots are sharply differentiated from the axes which bear them. They are simpler in stelar construction, have a restricted apical meristem and root cap which soon abort, if not colonised. If infected they continue to grow, branch dichotomously and form mycorrhizal systems. Though the colonization by mycorrhizal fungi restricts the elongation of roots, yet it stimulates the branching, and therefore, growth of roots is not discontinued. After infection the roots become swollen and the colour of the symbiotic fungus shows through them in different shades making them variously coloured. As per the branching, the ectomycorrhizal roots may be simple or monopodial as in spruce and fir, coralloid as in deodar or often exhibit repeated dichotomy as in pines or may be pyramid type as in *Eucalyptus* (Fig. 1).

Each short root, regardless of branching pattern is an ectomycorrhiza. Functionally when ectomycorrhizae are formed, the absorbing area of root surface is increased. Root hairs are absent and these roots are completely enveloped by a fungal mat or mantle along the length and apex. Mycorrhizal structures are not static organs but change ontogenetically from a juvenile to a mature state and finally senesce. Mycorrhizae may become inactive and are lost by attrition; alternatively, they may undergo cycles of dormancy and activity, which is a common feature of perennial root system of trees (Suvercha *et al.*, 1991).

2.1. Mantle

Mycorrhizal roots gradually become completely enveloped by a mat of fungal symbiont, called "Mantle". Mantle may be smooth or with rediating hyphae, which enter into soil to increase the absorptive surface of short roots. Different ectomycorrhizal fungi form distinctive mantles of varying thickness, texture and colour, with different hosts. For example, mantle formed by *Cenococcum graniforme* consists of horizontally arranged palisade cells alternating with groups of small pseudoparenchymatous cells. In *Scleroderma verrucosum, Hysterangium* sp. and *Pulveroboletus shoreae*, hyphae in the mantle may be less organised, forming a loose mantle where individual hyphae may be recognised or hyphae may be organised still maintaining their identity (Fig. 2a,b).

When the hyphae composing the mantle can no longer be recognised., it is then called a pseudoparenchyma or synnema. The synnematous mantle structures may have approximately isodiameteric cells with fairly straight walls or puzzle-like cells with wavy walls. The interhyphal space may be large or cemented with a matrix from the slimed hyphal walls. But when hyphae maintain their identity and can still be recognised, the structure is called plectenchyma or prosenchyma.

Mantle surface can range from thin to profuse and texture can vary from smooth cottony, velvety, warty to granular. The hyphae radiating from mantle surface may be simple or branched bearing simple or clamped septa; colour of these hyphae may be hyaline, black or in various shades such as orange, yellow and brown. The colour of mantle is mainly due to the colour of radiating hyphae.

The main tissues of the tuber sheath, may be reminiscent of the fruit bodies of the fungi and had both active and storage hyphae in them. It is believed that the interhyphal matrix, probably of carbohydrate polymers, that binds the hyphae of the sheath together differed in ascomycetous and basidiomycetous sheaths, being electron dense in the former and electron lucent in the latter. Others do not agree that there is a consistent difference of this sort.

In general the fine structure of the mantle is similar to that of basidiomycetous hyphae. The cell walls of the mantle hyphae is covered with an amorphous layer which on mild maceration revealed few layers of microfibrils, the inner one being more organised then the outer as in fungi of many groups. The inner layer of mantle was characterized by more closely interwoven hyphae and an increase in the number of cytoplasmic organelles. It is possible to recognise and identify mycorrhizae by structural peculiarities of the hyphal mantle.

The sheath varies in thickness but is typically 20-40 μm thick and comprises 20-30% of the volume of the rootlet. It is estimated that sheath contributed about 39.1 ± 0.5mg per 100gm of dry

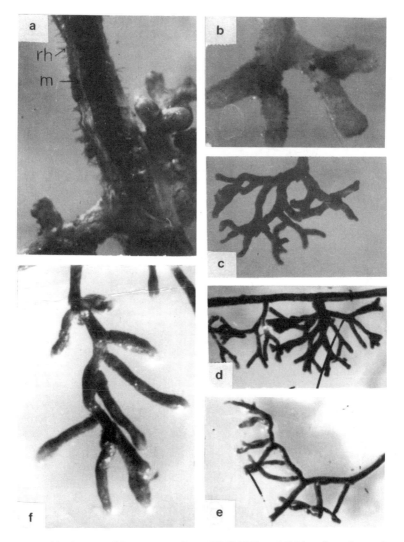

Fig. 1 : a. Mycorrhizal roots of *Pinus gerardiana* Wall (5X), exhibiting fungal mantle = m and radiating hyphae = rh.

b, c, d. Gross morphology of ectomycorrhizae of *Pinus gerardiana* Wall., dichotomies of short roots (b) and corralloid mycorrhizae (c and d), arising from the main root, 2.5 X.

e. Gross morphology of monopodial mycorrhizae of *Picea smithiana* (Wall.) Boiss., 2.5X.

f. Mycorrhizal roots of *Picea smithiana* (Wall.) Boiss. Monopodial mycorrhizae showing bulbous and rounded apices covered with hyphal sheath.

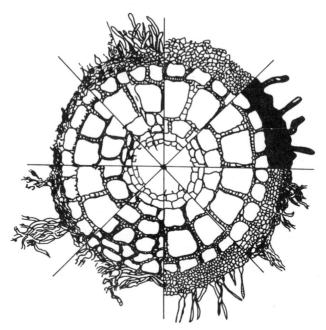

Fig. 2a : Diagram showing a range of structure of mycorrhizas in transverse section in *P. maritima* (adapted from Harley and Smith, 1983).

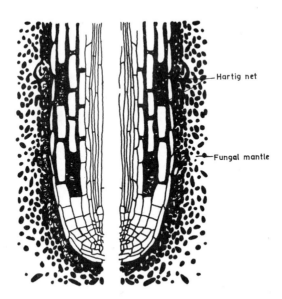

Fig. 2b : Vertical section of a root with a schematic representation of ectomycorhiza (modified from Kendrick and Berch, 1985).

weight of mycorrhiza in *Fagus*.

The variations in structures of the sheath, the ornamentation of its surface by hyphae, rhizomorphs, setae or cystidia like structures and its colour together with chemical, immunological and other tests and the structure of the Hartig net have been used to classify ectomycorrhizae (Zak, 1971;1973).

In any event, the fungal layer is a very significant part of the mycorrhizal organ which is frequently characteristic for the particular fungus host symbiosis. Hyphae extend from the mantle into the soil to increase the absorptive surface. Fungal mantle creates a unique and totally different type of obstruction to pathogens attempting penetration. The fungus mantles of mycorrhizae are formidable physical barrier to penetration by *Phytophthora cinnamomi*.

2.2.   Hartig's Net

The juvenile stage is the initial infection stage when the 'Hartig net' begins to develop around differentiated metabolically active cortical cells. The cellular interaction begins sub-apically, the distance to the merristem depending upon root elongation rates. When Hartig net development is initiated the cortical or epidermal cells have obviously already acquired their final dimensions (Fig. 2a, b).

When the metabolically active cortical cells become enveloped more or less completely by a living hyphal system, it is the mature stage of the mycorrhizal structures. The dormant stage is the stage when root cap cells and the cells lying in between the endodermis and the root cap become suberized and accumulate polyphenols. Thus, the apical meristerm is encircled by a protective layer and mycorrhizal fungi are barred from the meristem during dormancy. The dormant mycelium may regrow slowly or rapidly after dormancy. In the former the root tip does not break through the sheath, but "both host and fungal tissues may grow slowly in unison". In the latter, the root apex breaks through the hyphal sheath.

The Hartig net development starts when hyphae come in contact with unsuberized living cortical or epidermal cells and is characterized by changes of hyphal growth and morphology. The diameter of the hyphae may be greater or smaller than those of the hyphal sheath. Hyphae are oriented transversly to the root axis and begin to branch out irregularly, septation is rare, the hyphae penetrate in the direction of the endodermis-growth and longitudinal direction through the intercellular spaces is rather restricted. The transversal growth direction is surely an advantage for nutrient transport between the endodermis and the hyphal mantle exploiting the shortest way between the both. The transversal growth direction also ensures hyphal establishment in the intercellular spaces of cortex cells of a suitable developmental space.

Hartig net has been described to be made up of septate hyphae and has been considered to be essentially hyphal. Hartig net has been described to be "lames fungique" and examination of its structure by electron microscopically reveals that it consists of complicated fan-like or labyrinthine branch systems which provide a very large surface of contact between cells of the two symbionts. With regards to septation, it has been opined that the walls of hyphae in close contact have been erroneously interpreted as septate or incompletely septate. The double wall nature of the structures reveals, however, that they are apparently not septate and that incomplete septa mark beginning of the branches.

The fungal mantle may consist of 1-2 layers of plectenchyma of hyphae or the hyphae may be densely interwoven to form a more compact pseudoparenchymatous layer with isodiametric or puzzle-like cells. Setae, cystidia, laticifers and pigmented hyphae in sheath may be reminiscent of the fruit bodies of fungi and therefore, are valuable in tracing link between the fungal mantle and in identifying the fungal species involved.

3.   THE ECTOMYCORRHIZA FORMING FUNGI

Taxonomically the ectomycorrhiza forming fungi usually belong to Basidiomycotina; some are also members of Ascomycotina (especially the hypogeous ones) and a few others of Deuteromycotina. It is estimated that in all such ectotrophic mycorrhiza forming members number about 5000 and their host species about 2000 woody plant species. In North America alone 2100 species of fungi are estimated to form ectomycorrhizae with forest trees. Report of such species from most of the other countries are still incomplete because of the lack of proper surveys. Between 100-150 fungal species in India are supposed to be mycorrhizal by virtue of their circumstantial distributional evidence. This is least representative of the floral diversity of this vast landmass and therefore, reflects the neglect of systematic surveys.

Among the basidiomycetous fungi, the majority are members of Hymenomycetes, particularly belonging to families Boletaceae, Gomphidiaceae, Russulaceae, Strobilomycetaceae, Cantherellaceae and the genera, *Amanita, Armillaria, Astreaus, Boletus, Cortinarius, Hebeloma, Inocybe, Suillus, Tricholoma, Laccaria* and *Lactarius* and in the Gasteromycetes, they belong to the genera *Rhizopogon, Scleroderma* and *Pisolithus*. Most of these genera are cosmopolitan mycorrhiza forming. *Cortinarius* alone is estimated to have around 2000 species. In Aphylophorales only a few members are ectomycorrhizal; most of the members are saprobes with high cellulolytic and lignolytic capabilities. Certain orders in the Ascomycetes such as Eurotiales (*Cenococcum geophilum*) Tuberales (Truffles) and Pezizales have species that form mycorrhiza on trees. Hypogeous ascomycetes, especially species of *Elaphomyces* and Tuberales have been proved beyond doubt to be mycorrhiza forming. All such fungi were comprehensively listed by Trappe (1962) but since then many additions have been made; notable being the Phycomycetes. *Endogone flamiocrna*, *E. eucalypti* and *E. lactiflua* which form ectomycorrhizae with different Pinaceae and *Eucalyptus*. Mycorrhiza forming activity has been recorded in representatives of 73 basidiomycete genera distributed among 27 families in nine orders. These fungi are holobasidiomycete Agarics; Agarics which have evolved or are evolving into a geasteroid habit and have in many cases become hypogeous and non gilled hymenomycetes from among the club fungi, Chanterelles, tooth fungi and what are called the resupinate hymenomycetes. There are also ectomycorrhizal representatives of 16 unitunicate ascomycete genera from eight families and two orders. All but one of the ascomycete genera involved are hypogeous and all but one are related to Operculate Discomycetes.

Most of the ectomycorrhizal fungi produce macroscopic fruit bodies in nature during rainy and/or winter season. These fungi can be easily and accurately identified from their fruit bodies, even though they appear for a short and limited period. Many standard reference works are available for their identification.

Various techniques have been used for establishing the mycorrhizal association of trees and the fungal fruit bodies. The most commonly used method is the one of tracing the hyphal connections between the roots of plants and the fungal sporocarps. In most of these, the connections are ascertained if the fungal hyphae or mycelium is abundant and distinctive and attached to - the both fruit body and a mycorrhizal root. This method has been successfully used to link identified sporocarps to mycorrhizae. For example one could easily trace the brilliant yellow mycelial strand of *Pisolithus tinctorius* through the contrasting dark anthracite wastes to the base of the fungi basidiocarp. Such coupling is often easily established with sporocarps of hypogeous fungi and nestled among mycorrhizae for *Pinus strobus* and *Endogone lactiflua; Octaviania densa* and *Cedrus deodara*. In the same way mycelial connections between sporocarps of *Suillus sibricus* and *Pinus wallichiana* and in species of *Morchella* and grass, fern and Raspberry roots were also similarly established.

But mycelial connections between fruiting bodies and mycorrhizae have been regarded as inadequate evidence to establish identity of the fungal partner, as the mycelium in some cases may be merely attached superficially to and not actually joined to part of the fungal mantle. The connections can be ascertained easily microscopically by matching the colour, texture and other gross characters of the respective tissue.

An extension of the method of tracing rhizomorphs and hyphae to sporocarp, is the linking sporocarps to underlying mycorrhizae. In this method tissues of sporocarps and mycorrhiza are compared. This is simple and can be applied easily but is limited by the availability of sporocarps. In fact, no single method can be described as exclusive and in practice more than one method should be employed. In certain cases it has been possible to ascertain identification by either achieving hyphal fusion with known species or immunological or serological or chromatographic methods.

Another method developed in the identification of the fungal symbiont is the pure culture synthesis. The ectomycorrhizal fungi are isolated in prue culture and utilized to form a recognisable mycorrhizal organ under *in vitro* conditions. However, the identification of the fungi isolated from the ectomycorrhizae is more problematic. Such a mycorrhiza formed in a wholly artificial environment is morphologically rarely the same as its natural counterpart because of the high variability of many ectomycorrhizal fungi. Furthermore all mycorrhizal fungi may grow easily on laboratory media.

Some of the ectomycorrhizal fungi are host specific, others have a wide host range and at the same time a single tree may have several or many different ectomycorrhizal partners on its roots. The host specificity is the result of co-evolution of ectomycorrhizal fungi and their hosts and, therefore, such fungi are usually distributed in the range of their specific hosts. But in due course of time the original associate may be replaced or succeeded by other fungi and hence, the continuity of the range may be interrupted. The genera of families Gomphidiaceae and Rhizopogonaceae were orginally restricted to Northern hemisphere with their conifer hosts. Species of genera *Suillus, Fuscoboletinus* and *Boletopsis* are found associated only with Northern hemisphere Pinaceae, but other related genera are cosmopolitan. *Mesophellia* the only known mycorrhizal memebr of the Lycoperdales is restricted to Australia with *Eucalyptus*. The sporocarps of mycorrhizal fungi associated with a mixed stand of *Betula* spp., produced in a pattern, ordered in time and space, suggesting a succession with identifiable early-and late-stage fungi species of *Laccaria* were observed to be the early-stage fungi and *Amanita muscaria*, as late-stage fungus. There are some intermediate species as well so that a single tree species may have a very large number of potential mycorrhizal partners. Such a concept has an important bearing on the selection of fungi for inoculating young tree seedlings. It is estimated that Douglas fir alone may be able to form ectomycorrhizae with about 2000 different species of fungi. *Boletus edulis, B.hoarkii* and *Octaviania densa* have been observed to be specifically mycorrhizal with *Cedrus deodara*. Some other host specific ectomycorrhizal fungi are *Suillus grerillei-Larix; Suillus lakei-Pseudotsuga menziesii, Lactarius obseuratus-Alnus; Gomphidius vinicolor-Pinus* and *Cortinarius hemitrichus-Betula*. Some of the broad host range species are: *Amanita muscaria, Boletus edulis, Cantharellus cibarius, Cenococcum geophilum, Laccaria laccata, Pisolithus tinctorius* and *Thelephora terrestris*. List of mycorrhizal fungi is presented in Tables 1-4. Out of such a large number of ectomycorrhiza forming species, it is not easy to select out the best possible fungus easily for a chosen tree/trees. There is no single feature that makes a fungus 'ideal' for mycorrhiza forming capacity nor is there any fungus that fulfills all the requisite criteria. Still for any fruitful biotechnological exploitation, it should be possible to grow the potential fungus in pure culture and manipulate it with ease for mass multiplication. Further, such a culture should be compatible with the host and it should quickly enter into mycorrhizal association. The fungal strain should also show tolerance to extremes of pH, water stress or drought, toxicity and should possess competitive, ability to overgrow other fungal species when the seedlings are outplanted.

Table 1 :   Synopsis of orders and families containing ectomycorrhizal fungi (Modified from Miller, 1982)

I	**BASIDIOMYCOTINA**

1. Agaricales
Amanitaceae
Hygrophoraceae
Tricholomataceae
Entolomataceae
Cortinariaceae
Paxillaceae
Gomphidiaceae
Boletaceae
Strobilomycetaceae

2. Russulales
Russulaceae
Elasmomycetaceae

3. Gautieriales
Gautieriaceae

4. Hymenogastrales
Octavinadaceae
Hymenogastraceae
Rhizopogonaceae
Hydnangiaceae

5. Phallales
Hysterangiaceae

6. Lycoperdales
Mesophelliaceae
Lycoperdaceae

7. Melanogastrales
Melanogastraceae
Leucogastraceae

8. Sclerodermatales
Sclerodermataceae
Astraceae

9. Aphyllophorales
Corticiaceae
Cantharellaceae
Thelephoraceae

II.   ASCOMYCOTINA

1. Eurotiales
Elaphomycetaceae

2. Pezizales
Humariaaceae
Pezizaceae

3. Tuberales
Pseudotuberaceae
Hydootryaceae
Geneaceae
Eutuberaceae
Terfeziaceae

III.   ZYGOMYCOTINA

1. Endogonales
Endogonaceae

Table 2: Mycorrhiza forming species as recorded by Bakshi (1974)

Agaricus trisulphuratus Berk.
Amanita hemibapha (Berk. & Br.)
Amanita verna (Bull. ex. Fr..) Vitt.
Amanita sp.
Astraeus hygrometricus (Pers.) Morgan
Boletus sp. 1.
Boletus sp. 2.
Boletus sp. 3.
Boletellus sp.
Calvatia sp.
Cantherellus cibarius Fr.
Cenococcum graniforme (Sow.) Ferd. & Winge
Cortinarius sp.
Gautieria sp.
Geastrum fimbriatum Fr.
Hysterangium sp. 1
Hysterangium sp. 2
Lactarius scorbiculatus (Scop. ex. Fr.) Fr.
Mycelium radicis-actrovirens Melin
Pulveroboletus shoreae Singh & Singh
Rhizopogon flavum Petch
Rhizopogon sp.

Table 3 : Mycorrhizal associates of different tree species in Himachal Pradesh as recorded by Lakhanpal (1988)

***Abies pindrow* Royle**
    Amanita pantherina (DC ex Fr.) Secr.
    Amanita vaginata (Bull. ex Fr.) Vitt.
    Citocybe gibba (Fr.) Kummer
***Betula utilis* D. Don**
    Amanita fulva (Schaeff) Pers.
    Leccinum scabrum (Fr.) S.F. Gray
    L. oxydabile Singer
***Cedrus deodara* (Roxb.) Loud.**
    Agaricus silvaticus Schaeff. ex Secr.
    Amanita emilii Riel.
    A. flavoconia Atk.
    A. gemmata (Fr.) Bert.
    A. inaurata Secr.
    A. pantherina (DC ex Fr.) Secr.
    Amanita rubescens (Fr.) S.F. Gray
    Boletus edulis Bull. ex Fr.
    Boletus sp.
    Clitocybe dealbata (Fr.) Kummer
    C.dialatata Pers. ex Karst.
    Cortinarius cinnabarinus (Fr.) Fayod
    Cystoderma amianthianum (Fr.) Fayod
    Inocybe fastigata (Schaeff. ex Fr.) Quel.
    Lepiota depeolaria (Bull. ex Fr.) Quel.
    L. cristata (Fr.) Kummer.
    Leucopaxillus gigenteus (Fr.) Singer

Macrolepiota sp.

Russula densifolia (Secr.) Gillet

Picea smithiana (Wall.) Boiss.

Hygrophorus chrysodon (Batsch ex Fr.) Fr.

H. pudorinus (Fr.) Fr.

Lactarius deliciosus (Fr.) S.F. Gray

Leucopaxillus amareus (A. & S. ex Fr.) Kuhn.

Suillus sibricus (Singer) Singer

Pinus roxburghii Sarg.

Amanita berkeleyi (Hook. F.) Bas

A.emilii Riel.

A. gemmata (Fr.) Bert.

A. vaginata (Bull. ex fr.) Vitt.

Lactarius sanguifluus (Paulet ex Fr.) Fr.

Pinus wallichiana A.B. Jackson

Clitocybe clavipes (Pers. ex Fr.) Kummer

C. gibba (Fr.) Kummer

Hygrocybe conica (Fr.) Kummer

Laccaria amethystina (Bull. ex Merat) Murrill

L. laccata (Scop. ex Fr.) Berk. & Br.

Nematoloma fasciculata

Suillus granulatus (Fr.) Kuntze

S. glandulostipes Thierse & Smith

S. placidus (Bonorden) Singer

S. umbonatus Dick & Snell

Rhododendron arboreum Smith

Hygrophorus subalpinus Smith

Russula subgalachora

Quercus incana Roxb.

Agaricus angustus Fr.

Amanita rubescens (Fr.) S.F. Gray

A. umbonata Pomerleaus

Boletus sp.

B. gertrudiae Peck

B. vermiculosoides Smith & Thiers

Collybia fusipes (Bull. ex Fr.) Quel.

Gomphus clavatus (Fr.) S.F. Gray

Hygrocybe sp.

Lactarius hygrophoroides Berk. & Curt.

L. indicus

L. piperatus (Scop.) Fr.

L. zonarius (Bull. ex St-Amans) Fr.

Leucoagaricus rubrotinctus (Peck) Singer

Phylloporus rhodoxanthus (Schw.) Bres.

Leccinum luteum Smith, Thiers and Walting

Russula brevipes Peck

R. subflaviscens

R. lilacea Quel.

R. mucronoides

R. subgalachora

Strobilomyces annulatus Corner

S. mollis Corner

Stropharia rugusoannulata Farlow ex. Murr.

Quercus semicarpifolia Smith

Lepista nuda (Bull.) Fr.

Table 4 : Mycorrhizal associaties of different tree species as recorded by Sharma & Singh, 1990

Pinus roxburghii

Scleroderma verrucosum
S. arcolatum
S. dictyosporum
Rhizopogon sp.
Suillus spp.
Astrcus hygrometricus
Amanita vaginata
Laccaria laccata
Lactarius sanguiflus
Amanita spp.
Lepista nuda
Suillus sibricus
Boletus edulis
Cantharellus cibarius
Scleroderma texense
Rhizopogon rubescence
Thelephora terrestris

Cedrus deodara

Amanita spp.
Russula spp.
Clitocybe infundibuliformis
Lactarius deliciosus zonarius
Leucopaxillus gigonteus
Strobilomyces floccopus
Boletus edulis
B. erythropus
Thelephora terrestris
Canthrellus cibarius
Laccaria laccata

Quercus spp.

Russula spp.
Agaricus angustus
Amanita nuscaria var. *fluavivolvata*
Collybia spp.
Lactarius piperatus
L. zonarius
L.deliciosus
L. camphoratus
Scleroderma dictyosporum
Stropharia rugosoannulata
Boletus edulis
Thelephora terrestris
Rameria spp.

4. ROLE OF ECTOMYCORRIZAL FUNGI IN FORESTRY PRACTICES

Among various factors responsible for arousing interest in mycorrhizal studies one is denudation of the forests and realization of the fact that mycorrhiza play specific role in the establishment of seedlings on such sites. A large number of factors are responsible for the depletion of forests. One of the prime factors has been the increase in human and cattle population and undertaking of such developmental activities such as mining, building of communication links, construction of dams, highways and bywasy, etc. The top soil at such sites gets eroded and the subsoil becomes compacted, droughtly, stony and inherently low in fertility. Such degraded sites will usually not become rehabilitated through normal successional processes during the life time of man. Rehabilitation of such sites is suggested through a number of means, e.g. from a new growth medium, from the artificial introduction of new plant communities, or by altering the conditions under which new plant communities may develop. It is ideal, if all these efforts are undertaken simultaneously in order to expedite the rehaibilitation process.

The true nature and significance of mycorrhizal association in relation to the plant health was realized and understood by the middle of the 20th century. But its importance in forest plantations received appreciation towards the first half of this century when it was observed that the forest trees often fail to establish at new sites, if the ectomycorrhizal symbiont is absent; this was first observed in experimental plantations of exotic pines in different parts of the world, where plantations consistently failed until suitable ectomycorrhizal fungi had been introduced. In Western Australia *P. radiata* and *P. pinaster* failed to establish in nursery beds that lacked mycorrhizal fungi but the seedlings grew normally after soil from healthy pine stands was added to the beds. Even addition of fertilizer had no effect on such sites and on seedlings in terms of their establishment. But when forest soil was added, normal seedlings were produced because the forest soil had the propagules of mycorrhizal fungi. The early inoculations were, therefore, by introduction of soil that contained mycorrhizal propagules. Later, inoculation with pure mycelial cultures was introduced.

The seedlings of *Pinus strobus, P. patula, P. caribeae* and *P. taeda* were noticed to be chloretic when grown in non-mycorrhizal soil but when the seedlings were inoculated with the mycelia of different fungi, they grew well and contained almost twice as much NPK as the uninoculated ones. Eighty eight per cent increase in growth of several species of *Eucalyptus* inoculated with spores of the gasteromycetous fungus, *Scleroderma flavidum* has been reported. It is concluded that forest soil inoculum is preferable for most practical purposes and that pure culture inoculation is necessary only when a desirable species of mycorrhizal fungus is lacking in the soil or the existing species is insufficiently virulent.

5. SOURCES OF INOCULUM

There are four primary sources of artificial inoculum. The most commonly used and probably the most reliable source of ectomycorrhizal inoculum is soil taken from ectomycorrhizal host and incorporated roughly 10% by volume into nursery beds prior to sowing. This method, therefore, requires relatively large amounts of soil which at times may have to be transported to long distances. There is also a possibility of the introduction of pathogens and pests. Morever, the fungal symbiont is unknown. Still this is the most reliable method eliminating mycorrhizal deficiencies and is still used by nurserymen regularly.

Another method of artificial inoculation is by planting mycorrhizal 'nurse' seedlings into nursery beds. The fungi from these seedlings serve as a source for neighbouring young seedlings. The method also suffers from the drawbacks, e.g. introduction of unwanted pests and pathogen, and that the introduced fungi are unknown. Moreover the inoculum spreads slowly and unevenly.

The third method of inoculation involves use of spores or crushed sporocarps. The method has been used successfully in small experiments but on a large scale the use of this type of inoculum is limited by the availability of a short collecting season for the collection of sporocarps and spores in quantity and because after collection, their long term storage requirements are unknown.

The fourth type of inoculum is pure culture of ectomycorrhizal fungi. There is a general consensus that the use of pure mycelial cultures of selected fungi is biologically the most sound inoculum method since harmful organisms are excluded. Some difficulty may, however, be encountered in isolating, maintaining and in the production of sufficient inoculum of some mycorrhizal fungi; still this type of inoculum production offers the greatest potential for wide scale application, for which now the techniques have been developed.

Initially the pure culture of ectomycorrhizal fungi is obtained from fruit bodiies or from the ectomycorrhiza itself. The isolation from sporocarps is, however, easiest for most fungi and permits identification of the species. Isolations from ectomycorrhizae and rhizomorphs are difficult and often can not be identified. Since ectomycorrhizal fungi are to all intents and purposes, obligately biotrophic in the natural habitat, mycelial incolculum will not be able to grow far through the soil to find and colonize a hospitable root. Therefore, an initial inoculum must be delivered in a very precise manner- it must be placed in contact with or in the path of the young roots.

6.   USE OF ECTOMYCORRHIZA IN BIOTECHNOLOGY

6.1.   Isolation of Mycobiont

The use of pure mycelial cultures of ectomycorrhizal fungi has been repeatedly recommended as the most biologically sound method of inoculation. But large scale nursery application of pure culture was not possible because of the lack of large quantities of inoculum. The inoculum of *Suillus plorans* was produced on nutrient solution in small flasks for several days. The inoculum was transferred to 10 litre tank containing the same nutrient solution and aerated for 2-3 hrs daily for three-four months. The mycelium from the 10 litre container was poured into 5-litre flask containing sterilized peat moss and fresh nutrient solution. Then within next few months the fungus mycelium colonised the whole substrate and inoculum was ready for use. Inoculum removed from the culture tanks was packed in sterile polyethylene bags, transported to nursery, and applied within 2-3 days to nursery 10 cm. deep in furrows, in soil @ 3-4 litres inoculum per m² of soil surface. Several genera of fungi e.g. *Suillus, Amanita, Paxillus, Lactarius, Tricholoma, Leucopaxillus* and *Phlemacium* were utilized for inoculating *Pinus cembra* and *P. sylvestris* and reported differences in stimulation of the growth of pine seedlings inoculated with different species of mycorrhizal fungi.

Another recommended method is the growing of specific ectomycorrhial fungi in a liquid medium until adequate mycelium is obtained, which then is transferred to 1 litre flasks containing cooked sterlized cereal grains (wheat or white millet). Addition of gypsum (0.4-0.5g/100g of grain) was recommended to improve growth of some fungi. The grains become thoroughly colonized by the fungus when shaken and incubated at 20-22°C. The colonized grains were added to peat moss enriched with ammonium tartrate, asparagine, soybean meal, blood meal, malt extract glucose, for the final production of inoculum. The sterilized enriched peat moss in 10-15 lit. quantities, was placed in large transparent plastic bags and inoculated with cereal grain culture and incubated at 20-22°C. Within 3-6 weeks inoculum is ready for nursery inoculation. This method has been used to produce inoculum of *Suillus plorans, S. grevillei, Boletinus caripes, Amanita muscaria,* and *Hebeloma crustuleniforme.*

Basidiospores or pieces of tissues from the sporophores of *Amanita verna, Suillus granulatus, S. luteus, Hebeloma crustuliniforme, Russula* sp., *Scleroderma verrucosum* and *S. vulgare,* were used to raise culture on agar medium. The mycelium is transferred to liquid culture, incubated and added either to sterlized, germinated grains of cereals such as barley, a mixture of grain and chaff, or sterilized peat moss. All substrates are enriched with a liquid medium; and inoculated either with mycelial agar discs or liquid culture mycelium.

Pure culture of *R. luteolus* has been produced in a medium of vermiculite, chaff and corn meal in a ratio of 10:2:1 moistened with a liquid medium. The fungus was placed in bottles which contained 80g of medium and incubated at 25°C for one month. Sterile peat moss and vermiculite in the ratio of 2:1, moistened with glucose, ammonium tartrate nutrient solution with final pH at 3.8 were used for the inoculum production of *Cenococcum graniforme, Corticium bicolor, Rhizopogon roseolus* and *Suillus cothurnatus*. After 16 weeks of incubation inoculum was shifted to USDA, Forest Service Research Laboratory, Beltsville, Maryland for field use.

6.2. Multiplication of Mycobiont

During the past three decades, the role of ectomycorrhizal fungal inoculations in artificial regeneration programmes has been clearly illustrated. The pure culture isolations have been tried from the fruiting bodies, surface sterilized mycorrhizal roots and sclerotia. Mycelial cultures have been derived from a number of known genera; e.g. *Amanita, Astraceus, Boletus, Cortinarius, Fuscoboletinus, Hebeloma, Hymenogaster, Leccinum, Melanogaster, Paxillus, Rhizopogon, Scleroderma, Suillus* and *Tricholoma*. However, dramatic improvements in survival and growth of pine and other seedlings have been reported in recent times with two versatile fungi, *Pisolithus tinctorius* and *Thelephora terrestris*.

Pisolithus tinctorius (Pers.) Cooker and Couch (henceforth abbreviated as *Pt*), is a gasteromycetous, puffball like fungus, which was first found to colonize anthracite wastes in Pennsylvania. This was observed to be the most predominant fungus, forming large, brilliant golden yellow mycelial strands, running as far as 15 feet, through the contrasting anthracite wastes, from the seedling to the basidiorcarp.

Eversince then fructifications of *Pt* and its ectomycorrhizae have been reported on various tree species and on numerous types of adverse sites, such as exhibiting high soil temperature, extreme acidity, droughtiness, low fertility of high levels of toxic metals. Therefore, *Pt* was the type of fungus needed for rehabilitating the degraded sites because in comparison to other fungi, it was better adapted ecologically in so diverse and adverse sites. In anthracite wastes where this fungus grew, the soil temperature was between 35°C to 65°C at a depth of 6-7 cm. At such temperatures perhaps other symbiotic fungi could not establish, whereas, *Pt* could and hence it was dominant there. The temperature tolerance experiments conducted with the fungus have lent support to this observation. In pure culture, the fungus could tolerate temperatures upto 42°C and under aseptic conditions, it formed more ectomycorrhizae with loblolly pine (*Pinus taeda*) seedlings at a constant temperatures of 34°C than at other temperatures.

Prompted by these observations, extensive surveys were carried out in the U.S. and other countries and this fungus was discovered to have a host range of over 100 species of woody plants, was represented in over 33 countries of the world and 38 states of the United States.

The procedure involved in growing mycelium of the fungus for 3-4 months at room temperature in 1.5 litre; jars containing a mixture of vermiculite and peat moss moistened with liquid medium. The substrate permeated by the fungus is removed from the jars, leached in tap water, and dried to a moisture content of the 12-20 per cent. Mycelium which develops in the laminated structure of the vermiculite particles maintains viability and is protected from environmental stress and microbial saprophytic competitions in the nursery soil. Whereas fungus grown on other substrates, such as a sand perlite, peat moss and cereal grains has little or no protection from these factors. Vermiculite peat moss incoculum of *Pisolithus tinctorius* is widely used for the nursery inoculation in the United States and has been shown to form ectomycorrhizae in fumigated nursery soil on pine, oak and peacan seedlings in microplots and on pine seedlings in conventional base-root nurseries in Georgia, Florida and North Carolina, Mississippi and Missouri. Another equally important fungus which occurs naturally in the forest nurseries throughout the world is *Thelephora terrestris* (henceforth abbreviated at *Tt*), has not been tried for its mycorrhizal potential. It has been reported to be unable to function of harsh sites, but it is ecologically well adapted to the fertile and moist conditions of nursery soil.

Extensive studies have been carried out by USDA forest service and Institute for Mycorrhizal Research and Development (IMRD) on inoculum production and its use has been evaluated in microplots and green house conditions on various pine species in Brazil, South Korea and France. Inoculum formulations and inoculation techniques have been developed for use in seedling production system. The vermiculite peat moss inoculum of *Pt* has been used in fumigated nursery soils to form ectomycorrhizae on species of *Pinus, Quercus, Carya, Pseudoptsuga, Picea* and *Abies*, seedlings. The *Pt* inoculated seedlings were observed to be almost always larger than the control seedlings. The fumigation of the nursery soil prior to inoculation, reduces populations of soil microorganisms that otherwise usually colonise introduced inocula, it also suppresses feeder root pathogens that damage roots and thus reduce ectomycorrhizal development and it also checks indigenous competing ectomycorrhizal fungi such as *Tt*. The vermiculite based inoculum broadcast at a rate of about $11/m^2$ of soil surface and mixed in soil has been as effective at higher rates in forming abundant *Pt* ectomycorrhizae. The effectiveness of the inocula is influenced by fertility, type of container, growing medium, fungicide, inoculum storage and frequency of watering.

From small scale production of *Pt* inoculum, commercial production for large scale forest regeneration programmes was initiated in 1976 jointly by IMRD and States and Private Forestry, USDA Forest Service and Abbott Laboratories, North Chicago, Illinois and a comprehensive report was published on the testing of various formulations for ectomycorrhizal development of container grown tree seedlings of various species. They jointly developed a commercial formulation of *Pt* mycelial inoculum called 'Mycorrhiz', which weighed about 250-300 gl^{-s}, was packed in 50 l units, stores at 5°C and had a shelf life of 5-6 weeks.

The Abbott Laboratories stopped commercial production of *Pt* inoculum in September 1983, due to quality control problems and IMRD and State and Private Forestry have initiated expanded programme of research and nursery evaluation of various types of vegetative inocula of ectomycorrhizal fungi produced by Sylvan Spawn Laboratory, Pennsylvania. In these studies better formulations have been evolved for the production of vegetative inoculum of *Pt* which do not require leaching and drying before use. This way around 5 million pine seedlings were being produced in the eastern United States.

6.3. Ectomycorrhiza and Forest Plantations

In North America alone nearly 2 billion tree seedlings are grown annually in nurseries for artifical regeneration programmes. They recorded dramatic improvements in survival and growth of various pine species with Pt ectomycorrhizae on acid coal spoils in Appalachia, Kaolin spoils in Georgia, severely eroded sites in the copper basin of Tennessee, Borrow pits in south and north Carolina and a Prioirie soil in South Dakota. The tree volume was seen to increase upto four fold even on such hostile sites, improvements in seedlings performance were also reported in routine reforestation sites in Florida, Georgia, North Carolina, Oklahoma, Arkansas and Mississippi. In parts of Asia, Africa and South America tropical pine seedlings with *Pt* ectomycorrhizae also outperformed seedlings with ectomycorrhizae from soil inoculum.

Bare root *Pinus taeda* seedlings with abundant *Pt* ectomycorrhizae had a 400% greater plot volume index than seedlings with *Tt* ectomycorhizae after three years on an acid coal spoil (pH 4.5) in Kentucky. On plots where fertilizer starter tablets were used on all seedlings, volume for *Pt* seedlings were nearly 250% greater in the same test, short leaf pine seedlings with *Pt* ectomycorrhizae were over 400% larger without fertilizer tablets and over 100% larger with fertilizer tablets than *Tt* seedlings. Seedlings with *Pt* ectomycorrhizae also contained more foliar N and less foliar S, Fe, Mn, and Al than seedlings with *Tt* ectomycorrhizae. The fertilizer starter tablets stimulated seedlings growth for the first and second growing seasons but once the nutrients in the tablets were depleted growth increments slowed and N deficiency symptoms appeared on the foliage. This efficiency was less striking on *Pt* seedlings than on others. The same report, plot volume for *P. taeda* with *Pt* ectomycorrhizae, after four years on an acid coal spoil (pH 3.4) in Virginia, were over 180% more than those for seedlings with *Tt* ectomycorrhiza.

It is clear, therefore, that seedlings tailored with *Pt* ectomycorrhizae survive and grow faster on adverse sites than routine nursery seedlings with naturally occurring ectomycorrhizae. Such seedlings have been used successfully in redressing such disturbed and eroded sites as surface mines, strip mines, and borrow pits. Even in the forestation of undisturbed sites which might have naturally occurring ectomycorrhizal fungi, it has been emphasized that there is need for parallel introduction of ectomycorrhizal fungi because it is not necessary that naturally occurring fungal propagules colonise the seedlings immediately. The tree seedlings have been noticed to fail to survive on transplanting if they do not have an adequate complement of mycorrhizae from the nursery. Further, the containerized seedlings are better than bare root seedlings because in the latter, roots get damaged while transplanting. Therefore, what emerges from these studies is that any ectomycorrhizae on the roots of planting stock are better than non-mycorrhizae, and that ectomycorrhizae formed by certain species of fungi are more beneficial to seedlings on certain sites than ectomycorrhizae formed by other fungi. The inoculated seedlings exhibited better shoot/root ratio, stem diameter, shoot/root fresh weight ratios. In this way in the inoculated seedlings, the transplanting period was reduced by almost a year. The studies which were conducted under glass house conditions, are now being extended to the field, which if equally successful, will involve quite a saving in terms of time, money and energy (Fig. 3).

7. FUTURE PROSPECTS

The past few decades have witnessed a tremendous increase in food production due to the development of several high yielding varieties. As a consequence, the use of fertilizers has also increased manyfold because most of these cultivars requires high doses of fertilizer for giving higher yields. But currently the cost of fertilizer has become prohibitive and moreover, they are also in short supply, especially the phosphorus fertilizer. Therefore, it is imperative that some alternative is found by which the present day use of fertilizers is minimized. Mycorrhiza which has been termed as "biotic fertilizer", is perhaps one such choice which may provide an alternative to high phosphate application (Johnson and Menge, 1981).

The most important envisaged role of mycorrhiza is in the reclamation and reforestation programmes. In such programmes it needs to be realized that these programmes involve forest trees which unlike agricultural crops need to be monitored for atleast 5-10 years under field conditions for correct assessment of their survival and growth. During this time period, they will automatically be exposed to the stress and strain of the environment, which could consequently determine their growth and survivability. Various silvicultural aspects must also be considered in such programme, e.g. intensity of site preparation, soil quality, their genetics, fertilization, nutrient cycling, soils, diseases and insect pests, competing vegetation and allelopathy and tree spacing.

It is only *Pisolithus tinctorius* which has so far achieved commercially exploitable status. No other fungus has been so extensively investigated and field tested for possible benefits to artificial regeneration programmes. There is need to investigate many of the other known mycorrhizal fungi on similar lines for attaining the same biotechnological developments as *Pisolithus*. If some of these happen to be edible species, so much the better, as such species could be used to establish "Edible fungus orchards" the same way as has been possible with truffles as "Truffle Orchards" in France.

At present little is known about the basic genetics of ectomycorrhizal fungi; as such they afford little promise in manipulation to produce superior strains useful in agriculture and forestry. However, with a better understanding of their basic biology, these microorganisms can be utilized as they now exist, in biotechnology related to agriculture and forestry (Peterson *et al.,* 1984).

Finally in words of Hacskaylo (1983), "To derive the potentially vast benefits from mycorrhizal associations in field applications and to achieve genetic alternations so as to improve physiological benefits, fundamental characterstics and biochemical mechanisms need research to much higher level."

Fig. 3 : a. Six months old *Pinus gerardiana* Wall., inoculated (Mycorrhizal=m) and uninoculated (non-mycorrhizal=nm) seedlings.

        b. Two and half years old *Picea smithiana* (Wall.) Boiss., inoculated (mycorrhizal=m) and uninoculated (non-mycorrhizal=nm) seedlings.

        c. Six months old and

        d. Eight months old *Pinus geradiana* Wall., inoculated (I) and uninoculated (U) seedlings.

        e. Two and half years old and

        f. Six months old *Picea smithiana* (Wall.) Boiss., inoculated (I) and uninoculated (U) seedlings.

REFERENCES

Foster, R.C. and Marks, G.C. 1966, The fine structure of the mycorrhizas of *Pinus radiata*, D.Don. *Aust. J. Bio. Sci.* 18: 1027-1038.

Gerdemann, J.W. 1968, Vesicular-arbuscular mycorrhiza and plant growth, *Annu. Rev. Phytopathol.* 6: 397-418.

Hacskaylo, E. 1953, Pure culture synthesis of pine mycorrhizae in terra-lite, *Mycologia.* 45:971-975.

Hacskaylo, E. 1983, Researching the potential of forest tree mycorrhizae, in: *Tree Root System and their Mycorrhizas*, D. Atkonson, K.K.S., Bhat, P.A., Kason, M.P. Coutts and D.J. Read, eds., Martinus Nijhoff and W. Junk Publishers, London, pp.1-8.

Harley, J.L. 1986, Mycorrhizal studies: Past and Future, in: *Physiological and Genetical Aspects of Mycorrhizae*, P.V. Ginanazzi, and S. Gianinazzi, eds., INRA, Paris, pp. 25-33.

Hayley, J.L. and Smith, S.E. 1983, *Mycorrhizal Symbiosis*, Academic Press, London.

Kendrick, B. and Berch, S. 1985, Mycorrhizae: Applications in Agriculture and Forestry, in: *Comprehensive Biotechnology* Vol. 4. C.W. Robinson and J.A. Howell, eds., Pergamon Press, pp. 109-151.

Kottke, I. and Obserwinkler, F. 1986a, Mycorrhiza of forest tree-structure and function, *Trees* pp. 1-24.

Malloch, D.W., Pirozynzki, K.A. and Raven, P.H. 1980, Mycorrhizal symbiosis in vascular plants (A Review), *Proc. Natn. Acad. Sci*, USA. 77: 2213-2218.

Marx, D.H., Cordell, C.E., Kenney, D.S., Mexal, J.G., Artman, J.D., Riffle, J.W. and Molina, J.R. 1982, Commercial vegetative inoculum of *Pisolithus tinctorius* and inoculation techniques for development of ectomycorrhizae on bare root tree seedlings, *For. Sci.* 28: 1-101.

Marx, D.H. and Schenck, N.C. 1983, Potential of mycorrhizal symbiosis in agriculture and forest productivity, in: *Challenging Problems in Plant Health*, T. Kommendahl and P.H. Williams, eds., 75th Anniv. Publ. of Am. Phytopathol. Soc. pp 334-347.

Marx, D.H. and Ruehle, J.L. 1989, Ectomycorrhizae as biological tools in reclamation and revegetation of waste lands, in: *Mycorrhizae for Green Asia: Proc. Ist Asian Conf. on Mycorrhizae*, A. Mahadevan, N. Raman, and K. Natrajan, eds., Madras, pp 336-343.

Mason, P.A., Wilson, J., Last, F.T. and Walker, C. 1983, In: *Trees Root Systems and their Mycorrhizas*, D. Atkinson, K.K.S. Bhat, M.P. Coutts, P.A. Mason and D.J. Read., eds., Martinus Nighoff and W. Junk Publishers, London.

Mikola, P. 1973, Application of mycorrhizal symbiosis in forestry practices, in: *Ectomycorrhizae: Their Ecology and Physiology*, G.C. Marks and T.T. Korlowski, eds., Academic Press, New York, pp. 383-411.

Mukerji, K.G. and Mandeep 1998, Mycorrhizal relationship of Wetlands and Rivers associated plants, in : *Ecology of Wetland and Associated Systems*, S.K. Mazumdar, E.W. Miller, and F.J. Brenner, eds., The Penn. Acad. Sci., USA., pp. 240-257.

Peterson, R.L., Piche, Y. and Plenchette, C. 1984, Mycorrhizae and their potential use in the agricultural and forestry industries, *Biotech. Advs.* 2: 101-120.

Suvercha, Mukerji, K.G. and Arora, D.K. 1991, Ectomycorrhiza, in : *Handbook of Applied Mycology*, Vol. I, Plant and Soil, D.K. Arora, B. Rai, K.G. Mukerji and G.R. Knudsen, eds., Marcel Dekker Inc., USA, pp. 187-217.

Trappe, J.M. 1977, Selection of fungi for ectomycorrhizal inoculation in nurseries, *A. Rev. Phytopathol.* 15: 203-322.

Zak, B. 1971, Characterization and classification of Douglas fir mycorrhizae, in: *Mycorrhiae: Proc. I. Nacom*, E. Hacskaylo, ed., USDA For. Ser. Misc. Publ. 1189, US Government Printing Office, Washington, DC, pp. 38-53.

Zak, B. 1973, Classification of Ectomycorrhizae, in: *Ectomycorrhizae: Their Ecology and Physiology*, G.C. Marks and T.T. Kozlowski, eds., Academic press, New York and London, pp. 43-78.

MOLECULAR GENETICS OF ECTOMYCORRHIZAL FUNGI

Vandana Gupta and T. Satyanarayana

Department of Microbiology
South Campus, University of Delhi
New Delhi-110021, INDIA

1.   INTRODUCTION

The mycorrhizal associations have evolved as a survival mechanism for both fungi and plants. These associations are as old as land plants. Ectomycorrhizal plants have evolved along with many of their ectomycorrhizal fungal partners (Miller and Walting, 1987; Nicolson, 1975). It is very difficult to find plants without mycorrhiza in natural ecosystems and in disturbed habitats, and forest plantations, agricultural and horticultural crops growing at low phosphate nutrition (Harley and Smith, 1983). From genetics point of view, mycorrhizal fungi are very different than other biotrophic fungi in that most species of ectomycorrhizal fungi have evolved to have a wide host range in contrast to the narrow range for pathogenic biotrophic fungi.

Detailed knowledge about genetics of ectomycorrhizal fungi in the past was scanty due to the lack of readily manipulatable mononucleate life cycle stages, difficulty in growing them in axenic culture and lack of sexual reproduction in axenic culture. Since ectomycorrhizal fungi are known to aid in the survival and establishment of plants on waste and degraded lands, the development of new superior strains to increase forest productivity and profitability is required. For this purpose both classical and molecular genetic approaches are being developed. Key questions relating to the current and the future management of EM fungi are: (i) How widely has the existing genetic pool been examined? (ii) Is genetic variability within populations likely to be large? (iii) Can fungi with desirable characteristics relating to formation and functions be bred or engineered? and (iv) Fungi that do not form mycorrhiza be genetically manipulated to become mycorrhizal.

Identification of ectomycorrhizal fungi in the field is difficult because of the lack of asexual spores, and presence of a confusing range of colour and structure. Therefore, there is an emphasis on the molecular taxonomy of these fungi. In this chapter, an attempt has been made to review various molecular genetic aspects of ectomycorrhizal fungi.

2.   MOLECULAR METHODS IN TAXONOMY OF MYCORRHIZAL FUNGI

Molecular techniques have been used in taxonomic studies of fungi. Two major applications of molecular methods are systematic and the elucidation of evolutionary relationships and the delineation of species concept. Molecular methods employed in the classification of fungi include polypeptide analysis, isozyme analysis, immunochemical methods and DNA analysis. Systematics and methods of characterizing and classifying mycorrhizas are needed because it is always very difficult to define and identify each mycorrhiza as a function of both a known tree species and a known fungus. The different kinds of mycorrhizas are a confusion of colours, sizes and shapes (Zhu *et al.*, 1988). In many cases morpho-anatomical features of ectomycorrhizas are species specific as in the case of *Piloderma croceum* (Hang and Oberwinkler, 1987; Brand, 1991) or *Russula ochroleuca* (Agerer, 1996). In other groups (e.g. *Lactarius* spp.) such features only allow characterization of morphotypes, which do not necessarily represent single species (Kraigher *et al.*, 1995). A classical approach for identifying mycorrhizas is, therefore, to trace hyphal connections between sporocarps and mycorrhizal sheaths, which allow identification of many ectomycorrhizas (Agerer, 1995). This means of identification is restricted to direct sampling of mycorrhizas and sporocarps, and it is difficult or impossible when root density is so high that hyphae emerging from the stipe base cannot be attributed unambiguously to a single mycorrhizal type (Pritsch, 1996).

2.1.   Polypeptide Analysis

A few attempts have been made to identify ectomycorrhizal fungi by the protein pattern of different species (Burgess *et al.*, 1995b; Mouches *et al.*, 1981). Burgess *et al.* (1995b) compared and classified 85 Australian and 15 non-Australian *Pisolithus* isolates according to basidiospore and basidiome morphology, culture characteristics and separation of polypeptides using one dimensional SDS-PAGE. They found that one cluster was composed of all the non-Australian isolates collected beneath *Pinus*, whilst within Australia, isolates from the eastern, southern and western seaboards fell into distinct clusters. The analysis of polypeptide pattern could provide a meaningful classification system to assist in isolate selection for further studies.

2.2.   Isozyme Analysis

Isozyme analysis is now used by mycologists to resolve taxonomic disputes, identify unknown fungal taxa, 'fingerprinting' patentable lines, analyze the amount of genetic variability in population, trace the origin of pathogens, follow the segregation of loci and identify ploidy levels throughout the life cycle of an organism (Burdon and Marshall, 1983; Micales *et al.*, 1986). The technique is based on analyzing the variation in isozyme electrophoretic mobilities encoded by different alleles or separate genetic loci. Such variations are the result of variations in the amino acid contents of the molecule, which is dependent on the sequence of nucleotides in the DNA.

Cameleyre and Olivier (1983) studied variability between twelve isolates of the EM fungus *Tuber melanosporum* using isoelectric focussing of four enzymes including acid and alkaline phosphatases, phosphoglucomutases and esterases. Authors observed intraspecific variations that resulted in the separation of isolates into several groups. Zhu *et al.* (1988) studied the genetic variability of isozymes of eight different enzymes of various strains of *Suillus tomentosus* showing geographical clustering of isolates. Ho and Trappe (1987) have shown that the isolates representing six species and three host sections of *Rhizopogon* had acid phosphatase isozyme mobilities that differed greatly in different species but were more conserved with in a species. The isolates of *L. laccata* were divided into three host related groups based on similar analysis (Ho, 1987). While Sen and Hepper (1986) used selective enzyme staining following polyacrylamide gel electrophoresis to characterize VAM fungi like *Glomus*, and showed that the isozymic analysis allowed at least the characterization of species. Hepper *et al.* (1986) could identify the mycelium of the VAM fungi in roots of leak and maize on the basis of isozyme analysis. Hepper *et al.* (1988) also showed that electrophoretic analysis of isozymes could differentiate isolates of *Glomus clarum, G. monosporum* and *G. mosseae* into several clusters. Isozyme and somatic incompatibility analysis of 43 isolates of *Suillus collonitus* indicated high intraspecific variation between isolates from the same location and from different locations, and showed a high correlation between the results of electrophoretic isozyme analysis and those of somatic incompatibility reactions (Karkauri *et al.,* 1996). Host genotype and fungal diversity was characterized in Scots pine ectomycorrhiza from natural humus microcosms using isozyme analysis (Timonen *et al.,* 1997)

2.3.   Immunochemical Approach

Serological techniques have been used to characterize various fungi including ectomycorrhizal fungi. Cleyet-Marel *et al.* (1989) purified an acid phosphatase from *Pisolithus tinctorius* and used it to prepare polyclonal antibodies. These antibodies were applied in homologous and heterologous tests on the mycelium and mycorrhizas of several fungi. Immunochemical approach appeared to help to characterize and detect fungi in case a particular protein is used as a marker.

Immunochemical methods facilitate the study of difficult systematics of the arbuscular mycorrhiza (AM) forming fungi belonging to the order Glomales. Though the use of polyclonal antibodies rarely reach below the generic level, monoclonal antibodies (mAb) can differentiate AM fungal spores to the species and strain level. But these serological techniques need to be combined with an improved enrichment procedure for spores and hyphae of AM fungi, so that the risk of unspecific binding is reduced. The production of specific antibodies assumes that fungal species or isolates contain different antigens or groups of antigens, which can be extracted. Polyclonal antisera varied greatly in specificity in discriminating between VAM and non-VAM fungi (Wilson *et al.,* 1993), and also in distinguishing between VAM fungi at the generic level (Adwell *et al.,* 1985; Kough *et al.,* 1983). Wright *et al.* (1987) successfully attempted to raise mAb against soluble antigens from spores of *Glomus occultum*. These mAbs had a high specificity for their homologous antigens and gave significantly weaker signals with 29 other specific mycorrhizal and 5 non-mycorrhizal fungal isolates when tested in an ELISA. These mAbs were even able to differentiate their antigens from isolates of the same species from different location (Ormas-Shirey and Morton, 1990). Sanders *et al.* (1992) raised polyclonal antisera to soluble spore fractions of VAM fungi, *Gigaspora margarita* and *Acaulospora laevis*. They found a major cross reactivity between *A. laevis* and *G. margarita* and also to some extent with the heterologous antigens from spores of the other VAM fungi tested in a dot immunobinding assay. While an indirect ELISA gave more specific reactions. Fries and Allen (1991) have shown by using polyclonal Abs coupled to a fluorescent strain that inoculated AM fungi can survive in the field for upto two years, even in competition with indigenous AM fungi.

2.4.   Analysis of DNA

DNA analysis methods employed in the classification of fungi include G+C molar ratio, DNA complementarity, ribosomal RNA sequence comparison and sequence or structural polymorphism in DNA.

The G+C molar ratio is one of the simplest molecular methods, but does not give good resolution to many taxonomic problems because of a large overlap among unrelated species. For the determination of G+C value either thermal denaturation or cesium chloride density gradient equilibrium centrifugation can be used. It has been reported that the G+C contents of Ascomycetes and Deuteromycetes were close to 50%, whereas G+C contents of Basidiomycetes were consistently greater than 50%.

The extent of hybridization of denatured DNA from two isolates is a measure of the relatedness of the two genomes. Highly purified nuclear DNA consisting of non-repeated sequences gives more accurate results because the repeated sequences from an isolate can re-associate with each other giving obscure results.

Ribosomal RNA coding DNA sequences are highly repetitive in addition to being highly conserved, which makes them suitable for taxonomic and phylogenetic comparisons above the species level. Initially researchers compared the molecular weights of RNA subunits as taxonomic criteria (Lovett and Hasselby, 1971). Gradually ribosomal RNA-DNA complimantarity analysis was used (Bicknell and Dauglas, 1969). As the era of DNA sequencing began, workers have started comparing the actual rRNA sequences of the smaller subunits like 5S rRNA (Hasegawa *et al.*, 1985; Hori and Osagawa, 1987). Since RNA sequence determines the secondary structure of the molecule, Blanz and Gottschalk (1986) used the secondary structure to compare the taxa.

Molecular biology finds wide applications in the identification of genetic variability in the mycorrhizal fungal genome by DNA fingerprinting. This technique has the potential to identify species and isolates by observing DNA polymorphism resulting from mutations and chromosomal rearrangements. Restriction Fragment Length Polymor-phism (RFLP) analysis which detects length mutations and alterations in base sequences has been used for constructing DNA finger prints of *L. bicolor* and *L. laccata* (Martin *et al.*, 1991), *Hebeloma* (Marmeisse *et al.*, 1992a), *Cenococcum geophilum* (Lemke *et al.*, 1991; Lo Buglio *et al.*, 1991) and *Tuber* species (Henrion *et al.*, 1994a), and to distinguish EM fungi (Armstrong *et al.*, 1989). The PCR amplification of targeted genomic sequences from microorganisms followed by RFLP, allele specific hybridization, direct sequencing or single strand conformation polymorphism is increasingly used to detect and characterize rhizospheric microbes, including mycorrhizal fungi in natural ecosystems (Bruns *et al.*, 1991; Gardes *et al.*, 1991; Henrion *et al.*, 1992; Simon *et al.*, 1992a,b; Pritsch *et al.*, 1997). The PCR primers have been designed based on highly conserved regions of rDNA to amply the internal transcribed spacers (ITS) and the intergenic spacers (IGS) which are highly polymorphic non-coding regions providing a useful tool for taxonomic and phylogenetic studies (Gardes and Bruns, 1993; Pritsch *et al.*, 1997). The technique involves enzymatic amplification of DNA fragments spanning between terminal sequences recognized by specific oligonucleotide primers. Amplified products are then digested by restriction endonucleases giving rise to DNA fingerprints. PCR probes with high specificity and high sensitivity has been developed based on DNA sequences with low degree of conservation and which are highly repetitive. Henrion *et al.* (1994a) studied the variation within ITS and IGS of the ribosomal RNA genes of European species of *Tuber* by PCR coupled RFLP and concluded that PCR-RFLP analysis of rDNA spacer provide an efficient alternative for typing pure cultures and the fruit bodies, and a versatile tool for strain fingerprinting of ectomycorrhizas in ecosystems. The PCR-RFLP analysis suggested that the genetic variation occurred in the targeted rRNA genes within local geographical population of *Laccaria* spp. as well as among the isolates from a world wide distribution (Gardes *et al.*, 1991; Henrion *et al.*, 1992). Pritsch *et al.* (1997) compared restriction patterns made from sporocarps of 28 mycobionts using PCR/RFLP analysis of ITS regions. PCR/RFLP analysis were largely consistent with the morphotyping and complementarity of both methods, as earlier reported by Pritsch and Buscot (1996) and confirmed by Pritsch *et al.* (1997). Timonen *et al.* (1997) also analysed the PCR/RFLP of ITS from fungal RNA of Scotspine ectomycorrhiza from natural humus microcosms. Kreuzinger *et al.* (1996) amplified gene encoding glyceraldehyde-3-phosphate dehydrogenase for RFLP analysis for identification of ectomycorrhizal fungi. Gardes *et al.* (1991) used ITS of the nuclear ribosomal repeat unit and a portion of the mitochondrial large subunit rDNA as target sequences because of their high copy number and specificity of the primers used for fungal partner. The nucleotide sequence, their structure and evolution are relatively well known. Variation in the nucleotide sequences was 32% between *L. bicolor* and *T. terrestris*, 3 to 5% among three *Laccaria* spp. and 1 to 2% within *L. bicolor*. Karen *et al.* (1997) amplified ITS region using universal primers ITS1 and ITS4 of 44 species in 17 genera and subjected to RFLP analysis. Intraspecific polymorphism in the ITS region were found in seven species. A high interspecific polymorphism was observed. Erland (1995) used a similar method. Gardes *et al.* (1990) examined RFLPs in the nuclear rDNA of 29 isolates of *L. bicolor*, 8 of *L. laccata*, 3 of *L. proxima* and 2 of *L. amethystina*. Restriction endonucleases used were BamHI, Bgl II, EcoR1, Hind III and Pst 1, and the probe used was cloned rDNA from *Armillaria ostoyae*. They clearly showed that the four *Laccaria* spp. as well as four strains of *L. laccata* were distinguishable. The isolates of *L. bicolor* from a particular geographical area form relatively homogenous group distinguishable from the isolates of another geographical area. Henrion *et al.* (1994b) analyzed RFLPs of specifically amplified 17 S and 25 S nuclear rDNA from 26 isolates of four species of *Laccaria* (*L. laccata, L. bicolor, L. proxima, L. tortilis*). Analysis revealed interspecific and intraspecific polymorphism. The degree of variation observed was sufficient to discriminate several isolates of the same species. LoBuglio *et al.* (1991) studied the rDNA variability among 71 *Cenococcum geophilum* isolates of both geographically distinct and similar origins using RFLP analysis. They reported 32 unique phenotypes and grouped *C. geophilum* isolates into a broad range of clusters with similarity ranging from 100 to 44%. This wide variation in rDNA indicates that *C. geophilum* is either an extremely heterogenous species or a fungal complex representing a broader taxonomic rank than presently considered. The RFLP analysis of ITS region of rDNA of *Tylospora fibrillosa* suggested low genetic variability in this species and that RFLPs are not useful in discriminating the isolates of this species (Erland *et al.*, 1994). Similar techniques of PCR- RFLP analysis of ITS regions have been used successfully by Sweeney *et al.* (1996) for *L. proxima*, Gandeboeuf *et al.* (1997a) for economically important *Tuber melanosporum*, Erland (1995) for measuring abundance of *Tylospora fibrillosa* ectomycorrhizas in a South Swedish spruce forest, and by Anderson *et al.* (1998) to determine genetic variation in *Pisolithus* isolates from a defined region in New South Wales, Australia.

The use of different target sites such as mitochondrial DNA and tRNA genes and different endonucleases combinations would extend the scored RFLPs. The small size of the mtDNA and its mode of inheritance independent of the nuclear genome make this molecule attractive for studies of fungal biology, taxonomy and evolution (Taylor, 1995). The RFLP analysis of mtDNA has been used to estimate the relationships among and within the species of many groups of fungi including Basidiomycetes, Ascomycetes and Oomycetes (Gardes *et al.*, 1991). Mitochondrial DNA analysis was found to be useful for isolate identification. The RFLP analysis of mtDNA from 25 *L. bicolor*, 8 *L. laccata*, 3 *L. proxima* and 2 *L. amethystina* suggested that most of the *Laccaria* isolates have unique overall mitochondrial pattern. This variability could be used for isolate typing. In an attempt to develop DNA probes for identification of *Hebeloma* species and strains, Marmeisse *et al.* (1992a) used randomly cloned genomic sequences of *H. cylindrosporum* as hybridization probes. They found 2 sequences of *H. cylindrosporum* as hybridization probes, which gave unique RFLP patterns in all species tested (species specific). The third sequence, however, hybridized only with *H. cylindrosporum* DNA, and it was strain specific. The intraspecific genetic variability of *Tuber aestivum* was studied using PCR-RFLP analysis to examine the variation of the nuclear and mitochondrial ribosomal DNA (rDNA). The RFLPs were found in the nuclear

ITS and three alleles were detected in the six populations analyzed, and no variability was found in mitochondrial rDNA (Guillemaud *et al.*, 1996). Longato and Bonfante (1997) screened the genome of ecto- and endomycorrhizal fungi by using primers designed on microsatellite sequences (CT)8, (CA) 8, (GACA) 4, (TGTC) 4, (GTG) 5, with the aim (i) to establish the presence of these sequences in fungal genome, (ii) to assess their value in discriminating fungal symbionts belonging to closely related species or evolutionarily separate species, and (iii) to compare their efficiency with ITS and RAPD patterns. They demonstrated that microsatellite primers are reliable, sensitive and technically simple tools for assaying genetic variability in mycorrhizal fungi, and can be used to discriminate mycorrhizal symbionts with different taxonomic features. The sequence (GTC) 5 led to species- specific fingerprints in both truffles, which are closely related species in evolutionary terms. Interspecific variability in strains of the ectomycorrhizal fungus *Suillus* by PCR-RFLP of ITS region and intraspecific variability was revealed by RAPD amplification and using primers for microsatellites (Bonfante *et al.*, 1997). According to Tigano-Milani (1995), tRNA fingerprinting provides a valuable tool to develop molecular taxonomy for *Paecilomyces lilacinum*. Such an approach could be useful in the molecular taxonomy of ectomycorrhizal fungi. Henrion *et al.* (1994b) made an attempt to monitor the dissemination and persistence of *L. bicolor* introduced in nursery grown Dauglas fir by PCR of the rDNA intergenic spacer. Results indicated that test trees remained exclusively colonized by the inoculated isolates suggesting that the inoculated *L. bicolor* is more competent than the indeginous strains. This provides further illustration of the potential of exotic species for large scale application.

Randomly Amplified Polymorphic DNA (RAPD) is an alternative way of obtaining polymorphisms based on PCR (Tommerup, 1992; Tommerup *et al.*, 1992). Short oligonucleotide primers of arbitratry sequences anneal to complementary sequences occurring randomly in the genome. Amplified templates are separated by electrophoresis to reveal DNA fingerprints. This technique is very rapid and does not require target sequence information and can provide markers in genomic regions inaccessible to PCR/RFLP analysis. A successful distinction of isolates of *Laccaria* and *Hydnangium* was achieved using RAPD technique by Tommerup (1992) and Tommerup *et al.* (1992). Wyss and Bonfante (1992) have also used the same technique to distinguish the isolates of *Tuber*. Lanfranco *et al.* (1993) used RAPD fingerprints to reveal a high degree of interspecific variability. They also developed DNA probes for the identification of *Tuber* spp. using a 1.5 Kb fragment that consistently appeared when DNA was amplified with a specific primer (OPA 18). RAPD analysis has been used by Tommerup *et al.* (1995) to differentiate three basidiomycetous fungi, *Laccaria, Hydnangium* and *Rhizoctonia*, Anderson *et al.* (1998) to screen *Pisolithus* isolates for variation, Doudrick *et al.* (1995) for genetic analysis of homokaryons from a basidiome of *Laccaria bicolor* and by Gandeboeuf *et al.* (1997b) for grouping and identification of *Tuber* species. Tommerup *et al.* (1995), have reported that RAPD fingerprinting of *Laccaria, Hydnangium* and *Rhizoctonia* was reproducible in any laboratory provided the same set of reaction and thermocycle conditions are used. Although more sophisticated molecular techniques for identification of mycobionts will be available in future, the cited work suggests that we already have a tool that enables us to make more thorough analysis of the community structure of EM fungi than ever before (Karen *et al.*, 1997). Identification of ectomycorrhizal fungi using molecular techniques has been extensively reviewed by Gupta and Satyanarayana (1996).

Using an altogether different approach, Bruns and Gardes (1993) designed five probes for the identification of suilloid fungi by comparing partial sequence from mitochondrial large subunit rRNA gene (mt-Lr RNA). Out of five probes used, three (SI, RI and GI) were targeted at the genera *Suillus, Rhizopogon* and *Gomphidius*. Probe G2 was designated to recognize all these taxa, and any other member of suilloid group. Mehmann *et al.* (1994) studied the DNA sequence of single copy genes coding for chitin synthases. The presence of introns at conserved positions has a potential use in the identification of genera by analyzing

PCR-generated DNA fragment pattern using degenerate primers from short completely conserved amino acid stretches. Data confirmed the current taxonomic groupings.

Since endomycorrhizal fungi are unable to grow in pure culture and the spores that can be identified are produced in the soil outside the colonized roots, the identity of the endophyte inside the field collected colonized roots is difficult to determine with any degree of certainty. The development of DNA based methods could allow the identification of endomycorrhizal fungi in a sample and provide a useful tool to complement the expertise of fungal taxonomists. Recently, sequences of the gene coding for the small subunit rRNA (SSU) were obtained from 12 endomycorrhizal fungal species, and the comparisons with other 185 sequences showed that the Glomales formed a phylogenetically coherent group that originated about 400 million years ago (Bruns *et al.*, 1991; Simon *et al.*, 1993a). Twelve nuclear genes coding for SSU were sequenced from different species of VAM fungi (Simon *et al.*, 1992a,b; 1993b). These were aligned and the regions showing potential for the design of taxon specific probes or primers were identified by Simon *et al.* (1993b), and four taxon specific primers (VAACAU, VAGIGA, VAGLO and VALETC) were designed and synthesized. Simon *et al.* (1992a,b) used these in conjunction with Glomales specific VANS 1 primer described. These taxon specific primers were used to amplify portions of the nuclear gene coding for the small subunit rRNA. By coupling the sensitivity of the PCR and the specificity afforded by taxon-specific primer, a variety of samples can be analyzed including small amounts of colonized roots. The amplified products are then subjected to single-strand conformation polymorphism analysis to detect sequence differences. Advantage of this approach is the possibility of directly identifying the fungi inside field collected roots, without having to rely on the fortuitous presence of spores (Simon *et al.*, 1993b). Simon *et al.* (1992b) devised a rapid quantitation method of endomycorrhizal fungus *Glomus vasiculiferum* colonizing leak roots (*Allium porum*).

3.   GENETIC TRANSFORMATION OF MYCORRHIZAL FUNGI

In order to successfully transform any organism, three requirements must be met. A means to introduce the vector into the organism (transformation), a marker for the selection of transformants and a means to allow the DNA to replicate. In fungal systems, the protoplasts can be transformed by using a combination of polyethylene glycol (PEG) and Ca^{2+} ions. Polyethylene glycol and Ca^{2+} ions help neutralizing negative charge on DNA and plasmalemma, hence reducing the electrostatic repulsion resulting in protoplast aggregation, localized membrane fusion and DNA uptake. This technique has been successfully used for a number of Ascomycetous and Basidiomycetous fungi. *Aspergillus* spp. have been transformed by incubation of protoplasts with DNA containing genes for utilization of novel nitrogen sources, for example acetamidase (Tilburn *et al.*, 1983), genes for amino acid synthesis such as tryptophan synthase (Yelton *et al.*, 1984) and genes for drug resistance like hygromycin phosphotransferase. A number of Basidiomycetous fungi including *Coprinus cinereus* and *Schizophyllum commune* were successfully transformed to prototrophy by complementation of auxotrophic mutations such as trp C genes and regulatory signals (Binninger *et al.*, 1986; Monroz-Rivas *et al.*, 1986). An adenine auxotroph of *Phanerochaete chrysosporium* was transformed with an adenine biosynthetic enzyme gene from *S. commune* (Alice *et al.*, 1989).

Another means of transformation include electroporation and gene gun. Electroporation gives high frequency transformation of both intact cells and protoplasts. This simple and rapid technique involves a brief voltage pulse, which reversibly permeabilizes cell membranes facilitating entry of DNA molecules into the cells. Though the transformation frequency is same as the PEG method, electroporation is easier to handle, reproducible and can also be applied to intact cells in addition to protoplast (Chakraborty and Kapoor, 1990; Goldman *et al.*, 1990). Gene gun or balistic technique has been developed and optimized for intact plant tissues (Klein *et al.*, 1987). In this technique DNA coated on tungsten or gold microprojectiles is shot into cells followed by insertion of foreign DNA into the genome. It is an easy technique to handle and is not very expensive for vector independent DNA delivery into a variety of cells. It has been applied to fungal cells (Armaleo *et al.*, 1990).

Selection markers used for transformation of fungal systems include four types. First type includes those in which a fungal promoter has been spliced to a prokaryotic gene encoding resistance to a fungicide compound, for example aminoglycoside and hygromycin phosphotransferase (*aph* and *hph*) and bleomycin binding protein *ble* conferring resistance to G 418, hygromycin and bleomycin, respectively (Kolar *et al.*, 1988; Turgeon *et al.*, 1987). Another type is that of fungal benomyl resistant beta-tubulin genes, cloned from mutants that are resistant to this fungicide (Orbach *et al.*, 1986). Third type that has found widespread use in transformation is the acetamide gene of *A. nidulans*. This gene allows heterologous organisms that can use acetate as a sole carbon source to grow on acetamide (Kelley and Hynes, 1985). Still another such technique includes selection by complementation of mutants. Prerequisite for such a system is a host harboring mutation in readily complementable genes, for example fungi with lesions in orotodylate decarboxylase, nitrate reductase and ATP sulfurylase (Buxton *et al.*, 1989; Malardier *et al.*, 1989; Van Hartingsveldt *et al.*, 1987). Prospects and procedures for DNA mediated transformation of ectomycorrhizal fungi has been reviewed earlier (Barrett, 1992; Lemke *et al.*, 1991; Martin *et al.*, 1994; Tigano-Milani *et al.*, 1995). The very first report of genetic transformation of a mycorrhizal fungi is that of Barrett *et al.* (1990). They successfully transformed *L. laccata* protoplasts at a frequency of 5 tranformants per μg DNA. Selection marker used was hygromycin B (HmB) resistance using *E. coli* aminocyclitol phospho-transferase (*aph*) gene. The promoter used was of glyceraldehyde-3-phosphate dehydrogenase, and flanked by tryptophan (*trp* C) terminator from *Aspergillus nidulans*. Southern blot hybridization revealed that HmB resistant transformants of *L. laccata* had integrated vector sequences. This study provided an evidence for the ability of promoter and termination signals of Ascomycetous origin to function in Basidiomycetes. Another evidence for transformation of ectomycorrhizal fungus *Hebeloma cylindrosporum* came from the study of Marmeisse *et al.* (1992b). The pAN 71 plasmid containing *E. coli* hygromycin B phosphotransferase gene was used to transform protoplasts of *H. cylindrosporum*. Plasmid containing either tryptophan biosynthesis gene and/or NADP-glutamate dehydrogenase gene from *Coprinus cinereus* were successfully co-transformed in *H. cylindrosporum* genome, and were stably maintained. All trasformants retained their ability to form mycorrhizas. Successful transformation of *L. laccata* and *H. cylindrosporum* suggest that there is a possibility of transforming other ectomycorrhizal fungi. Since other techniques such as electroporation and gene gun can be applied for intact cells for transformation, the range for ectomycorrhizal fungi can be increased.

Occurrence of two mitochondrial, linear double stranded (DNA) plasmids have been reported in *Hebeloma circinans* (Schrunder *et al.* 1991). These investigators have suggested that there is a possibility to identify sequences of origin of replication from these plasmids and to construct tranformation vectors utilizing these sequences.

The tranformation systems described so far have been developed for fungi that are well characterized and can be cultured in the laboratory in defined media. Since the extraction of DNA and cloning of certain genes in *E. coli* appears to be possible for a wide variety of fungi, theoretically it should be possible to develop systems in any fungus that can be readily cultured *in vitro*. This makes development of gene cloning systems in the culturable EM fungi possible. The most obvious drawback to developing such systems for VAM fungi is the lack of a defined culture medium and problems in obtaining amounts of pure fungal material for nucleic acid extraction. Though there is a possibility of introducing DNA into germinating spores, the presence of an introduced gene can be selected by the limited growth occurring on defined medium. Since the spores are multinucleated, a strong selection for the introduced gene would be necessary to ensure its maintenance during subsequent propagation of the fungus in association with a suitable plant.

The applications of gene cloning in mycorrhizal fungi include the possibility of introducing novel genes determining products that would enhance the growth or health of the plant partner, introduction of fungicide resistance in the mycorrhizal fungi and it also provides the information about the fundamental biology of mycorrhizal associations.

Bills *et al.* (1995) described transformation of EM fungus *Paxillus involutus* by particle bombardment. Transformation was determined using the *hph* gene as the selectable marker and the (-glucuronidase gene (GUS) as a reporter gene. The hygromycin resistant transformants were obtained which were mitotically stable and maintained both the *hph* and GUS genes in the fungal genome in integrated form. Synthesis experiments showed that transformed *P. involutus* formed ectomycorrhiza with *Pinus resinosa*. These provided the first report of a successful transformation of an EM fungus using particle bombardment.

4. FUNGUS ROOT INTERACTIONS

The interaction between mycorrhizal fungi and roots involves a series of events ultimately leading to an integrated, functional structure, and these events are not fully understood as yet at molecular level. Endomycorrhizas are the most widespread and most ancient type of plant-fungus associations. Majority of terrestrial plants has evolved with compatibility systems towards the fungal symbionts. Cellular interactions leading to mycorrhiza formation are complex. Some events are shared with nodulation processes but molecular modifications specific to arbuscular mycorrhiza formation are also found. It has been suggested that defense gene expression in plants is controlled during establishment of a successful symbiosis. Modifications such as changes in wall metabolism and protein expression are also induced in fungal symbiont. Since endomycorrhizal fungi cannot be grown, cloning of DNA from these fungi opens up the possibility of identifying genes involved in symbiosis establishment.

4.1. Fungal Induced Plant Responses

Host roots show little cellular response to invasion by an arbuscular mycorrhizal fungus until arbuscule formation occurs in the cortical cells. This results in the induction of increased abundance of cell organelles, enhanced nuclear activity, extensive proliferation of membrane systems and accumulation of a number of host derived molecules in the resulting interface formed between the symbiont cells (Bonfante-Fasolo, 1994; Gianinazzi-Pearson, 1995). It has been observed that very weak activation of plant defense genes occurs which is transient and very localized during mycorrhiza establishment (Bonfante-Fasolo, 1994; Gianinazzi-Pearson *et al.*, 1992; Harrison and Dixon, 1994; Lambais and Mehdy, 1995). It has been proposed that in mycorrhizal plants specific host genes may play a regulatory role towards defense genes, so that their expression is suppressed or maintained at a low level.

Some events in symbiosis establishment by AM fungi and rhizobia are determined by common plant genes, that is why modifications in similar molecular components (glycoconjugates and nodulins) occur during the two types of root infections (Perotto *et al.*, 1994; Wyss *et al.*, 1990). Now the presence of symbiosis specific plant genes has been confirmed because of : (i) The detection of new proteins or polypeptides (endomycorrhizins) indicating *de novo* gene expression, (ii) mycorrhiza induced modifications in gene expression as confirmed by analysis of mRNA translation products and differential RNA display banding pattern, and (iii) disappearance of certain root polypeptides but their precise role in mycorrhiza development and functioning is still unknown.

5. CHANGES INDUCED IN FUNGAL BEHAVIOUR

Spore germination and hyphal elongation is stimulated by factors secreted by host plants, whilst root exudates of non-host plants do not have effect (Giovannetti *et al.*, 1994; Glenn *et al.*, 1988). Root exudates also elicit morphogenetic modification in the mycelial growth leading to hyphal branching and appressoria formation (Giovennetti *et al.*, 1994; Glenn *et al.*, 1988). When the hyphae reach the cortical cells, extensive proliferation of mycelium and differentiation of arbuscules is induced, accompanied by modifications in cell wall structure and molecular composition. Investigation into the structure and regulation of fungal genes require the establishment of representative genomic libraries of AM fungi. The development of the mycorrhizal association involves considerable morphogenetic and physiological changes in both symbionts which are triggered by a complex cascade of signals. Rapid growth in molecular techniques is facilitating the possibility of analysing temporal and spatial gene expression in the two partners.

6. ECTOMYCORRHIZAL ASSOCIATIONS

Physiological, morphological and anatomical studies have demonstrated that the metabolism of free-living plants and EM fungi are altered while entering into symbiosis (Harley and Smith, 1983). However, the molecular events associated with the establishment and maintenance of these associations are poorly understood. Compatibility in EM associations is genetically controlled in plants (Walker *et al.*, 1986) and in fungi (Kropp and Fortin, 1988). A significant contribution to this field is the observation that EM root formation leads to differential gene expression (Hilbert *et al.*, 1991a). The protein profile of *Eucalyptus globulus* and EM fungus *Pisolithus tinctorius* association are different than the proteins of these two organisms living axenically. Specific proteins called 'ectomycorrhizins' were uniquely associated with EM roots (Hilbert *et al.*, 1991a,b). Duchesne (1989) obtained similar results while studying the protein profiles of *Pinus resinosa* and *Paxillus involutus* association. He concluded that metabolism of *Pinus resinosa* roots and *P. involutus* change even prior to physical contact between these two organisms indicating gene expression in EM plants, and fungi are altered by the onset of the symbiotic stage.

The EM association between *Eucalyptus* and *Pisolithus tinctorius* has been a useful model for the study of many diverse, basic aspects of the biology of symbiosis such as early stages of many diverse, basic aspects of the biology of symbiosis such as early stages of EM formation and the discovery of new gene products involved in the establishment of symbiosis and its morphogenesis (Martin and Tagu, 1995), because of its amenability to manipulation *in vitro* (Malajczuk *et al.*, 1990). Symbiosis induced gene expression is well studied for *Eucalyptus-Pisolithus* associations (Martin and Tagu, 1995; Nehls and Martin, 1995). Changes in gene expression leads to appearance and disappearance of new proteins, and changes in metabolic organization in fungal and plant cells (Burgess *et al.*, 1995a; Hilbert and Martin, 1988). The role and function of several symbiosis related (SR) proteins have recently been characterized by molecular and biochemical approaches. The SR proteins correspond to pathogenesis-related proteins, products of auxin-induced genes, enzymes (Nehls and Martin, 1995) and fungal hydrophobins (Tagu *et al.*, 1996).

The key event in mycorrhiza formation is the adhesion of hyphae, which involves cellular differentiation including dramatic changes in the structure and composition of the partner cell walls. The knowledge of these processes is mostly based on cytological observations but several cell wall proteins (CWP) and genes coding for CWP and their regulation have recently been described (Tagu and Martin, 1996).

6.1.   Morphological Alterations During Early Stages of Ectomycorrhiza Formation

Initial changes in host plant include stimulation of lateral root formation by fungal auxins, radial elongation of epidermal cells and arrest of cell division of ensheathed roots (Peterson and Bonfante, 1994). Changes in fungi include, shortening of hyphal cells, with few clamp connections and branching leading to pseudoparenchymatous mode of growth (Horan and Chilvers, 1990). Hyphae in the mantle are highly branched, glued together, and embedded in an abundant extracellular matrix composed of polysaccharides, cysteine rich proteins and glycoproteins. Mycelium in the Hartig's net has multilobed hyphal fronts with a complex fan like structure (Scheidegger and Brunner, 1995). The development of interface between the two partners is remarkably uniform regardless of the species of host or fungus, which contain a complex matrix of polysaccharides and proteins. Insight into the cell wall formation and alteration during the symbiosis shows a high level of up regulated 32 kDa symbiosis related acidic polypeptides and hydrophobins, and decreased content of mannoprotein gp95, which simultaneously regulates the molecular architecture of protein network in a manner to allow new development fates for both fungal cell cohesion and root colonization by the fungus (Tagu and Martin, 1996).

6.2   Molecular Cloning of Symbiosis Related Fungal Genes

It has been well documented that there is a shift in the expression of a set of developmentally related genes during ectomycorrhiza formation both in fungal as well as plant symbiont. Several clones were obtained by differential screening of a cDNA library of *E. globulus - P. tinctorius* mycorrhiza using cDNA probes prepared either from mRNA isolated from free-living mycelium or *P. tinctorius* colonized roots (Tagu *et al.,* 1993), that represented about one third of the cDNA clones screened. This confirmed early contentions based on protein analysis, that early development steps of ectomycorrhiza formation induce major shift in fungal gene expression (Hilbert and Martin, 1988). This has been confirmed by the analysis of mRNA populations, and the drastic decreased expression of plant genes is noteworthy.

7.   CONCLUSIONS

Attempts have been made to develop nucleic acid probes for rapid and accurate identification of EM fungi associated with ectomycorrhizal roots in order to understand population dynamics. The PCR-RFLP of nuclear rRNA and mitochondrial rRNA genes, alone or in conjunction with RAPD, or RAPD alone, and the analysis of microsatellite DNA have been shown to be useful in the identification of ectomycorrhizal fungi. Further efforts are also continuing to develop transformation systems for various ectomycorrhizal fungi so as to bring desirable changes in the genome of these fungi to improve their efficacy as symbionts and to understand molecular events during the establishment of symbiosis.

REFERENCES

Adwell, F.E.B., Hall, I.R. and Smith, J.M.B. 1985, Enzyme linked immunosorbant assay as an aid to taxonomy of the Endogonaceae, *Trans. Bri. Mycol. Soc.* 84: 399-402.
Agerer, R. 1996, Studies on ectomycorrhizae, III, Mycorrhizae formed by four fungi in the genera *Lactarius* and *Russula* on spruce, *Mycotaxon* 27: 1-59.
Agerer, R. 1995, Anatomical characteristics of identified ectomycorrhizas: an attempt towards a natural classification, in: *Mycorrhiza: Structure, Function, Molecular Biology and Biotechnology,* A. Varma and B. Hock, eds., Berlin, Heidelberg, New York Springer-Verlag, pp. 605-734.

Alice, M., Kornegay, J.R., Pribnow, D. and Gold, M.H. 1989, Transformation of complementation of an adenine auxotroph of lignin-degrading basidiomycete *Phanerochaete chrysosporium*, *App. Environ. Microbiol.* 55:406-411.

Anderson, I.C., Chambers, S.M. and Cairney, J.W.G. 1998, Molecular determination of genetic variation in *Pisolithus* isolates from a defined region in New South Wales, Australia, *New Phytol.* 138:151-162.

Armaleo, D.Y.G.N., Klein, T.M., Shark, K.B., Sanford, J.C. and Johnson, S.A. 1990, Biolistic nuclear transformation of *Saccharomyces cereviseae* and other fungi, *Curr. Genet.* 17:97-103.

Armstrong, J.L., Fowles, N.L. and Rygiewiz, P.T. 1989, Restriction fragment length polymorphisms distinguish ectomycorrhizal fungi, *Plant Soil* 116: 1-7.

Barrett, V. 1992, Ectomycorrhizal fungi as experimental organisms, in: *Hand Book of Applied Mycology* Vol.I, Soil and Plants, D.K. Arora, B. Rai, K.G. Mukerji, and G.R. Knudsen, eds., Mascal Dekker Inc., USA, pp.217-229.

Barrett, V., Dixon, R.K. and Lemke, P.A. 1990, Genetic transformation from selected species of ectomycorrhizal fungi, *App. Microbiol. Biotechnol.* 33:313-316.

Bicknell, J.N. and Dauglas, H.C. 1969, Conservation of ribosomal RNA sequences in yeast, filamentous fungi and plants, Bacteriological Proceedings of 69th Annual Meeting of the American Society for Microbiology, Florida, pp. 39 (Abstr.)

Bills, S.N., Richter, D.L. and Podila, G.K. 1995, Genetic transformation of the ectomycorrhizal fungus *Paxillus involutus* by particle bombardment, *Mycol. Res.* 99: 557-561.

Binninger, D.M., Skrzynia, C., Pukkila, P.J. and Casselton L.A. 1986, DNA-mediated transformation of the basidiomycetes *Coprinus cinereus*, *EMBO J.* 6:835-840.

Blanz, P.A. and Gottschalk, M. 1986, Systematic position of *Septobasidium grophiola* and other basidiomycetes as deduced on the basis of their 5S ribosomal RNA nucleotide sequence system, *App. Microbiol.* 8:121-127.

Bonfante-Fasolo, P. 1994, Ultrastructural analysis reveals the complex interactions between root cell and arbuscular mycorrhizal fungi, in: *Impact of Arbuscular Mycorrhizas on Sustainable Agriculture and Natural Ecosystems*, S. Gianinazzi and H. Schuepp, eds., Birkhauser, Basel, pp. 73-87.

Bonfante, P., Lanfranco, L.O., Cometti, V. and Gure, A. 1997, Inter- and intraspecific variability in strains of the ectomycorrhizal fungus *Suillus* as revealed by molecular techniques, *Microbiol. Res.* 152:287-292.

Brand, F. 1991, Ektomykorrhizen an *Fagus sylvatica*, Charakterisierung und Identifuzierung, okologische Kennzeichnung und unsterile Kultivierung, Libri Botanici 2, Eching: IHW- Verlag, 1-229.

Bruns, T.D. and Gardes, M. 1993, Molecular tools for the identification of ectomycorrhizal fungi - taxon specific oligonucleotide probes for Suilloid fungi, *Mole. Ecol.* 2:233-242.

Bruns, T.D., White, T. J. and Taylor, J.W. 1991, Fungal molecular systematics, *Ann. Rev. Ecol. Syst.* 22: 525-564.

Burdon, J.J. and Marshall, D.R. 1983, The use of isozymes in plant disease research, in: *Isozymes in Plant Genetics and Breeding*, S.D. Tanksley and T.J. Orton, eds., Vol. A., Elsevier, New York, pp. 401-412.

Burgess, T., Laurent, P., Dell, B., Malajczuk, N. and Martin, F. 1995a, Effect of the fungal isolate aggrresivity on the biosynthesis of symbiosis-related polypeptides in differentiating eucalypt ectomycorrhiza, *Planta* 196: 408-417.

Burgess, T., Malajczuk, N. and Dell, B. 1995b, Variation in *Pisolithus* based on basidiome and basidiospore morphology; Cultural characteristics and analysis of polypeptides using ID SDS PAGE, *Mycol. Res.* 99: 1-13.

Buxton, F.P., Gwynne, D.I. and Davies, R.W. 1989, Cloning of a new bidirectionally selectable marker for *Aspergillus* strains, *Gene* 84:329-334.

Cameleyre, I. and Olivier, J.M. 1983, Evidence for intraspecific isozyme variations among French isolates of *Tuber melanosporum* (Vitt.), *FEMS Microbiol. Lett.* 110: 159-162.

Chakraborty, B.N. and Kapoor, M. 1990, Transformation of filamentous fungi by electroporation, *Nuc. Acids Res.* 18:6737

Cleyet-Marel, J.C., Bausquet, N. and Mousain, D. 1989, The immunochemical approach for the characterization of ectomycorrhizal fungi, *Agri. Ecosy. Environ.* 28: 79-83.

Doudrick, R.L., Raffle, V.L., Nelson, C.D. and Furnier, G.R. 1995, Genetic analysis of homokaryones from a basidiome of *Laccaria bicolor* using random amplified polymorphic DNA (RAPD) markers, *Mycol. Res.* 99:1361-1366.

Duchesne, L.C. 1989, Protein synthesis in *Pinus resinosa* and the ectomycorrhizal fungus *Paxillus involutus* prior to ectomycorrhiza formation, *Trees* 3: 73-77.

Erland, S. 1995, Abundance of *Tylospora fibrillosa* ectomycorrhizas in a South Swedish spruce forest measured by RFLP analysis of the RCR- amplified rDNA ITS region, *Mycol. Res.* 99:1425-1428.

Erland, S., Henrion, B., Martin, F., Glover, L.A. and Alexander, I.J. 1994, Identification of the ectomycorrhizal basidiomycete *Tylospora fibrillosa* Donk by RFPL analysis of PCR amplified ITS and IGS regions of ribosomal DNA, *New Phytol.* 126:525-532.

Fries, C.F. and Allen, M.F. 1991, Tracking the fates if exotic and local VA mycorrhizal fungi: methods and patterns, *Agricul. Ecosys. Environ.* 34: 87-96.

Gandeboeuf, D., Dupre, C., Roeckel-Drevet, P., Nicolas, P. and Chevalier, G. 1997a, Typing *Tuber* ectomycorrhizae by polymerase chain amplification of the internal transcribed spacer of rDNA and the sequence characterized amplified region markers, *Can. J. Microbiol.* 43:723-728.

Gandeboeuf, D., Dupre, C., Roeckel-Drevet, P., Nicolas, P. and Chevalier, G. 1997b, Grouping and identification of *Tuber* species using RAPD markers, *Can. J. Bot.* 75:36-45.

Gardes, M. and Bruns, T.D. 1993, ITS primers with enhanced specificity for Basidiomycetes: application to the identification of mycorrhizae and rusts, *J. Mol. Ecol.* 30:17-121.

Gardes, M., Fortin, J.A., Mueller, G.M. and Kropp, B.R. 1990, Restriction fragment length polymorphisms in the nuclear ribosomal DNA of four *Laccaria* spp. (*L. bicolor, L. laccata, L. proxima, L. amethystina*), *Phytopathol.* 80:1312-1317.

Gardes, M., White, T.J., Fortin, J.A., Bruns, T.D. and Taylor, J.W. 1991, Identification of indigenous and introduced symbiotic fungi in ectomycorrhizae by amplification of nuclear and mitochondrial ribosomal DNA, *Can. J. Bot.* 69: 180-190.

Gianinazzi-Pearson, V., Tahiri-Alaoni, A., Antoniw, J.F., Gianinazzi, S. and Duma-Gandot, E. 1992, Weak expression of the pathogenesis related PR-b1 gene and localisation of related proteins during symbiotic endomycorrhizal interaction in tobacco roots, *Endocytobiosis Cell Res.* 8: 177-185.

Gianinazzi-Pearson, V. 1995, Morphological compatibility in interactions between roots and arbuscular endomycorrhizal fungi: molecular mechanisms, genes and gene expression, in: *Pathogenesis and Host Parasite Specificity in Plant Disease*, Vol. II., K. Kohmots, R.P. Singh and U.S. Singh, eds., Pergamon Press, Elsevier Science, Oxford.

Giovannetti, M., Shrana, C., Citternesi, A.S., Avio, L., Gollotte, A., Gianinazzi-Pearson, V. and Gianinazzi, S. 1994, Recognition and infection process, basis for host specificity of arbuscular mycorrhizal fungi, in: *Impact of Arbuscular Mycorrhizas on Sustainable Agriculture and Natural Ecosystems*, S. Gianinazzi, and H. Schuepp, eds., Birkhauser, Basel, pp.61-72.

Glenn, M.G., Chew, F.S. and Williams, P.H. 1988, Influence of glucosinolate content of *Brassica* (Cruciferae) roots on growth of vesicular-arbuscular mycorrhizal fungi, *New Phytol.* 110:217-225.

Goldman, G.H., Van Montague, M. and Herrera-Estrella, A. 1990, Transformation of *Trichoderma harzianum* by high voltage electrical pulse, *Curr. Genet.* 17:169-174.

Gupta, V. and Satyanarayana, T. 1996, Molecular and general genetics of Ectomycorrhizal fungi, in: *Concepts in Mycorrhizal Research*, K.G. Mukerji, ed., Kluwer Academic Publishers, Netherlands, pp. 343-361.

Guillemaud, T., Raymond, M., Callot, G., Cleyet-Marel, J.C. and Fernandez, D. 1996, Variability of nuclear and mitochondrial ribosomal DNA of a truffle species (*Tuber aestivum*), *Mycol. Res.* 100:547-550.

Harley, J.L. and Smith, S.E. 1983, *Mycorrhizal Symbiosis*, Academic Press, New York.

Harrison, M.J. and Dixon, R.A. 1994, Spatial patterns of expression of flavonoid/isoflavonoid pathway genes during interactions between roots of *Medicago trunculata* and the mycorrhizal fungus *Glomus versiforme*, *Plant* J. 6: 9-20.

Hasegawa, M., Iida, Y., Yano, T., Takaiwa, F. and Iwabuchi, M. 1985, Phylogenetic relationship among eukaryotic kingdoms inferred from ribosomal RNA sequences, *J. Mol. Evol.* 22:32-38.

Hang, I. and Oberwinkler, F. 1987, Some distinctive types of spruce mycorrhizae, *Trees* 1:172-188.

Henrion, B., Chevalier, G. and Martin, F. 1994a, Typing truffle species PCR amplification of the ribosomal DNA spacers, *Mycol. Res.* 98: 37-43.

Henrion, B., Di Battista, C., Bouchard, D., Varielles, D., Thompson, B.D., Le Tacon, F. and Martin, F. 1994b, Monitoring the persistence of *Laccaria bicolor* as an ectomycorrhizal symbiont of nursery grown Dauglas fir by PCR of the rDNA intergenic spacer, *Mol. Ecol.* 3:571-580.

Henrion, B., LeTacon, F. and Martin, F. 1992, Rapid identification of genetic variation of ectomycorrhizal fungi by amplification of ribosomal RNA genes, *New Phytol.* 122: 289-298.

Hepper, C.M., Sen, R. and Maskall, C.S. 1986, Identification of vesicular arbuscular mycorrhizal fungi in roots of leak (*Allium porrum* L.) and maize (*Zea mays* L.) on the basis of enzyme mobility during polyacrylamide gel electrophoresis, *New Phytol.* 102: 529-539.

Hepper, C.M., Sen, R., Azcon-Aguiliar, C. and Grace, C. 1988, Variation in certain isozymes among different geographical isolates of the vesicular-arbuscular mycorrhizal fungi *Glomus clarum, Glomus monosporum* and *Glomus mosseae*, *Soil Biol. Biochem.* 20: 51-59.

Hilbert, J.L. and Martin, F. 1988, Regulation of gene expression in ectomycorrhizas 1, Protein changes and the presence of ectomycorrhiza specific polypeptides in the *Pisolithus-Eucalytus* symbiosis, *New Phytol.* 110: 339-436.

Hilbert, J.L., Costa, G. and Martin, F. 1991a, Regulation of gene expression in ectomycorrhizas, Early ectomycorrhizas and polypeptide cleansing in eucalypt ectomycorrhizas, *Plant Physiol.* 97: 977-984.

Hilbert, J.L., Costa, G. and Martin. F. 1991b, Ectomycorrhizin synthesis and polypeptide changes during the early stage of eucalypt mycorrhiza development, *Plant Physiol.* 97: 977-984.

Ho, I. 1987, Enzyme activity and phytohormone production of a mycorrhizal fungus *Laccaria laccata*, *Can. J. For. Res.* 17: 855-858.

Ho, I. and Trappe, J.M. 1987, Enzymes and growth substances of *Rhizopogon* species in relation to mycorrhizal hosts and infrageneric taxonomy, *Mycologia* 79:553-558.

Horan, D.P. and Chilvers, G.A. 1990, Chemotropism : the key to ectomycorrhizal formation, *New Phytol.* 116: 297-301.

Hori, H. and Osagawa, S. 1987, Origin and evolution of organisms as deduced from 5S ribosomal RNA sequences, *Mol. Biol. Evol.* 4:445-472.

Karen, O., Hogberg, N., Dahlberg, A., Jonsson, L. and Nylund, J. 1997, Inter- and intraspecific variation in the ITS region of rDNA of ectomycorrhizal fungi in Fennoscandia as detected by endonuclease analysis, *New Phytol.* 136:313-325.

Karkouri, K.E., Cleyet-Marel, J.C. and Mousain, D. 1996, Isozyme variation and somatic incompatibility in populations of the ectomycorrhizal fungus *Suillus collinitus, New Phytol.* 134:143-153.

Kelley, M.K. and Hynes, M.J. 1985, Transformation of *Aspergillus niger* by *Amd* S gene of *Aspergillus nidulans, EMBO* J. 4:475-497.

Klein, T.M., Wolf, E.D., Wu, R. and Sanford, J.C. 1987, High velocity microprojectiles for delivering nucleic acids into living cells, *Nature* 327:70-73.

Kolar, M., Punt, P.J., Van Den Hondel, C.A.M.J.J. and Schwab, H. 1988, Transformation of *Penicilium chrysogenum* using dominant selection markers and expression of an *Escherichia coli Lac* Z fusion, Gene 62:127-134.

Kough, J., Malajczuk, N. and Linderman, R.G. 1983, Use of an indirect immuno-fluorescent technique to study the vesicular arbuscular fungus *Glomus epigaeum* and other *Glomus* species, *New Phytol.* 94: 57-62.

Kraigher, H., Agerer, R. and Javornik, B. 1995, Ectomycorrhizae of *Lactarius ligniuotus* on Norway spruce characterized by anatomical and molecular tools, *Mycorrhiza* 5:175-180.

Kreuzinger, N., Podeu, R., Gruber, F., Gobl, F. and Kubicek, C.P. 1996, Identification of some ectomycorrhizal basidiomycetes by PCR amplification of their GDP (glyceraldehyde-3-phosphate dehydrogenase) genes, *App. Environ. Microbiol.* 62:3432-3438.

Kropp, B.R. and Fortin, J.A. 1987, The incompatibility system and ectomycorrhizal performance of monokaryons and reconstituted dikaryons of *Laccaria bicolor, Can. J. Bot.* 66: 289-294.

Lanfranco, L., Wyss, P., Marzachi, C. and Bonfante, P. 1993, DNA probes for identification of the ectomycorrhizal fungus *Tuber magnatum* Pico, *FEMS Microbiol. Lett.* 114: 245-252.

Lambais, M.R. and Mehdy, M.C. 1995, Differential expression of defence-related genes in arbuscular mycorrhiza, *Can. J. Bot.* 73: S533-S540.

Lemke, P.A., Barrett, B. and Dixon, R. K. 1991, Procedures and prospects for DNA mediated transformation of ectomycorrhizal fungi, in: *Methods in Microbiology*, Vol. 23, J.R. Norris, D.J. Read and A.K. Verma, eds., Academic Press, London, pp. 281-293.

Lo Buglio, K.F., Rogers, S.O. and Wang, C.J.K. 1991, Variation in ribosomal DNA among isolates of the mycorrhizal fungus *Cenococcum geophilum, Can. J. Bot.* 69: 2331-2343.

Longato, S. and Bonfante, P. 1997, Molecular identification of mycorrhizal fungi by direct amplification of microsatellite regions, *Mycol. Res.* 101:425-432.

Lovett, I.S. and Hasselby, J.A. 1971, Molecular weights of the ribosomal RNA of fungi, *Archeives Microbiol.* 80:191-204.

Malajczuk, N., Garbay, J. and Lapeyrie, F. 1990, Infectivity of pine and eucalyptus isolates of *Pisolithus tinctorius* on roots of *Eucalyptus urophilla in vitro*, 1, Mycorrhiza formation in Model Systems, *New Phytol.* 111: 627-631.

Malardier, L.M., Dadoussi, J., Julian, J., Roussel, F., Scazzocchio, C. and Brygoo, Y. 1989, Cloning of the nitrate reductase gene (nia D) of *Aspergillus nidulans* and its use for transformation of *Fusarium oxysporium, Gene* 78:147-156.

Marmeisse, R., Debaud, J.C. and Casselton, L.A. 1992a, DNA probes for species identification in the ectomycorrhizal fungus *Hebeloma, Mycol. Res.* 96: 161-165.

Marmeisse, R., Gay, G., Debaud, J.C. and Casselton, L.A. 1992b, Genetic transformation of the ectomycorrhizal fungus *Hebeloma cylindrosporum, Curr. Gene*, 228: 41-45.

Martin, F. and Tagu, D. 1995, Ectomycorrhiza development: a molecular perspective, in: *Mycorrhiza : Structure, Function, Molecular Biology and Biotechnology*, A.K. Verma and B. Hock, eds., Springer-Verlag, Berlin, Heidelberg, pp. 29-58.

Martin, F., Zaiou, M., Le Tacon, F. and Rygiewicz, P. 1991, Strain specific difference in ribosomal DNA from the ectomycorrhizal fungi *Laccaria bicolor* (Maire) Orton and *Laccaria laccata* (Scop exfr), *Brit. Ann. Sci. For.* 48:133-142.

Martin, F., Tommerup, I.C. and Tagu, D. 1994, Genetics of ectomycorrhizal fungi : progress and prospects, *Plant Soil* 159: 159-170.

Martin, F., Laurent, P., DeCarvalho, D., Burgess, T., Murphy, P., Nehls, U. and Tagu, D. 1995, Fungal gene expression during ectomycorrhiza formation, *Can. J. Bot.* 73: S541-S547.

Mehmann, B., Brunner, I. and Braus, G.H. 1994, Nucleotide sequence variation of chitin synthase gene among ectomycorrhizal fungi and its potential use in taxonomy, *Appl. Environ. Microbiol.* 60:3105-3111.

Micales, J.A., Bonde, M.R. and Peterson, G.L. 1986, The use of isozyme analysis in fungal taxonomy and genetics, *Mycotaxon* 27: 404 449.

Miller, O.K. and Walting, R. 1987, In: *Evolutionary Biology of Fungi,* A.D.M. Rayner, C.M. Brasied and D. Moore, eds., Cambridge University Press, Cambridge, pp. 435-448.

Monroz-Rivas, A.M., Specht, C.A., Drummond, B.J., Froelinger, E., Novotony, C.P. and Ullrich, R.C. 1986, Transformation of the basidiomycete, *Schizophyllum commune, Mol. Gen. Genet.* 205:103-106.

Mouches, C.P., Duthil, N., Poitou, J., Delmas and Bove, J.M. 1981, Characterisation des especes truffieres par analyse de leurs protienes en gel depolyacrylamide et application de ces techniques a la taxonomie des champignons, *Mushroom Sci.* 11: 819-831.

Nehls, U. and Martin, F. 1995, Changes in root gene expression in ectomycorrhiza, in: *Biotechnology of Ectomycorrhizae: Molecular Approaches,* V. Stocchi, P. Bonfante and M. Nuti, eds., Plenum Press, New York, London, pp. 125-137.

Nicolson, T.H. 1975, In: *Endomycorrhizas,* F.E. Sanders, B. Mosse and P.B. Tinkler,eds., Academic Press, London pp. 25-34.

Oramas-Shirey, M. and Morton, J.S. 1990, Immunological stability among different geographical isolates of the arbuscular mycorrhizal fungus *Glomus occultum,* Abstracts of the 90th Annual meeting of American Society for Microbiology, Washington, pp.311.

Orbach, M.J., Porro, E.B. and Yanafsky, C. 1986, Cloning and characterization of the gene for B-tubulin from a benomyl-resistant mutant of *Neurospora crassa* and its use as a dominant selectable marker, *Mol. Cell Biol.* 6:2452-2461.

Perotto, S., Brewin, N. J. and Bonfante-Fasolo, P. 1994, Colonization of pea roots by the mycorrhizal fungus, *Glomus versiforme* and by *Rhizobium* : immunological comparison using monoclonal antibodies as probes for cell surface components, *Mol. Plant Microbe Interac.* 7: 91-112.

Pritsch, K., Boyle, J.C., Munch, J.C. and Buscot, F. 1997, Characterization and identification of black alder ectomycorrhizas by PCR/RFLP analysis of the rDNA internal transcribed spacers (ITS), *New Phytol.* 137:357-369.

Pritsch, K. 1996, Untersuchungen zur Diversitat und Okologie von Mycorrhizen der Schwarzerle [*Alnus glutinous* (L.) Gaertn.], Dissertation, University of Tuebingen, 1-197.

Pritsch, K. and Buscot, F. 1996, Biodiversity of ectomycorrhizas from morphotypes to species, in: *Mycorrhizas in Integrated Systems from Genes to Plant Development,* C. Azcon-Aguilar and J.M. Barea, eds., Proceedings of the Fourth European Symposium on Mycorrhizas, Europian Commission report 16728, Luxembour : Office for Official Publications of the European Communities, 9-31.

Scanders, I.R., Ravolanirina, F., Gianinazzi-Pearson, V., Gianinazzi, S. and Lemoine, M.C. 1992, Detection of specific antigens in the vesicular arbuscular mycorrhizal fungi *Gigaspora margarita* and *Acaulospora laevis* using polyclonal antibodies to soluble spore fractions, *Mycol. Res.* 96:477-480.

Scheidegger, C. and Brunner, I. 1995, Electron microscopy of ectomycorrhiza: Methods, applications and findings, in: *Mycorrhiza: Structure, Function, Molecular Biology* and *Biotechnology,* A.K. Verma, and B. Hock, eds., Springer Verlag, Heidelberg, Berlin.

Schrunder, J., Debaud, J.C. and Mainhardt, F. 1991, Adenoviral like genetic elements in *Hebeloma circinans,* in: *Mycorrhizas in Ecosystems-Structure and Function,* Abstracts, 3rd ESM Sheffield, U.K.

Sen, R. and Hepper, C.M. 1986, Characterization of vesicular arbuscular mycorrhizal fungi (*Glomus* spp.) by selective enzyme staining following polyacrylamide gel electrophoresis, *Soil Biol. Biochem.* 18:29-34.

Simon, L., Lalonde, M. and Bruns, T.D. 1992a, Specific amplification of 18S fungal ribosomal genes from vesicular arbuscular endomycorrhizal fungi colonizing roots, *App. Environ. Microbiol.* 58: 291-295.

Simon, L., Levesque, R.C. and Lalonde, M. 1992b, Rapid quantitation by PCR of endomycorrhizal fungi colonizing roots, *PCR Methods Application* 2: 76-80.

Simon, L., Bousquet, J., Levesque, R.C. and Lalonde, M. 1993a, Origin and diversification of endomycorrhizal fungi and coincidence with vascular land plants, *Nature* 363:67-69.

Simon, L., Levesque, R.C. and Lalonde, M. 1993b, Identification of endomycorrhizal fungi colonizing roots by fluorescent single-strand conformation polymorphism-polymerase chain reaction, *App. Environ. Microbiol.* 59:4211-4215.

Southern, E.M. 1975, Detection of specific sequences among DNA fragments separated by gel electrophoresis, *J. Mol. Biol.* 98:503-517.

Sweeney, M., Harmey, M.A. and Mitchell, D.T. 1996, Detection and identification of *Laccaria* species using a repeated DNA sequence from *Laccaria proxima, Mycol. Res.* 100:1515-1521.

Tagu, D. and Martin, F. 1996, Molecular analysis of cell wall proteins expressed during the early steps of ectomycorrhiza development, *New Phytol.* 133:73-85.

Tagu, D., Nasse, B. and Martin, F. 1996, Cloning and characterization of hydrophobins-encoding cDNAs from the ectomycorrhizal basidiomycete *Pisolithus tinctorius*, *Gene* 168: 93-97.

Tagu, D., Python, M., Cretin, C. and Martin, F. 1993, Cloning symbiosis-related cDNAs from eucalypt ectomycorrhizas by PCR assisted differential screening, *New Phytol.* 125: 339-343.

Taylor,J.W. 1986, Fungal evolutionary biology and mitochondrial DNA, *Exp. Mycol.* 10:259-269.

Tigano-Milani, M.S., Samson, R.A., Martin, I. and Sobral, B.W.S. 1995, DNA markers for differentiating isolates of *Paecilomyces lilacinus, Microbiology* 141: 239-245.

Tilburn, J., Scazzocchio, C., Taylor, G.G., Zabicky Zissman, J.H., Mockington, R.A. and Davies, R.W. 1983, Transformation by integration in *Aspergillus nidulans, Gene* 26:205-221.

Timonen, S., Tammi, H. and Sen, R. 1997, Characterization of the host genotype and fungal diversity in Scots pine ectomycorrhizal from natural humus microcosms using isozyme and PCR- RFLP analyses, *New Phytol.* 135:313-323.

Tommerup, I.C. 1992, Genetics of eucalypt ectomycorrhizal fungi, in: *Internatioanl Symposuim on Recent Topics in Genetics, Physiology and Technology of Basidiomycetes*, M. Miyaji, A. Suzuki and K. Nishimura, eds., Chiba University, Chiba, Japan, pp.74-79.

Tommerup, I.C., Barton, J.E. and O'Brian, P.P. 1992, RAPD fingerprinting of *Laccaria, Hydnangium* and *Rhizoctonia* isolates, in: *International Symposium on Management of Mycorrhizas in Agriculture, Horticulture and Forestry*, the University of Western Australia, Nedlands (Abstracts), pp.161.

Tommerup, I.C., Barton, J.E. and O'Brian, P.P. 1995, Reliability of RAPD fingerprinting of three basidiomycete fungi, *Laccaria, Hydnangium* and *Rhizoctonia, Mycol. Res.* 99: 179-186.

Turgeon, G.G., Garber, R.C. and Yoder, O.C. 1987, Development of a fungal transformation system based on selection of sequences with promoter activity, *Mol. Cell Biol.* 7:3297-3305.

Walker, C., Biggin, P. and Jardine, D. C. 1986, Differences in mycorrhizal status among clones of Sitka spruce, *For. Eco. Manag.* 14: 275-283.

Wilson, J.M., Trinick, M.J. and Parker, C.A. 1983, The identification of vesicular-arbuscular mycorrhizal fungi using immunofluorescence, *Soil Biol. Biochem.* 15:439-445.

Wright, S.F., Morton, J.B. and Sworobuk, J.E. 1987, Identification of a vesicular arbuscular mycorrhizal fungus by using monoclonal antibodies in an enzyme linked immunosorbent assay, *App. Environ. Mircorbiol.* 53: 2222-2225.

Wyss, P., Mellor, R. B. and Wiemken, A. 1990, Vesicular-arbuscular mycorrhizas of wild type soybean and non-nodulating mutants with *Glomus mosseae* contain symbiosis-specific polypeptides (mycorrhizins), immunologically cross-reactive with nodulins, *Planta* 182: 22-26.

Wyss, P. and Bonfante, P. 1992, Identification of mycorrhizal fungi by DNA fingerprinting using short arbitrary primers, in: *International Symposium on Management of Mycorrhizas in Agriculture, Horticulture and Forestry*, The University of Western Australia, Nedlands (Abstract). pp. 154.

Van Hartingsveldt, W., Mattern, I.E., Van Zeul, C.M.J., Pouwels, P.H. and VanDen Hondel, C.A.M.J.J. 1987, Development of a homologous transformation system for *Aspergillus niger* based on the pyr G gene, *Mol. Gen.Genet.* 206:71-75.

Yelton, M.M., Hamer, J.E. and Timerlake, W.E. 1984, Transformation of *Aspergillus nidulans* by using Trp C plasmid, *Proc. Natl. Acad. Sci. USA*, 81:1470-1474.

Zhu, H., Higginbotham, K.O., Dancik, B. and Navratil, S. 1988, Intraspecific genetic variability of isozymes in the ectomycorrhizal fungus *Suillus tomentosus, Can. J. Bot.* 66: 588-594.

PLANT MINERAL NUTRITION THROUGH ECTOMYCORRHIZA

C. Manoharachary and P. Jagan Mohan Reddy

Department of Botany
Osmania University
Hyderabad-500 007
A.P. INDIA

1.   INTRODUCTION

Mycorrhizae are the structures formed by the association of root in plant with fungi. Such root fungus associations were found to be normal on the root system of trees such as pine, oak, etc. The studies of the German Botanist Frank (1885) in the last century had initiated world wide interest on mycorrhizae. Mycorrhizae are found associated with many plants under natural conditions. The fungi are restricted for essential nutrients and growth factors on their hosts. Such root fungus contact and dependency is of symbiotic nature.

Ectotrophic mycorrhizae are those in which fungus completely encloses the rootlet in a sheath or mantle of tissue formed of compact hyphal cells and penetrates only between the cells of the root cortex. This kind of mycorrhizae are associated with many forest trees (Suvercha *et al.*, 1991). According to Trappe (1962) more than 90% of plants are mycorrhizal. Mycorrhizae play significant role in plant productivity and energy systems besides helping as biocontrol agents.

2.   ECTOMYCORRHIZAE

The ectmycorrhizae a common symbiotic association between roots of higher plants and fungi, plays an important role in forest ecosystem. The plants deliver photosynthate to the mycorrhizal fungi and plants receive mineral nutrients and water through mycorrhizal roots. Most of the temperate, boreal and subarctic forest trees (*Pinus, Picea, Larix, Abies*) amentiferous deciduour trees (*Alnus, Betula, Fagus, Quercus, Populus, Salix*) tropical and and semi-arid tropical plants (*Casuarina, Eucalyptus*) harbour ectomycorrhizal association, whose mycorrhizal roots function as nutrient absorbing organ in the plants (Gerdemann, 1975). Ectomycorrhizal associations are generally formed by higher basidiomycetes, ascomycetes, deuteromycetes and some members of zygomycotina (Fontana, 1962; Richards and Wilson, 1963; Slankis, 1958; Trappe, 1962).

2.1. Development

Mycorrhizae have been classified into ectomycorrhiza, ectoendomycorrhiza, endomycorrhiza, arbutoid, monotropoid, ericoid and orchidaceous type (Bhandari and Mukerji; 1993; Harley and Smith, 1983). Fontana (1962) surveyed mycorrhizal association in 14 tree species from plain hills and mountains of Italy which possessed ectotrophic mycorrhiza. Richard and Wilson (1963) noticed the variation in the development of mycorrhiza by adding some nutrients but they could not find any correlation between the concentration of carbohyrates and mycorrhizal development. Miller (1982) has identified a kind of cytokinin responsible for fruiting body formation of *Rhizopogon roseolus* and this kind of cytokinin has been found responsible for the change in the morphology of ectomycorrhizal roots. All the mycorrhizal types share a fundamental features that are associated with the interactions between symbionts features like chemotropism recognition, compatibilities and alterations during morphogeneses of the symbionts are of much significance. The events concerning between modulations of genotypes of various ecto and VA mycorrhiza formation offers new possibilities of exploring factors controlling early events in the establishment of this symbiosis. This kind of information has become an important tool in understanding chemotropism, recognition, compatibility, changes in fungal cyto-skeleton, enzyme production and hormone production in the establishment of a mycorrhiza. Peterson and Farquhar (1994) have reviewed various aspects of mycorrhizal types, formation and establishment.

2.2. Taxonomy and Morphogenesis

Most of the fungi belong to Basidiomycotina representing *Amantia, Boletus, Cenococcum, Laccaria, Inocybe, Pisolithus, Russuia, Scleroderma, Suillus, Thelephora*, etc. Observations of Marks and Foster (1973) and the subsequent literature indicated a mycorrhizal infection zone, which is the initial site of interaction between the symbionts and further studies indicated that the initial site of contact being root cap. Horan and Chilvers (1990) have shown that the root apex was shown to be the source of chemotrophic stimulus to hyphae. Piche *et al.* (1986) have found that the contact of hyphae with the root surface appears to involve the formation of polysaccharides that affect the adhesion of hyphae to the root surface (Kannan and Natrajan, 1987). Giellant *et al.* (1993) reported that a lectin purified from *Lactarius deterrimus* preferentially bound to root hairs and tips of young lateral roots of *Picea abies* (L) Kast and that this binding was prevented by pretreatment of roots with B-D-galactosyl (1-3) D-N-acetyl glucosamine. Thus the possibility of a lectin-polysaccharide recognition system in mycorrhizal association is an important event in ectomycorrhizal association which needs more experimental evidences. Fungal hyphae continue to proliferate to form a mantle which consists of a compact hyphae, separating living root cells from the soil and these hyphae store glycogen, proteins and lipids. The hyphae of hartigs net undergo branching thus expanding the surface area for nutrient exchange. Martin and Tagu (1994) have emphasised the role of gene expression during ectomycorrhizal formation.

2.3. Significance

Ectomycorrhizae help in the phosphorus uptake especially in soils of its low availability. The application of ectomycorrhizal fungi may help in savings upto 50% of phosphorus if suitable methodologies are employed to derive maximum mycorrhizal response. The other benefits made available are increased nutrient and water absorption, nutrient mobilization, increase in feeder root longevity, accmulation of nitrogen, phosphorus, potassium, calcium, zinc and their translocation to the host tissue, minerals get mobilized in F and H horizons before they reach sub soil system, help in the degradation of complex minerals in soil and make them available to the host plant, afford protection to delicate root tissue, provide host plant with growth promoters, play significant role in nutrient cycling besides the entry of mycorrhizal fungal biomass into soil organic matter pool.

However, the ectomycorrhizal formation and their role are delimited/hampered in the presence of high P:N ratio, high soil fertility, high temperatures, alkalinity, water stress and their related parameters.

2.4. Ectomycorrhizae Research in India

Taxonomy of ectomycorrhizal fungi reached a sound footing because of the concerted research efforts made by Bakshi (1966; 1974); Bakshi and Thaper (1966); Bakshi *et al*,(1966). Mishra *et al.* (1981) and Natarajan (1977) who have consolidated the ectomycorrhizal association in some forest localities. Bakshi (1974) has made some attempts to get pure culture synthesis of some ectomycorrhizal fungi and their formation in *Pinus patula*. Kannan and Natarajan (1987) have established ectomycorrhizal association between *Pinus patula* and *Scleroderma citrinum*. Bakshi (1974) has studied the nutritional requirements of different mycorrhizal fungi and found the effect of different sources of carbon, nitrogen and vitamins.

2.5. Ectomycorrhiza in Forestation and Stabilization

Forestation and stabilization of waste lands viz. coal spoils, borrow pits, eroded soils and routine forestation sites have been more successful when tree seedlings tailored in the nursery with the ectomycorrhizal fungus *Pisolithus tinctorius* was used. Marxan Foster, (1973), Marks, (1975) and Marx *et al.* (1992) found that seedlings with ectomycorrhizae could survive and grow better than the non-mycorrhizal seedlings. *Pisolithus tinctorius* enhanced the growth of *Pinus taeda* and *Pinus strobus*. Trappe (1962) provided an information about the selection of ectomycorrhizal fungi in forest nurseries practice. The preliminary screening of mycorrhizae will be helpful for experimental inoculation of seedlings and planting them at different sites. *Pisolithus tinctorius* has evidenced better role in the establishment of pine seedlings even under extreme soil conditions like high temperature, drought, and low pH of the soil. Ectomycorrhizal mycelial strand can also form a link from plant to plant and thus provide pathway for the transfer of assimilates between individuals and for making available the sufficient quantities of water (Bjorkman *et al.,* 1967; Mukerji and Mandeep, 1998). Ectomycorrhizal mycelial slurries have proved to have good potenlial for use in inoculation of containerized tree seedlings besides being biofertilizer.

2.6. Nutrition in Plants Colonized by Ectomycorrhizal Fungi

Mycorrhizal fungi derive their organic nutrition (carbohydrates, vitamins, amino acids) from the primary cortical tissues of fine roots. In return the fungus benefits the tree chiefly by increasing solubility of nitrogen, phosphorus, potassium, calcium and other nutrients there by making them easily absorbable (Tables 1 & 2).

Table 1. Growth of one year old *Pinus* sp. from Araku, Waltair, A.P.

Treatment	Seeding height (cm)	% of Mycorrhizal roots	Total dry wt.
Soil	8.0	60.0	0.8
Soil + Inoculum *(Pisolithus tinctorius)*	12.0	80.0	0.88

Table 2. Macro and micro elemental composition (in mg) of one old *Pinus* sp.

Treatment		N	P	K	Zn
Soil	Shoot	6.1	0.4	0.6	0.09
	Root	1.2	0.1	0.62	0.13
Soil + Inoculum	Shoot	10.0	0.9	0.7	0.13
(Pisolithus tinctorius)	Root	1.6	0.18	0.7	0.14

Trees with abundant ectomycorhizae have a much larger physiologically active root fungus area for nutrient and water absorption than trees with few or no ectomycorrhizae.

The surface area increases due to branching habit of most ectomycorrhizae and is also due to extensive vegetative growth of hyphae of the fungal symbionts from the ectomycorrhizae into the soil. The extra matrical hyphae function as additional nutrient and water absorbing entities and assure maximum nutrient absorption from the soil by the host. Ectomycorrhizal plants are able to absorb and accumulate nitrogen, phosphorous, potassium and calcium in the fungus mantle more reapidly and for larger periods of times than non-mycorrhizal feeder roots (Lei *et al.,* 1990). Harley (1969) has reported that ectomycorrhizae remain active for periods ranging from several months to 3 years and such ectomycorrhizal association increase the tolerance of trees to drought, high soil temperatures, soil toxins and extremes of soil acidity besides paving way for biological control of pathogens. Ectomycorrhizal fungi produce some growth regulators and enzymes which change the morphology of roots regulate nutrient absorption respectively (Piche *et al.,* 1986). Bjorkman *et al.* (1967) have emphasized the role of mycorrhizal fungi in the liberation of nutrients from complex compounds in the forest soil. Fertilizer application lead to reduction of ectomycorrhizal infection (Piche *et al.,* 1986). In general ectomycorrhizae have been found to increase the uptake of nitrogen, phosphorus and potassium but few other elements have been investigated. Selected ectomycorrhizal inoculation increased calcium content of *Pinus radiata* (Harley, 1969; 1989; Harley and Smith, 1983). Bowen (1973) has pointed out that the studies on uptake of trace elements by ectomycorrhizal trees need to be strengthened as there are few concerted efforts made in this direction. Berry and Marx (1986) have stated that the use of inoculation of *Pisolithus tinctoriuin, Pinus taeda* and *P. echinata* lead to efficient uptake of Bo, Cu, Fe, Mo, Mn and Zn from sewage sludge. Mycorrhizal plants have shown increased growth and enhanced uptake of phosphorus. Mycorrhizal roots seem likely to increase 'P' uptake by exploring large volumes of soil thereby making positionally unavailable nutrients available. This is achieved by decreasing the distance for diffusion of phosphate ions all by increasing the surface area for absorption (Bjorkman, 1942; Bonsqnest *et al.,* 1986).

2.7. Phosphorus Uptake and Phosphotase Activity

It is an established phenomenon that the mycorrhizal fungi play a significant role in phosphate absorption. In general mycorrhizal effect decreases with increased supplies of soluble phosphate and therefore non-mycorrhizal plants show greater responsiveness towards fertilizer applications. One of the applications of mycorrhizal fungi therefore appears in greater utilization of unprocessed phosphate rocks that contain sparingly soluble phosphorus. Mycorrhizal association provides greater exploration of 'P' absorbing area, resulting in increased flow of phosphorus into the plant. Mycorrhizal combination with rock phosphate may improve greatly the formation and activity of root nodules of tree legumes in 'P' deficient soil. However, studies of ectomycorrhizal association in relation to root nodules are meagre. Harley (1969) suggested that production of phosphotases by ectomycorrhizal fungi is important in the solubilization of organic phytates, which constitute a large fraction of total phosphate in humic-soils. These enzymes are many times more active than those on non-mycorrhizal roots (Gianinazzi-Pearson and Gianinazzi, 1989; Marx *et al.,* 1982).The evidence of such induction of phosphatase activity in response to the lack of inorganic phosphate by ectomycorrhizal fungi have been brought out by some workers. (Alexander and Hardy, 1981; Bonsquest *et al.,* 1986; Miller, 1982). It has been shown that the ectomycorrhizae produce large amounts of calcium oxalate which help in the chelation of Fe and Al and there by release 'P' for plant uptake (Lei *et al.,* 1990; Trappe, 1987; Trebby *et al.,* 1989). Acid phosphatases (Ec 3.1 3.2) are generally considered to be involved in the phosphate nutrition of plants and to be key enzymes in the utilization of complexed phosphate esters by ecto and ericoid mycorrhizal systems.

Ectomycorrhizal plants are capable of increasing 'P' uptake especially in low fertility soils. In experiments using ^{32}P it was found that the mantle of ectotrophic mycorrhizae accumulates and stores 'P'. This accumulation remains consistent unless there is a dramatic change in the plants 'P' status. A phosphorus deficiency in the plant may stimulate the release by the mantle to the plant some of its accumulated 'P'. However, whether or not an ectotrophic mycorrhizae can store other nutrients is questionable. The association of *Boletus* on *Pinus* sp. has resulted in the accumulation of phosphorus. Ectomycorrhizal association has resulted in increased dry weights and biomass over non-mycorrhizal seedlings. Some workers (Berry and Marx, 1976; Harley, 1989; Marx, 1975; Marks and Foster, 1973; Richards and Wilson, 1963; Slankis, 1958), have inoculated *Quercus* sp. and *Pinus* sp. with eleven isolates of ectomycorrhizal fungi and observed that mycorrhizae increased foliar nutrients. Bowen (1973) reviewed the role of mycorrhizal fungi in the uptake of phosphorus by plants. Koide (1991) gave a comprehensive review of nutrient supply, nutrient demand and plant response to mycorrhizal infection. Some workers have suggested that nitrogen deposition from the atmosphere may damage the function of mycorrhiza even before root tip studies reveal any decline in the symbiotic state.

2.8.   Future Lines of Research

(i)   The mycofloristics of ectomycorrhizal fungi are incomplete from tropical region, hence relevant studies have to be taken up through concerted efforts.

(ii)   Indigenous and host specific ectomycorrhizal fungal strains have to be identified and have to be multiplied if they are utilized to the advantage of host.

(iii)   Critical role of temperature, soil conditions, sporulation, survivability, active state and other related aspects need to be studied indepth.

(iv)   Proper understanding of molecular genetics of ectomycorrhizal fungi including that of understanding gene expression, protoplast fusion and other related strategies would lead to improvement of the fungal component and thereby their practical applications. These studies may open up possibilities of introduction of the new traits in ectomycorrhizal fungi to the advantage of the host plant.

(v)   No information is available about the ecology of ectomycorrhizal fungi and it is necessary to protect the biodiversity of ectomycorrhizal fungi through critical taxonomic considerations and also to maintain germplasm of beneficial ectomycorrhizal fungi.

3.   CONCLUSIONS

Mycorrhizal plants suggest saving of 50% of phosphorus if proper management is observed by deriving maximum mycorrhizal response. Ectomycorrhizal inoculation is a common practice and it is of immense utility in forestry, silviculture and agroforestry operations in the tropics Experimental synthesis of ectomycorrhizae for field inoculation has undergone continuous change to meet the situations and then can be introduced in ectomycorrhizal fungi adopted to desired soil and enviromental conditions. The future success of ectomycorrhizal fungi utilization mostly depend on its molecular genetics and gene expression in particular so that the new traits can be introduced in ectomycorrhizal fungi to the advantage of the host plant.

ACKNOWLEDGEMENTS

The authors are thankful to Prof. K.G. Mukerji, Department of Botany, University of Delhi, Delhi, for his valuable suggestions and encouragement.

REFERENCES

Alexander, I.J. and Hardy, K. 1981, Surface phosphotase activity of sikka spruce mycorrhizals from a serpentise soil, *Soil. Biol. Biochem.* 13 : 301-305.

Bakshi, B.K. 1966, Mycorrhiza in *Eucalyptus* in India, *Ind. For.* 92 : 19-20.

Bakshi, B.K. 1974, Mycorrhiza and its role in forest, Forest Research Institute, Dehradun, PL-480 Project Report.

Bakshi, B.K and Thapar, H.S. 1966, Studies on mycorrhiza of chirpine (*Pinus roxburghii*), *Proc. Nat. Inst. Sci.* 32 : 6-20.

Bakshi, B.K., Thapar, H.S. and Singh, B. 1966, D N Type of mycorrhiza in six species of indian conifers, *Proc. Nat. Inst. Sci.* 32 : 1-15.

Berry, C.R. and Marx, D.W. 1976, Effect of *P. tinctorius* ectomycorrhizae on growth of loblolly and Virginia pines in Tennessee copper bagin, USDA *Forest Serv. Res.* Note SE-264.

Berry, A.M., McIntyre, L. and McCully, M.F. 1986, Fine structure of root hair infection leading to modulation in the *Frankia-Alnus* Symbiosis, *Can. J. Bot.* 64 : 292-305.

Bhandari, N.N. and Mukerji, K.G. 1993, *The Haustorium*, John Wiley and Sons Ltd., U.K.

Bjorkman, E. 1942, Uber dieBedingum-gen der mykor-rhizabildung bei kitter and fichte, *Symp. Bot. Upsal.* 6, No. 2.

Bjorkman, E., Lunderberg G. and Nommik, H. 1967, Distribution and balance of ^{15}N lablelled fertilizer nitrogen applied to young pine trees, *For. Sci.* 48 : 1-23.

Bousquest, N., Mousin, D. and Salac, A 1986, Use of phytase by ectomycorrhizal fungi, in : *Mycorrhizae Physiology and Genetics*, V. Gianinazzi-Pearson and S. Gianizazzi, eds., *Proc. Inst. Eur. Sym. Mycorrhizae*, Dijon INRA, Paris, pp.363-368.

Bowen, G.D. 1973, Mineral nutrition of ectomycorrhizae, in : *Ectomycorrhizae-Their Ecolgy and Phsiology*, G.C. Marks and T.T Kozlowshi, eds., Academic Press. N.Y. pp. 151-205.

Brownlee, C., Duddridge, J.A., Malibai, A. and Read, D.J. 1983, The structure and function of mycelial systems of ectomycorrhizal roots with special reference to their role in forming inter-plant connections and providing pathways for assimilates and water transport, *Plant Soil* 73 : 433-443.

Fontana, A. 1962, Researches on the mycorrhizae of the genus *Salix, Allionia* 8 : 67-85.

Frank, A.B. 1885, Liberdie out Wurzelhm biose beruuhende Evanurung Sewise, Baumedurch untervividche pits, *Ber. dt. bot. Ges.* 3 : 128-145.

Gerdemann, J.W. 1975, Vesicular-arbuscular mycorrhiza and plant growth, *Ann. Rev. Phytopathol.* 6 : 397-418.

Gianinazzi-Pearson, V. and Gianinazzi, S. 1989, Phosphorus metabolism in mycorrhizas, in : *Nitrogen, Phosphorus and Sulphur Utiliztion by Fungi*, L.Boddy, R. Marchant and D.J. Read, eds., Cambridge University Press, Cambridge, pp.227-241.

Giellant, M., Guillot, J., Damez, M., Dusser, P. and Didler, E. 1993, Characterization of a lectin from *Lactarius deterrimus, Pl. Physiol.* (Lancaster) 101 : 513-522.

Harley, J.C. 1969, *The Bioloy of Mycorrhizae* (2nd ed.), Leonard Hill, London.

Harley, J.L. 1989, The significance of mycorrhiza, *Mycol. Res.* 92 : 129-139.

Harley, J.L. and Smith, S.E.1983, *Mycorrhizal Symbiosis*, Academic Press, London.

Henderson, G.A. and Stone, E.L. 1970, Interactions of phosphorous availability, mycorrhizae, and soil fumigation on coniferreadings, *Agron. Abstr.*, p.134.

Horan, D.P. and Chilvers, G.A. 1990, Chemotropism - the key to ectomycorrhizal formation, *New Phytol.* 116 : 297-301.

Kannan, K. and Natarajan, K. 1987, Pure culture synthesis of *Pinus patula* ectomycorrhizae with *Scleroderma citirinum, Curr. Sci.* 56 : 1066-1068.

Koide, R.T. 1991, Nutrient supply, nutrient demand and plant response to mycorrhizae infection, *New Phytol.* 117 : 365-368.

Kucey, R.M.N., Janzen, H.H and Leggett, M.E. 1989, Microbially mediated increases in plant available phosphous, *Adv. Agron.* 42 : 199-229.

Lei, J., Lapeyrie, F., Malajczuk, N. and Dexheimer, J. 1990, Infectivity of pine and eucalypt isolates of *Pisolithus tintorius* on roots of *Eucalyptus urophylla in vitro* 11, Ultrastructural and biochemical changes at the early stage of mycorrhiza formation, *New Phytol.* 116 : 115-122.

Malazenuk, N. and Cromack, K. 1982, Accumulation of calcium oxalate in the mantle of ectomycorrhizae roots of *Pinus radiata* and *Eucalyptus marginata, New Phytol*. 92 : 527-531.

Marks, G.C. and Foster, R.C. 1973, Structure, morphogenesis and ultrastructure of ectomycorrhizae, in : *Ectomycorrhizae: Their Ecology and Physiology*, G.C. Marks and T.T. Kozlowski, eds., Academic Press, London, pp. 1-41.

Martin, F. and Tagu, D. 1994, Ectomycorrhizae development : a molecular perspective, in : *Mycorrhiza, Structuture, Function, Molecular Biology and Biotechnology*, A. Varma and B. Hock, eds., Springer-Verlag, Heidelberg, Germany.

Marx, D.H. 1975, Mycorrhizae and establishment of trees on strip mined land, *The Ohio J. Sci*. 75 : 288-297.

Marx, D.H., Ruhle, J.L., Kenney, D.S., Cardell, C.E., Riffle, J.W., Molina, M.J., Pawu, K. W.H., Navratil, S., Tinus, R.W. and Goodwin, O.C. 1982, Inoculum of *Pisolithus tinctorius* and inoculation techniques for development of ectomyorrhizae on container grown tree seedlings, *For. Sci*. 28 : 373-400.

Miller, O.K. Jr. 1982, Taxonomy of ecto and ecto-endomycorrhizal fungi, in : *Methods and Principles of Mycorrhizal Research*, N.C. Schenck, ed., Am. Phytopath. Soc. Minn., pp. 91-101.

Mishra, R.R., Sharma, G.D. and Tiwari, B.K. 1981, Report of *Fusarium* with in *Pinus kesiya* Royle, *Curr. Sci*. 50 : 36.

Mousin, D., Bousquet, N. and Polard, C. 1988, Comparison des activities phosphatases d' Homobasidiomycetes ectomycorrhiziens exculture *in vitro, Eur. J. For. Path*. 18: 299-309.

Mukerji, K.G. and Mandeep 1998, Mycorrhizal relationships of wetlands and rivers associated plants, in : *Ecology of Wetlands and Associated System*, S.K. Mazumdar, E.W. Miller and F.J. Brenner, eds., the Penn. Acad. Sci. Vol. XXV., pp. 240-257.

Natarajan, K. 1977, Indian Agaricales, *Kavaka* 5: 35-42.

Peterson, R.L. and Farquhar, M.L. 1994, Mycorrhiza-Integrated development between roots and fungi, *Mycologia* 3 : 311-326.

Piche, Y., Ackerley, C.A. and Peterson, R.I. 1986, Structural characteristics of ectendomycorrhizas synthesized between roots of *Pinus resinosa* and the F-strain fungus, *Wilcoxina mikolae* var. *mikolae, New Phytol*. 104 : 447-452.

Richards, B.N. and Wilson, G.L. 1963, Nutrient supply and mycorrhiza development in caribbean pine, *For. Sci*. 9 : 405-412.

Slankis, V. 1958, Mycorrhiza of forest trees, *Proc. Ist NAAFS Conf.*, Agricultural Experiment Station, Michigan Stae University, East Lansing, pp. 130-137.

Suvercha, Mukerji, K.G. and Arora, D.K. 1991, Ectomycorrhiza, in : *Hand Book of Applied Mycology*, Vol. I, Soil and Plants, D.K. Arora, B. Rai, K.G. Mukerji and G.R. Knndsen, eds., Marcel Dekker Inc., New York, pp. 187-215,

Trappe, J.M. 1962, Fungus associates of ectotrophic mycorrhizae, *Bot. Rev*. 28 : 538-606.

Trappe, J.M. 1987, Phylogenetic and ecological aspects of mycotrophy in the angiosperms from an evolutionary stadpoint, in : *Ecophysiology of VA Mycorrhizal plants,* N.S.Safir, ed., CRC, Boca Raton, Fla., pp. 5-25.

Trebby, M., Manschrer, H. and Ramheld, V. 1989, Mobilization of iron and other micronutrient cations from a calcareas soil by plant-borne microbial and synthetic metal chelators, *Plant Soil* 114 : 217-226.

MYCORRIZOSPHERE : INTERACTION BETWEEN RHIZOSPHERE MICROFLORA AND VAM FUNGI

M. Bansal, B.P. Chamola, N. Sarwar, and K.G. Mukerji

Applied Mycology Laboratory
Department of Botany
University of Delhi
Delhi-110007, INDIA

1.   INTRODUCTION

Mycorrhizosphere includes the region around the mycorrhizal roots. The concept of mycorrhizosphere is based on the fact that mycorrhizae exert a strong influence on the microflora in the rhizosphere (Bansal and Mukerji 1994; Fitter and Garbaye, 1994; Garbaye, 1991; Mukerji *et al.*, 1998; Paulitz and Linderman, 1991). As a result, there are taxonomic and functional differences in populations of bacteria, fungi and nematodes associated with rhizospheres of vesicular arbuscular mycorrhizal (VAM) and non-VAM plants. How, the symbiotic association with mycorrhizal fungi alters the rhizosphere microflora is one of the fundamental questions in mycorrhizal research.

During the past decade, our knowledge on interaction of VAM fungi with rhizosphere microflora has expanded. Reviews are appearing in literature about potential of VA mycorrhizal fungi in biological control of plant diseases (Azcon Aguilar and Barea, 1996; Fitter and Garbaye, 1994; Fitter and Miller *et al.*, 1985; Jalali and Jalali, 1991; Miller *et al.*, 1985; Morandi, 1996; Schenck, 1981). However, there is a need to understand the mechanisms of interactions in mycorrhizosphere for effective application of VAM fungi as a tool in increasing the crop yield. In this paper work done during the last decade in the field of interaction mechanisms is reviewed.

2.   TYPES OF INTERACTIONS

Interactions among VAM fungi and other soil microorganisms are reciprocal i.e. mycorrhizal fungi affect the other soil microbes and other microbes inturn affect the mycorrhizal association. The interactions can be grouped as follows :

2.1.   Positive/Synergistic

VAM fungi are known to support the growth of various groups of bacteria viz., symbiotic bacteria free-living di-nitrogen fixers, plant-growth-promoting bacteria, and phosphorus-solubilising bacteria. There may not only be better germination of VAM fungal spores in presence of bacteria but more vigorous hyphae may be formed from the germinated spores (Mayo *et al.*, 1986). Certain

Mycorrhizal Biology, edited by Mukerji *et al.*
Kluwer Academic/Plenum Publishers, 2000

groups of saprophytic fungi are also known to grow better in presence of VAM fungi. Important references for synergistic interaction of VAM fungi with various rhizosphere microorganisms are given in Table 1.

Table 1: Synergistic interactions between VAM fungi and other soil microorganisms

Name of VAM fungi	Rhizosphere microorganism	Reference
	ACTINOMYCETES	
Glomus fasciculatum	*Frankia*	Gardener *et al.* (1984)
G. fasciculatum	*Streptomyces cinnamomeous*	Krishna *et al.* (1982)
	SYMBIOTIC BACTERIA	
Glomus mosseae	*Rhizobium leguminosarum*	Reinhard *et al.* (1993)
G. versiforme	*R. loli*	Valdes *et al.* (1993)
	FREE LIVING NITROGEN FIXER	
Glomus fasciculatum	*Beijerinckia mobilis*	Manjunath *et al.* (1981)
G. fasciculatum	*Azotobacter chroococcum*	Bagyaraj and Menge (1978)
G. fasciculatum	*Azospirillium brasilense*	Subba Rao *et al.* (1985)
G. margarita		
G. versiforme	*Corynebacterium* sp.	Mayo *et al.* (1986)
	PLANT GROWTH PROMOTING BACTERIA	
Glomus versiforme	*Pseudomonas* sp.	Mayo *et al.* (1986)
		Azcon *et al.* (1978)
	PHOSPHATE SOLUBILISING BACTERIA	
Endogone sp.	*Agrobacterium* sp.	Azcon *et al.* (1978)
	Pseudomonas sp.	
G. macrocarpum	*Bacillus megaterium*	Raj *et al.* (1981)
	Pseudomonas fluorescence	
	FUNGI	
Glomus macrocarpum	*Cladosporium* sp.	Bansal and Mukerji (1994b)
	Gliocladium virens	Paulitz and Linderman (1991)
G. intraradices	*Fusarium oxysporum* f. sp. *chrysanthemi*	Arnaud *et al.* (1995)

Positive interactions of VAM fungi with various dinitrogen fixing and phosphorus solubilising bacteria and the saprophytic fungi is the basis of application of VAM fungi as biofertiliser (Bagyaraj and Menge, 1978; Manjunath *et al.*, 1981; Raj *et al.*, 1981; Valdes *et al.*, 1993) and biocontrol agents (Bansal and Mukerji, 1994; Mukerji, 1999; Mukerji *et al.*, 1997) respectively.

2.2. Negative/Antagonistic

Most of the negative interactions of VAM fungi deal with plant pathogenic fungi and nematodes (Table 2). There are several reoports of VAM fungi inhibiting the occurrence of various pathogenic fungi (Dehne, 1982; Jalali and Jalali, 1991; Miller *et al.*, 1985) but there are relatively few reports of soil fungi inhibiting the growth and multiplication of VAM fungi. Soil microorganisms inhibit the germination and growth of VAM fungi. This is apparent from the fact that there is suppression of VAM fungal spore germination and growth responses in some non-sterile soils (Paulitz and Linderman,

1991) Recently the work of Pinochet *et al.* (1995) showed that nematodes may reduce spore and vesicle production of VAM fungi.

Table. 2 : Antagonistic interactions of VAM fungi with rhizosphere microorganisms

VAM Fungus	Soil Microorganism	Reference
	FUNGI	
Glomus mosseae	*Phytophthora nicotianae* var. *parasitica*	Trotta *et al.* (1996)
Gl. macrocarpum	*Fusarium* sp.	Bansal and Mukerji (1994b)
G. intraradices	*Fusarium oxysporum* f.sp. *radicis-lycopersici*	Caron *et al.* (1986) Dugassa *et al.* (1996)
G. intraradices	*Ordium lini*	Dugassa *et al.* (1996)
G. fasciculatum	*Pythium ultimum*	Kave *et al.* (1984)
G. fasciculatum	*Aphanomyces euteiches*	Rosendahl (1985)
Glomus sp.	*Verticillium albo-atrum*	Jalali and Jalali (1991)
Glomus sp.	*Rhizoctonia solani*	Jalali (1986)
G. fasciculatum	*Phytophthora perasitica*	Davis and Menge (1981)
G. fasciculatum	*P. fragariae*	Norman *et al.* (1996)
G. etumicatum *G. fistulosum*	*P. fragariae*	Mark and Casells (1996)
Gigaspora margarita	*Meloidogyne hapla*	Grandison and Copper (1986)
Glomus intraradices	*Pratylenchus vulnus*	Pinochet *et al.* (1995)
G. etunicatum	*Rhodopholus similis*	O'Bannon and Nemec (1979)
G. manihotis	*M. incognita*	Palacino and Leguizamon (1991)
	BACTERIA	
G. mosseae	*Azotobacter chrococcum*	Paulitz and Linderman (1991)

Mycoparasitism of VAM fungi has been reported for over 40 years. Mycoparasites include *Fusarium* sp., *Verticillium* spp., *Penicillium* spp., *Trichoderma* spp., *Stachybotyris* spp., *Phlyctochytrium* sp. and *Humicola* sp., which contaminate the outside of VAM fungal spores and survive there as such (Bhattacharjee *et al.,* 1982; Koske, 1981; Sylvia and Schenck, 1983). The negative/antagonistic interactions of VAM fungi with various soil borne plant pathogens is also the reason for their use as biocontrol agents (Caron *et al.,* 1986; Kaye *et al.,* 1984; Pinochet *et al.,* 1995).

3.   MECHANISMS OF INTERACTION

Mechanisms involved in interaction of VA mycorrhizal fungi and other rhizosphere microorganisms are grouped as follows:

3.1.   Physical Mechanism

This includes direct physical competition between endomycorrhizal fungi and other rhizosphere microorganisms so as to occupy a more suitable space in the altered root architecture (Davis and Menge, 1981). VAM fungi physically limit the number of available colonisation sites in roots and reduce the occurrence of other microorganisms. Localised morphological liginification of endomycorrhizal cell walls is suggested to increase the resistance against wilt diseases in tomato and cucumber (Dehre *et al.*, 1978). This hypothesis is supported by localised occurrence of arbuscular mycorrhizal effects i.e. the inhibition/stimulation of rhizosphere microorganisms is restricted to the site of mycorrhizal development (Rosendahl, 1985). However Norman *et al.* (1996) observed that root architecture does not play a role in the differences of genetic susceptibility to the pathogen.

3.2. Physiological Mechanism

Physiological mechanism of interaction can be explained in two ways:

3.2.1. Direct physiological effects

When a plant becomes endomycorrhizal there are significant changes in its physiology. VAM fungi directly influence the metabolic state of a cell by altering energy input and expenditure. Such changes in metabolism of a cell result in, altered rate of photosynthesis, respiration, enzyme activity and partitioning of photosynthate. (Gianinazzi-Person and Gianinazzi, 1994; Dugassa *et al.*, 1996; Dumas-Gaudot *et al.*, 1996) These changes in the physiology are reflected in more or less permanent alterations of biochemicals of cells, which in turn influence the composition of microorganisms in the rhizosphere.

Qualitative and quantitative differences in photosynthate are reported in mycorrhizal and non-mycorrhizal plants. Higher concentration of amino acids especially arginine (Baltruschat and Schonbeck, 1972; Graham *et al.*, 1981); soluble carbohydrate and sugars (Amijee *et al.*, 1993; Schenck, 1981), phenol metabolism and lignification (Dehne *et al.*, 1978; Moraudi; 1996) chitinolytic (Dehne *et al.*, 1978; Dumas-Gaudot *et al.*, 1996; Hodge *et al.*, 1995), acid phosphatase (Kieliszewska-Robiska, 1992) and peroxidase and esterases activity (Spanu and Bofante-Fasolo, 1988) has been reported in mycorrhizal plants. Increased glutamine synthetase, glutamate synthetase and glutamate dehydrogenase activity, succinate dehydrogenase and malate dehydrogenase activity (Saito, 1995) are also associated with mycorrhizal plants. The colonisation of roots by VAM fungi induces biochemical changes within host tissue which could be grouped as follows :

3.2.1.1. Stimulation of phenyl propanoid pathway

Plant phenolics in particular flavonoids/isoflavonoids are very important in biological control because of their anti microbial activity phytolaxenins play an important role in plant defense mechanisms (Bailey, 1982; Hallbrock and Scheel, 1989). It has been observed that phytoalexins and associated molecules accumulate in roots after mycorrhizal colonisation, but less intensively and more slowly in pathogenic interactions (Harrisson and Dixon, 1993). Following mycorrhizal colonisation enzyms of phenyl propanoid metabolism have been shown to be activated differently. Some flavonoids and isoflavonoids have been reported to stimulate *in vitro* germination of mycorrhizal fungi or *in vitro* mycorrhizal colonisation (Morandi, 1996) but their biological significance in signalling between two symbiotic partners and in biocontrol of plant diseases by vesicular arbuscular mycorrhizal fungi have not yet been illucidated.

3.2.1.2. Enhancement of certain hydrolases activity

Plant hydrolytic enzymes i.e. chitinases and ß-1,3 glucanases are proposed as potential anti fungal compounds in disease control (Boller *et al.*, 1983; Dumas - Gaudot *et al.*, 1996; Mauch *et al.*, 1988). Chitenases (Poly 1,4-N-acetyl-ß, glucosanine) glucanohydrolases, E.C. 2.3.1.14), catalyses the hydrolysis of chitin, a linear homopolymer of ß-1,4 linked N-acetyl glucasamine residue. ß-1-3 glucanases (E.C. 3.2.1.39) degrade 1,3,B-D glycosidic linkages in ß-D-glucans and sometimes even 13:1,6-ß glucans. Since the cell walls of many fungi contain chitin and or ß-D glucans as a major structural component, chitinases and ß-1.3-glucanases were proposed as potential antifungal compounds (Mauch *et al.*, 1988). This hypothesis has been reinforced by both *in vivo* experiments (Arlorio *et al.*, 1992; Brockaert *et al.*, 1988; Mauch *et al.*, 1988) and using transgenic plants over expressing either plant or microbial chitinases or glucanase genes (Broglie *et al.*, 1991).

3.2.1.3. Activation of defence related genes

Low priming of defence related pathways in roots by VAM fungi may provoke more rapid responses to subsequent pathogen so contribute to bioprotective effects through mechanisms similar to preimmunisation after pre-infection by hypo virulent viruses, bacteria or fungi (Gianinazzi *et al.,* 1996; Harrisson and Dixon, 1994; Ryals *et al.,* 1994).

3.2.1.4. Synthesis of proteins of unknown functions

These physiological changes in roots have a direct bearing on changes in rhizosphere microflora eg. arginine in root extract is known to reduce chlamydospore production in *Thelaviopsis basicola* (Dehne *et al.,* 1978). Similarly higher chitinolytic activity is implicated in reduced fungal infection. Deposition of lignin in the cell walls of endodermis and stele is responsible for restriction of endophyte to the cortex (Miller *et al.,* 1985). Accumulation of phenols in host cells during fungal colonisation increase the resistance against fungal penetration and acts as physical barrier (Krishna *et al.,* 1982 Morandi, 1996). Increased concentration of soluble carbohydrate and reducing sugars in mycorrhizal plants is indirectly linked to decreased pathogen activity (Amijee *et al.,* 1993; Bansal and Mukerj, 1994; Schenck, 1981). VAM fungal spores of *Glomus mosseae* also contain certain hydrolytic enzymes (pectinase, cellulase, hemicellulase) that are involved in fungal penetration in host root (Garcia Garrido *et al.,* 1992). Higher ATPase activity is associated with energisation of cells and is involved in transport of phosphorus (Gianinazzi-Pearson *et al.,* 1991).

Endomycorrhizal (VAM) symbiosis causes a modification of free sterols which could influence the development of biotrophic foliar pathogen feeding *via* haustoria (Bhandari and Mukerji, 1993; Dugassa *et al.,* 1996; Losel, 1991). The nutrient transfer between host and fungal cells may be mediated by sterols. The higher amount of sterols could stimulate the pathogen development e.g. some members of Pythiaceae are highly dependent on external sterols (Hendrix, 1970).

The VAM fungus after establishing inside a fine root, alters photosynthate leakage from root (Graham *et al.,* 1981). The influence of root exudates on microbial population in vicinity of roots is well known (Bansal and Mukerji, 1996; Curl and Truelove, 1983). Quantitative decrease in number of fungal colonies in the rhizosphere of *Leucaena leucocephala* is positively correlated with quantitative decrease in exudation of amino acids and sugars from mycorrhizal roots (Bansal and Mukerji, 1994). There is also possibility of exudation of antifungal substances from mycorrhizal roots. In case of nematodes also VAM fungi alter the host root exudates and change their chemotactic attraction (Pinochet *et al.,* 1995). VAM fungi and rhizobacteria are also suggested to interact *via* root exudation (Prikryl and Vancura, 1980).

VAM linseed plants show a higher CO_2 assimilation, besides direct damage, a biographic pathogen causes indirect damages through changes in assimilated allocation. It induces a sink in infected tissue which competes for assimilates in the sink, i.e. active developing plant parts such as shoot apex. This damage may be correlated with pathogen activity, which can be measured in terms of sporulation rates. Although the sporulation of the mildew fungus is estimated in AM linseed plants, these plants contained even more sucrose at the shoot apex. than non-AM ones. Consequently the AM plants suffered less from pathogen attack in terms of fresh weight production. This is an evidence that positive correlation between disease interrity and its damages, enabling the host plants to maintain their productivity, this AM effect inducing tolerance which is based on a stronger sink activity of shoot apex (Duggass *et al.,* 1996; Trotta *et al.,* 1996). This effect could be multiple. The increased concentration of the growth promoting, phytohormones such as cytokinins, auxins and gibberellins are one of the significant effects of VAM/AM association, these phytohormones increased photosynthetic activity (Druge and Schonbeck, 1992) and translation in the plant sinks (Miyamoto *et al.,* 1993).

3.2.2. Indirect effect of improved phosphorus nutrition

Researchers supporting this mechanism hypothesize that the influence of VAM fungi on rhizosphere microflora is not direct. Rather it is an outcome of improved phosphorus nutrition in mycorrhizal plants. Mycorrhiza through improved phosphorus uptake, indirectly alters the activity of microorganisms that responds to qualitative and quantitative changes in root metabolites in and outside root (Graham, 1988). The effect of mycorrhizal root colonisation of *Verticillium* (in case of cotton) and *Phytophthora* (on *Citrus*) are attributed to improved P-nutrition (Davis and Menge, 1981). Similarly increased resistance of mycorrhizal plants to nematodes is said to be due to improved phosphorus nutrtion (Smith, 1988).

Some studies indicate that increased growth response of diseased plants colonised by VAM fungi can be mimicked by adding phosphorus exogenously (Davis and Menge, 1981; Smith, 1988). Graham and Egel (1988) studied the role of P-nutrition in interaction between *G. intraradices* and *Phytophthora parasitica*. They observed that VAM fungus did not increase the resistance or tolerance unless it gave the plant increased P-nutrition over non-mycorrhizal plants.

However, not in all cases are effects of VAM fungi linked to phosphorus nutrition. For example, an increase in phosphorus level in stem or leaves of tomato had no influence on the population of *Fusarium oxysporum* (Paulitz and Linderman, 1991). Similarly in case of *Leucaena leucocephala* the VAM induced alterations in the rhizosphere could not be mimicked by exogenously supplied phosphorus (Bansal and Mukerji. 1994). Recently Trotta *et al.* (1996) studying the mechanism of interaction of soil and root borne pathogen *Phytophthora nicotianae* and *Glomus mosseae* observed the suppression of former.

3.3. Direct Mechanisms

There are various examples in literature which show direct stimulatory or antagonistic relations between VA mycorrhizal fungi and other microorganisms (Tilak *et al.*, 1987). Involvement of various water soluble, volatile or diffusible substances from rhizosphere bacteria in stimulation of VAM fungal hyphal growth and formation of secondary vesicles is suggested (Azcon, 1987).

Direct spatial association of VAM fungi (*G. fasciculatum, G. intraradices, G. mosseae* and *Gigaspora gilmori*) with some free living di-nitrogen fixers (*Azospirillum, Pseudomonas*) is reported (Tilak *et al.*, 1987). *Azotobacter brasilense* has also been isolated from root cortex with VAM fungi besides rhizosphere (Tilak and Subba Rao, 1987). This supports direct interaction between a VAM fungus and *Azospirillum* within the plant.

Involvement of growth hormones viz. auxins, gibberellins, cytokinins in interaction of bacteria with VAM fungi is also known (Azcon *et al.*, 1978). *Azotobacter vinelandii* and *A. beijerinckii* are known to produce phyto hormones like auxins, gibberellins and cytokinins in culture. Their effects on AM fungi can be simulated by exogenous addition of combination of gibberellic acid, kinetin and IAA (Azcon *et al.*, 1978). The growth promoting substances might also act directly on VAM fungi or these might produce morphological changes in plants which in turn influence mycorrhizal symbiosis (Paulitz and Linderman, 1991).

Direct parasitism (Mycoparasitism) of VAM fungi may occur (Bhattactharjee *et al.*, 1982; Sylvia and Schenck, 1983), where spores of VAM fungi are found to be parasitized by variuos other fungi including chytridiaceous fungi, *Fusarium* sp., *Penicillium* sp., *Trichoderma* sp., *Humicola* sp., etc. These fungi may play an important role in limiting the population of AM fungi. A number of bacteria have also been isolated from crushed surface of partially disinfected spores including *Pseudomonas, Streptococcus,* etc. (Paulitz and Linderman, 1991). Bacteria are also seen on inner spore membrane of *Glomus microcarpum* including anaerobic, fermentative and heterotrophic bacteria (Singh and Verma, 1985). Similarly presence of bacteria like organelles (BLOs) (Mac Donald *et al.*, 1982) and Spiroplasma like oranisms (SLOS) are reported in AM fungi (Tzean *et al.,* 1983). The mode of interaction of these microorganisms is not known but the direct microbial interaction/association is confirmed.

4.   LIMITATIONS AND PROSPECTS FOR FURTHER RESEARCH

(i)   The final effect of VAM fungi as biofertiliser/biocontrol agent can be seen in as series of events. This includes spore germination, growth of mycelium through soil, stimulation and attachment of infective hyphae to root, penetration of root, secondary spread inside the root, production of vesicles, arbuscules and spores. Most of the studies to date have given more emphasis on the root colonisation as a fungal parameter measure. Since VAM fungi are obligate symbionts and cannot be grown independently of plant, this makes it very difficult to study the exact mechanisms and stage of interaction involved.

(ii)   As VAM fungi cannot be cultured axenically the effect can only be observed on plant root. In order to work out the mechanism involved in interactions and to harness their potential as viable biocontrol/biofertilizer, workable technique for commercial production of VA mycorrhizal fungi needs to be designed.

(iii)   The mere presence of mycorrhizal fungi on roots of a plant does not ensure a positive effect on a disease or suppression of a pathogen. Rather the effects are very specific, both on host and companion microorganisms. Therefore, the mycorrhizal fungi on roots of a plant need to be very specifically managed for each VAM microbe combination. Standardisation of a particular VAM-pathogen or VAM fungus-biofertilizer partner combination for plant is also very important.

(iv)   There is an apparent difficulty in determining the biologically significant differences in population density of microbes in the rhizosphere. Statistical differences in many cases are not obvious.

(v)   The influence of VAM fungi on other rhizosphere microorganisms has not been demonstrated under field conditions despite the promising results in green house studies. Little effort has been made to use natural systems in evaluating the VAM fungi and their interactions with other microorganisms. For this reason the results from studies to date are very difficult to correlate with field situation.

5.   CONCLUSIONS

How do VAM fungi change the spectrum of microflora in mycorrhizosphere is still not clearly undrstood. Interactions among VAM-rhizosphere microorganism-host appear to be complex and seem to vary with each combination. Also, the proposed mechanisms still await experimental confirmation. Nonetheless generalisations can be made from research results of the past three decades. In general, indirect physiological mechanisms are more widely accepted and applied. In physiological mechanism, the interactions *via* root exudates need to be explored in large number of host-VAM fungus-rhizosphere microorganisms under different conditions.

In mycorrhizosphere, the physiological mechanisms alone may not be determing the microbial make up but the microbial interactions (positive and negative) may also be playing pivotal role. Production of several metabolites (growth hormones, toxins, antibiotics, etc.) by the associated micro-organisms are also important.

REFERENCES

Amijee, F.D., Stribley, P. and Tinker, P.B. 1993, The development of endomycorrhizal root systems VIII, Effects of ~oil phosphorus and fungal colonisation on the concentration of soluble carbohydrates in roots, *New Phytol.* 123: 297-306.

Arlorio, M., Ludwig, A., Boller, T. and Bonfante, P. 1992, Inhibition of fungal growth by plant chitinases and B-1, 3 gluec anases, amorphological study, *Protoplasma* 171: 34-43.

Azcon, R. 1987, Germination of hyphae growth of *Glomus mosseae in vitro*, Effect of rhizosphere bacteria and cell free culture media, *Soil Biol. Biochem.* 19 : 417-419.

Azcon, R., Azcon Aguilar, C. and Barea, J.M. 1978, Effects of plant hormones present in bacterial culture on formation and responses to VA endomycorrhiza, *New Phytol.* 80: 359-364.

Azcon-Aguilar, C. and Barea, J.M. 1996, Arbuscular mycorrhizas of biological control of soil-borne plant pathogen, on overview of mechanics involved, *Mycorrhiza* 6: 457-464.

Bagyaraj, D.J. and Menge, J.A. 1978, Interaction between a VA mycorrhiza and *Azotobacter* and their effects on rhizosphere microflora and plant growth, *New Phytol.* 80: 567-573.

Bailey, J.A. 1982, Physiological and biochemical events associated with the expression of resistance to disease, in : *Active Defence Mechanisms in Plants*, R.K.S. Woods, ed., Plenum Publishing Corporation, Bristol, U.K. pp.39-65.

Baltruschat, H. and Schonbeck, F. 1972, Influence of endotrophic mycorrhizas on chlamydospore production of *Thielviopsis basicola* in tobacco root, *Phytopath.* 74: 358-361.

Bansal, M. and Mukerji, K.G. 1994, Positive correlation between root exudation and VAM induced changes in rhizosphere mycoflora, *Mycorrhiza* 5: 39-44.

Bansal, M. and Mukerji, K.G. 1996, Root exudates in rhizosphere biology, in : *Concept in Applied Microbiology and Biotechnology*, K.G. Mukerji and V.P. Singh, eds., Aditya Book, New Delhi, India, pp. 98-120.

Bhandari, N.N. and Mukerji, K. G. 1993, *The Haustorium,* Research Studies Press Ltd, U.K. and John Willey & Sons Inc., New York, U.S.a. p. 308.

Bhattacharjee, M., Mukerji, K.G., Tewari, J.P. and Skorapad, W.P. 1982, Structure and hyperparasitism of a new species of *Gigaspora, Trans. Bri. Mycol. Soc.* 78: 184-188.

Boller, T., Gehri, A., Mauch, F. and Vogeli, U. 1983, Chitenase in bean leaves : induction by ethylene, purification, properties and possible function, *Planta* 157: 22-31.

Broekaert, W.A., Van Parijs, Allen, A.K. and Peumans, W.A. 1988, Comparison of some molecular enzymatic and antifungal properties of chitinases from tobacco and wheat, *Physiol. Mol. Plant Pathol.* 33: 319-331.

Caron, M., Fortin, J.A. and Richard, C. 1986, Effect of *Glomus intraradices* on infection by *Fusarium oxysporum* f.sp. *radicis lycopersici* in tomatoes over 12 week period, *Can. J. Bot.* 64: 552-556.

Curl, E. A. and Truelove, B. 1983, The *Rhizosphere*, Springer Verlag, Heidelberg.

Davis, R.M. and Menge, J.A. 1981, *Phytophthora parasitica* inoculation and intensity of vesicular arbuscular mycorrhizae in *Citrus, New Phytol.* 87: 705-715.

Dehne, H.W. 1982, Interaction between vesicular-arbuscular mycorrhizal fungi and plant pathogens, *Phytopath.* 78: 1115-1118.

Dehne, H.W., Schonbeck, F. and Baltruschat, H. 1978, The influence of endotrophic mycorrhizas on plant diseases, 3, Chitinases activity and ornithine-cycle, *Pflanzenkranth Pflanzenschutz* 85: 666-678.

Druge, U. and Schonbeck, F. 1992, Effect of vesicular-arbuscular mycorrhizal infection on transpiration, photosynthesis and growth of flora *Linum usitatissimum* L., in relation to cytokinin levels, *J. Plant Physiol.* 141: 40-48.

Dugassa, G.D., Alten, N. and Schoenbeck, F. 1996, Effect of arbuscular mycorrhiza AM, on health of *Linum usitatissimum* L. infected by fungal pathogens, *Plant Soil* 185: 171-182.

Dumas-Gaudot, E., Gusllaume, P., Tahiri - Alaoui, A., Gianinazzi-Pearson, V. and Gianinazzi, S. 1994, Changes in polypeptide pattern in tobacco roots colonised by two *Glomus* species, *Mycorrhiza* 4: 215-221.

Dumas-Gaudot, E., Slezack, S., Dassi, B., Pozo, M.J., Gianinazzi-Pearson, V. and Gianinazzi, S. 1996, Plant hydrolytic enzymes chitinases & β 1,3 glucanases, in root reactions to pathogenic and symbiotic microorganism, *Plant Soil* 185: 211-221.

Fitter, A.H. and Garbaye, J. 1994, Interaction between mycorrhizal fungi and other soil organisms, *Plant Soil* 159: 123-132.

Garbaye, J. 1991, Biological interactions in the mycorrhizosphere, *Experientia* 47: 370-375.

Garcia-Garrido, J.M., Garicia Romera, I. and Ocampo, J.A. 1992, Factors stimulate ^{32}P uptake and plasmalemma ATPase activity in vesicular arbuscular mycorrhizal fungus *Gigaspora margarita, New Phytol.* 118: 289-294.

Gardener, I.C., Clelland, D.M. and Scott, A. 1984, Mycorrhizal improvement in non-leguminous nitrogen fixing associations with particular reference to *Hippophae rhamnoides, Plant Soil* 78: 189-199.

Gianinazi-Pearson, V. and Gianinazzi, S. 1989, Cellular and genetical aspects of interactions between hosts and fungal symbiosis in mycorrhizae, *Genome* 31: 336-341.

Gianinazzi-Pearson, V., Smith, S.R., Gianinazzi, S. and Smith, F.A. 1991, Enzymatic studies on metabolism of vesicular-arbuscular mycorrhizas, V, Is H$^+$ ATPase a component of ATP hydrolysing enzyme activities in host fungus interface? *New Phytol.* 117: 67-75.

Gianinazzi-Pearson, V., Dunglas-Gandot, E., Gollotte, A., Tahiri-Alaovi, A. and Gianinazzi, S. 1996, Cellular and molecular defense-related root response to invasion by arbuscular mycorrhizal fungi, *New Phytol.* 133: 45-48.

Graham, J.H. 1988, Interaction of vesicular-arbuscular mycorrhizal fungi with soil-borne pathogens on take-all disease of wheat, *Phytopath.* 72: 95-98.

Graham, J.H. and Egel, D.S. 1988, *Phytophthora* root rot development on mycorrhizae and phosphorus fertilised non-mycorrhizal sweet orange seedlings, *Plant Disease* 72: 942-946.

Graham, J.H. and Menge, J.A. 1982, Influence of vesicular arbuscular mycorrhizae and soil phosphorus on take all disease of wheat, *Phytopath.* 72: 95-98.

Graham, J.H., Leonard, R.T. and Menge, J.A. 1981, Membrane mediated decrease in root exudation responsible for phosphorus inhibition of vesicular-arbuscular mycorrhiza formation, *Plant Physiol.* 68: 548-552.

Grandison, G.C. and Cooper, K.M. 1986, Interaction of vesicular arbuscular mycorrhizae and cultivars of alfalfa susceptible and resistant *Meliodogyne hapla, Nematol.* 18: 141-154.

Hallbrock, K. and Scheel, D. 1989, Physiology and molecular biology of phenyl propanoid metabolism, *Ann. Rev. Plant Physiol. Mol. Biol.* 40, 347-369.

Harrison, M.J. and Dixon, R.A. 1993, Isoflavonoid accumulation and expression of defence gene transcripts during establishment of vesicular-arbuscular mycorrhizal assciation in roots of *Medicago trancotula* and the mycorrhizal fungus *Glomus versiforme, Plant J.* 6,9-20.

Hendrin, J.W. 1970, Sterols in growth and reproduction of fungi, *Ann. Rev. Phytopathol.* 8: 111-130.

Hodge A., Alexander, I.J. and Gooday, G.W. 1995, Chitinolytic activities of *Eucatyptus pilularis* and *Pinus sylvestris* root system challenged with mycorrhizal and pathogenic fungi, *New Phytol.* 131: 255-261.

Jalali, B.L. 1986, Vesicular arbuscular mycorrhizae and current status, in : *Vistas in Plant Pathology,* A. Varma and J.P. Varma, eds., Malhotra Publishing House, New Delhi, pp. 437-450.

Jalali, B.L. and Jalali, I. 1991, Mycorrhizae in plant disease control, in : *Handbook of Applied Mycology,* Vol. I, Soil and Plant, D.K. Arora, B. Rai, K.G. Mukerji and G.R. Knudsen, eds., Marcel Dekker Inc., New York, pp.131-154.

Kaye, J.W., Pfleger, F.L. and Steward, E.L. 1984, Interaction of *Glomus fasciculatum* and *Pythium ultimum* on green house grown poinsettia, *Can. J. Bot.* 62: 1575-1579.

Kieliszewska-Robiska, B. 1992, Acid phosphatase activity in mycorrhizal and non-mycorrhizal scots pine seedling in relation to nitrogen and phosphorus nutrition, *Acta Soc. Bota. Pol.* 61: 253-254.

Koske, R.E. 1981, *Gigaspora gigantea* : observations on spore germination of VA mycorrhizal fungus, *Mycologia* 73: 283-300.

Krishna, K.R., Balakrishna, A.N. and Bagyaraj, D.J. 1982, Interactions between a vesicular-arbuscular mycorrhizal fungus and *Streptomyces cinnamomeous* and their affects on finger millet, *New Phytol.* 92: 401-405.

Losel, D.M. 1991, Synthesis and functioning of membrane lipids in fungi and unfeated plants, *Pesti. Sci.* 32: 353-362.

Mac Donald, R.M., Chandler, M.R. and Mosse, B. 1982, The occurrence of bacterium like organelles in the vesicular arbuscular mycorrhizal fungi, *New Phytol.* 90: 656-663.

Manju Nath, A., Mohan, R. and Bagyaraj, D.J. 1981, Interaction between *Beijerinckia mobilis, Aspergillus niger* and *Glomus fasciculatum* and their effects on growth of onion, *New Phytol.* 87: 723-727.

Mauch, F., Mauch-Mani, B. and Boller, T. 1988, Antifungal hydrolases in pea tissue, II, Inhibition of fungal growth by combinations of chitimases and β-1,3 glucanase, *Plant Physiol.* 88: 936-942.

Mayo, K., Davis, R.E. and Motta, J. 1986, Stimulation of germination of spores of *Glomus versiforme* by spore associated bacteria, *Mycologia* 78: 426-431.

Miller, J.C., Rajapakse, S. and Garbaer, R.K. 1985, Vesicular arbuscular mycorrhizae in vegetable crops, *Hort. Scie.* 21: 974-984.

Miyomoto, K., Veda, J. and Kami Saka, S. 1993, Gibberellin enhanced sugar accumulation in growing subhooks of eliolated *Pisum sativum* seedlings, Effects of gibberellic acid, indole acetic acid and cyclohemimide on invertage activity, sugar accumulation and growth, *Physiol. Plant.* 88: 301-306.

Morandi, D. 1996, Occurrence of phytoalexins and phenolic compounds in endomycorrhizal interaction and their potential role in biological control, *Plant Soil* 185: 241-251.

Mukerji, K.G. 1999, Mycorrhiza in control of plant pathogens : Molecular approaches, in : *Biotechnological Approaches in Biocontrol of Plant Pathogens,* K. G. Mukerji, B.P. Chamola and R.K. Upadhyay, eds., Planum Publishing Co. Ltd., New York, pp. 135-155.

Mukerji, K.G., Chamola, B.P. and Sharma, M. 1997, Mycorrhiza in control of Plant Pathogens, in : *Management of Threatening Plant Diseases of National Importance,* V.P. Agnihotri, A.K. Sarbhoy and D.V. Singh, eds., Malhotra Publ. House New Delhi, pp. 298-314.

Mukerji, K.G., Mandeep and Varma, A. 1998, Micorrhizosphere microorganisms : sareening and evaluation, in : *Mycorrhiza Manual,* A. Varma, ed., Springer Verlag, Berlin, pp. 85-97.

O'Bannon, J.H. and Nemec, S. 1979, The response of *Citrus limon* seedlings to a symbiont *Glomus etunicatum,* and a pathogen, *Radopholus similis, J. Nematol.* 11: 270-275.

Norman, J.R., Atkinson, D. and Hooker, J.E. 1996, Arbuscular mycorrhizal fungal-induced alterations to root architecturer in strawberry and induced resistance to root pathogen *Phytophthora fragariae, Plant Soil* 185: 191-198.

Pacovsky, R.S. and Fuller, G. 1988, Mineral and lipid composition of *Glycine - Glomus - Bradrrhizobium* symbiosis, *Physiol. Plantarum* 72: 733-746.

Palacino, J.H. and Leguizamon, J. 1991, Interaction between *Glomus manihotis* and *Meloidogyne incognita* in yellow and red pitaya, *Fitopatologia Colombiana* 15 1,, 9-17.

Pinochet, J., Calvet, C., Camprubi, A. and Fernandez, C. 1995, Interaction between the root leision nematode *Pratylenchus vulnus* and the mycorrhizal association of *Glomus intraradices* and Santa Lucia 64 cherry stock, *Plant Soil* 170: 323-329.

Paulitz, T. C. and Linderman, R.G. 1991, Mycorrhizal interactions with soil organisms, in : *Handbook of Applied Mycology*, Vol.1, Soil and Plants, D.K. Arora, B.Rai, K.G.Mukerji and G.R. Knudsen, eds., Marcel Dekker Inc., New York, pp. 77-129.

Prikryl, Z. and Vancura, V. 1980, Root exudation as dependent on growth, concentration gradient of exudates and the presence of the bacteria, *Plant Soil* 57, 69-83.

Raj, J., Bagyaraj, D.J. and Manju Nath, A. 1981, Influence of soil inoculation with vesicular arbuscular mycorrhizal fungi and phosphate dissolving bacterium on plant growth and ^{32}P uptake, *Soil Biol. Biochem.* 13: 105-108.

Reinhard, S., Martin, P. and Marschner, H. 1993, Interaction in the tripartite symbiosis of pea *Pisum sativum*, *Glomus* and *Rhizobium* under non limiting phosphorus supply, *J. Plant Physiol.* 141: 7-11.

Rosendahl, S. 1985, Interactions between the vesicular arbuscular mycorrhizal fungus *Glomus fasciculatum* and *Aphanomyces euteiches* root rot of peas, *J. Phytopathol.* 114 : 31-40.

Ryals, J., Uknes, S. and Ward, E. 1994, Systemic acquired resistance, *Plant Physiol.* 104: 1109-1112.

Saito, M. 1995, Enzyme activities of internal hyphae and germinated spores of an arbuscular mycorrhizal fungus, *Gigaspora margarita* Becker & Hall, *New Phytol.* 129: 425-431.

Samra, A., Dumas - Gaudot, E., Gianinazzi - Pearson, V. and Gianinazzi, S. 1996, Studies of *in vivo* polypeptide synthesis in non-mycorrhizal and arbuscular mycorrhizae *Glomus mosseae*, pea roots, in : *Mycorrhizas in Integrated Systems from Gene to Plant Development*, C. Azcon - Aguilar and J.M. Barea, eds., Kluwer Academic Publishers, Dordrecht, the Netherlands, pp.263-266.

Saxena, A.K. and Tilak, K.V.B.R. 1996, Interaction of soil microorganisms with vesicular arbuscular mycorrhiza, in : *New Approaches in Microbial Ecology*, J.P. Tewari, G. Saxena, N. Mittal, I. Tewari and B.P. Chamola, eds., Aditya Books, New Delhi, India, pp.187-204.

Schenck, N.C. 1981, Can mycorrhizae control root diseases, *Plant Diseases* 65: 232-234.

Schenck, N.C. 1987, Vesicular-arbuscular mycorrhizal fungi and the control of fungal diseases, in: *Innovative Approaches to Plant Disease Control*, I. Chet, ed., John Wiley, New York, pp.179-198.

Sekhon, B.S., Thapar, S., Notwal, A. and Singh, R. 1990, Effect of foliar application of urea on enzymes and metabolties of nitrogen metabolism in mycorrhizal moong plants under different phosphorus levels, *Plant Physiol. Biochem.* 28 : 393-398.

Singh, K. and Varma, A.K. 1985, Association of bacteria in the lumen of vesicular arbuscular mycorrhizal spores extracted from rhizosphere of xerophytes, *Trans. Mycol. Soc. Japan* 26 : 511-516.

Smith, G.S. 1988, The role of phosphorus nutrition in interactions of vesicular-arbuscular mycorrhizal fungi with soil borne nematodes and fungi, *Phytopath.* 18, 371-374.

Subba Rao, N.S., Tilak, K.V.B.R. and Singh, C.S. 1985, Synergistic effect of vesicular-arbuscular mycorrhizas and *Azospirillum brasilease* on the growth of barley, *Soil Biol. Biochem.* 17: 119-121.

Sylvia, D.M. and Schenck, N.C. 1983, Soil fungicides for controlling chytridiaceous mycoparasits of *Gigaspora margarita* and *Glomus fasciculatum*, *App. Environ. Microbiol.* 45: 1306-1309.

Spanu, P.G. and Bonfante-Fasolo, P. 1988, Cell wall bound peroxidases activity in roots of mycorrhizal *Allium porum*, *New Phytol.* 109: 119-129.

Tilak, K.V.B.R. and Subba Rao, N.S. 1987, Association of *Azospirillium brasilense* with pearl millet *Pennisetum americanum* (L) Lecke, *Biol. Ferti. Soils* 4: 97-102.

Tilak, K.V.B.R., Li C.T. and Ho, I. 1987, Occurrence of nitrogen fixing *Azospirillium* on surface sterilised mycorrhizal roots of green onion *(Allium cepa)*, in : *7th North American Conference on Mycorrhizae*, pp. 222.

Trotta, A., Varese, G.C., Gnavi, E., Fusconi, A., Sampo, S. and Berta, G. 1996, Interactions between soil borne root pathogen *Phytophthora micatinae* var. *parasitica* and arbuscular mycorrhizal fungus *Gloms mosseae* in tomato plants, *Plant Soil* 185: 199-209.

Tzean, S.S. and Chu, C.L. 1983, Spiroplasma like organisms in a vesicular arbuscular mycorrhizal fungus and its mycoparasite, *Phytopath.* 73: 989-991.

Valdes, M., Reza-Aleman, F. and Furlan, V. 1983, Response of *Leucaena esculenta* to Endomycorrhizae and *Rhizobium* inoculation, *World J. Microb. Biotech.* 9: 97-99.

THE ROLE OF ROOT EXUDATES IN ARBUSCULAR MYCORRHIZA INITIATION

Carol I. Mandelbaum[1] and Yves Piche[2]

[1] Department of Botany, University of Texas at Austin
Texas 78713-7460
USA
[2] Centre de Echerhe on Biologie Forestiere
Faculte de Foresteriet et de Geometigue
Pavillon C.E. Foresteriet et de Geomatique
Pavillon C.E. Marchand, University Lavel
Sté - F`oy, Quebec G1K 7P4, Canada

1. INTRODUCTION

The rhizosphere is a complex environment, the home to numerous microorganisms and ecosystems, the site of many chemical and physical reactions, and the substrate that supports terrestrial plants *via* anchorage of the roots. Plant roots continually elongate through this heterogeneous area, absorbing water and minerals vital for sustaining plant growth and development, and exuding various substances into the rhizosphere (Fig. 1). Plant root exudate constituents include actively secreted polysaccharides that form a layer of mucilage around the root, enzymes such as acid phosphatases; volatile compounds including ethylene and CO_2; low molecular weight metabolites such as sugars, organic acids, amino acids including phytosiderophores, and phenolics; cells sloughed off of the root cap, and dead cell lysates (Marschner, 1995; Rovira, 1969). Plant root exudate constituents vary between different plants, and exudate composition changes in the same plant at different ages or when grown under different environmental conditions (Hale *et al.*, 1971; Marschner, 1995; Rovira, 1969). Exudation levels of particular constituents are not always the same along the plant root axis, and is generally greater in the apical regions (Marschner, 1995; Rovira, 1969). Plant root exudates alter the physical and chemical conditions of the rhizosphere by changing pH levels and mineral availability *via* desorption and chelation, and also influence the growth and interactions of numerous microorganisms that populate the rhizosphere (Marschner, 1995; Rovira, 1969). One such important rhizosphere inhabitants are fungi from the order Glomales, which form the arbuscular mycorrhizae. Arbuscular mycorrhizal (AM) fungi are estimated to colonize greater than 80% of land plant species (Bonfante and Perotto, 1995) and are apparently responsive to the wide diversity of compounds exuded by different plant roots (Rovira, 1969). AM symbiosis enhances mineral uptake for the plant, particularly of phosphorus (P), and confers other benefits to plant health, such as increased uptake of additional minerals, drought tolerance, and resistance to root pathogens (Newsham *et al.*, 1995).

Mycorrhizal Biology, edited by Mukerji *et al.*
Kluwer Academic/Plenum Publishers, 2000

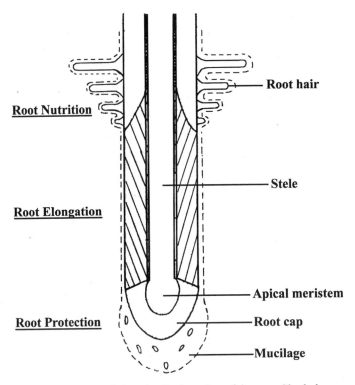

Fig. 1. Schematic representation of a longitudinal section of the root. Shaded areas represent the site of greatest AM colonization. White areas represent zones of mostly non-symbiotic infection. (Adapted from Marks and Foster, 1973, and Chaboud and Rougier, 1981).

The AM plant-fungal interaction consists of two distinct stages: first, pre-colonization stages of fungal development, attraction and attachment to a host plant root, dependent primarily on spore reserves and the close proximity of the root; and second, external and internal fungal development during colonization, dependent on processes that occur within the plant root (Bécard and Piché, 1989a). Many of the initial developmental stages for AM fungi occur without the addition of plant material or other rhizosphere constituents. AM-fungal spores germinate on water agar (Hepper and Smith, 1976; Mosse and Hepper, 1975), although specific requirements for germination of individual species vary, and may require different periods of dormancy or cold storage (Hepper and Smith, 1976), or specific ranges of moisture (Daniels and Trappe, 1980), pH (Green et al., 1976), or temperature (Tommerup, 1983). At the present time general conclusions regarding germination conditions are hard to draw, since few studies have examined more than one or two fungi. After spore germination hyphae grow predominantly unbranched and aseptate, supported only by spore reserves (Bonfante and Perotto, 1995; Powell, 1976). Thereafter, without plant stimulus, hyphal apices retract cytoplasm from the tips, produce retraction septa, and no longer grow at a measurable rate (Bonfante and Perotto, 1995). Numerous external factors stimulate the rate of spore germination and the length and duration of hyphal growth, e.g. soil extracts (Mosse, 1959), soil microorganisms (Azcon, 1987; Azcon-Aguilar et al., 1986), certain soil phosphorus concentrations (Miranda and Harris, 1994), suspension cell cultures and exudates (Carr et al., 1985; Paula and Siqueira, 1990), plant root volatiles (Balaji et al., 1995; Bécard and Piché, 1984 a,b; 1990; Carr et al., 1985) plant root exudates (Bécard and Piché, 1989b; Elias and Safir, 1987; Gemma and Koske, 1988; Gianinazzi-Pearson et al., 1989; Graham, 1982; Mosse, 1959; Poulin et al., 1993; Schreiner and Koide, 1993C; Tawaraya et al., 1996), isolated flavonoids (Balaji et al., 1995; Bécard et al., 1992; Chabot et al., 1992; Gianinazzi-Pearson, 1989; Kape et al., 1992; Nair et al., 1991; Poulin et al., 1993; Siqueira et al., 1991; Tsai and Philips, 1991), and other isolated phenolics (Douds et al., 1996). Exposure to host plant roots or root exudates switches AM-hyphal germ tube growth from mostly linear to a highly branched pattern (Gemma and Koske, 1988; Giovannetti et al., 1993b; Graham, 1982; Mosse, 1988; Powell, 1981; Schreiner and Koide, 1993c). Even with the stimulation provided by host plant roots and their exudates further fungal development cannot occur without penetration of the plant root and colonization (Bécard and Piché, 1989b). AM-fungal spore formation does not occur without root colonization, and isolated hyphal fragments do not regrow or colonize plant roots.

Initiation of AM colonization begins with hyphae growing toward the plant root and extensively around it, attaching to the root, hyphal tips swelling to form appressoria on the root, and infective penetration pegs emerging from the appressoria. The formation of appressoria is considered by some to be the point of host recognition and initiation of functional AM interactions (Giovannetti et al., 1993a; 1994). AM-fungal hyphae do not grow around, attach to or form swellings on various threads similar in shape and diameter to plant roots, suggesting that biological factors are necessary to elicit the change in growth pattern (Giovannetti et al., 1993a); however, this study did not examine finer physical structures such as ridges or indentations, which elicits appressoria formation for some pathogenic fungi (Manocha and Chen, 1990). AM fungi appear to colonize the plant root more readily behind the root tips and at the zone of lateral root formation (Fig.1) (Bécard and Piché, 1992; Marks and Foster, 1973; Smith and Walker, 1981).Greater quantities of exudate constituents are produced at apical regions of a root (Marschner, 1995), and the emergence of lateral roots may cause an increase in the amounts of cell wall constituents and lysates exuded in this area (Douds et al., 1996).

Thus, root exudates are likely to play a role in the initiation of AM-interactions. However, whether specific root exudate components are essential for the processes preceding the endophytic stages of AM interactions is debatable. The role of root exudates in AM initiation could be general, with root exudates serving as chemotropic attractants or providing a nutritive source for AM fungi. Root exudates may contain specific signals for host plant recognition, although AM fungi usually infect wide ranges of host plants. Also, plant root exudates could contain anti-fungal compounds that inhibit colonization. The role of root exudates in the reduction of AM colonization of plants grown in a high phosphorus (P) environment is also unclear. The cellular and molecular basis of AM-fungal chemotropism, growth stimulation, and morphogenesis is only just beginning to be addressed in the literature, hindered by the difficulty in growing substantial amounts of AM fungi in pure culture.

In view of the lack of experimentation involving AM-interactions and the root mucilage, this review considers the relatively stable, low molecular weight exudate components that could accumulate in the rhizosphere. As most studies of root exudates are performed under axenic conditions, we recognize that this review excludes mention of exudates from other microorganisms and considers plant root exudates that may not actually accumulate in the rhizosphere due to degradation or consumption by other rhizosphere inhabitants. Also, since exudation is usually greater under non-sterile conditions and is greater from roots grown in solid medium, estimations of exudate quantity when collected axenically and in liquid are probably underestimated (Marschner, 1995). Even if the conditions for exudate collection are adjusted to maximize overall quantity, the composition of exudate constituents may be altered by the different methods utilized. Therefore, it is difficult to draw conclusions about what occurs in the field from laboratory studies of root exudation and its effect on mycorrhizal colonization.

2. VOLATILE EXUDATES, GERM TUBE CHEMOTROPISM AND CARBON DI-OXIDE (CO_2)

Some of the strongest evidence about the influence of root exudates on AM-fungi pre-colonization development involves the role of volatile exudate constituents, particularly CO_2. In studies using the AM fungus *Gigaspora margarita*, which displays a strong negatively geotropic growth pattern in the germ tubes (Watrud *et al.*, 1978), aerial plant roots induce a positive chemotropic response of germ tube growth, presumably *via* volatile root exudates (Gemma and Koske, 1988; Koske, 1982). Also, volatile exudates from plant roots (Balaji *et al.*, 1995; Bécard and Piché, 1989 a,b; 1990) and cell cultures (Carr *et al.*, 1985) cause a general stimulation of AM-fungal hyphal growth.

The addition of potassium permanganate impregnated silica gel to a closed growth system suppressed AM colonization in clover, indicating that oxidizable, volatile compounds enhance AM colonization rates (St. John *et al.*, 1983). Experiments with $KMnO_4$ and KOH traps illustrate the importance of root volatiles in AM-hyphal growth: $KMnO_4$ and KOH traps eliminated the chemotropism of *Gigaspora gigantea* germ tubes towards plant roots (Gemma and Koske, 1988), and KOH traps completely stopped *G. gigantea* hyphal growth (Bécard and Piché, 1989a). KOH traps eliminate CO_2 from the air; thus, CO_2 is implicated as one volatile exudate constituent stimulating AM-fungal chemotropism and enhanced growth (Bécard and Piché, 1989a). Volatile root exudate constituents such as low molecular weight ketones and aldehydes also would be bound in KOH and $KMnO_4$ traps, and might stimulate AM-fungal germination and growth in addition to CO_2 (Balaji *et al.*, 1995; Bécard and Piché, 1989a).

The stimulation of AM-hyphal growth by root exudates (Bécard and Piché, 1989b) and flavonoids (Bécard *et al.*, 1992; Poulin *et al.*, 1993) is enhanced synergistically by the addition of CO_2. Therefore, the effects of isolated compounds on AM-fungal growth cannot be accurately determined unless tested under constant CO_2 levels, preferably in an optimal range (Bécard *et al.*, 1992). Optimal growth enhancement of *Gigaspora margarita* by flavonoids is achieved in the presence of 1.0%-2.5% CO_2; 5.0% CO_2 and greater reduces hyphal growth of *G. margarita* (Poulin *et al.*, 1993). However, optimal CO_2 concentrations may be different for other AM fungi.

Soil CO_2 levels are variable and are influenced by differences in soil moisture and temperature (Glinski and Stepniewski, 1985), and probably influence the seasonal variations in AM colonization patterns (Allen, 1983; Johnson - Green *et al.*, 1995; Koide and Schreiner, 1992). The lack of AM interactions in plants in waterlogged soils may be due to an absolute requirement for CO_2 (Tester *et al.*, 1987). Increases in soil CO_2 levels are correlated with enhancements of AM plant growth (Saif, 1984) and levels of colonization (Morgan *et al.*, 1994), but the effect of soil CO_2 on AM interactions can vary with different plant hosts (Monz *et al.*, 1994). Bécard and Piché (1989), proposed that the incorporation of radiolabeled $^{14}CO_2$ from plant root volatiles in the hyphae of *G. margarita* supports the hypothesis that CO_2 is utilized as a carbon source by germinating spores (Bécard and Piché, 1989a). Nonetheless, while CO_2 and other root volatile constituents provide stimulation of the AM-fungal pre-colonization stages, the gases alone do not support continued fungal development (Bécard and Piché, 1989a).

3. QUANTITY OF ROOT EXUDATES AND INHIBITION OF AM COLONIZATION IN PLANTS GROWN AT HIGH PHOSPHORUS

If root exudates serve as a source of food and nutrition for AM fungi during pre-colonization processes, increases in total exudate quantity should have a stimulatory effect, and plants that do not become colonized are expected to have lower overall levels of exudation. A number of factors are known to alter the levels of root exudation, such as soil temperature (Graham *et al.*, 1982), plant development (Johnson *et al.*, 1982a), and light exposure (Johnson *et al.*, 1982b). Overall root exudation increases and exudate constituent composition is altered when a plant undergoes a nutrient deficiency (Marschner, 1995). AM-fungal colonization seems to be particularly sensitive to phosphorus levels:when a plant is grown at extremely low P, AM colonization is inhibited, at limited P levels colonization increases, and beyond threshold P levels colonization is again inhibited (Bolan *et al.*, 1984). The inhibition of AM colonization at high P levels has been correlated to the concentration of P within the plant roots (Jasper *et al.*, 1979; Menge *et al.*, 1978), although high concentrations of P are found to have a direct inhibitory effect on *Glomus etunicatum* and *Scutellospora heterogama* (Miranda and Harris, 1994). The inhibition of AM colonization at higher P levels was correlated with root exudation by Tawaraya *et al.* (1996), with the hyphal growth stimulation of *Gigaspora margarita* by root exudates from onion (*Allium cepa*) decreasing correspondingly higher P levels.

Plant growth in increased P levels reduced exudation levels of amino acids and sugars in two studies that correlated the reduction with changes in membrane permeability, measured by phospholipid levels (Ratnayake *et al.*, 1978) and ^{86}Rb efflux (Graham *et al.*, 1981). Decreases in root exudation of high-P grown plants have also been correlated with decreases in the total amount of AM colonization (Graham *et al.*, 1981; Thomson *et al.*, 1986). The theory that increased exudation rates alone (not just P concentrations in the growth medium) increase infectivity was supported by comparisons of root exudation levels with AM colonization of different wheat cultivars (Azicon and Ocampo, 1981), and of AM colonization of *Chrysanthemum morifolium* during early flower bud development (Johnson *et al.*, 1982a). Several non-host plants for AM fungi exude lower levels of amino acids, sugars, and carboxylic acids when compared to host plants (Schwab *et al.*, 1984). Additionally, foliar application of the herbicide simizine, which increases root exudation of amino acids and sugars,stimulates AM colonization of the non-host plant *Chenopodium quinoa* (Schwab *et al.*, 1982). However, in one study no relationship between AM colonization and root exudation was found, and the non-host plants *Rhaphanus raphanistrum* and *Brassica oleracea* exuded higher levels of sugars than did several host plants (Azicon and Ocampo, 1984).

The influence of root exudation levels on AM colonization has not been linked to the exudation of any specific substance or to an overall increased nutritive source for pre-colonization stages of growth. Although total exudation levels of amino acids, sugars, and carboxylic acids are reduced for high-P grown host plant roots compared to low-P grown host plant root (Schwab *et al.*, 1983), and for non-host plant roots compared to host plant roots (Schwab *et al.*, 1984), no consistent qualitative differences in these exudate constituents were detected in either case. If AM colonization is enhanced by an increase in total root exudation and not by the exudation of specific compounds, it would follow that the fungi could be stimulated by a broad range of organic compounds and would be easier to maintain in pure culture; however, this is not the case (Schwab *et al.*, 1991). Elias and Safir, (1987) showed that root exudates from *Trifolium repens* grown under low-P availability stimulated *Glomus fasciculatum* hyphal growth independently of concentration, and high-P grown *T. repens* root exudates were not stimulatory at any concentration (Elias and Safir, 1987). These authors suggested that specific exudate constituents are required for AM-fungal stimulation and subsequent colonization, and that these constituents are lacking in the root exudates of plants grown in high-P concentrations (Elias and Safir, 1987). Furthermore, root exudates are reported to provide less stimulation of AM-fungi as the plant ages (Elias and Safir, 1987; Poulin *et al.*, 1993), and plant age is correlated to changes in exudate composition as well as quantity (Hale *et al.*, 1971).

Because there is much evidence that increases in root exudation increases colonization rates, together with the observations that regions along the axis of plant roots with the greatest exudation rates are also the areas where AM colonization is most likely to occur (Schwab *et al.*, 1991), it seems obvious that total root exudate levels influences AM-fungal colonization development. Regardless of whether one or several root exudate constituents stimulate the pre-colonization stages of AM-fungal development, increases in total exudation will enhance AM-fungal growth. If an AM fungus requires a specific signal molecule to initiate the colonization process, as membrane permeability and total exudation increase the quantity of signal exuded will also increase and the effects on fungal development would be concentration dependent, at least until a threshold value is obtained. However, the threshold value for a specific signal molecule may be so low that it is not easily detected from the concentrations obtained from the collection procedures used. Also, increases in root exudation may influence AM interactions indirectly, by affecting other rhizosphere microorganism populations.

The questions remain if AM fungi utilize carbon from the plant root exudates during the pre-colonization stages of development, either as CO_2, sugars or carboxylic acids, and if this carbon input is necessary to facilitate colonization. It is likely that growth in high-P also affects AM colonization through physiological changes within the plant root and not to just through alterations in fungal pre-colonization stages. Changes in P concentration in the plant root alters the rate of formation of effective entry points by *Glomus mosseae* on leeks (Amjee *et al.*, 1989), possibly due to recognition events that occur at the plant surface (Schwab *et al.*, 1991), or from a change in plant response to fungal attachment. Growth at high P concentrations changes carbon availability and allocation within the host plant (Graham *et al.*, 1981), which could limit AM-fungal development during colonization. Furthermore changes in soil P levels are reported to have direct effects on AM-fungal pre-colonization development (Miranda and Harris, 1994).

4. ANTI-FUNGAL ROOT EXUDATES AND AM NON-HOST PLANTS

Although mycorrhizal interactions occur in the majority of terrestrial plants, there are a few plant taxa that are non-host for these fungi. In this review we consider reports of non-host plants that do not form mycorrhizal associations of any kind, and not plants that support mycorrhizae other than AM. A lack of AM colonization in non-host taxa could be attributed to the environment in which they are found, as with plants that grow in waterlogged or saline soils (Tester *et al.*, 1987), due to the influence of root exudates on pre-colonization stages of AM development, or the result of events that occur after root penetration and the initiation of colonization. AM-symbiosis is believed to have been present when plants were evolving the ability to exist in a terrestrial environment (Simon *et al.*, 1993), and occurs throughout all divisions of vascular plants (Newman and Reddell, 1987). Given this evidence, it would appear those plant species that are non-host for AM-interactions have either lost the ability to form a functional AM-interaction, or have evolved a trait that excludes the fungus.

4.1. Non-host Mustards and Isothiocyanates

Plants in the Brassicaceae (mustards) generally do not form mycorrhizae (Tester *et al.*, 1987), possibly due to anti-fungal constituents released from the roots of these plants. The Brassicaceae plants can become colonized to a limited extent when grown alone (Glenn *et al.*, 1985; 1988; Tommerup, 1984), or grown together with host plants (Demars and Boerner, 1994; Hirrell *et al.*, 1978; Mukerji and Kocher, 1983; Ocampo *et al.*, 1980). However, when AM fungi are grown with Brassicaceae species, hyphal tips near the surface of the root retract cytoplasm before contact (Glenn *et al.*, 1985; Schreiner and Koide, 1993c), and fewer penetration pegs form (Ocampo *et al.*, 1980; Tommerup, 1984). In reports of Brassicaceae plants colonization by AM fungi colonized cells have either reduced or no arbuscules (Demars and Boerner, 1994; 1996; Glenn *et al.*, 1985; Hirrell *et al.*, 1978; Ocampo *et al.*, 1980; Tommerup, 1984), and colonization occurs in older, dying roots (Glenn *et al.*, 1985; Hirrell *et al.*, 1978). Reports of growing mustards with host plants are conflicting; in some studies growth with mustards reduced AM colonization levels of host plants (Black and Tinker, 1979; El.Atrach *et al.*, 1989; Hayman *et al.*, 1975), in others growth with mustards caused no significant change in AM colonization levels of host plants (Ocampo, 1980; Powell, 1981), and in one report pre-cropping with a mustard stimulated AM colonization of subsequently planted host crops (Ocampo and Hayman, 1981).

Isothiocyanates produced by mustard plants are anti-fungal, (Drobnica *et al.*, 1967; Walker *et al.*, 1937) and are potential inhibitors of AM colonization for Brassicaceae plants (Schreiner and Koide, 1993 a,b). Isothiocyanates from Brassicaceae plants form as a result of the reaction of glucosinolates with the myrosinase enzyme in myrosin cells. A mixture of isolated glucosinolates with myrosinase enzyme inhibits growth of the AM fungus *Glomus mosseae* (Vierheilig and Ocampo, 1990a) and *G. etunicatum* (Schreiner and Koide 1993b). No correlation was detected between glucosinolate levels in the roots of *Brassica* cultivars and the degree of penetration by *Glomus mosseae* (Glenn *et al.*, 1985; 1988); however, glucosinolates are generally non-toxic to fungi, and levels of toxic isothiocyanates were not measured in these studies (Koide and Schreiner, 1992). Koide and Schreiner, (1992), further suggested that the limited pattern of colonization sometimes observed in mustards is due to penetration by AM fungi in cortical cells other than the myrosin cells, which are found scattered throughout the root tissue. Usually myrosin cells produce isothiocyanates only after mechanical damage, because glucosinolates are compartmentalized in the vacuole (Koide and Schreiner, 1992). Nonetheless, isothiocyanates accumulate in the rhizosphere around *Carica papaya* and *Brassica kaber* roots (Koide and Schreiner, 1992; Tang and Takenaka, 1983).

Different mustard plant species vary in their glucosinolate/isothiocyanate composition, whether the isothiocyanates produced occur in volatile or non-volatile forms, and consequent anti-fungal effect (Drobnica *et al.*, 1967). Studies by Schreiner and Koide (1993a,b) provided evidence that the inhibitory activity of *Brassica kaber* and *B. nigra* root exudates on the growth of AM fungus *Glomus etunicatum* was due to isothiocyanates in the root exudates (Schreiner and Koide, 1993a, b). For *B. kaber*, which produces a glucosinolate that generates a non-volatile isothiocyanate, the greatest inhibitory activity is in the soluble root exudates; for *B. nigra*, which produces a glucosinolate that releases volatile isothiocyanates, the greatest inhibitory activity is in the volatile portion of the root exudates (Schreiner and Koide, 1993b). Furthermore, growth inhibition of *G. etunicatum* by *B. kaber* and *B. nigra* exudates is "rescued" by the addition of the amino acids lysine, arginine, and glutathione, which are reactive with isothiocyanates (Schreiner and Koide, 1993b).

Volatile root exudates from several mustard plant species inhibit spore germination of *Glomus* species (El-Atrach *et al.*, 1989; Schreiner and Koide, 1993a; Vierheilig and Ocampo, 1990 a,b), and root volatiles from the mustard plant kolhrabi fail to elicit a chemotropic growth response from *Gigaspora gigantea* germ tubes (Gemma and Koske, 1988). Conversely, Schreiner and Koide (1993c) reported that volatiles released from Ri-T-DNA transformed roots of *Brassica kaber* and *B. nigra* stimulated germination and growth of *Glomus etunicatum*, although when the hyphal tip was within 50 to 100 µm of the roots cytoplasm retracted from the hyphal apices and fungal growth was inhibited. The authors attributed the initial growth stimulation to the release of CO_2 by the roots. This example illustrates how different constituents of volatile root exudates may have opposing effects on AM-fungi, and that exudate constituents may be released in different proportions under various growth conditions. Studies monitoring exudate affects *in vitro* do not necessarily reflect the balance of constituents surrounding an intact root system in the field or the substances that actually encounter an AM fungus. Furthermore, the inhibition of spore germination and AM colonization attributed to isothiocyanates may not be the direct result of interactions with the AM fungus, but instead due to the reaction of isothiocyanates with other chemicals in the rhizosphere (Schreiner and Koide, 1993b).

4.2. *Lupinus* Species

An unusual non-host taxon is the genus *Lupinus* in the host family Fabaceae (Trinick, 1977). Root exudates of *Lupinus albus* do not affect on spore germination or hyphal growth of *Gigaspora margarita* (Gianinazzi-Pearson *et al.*, 1989) or *Glomus mosseae* (Giovannetti *et al.*, 1993a) when compared to controls. When *Lupinus cosentinii* was grown together with white clover the appressorium that formed on the clover roots were abnormally large, irregularly shaped, and many of the penetration pegs aborted; this phenomenon was attributed to anti-fungal compounds in *L. cosentinii* root exudates (Morley and Mosse, 1976). Similarly, functional appressorium from *Glomus mosseae* do not form on roots of living *Lupinus albus,* and hyphal swellings that form on excised roots of the same plant contained retracted cytoplasm, similar to the reaction seen with Brassicaceae plants (Avio *et al.*, 1990; Giovannetti *et al.*, 1993a). Interestingly, when shoots of *Lupinus albus* were grafted onto pea roots colonization by *Glomus intraradices* or *G. mosseae* was blocked, indicating that inhibitory factors present were produced or directed by the scion (Gianinazzi-Pearson and Gianinazzi, 1992).

Lupinus species root exudates contain isoflavonoids, particularly prenylated isoflavonoids (Gagnon *et al.*, 1995). Prenylation of isoflavonoids may enhance their fungitoxicity (Dixon *et al.*, 1983), and isoflavonoids from *Lupinus* species are anti-fungal (Harborne *et al.*, 1976; Ingham *et al.*, 1983). However, the effects of isolated compounds from *Lupinus* root exudates on AM fungi have not been investigated, and isoflavonoids may stimulate some AM fungi (discussed in section 5). Also, some annual legumes such as *Lupinus albus* respond to P deficiency with enhanced formation of proteoid roots, and most legumes respond to P deficiency with increased exudation of organic acids (Marschner, 1995). The proteoid roots of *Lupinus albus* under P deficiency exude relatively high amounts of citric acid in a localized area, leading to acidification and greater mobilization of minerals in a limited volume of soil (Marschner, 1955). It may be that increased acidification in the vicinity of *Lupinus* roots limits AM-fungal growth. However, it has also been suggested that *Lupinus* root exudates do not inhibit AM-fungal growth, but rather lack signal molecules essential for AM-fungal recognition and appressorium formation (Giovannetti *et al.*, 1993a).

4.3. Caryophyllales Plants

A large number of presumably non-host plant taxa are in the order Caryophyllales, a group evolutionarily distinct from other dicots (Behnke and Marby, 1994; Tester *et al.*, 1987). In particular, plants in the family Chenopodiaceae have been the focus of much attention as a non-host group. Colonization is reported in supposedly non-host members of the Caryophyllales both when grown alone (Allen, 1983; Ocampo *et al.*, 1980), or with companion plants (Hirrell *et al.*, 1978; Miller *et al.*, 1983; Ocampo *et al.*, 1980). Compared with plants in the Brassicaceae, AM colonization of plants in the Caryophyllales is often more extensive (Hirrell *et al.*, 1978; Ocampo *et al.*, 1980). Caryophyllales plants might be better defined as `weakly mycorrhizal' rather than non-mycorrhizal.

In one report root exudates of spinach (Chenopodiaceae) inhibited spore germination of *Glomus mosseae* (Vierheilig and Ocampo, 1990b), but in other studies Caryophyllaceous plant roots did not inhibit AM-fungal growth below the level of controls (Bécard and Piché, 1990; Giovanneti *et al*; 1993b; 1994; Schreiner and Koide, 1993c; Schwab *et al*., 1984). On the contrary, root exudates of beet (Chenopodiaceae) elicit germ tube chemotropism in *Gigaspora gigantea* (Gemma and Koske, 1988), and transformed root cultures of beet stimulate germination and growth of *Glomus etunicatum* (Schreiner and Koide, 1993c) and *Gigaspora margarita* (Bécard and Piché, 1990). Since none of these studies were performed at controlled CO_2 levels, the possibility that CO_2 was the only stimulatory exudate constituent cannot be ruled out. In light of these reports it would appear that Caryophyllales plants do not inhibit AM-fungal pre-colonization stages through the exudation of anti-fungal compounds. Nonetheless, we cannot exclude the possibility that anti-fungal compounds are exuded at concentrations so low under experimental conditions that their effects are masked, or that the effects of exudate constituents vary for different fungal isolates. For example, crude cell-wall extracts of Ri-T-DNA transformed beet roots inhibit hyphal growth of *Gigaspora gigantea,* but the same extracts stimulate hyphal growth of *G. margarita* (Douds *et al*., 1996).

Other theories have been proposed as to why Caryophyllales plants are predominantly non-host to AM fungi. It has also been suggested that for both the Chenopodiaceae and the Brassicaceae a lack of AM-colonization correlated to a relatively low release of border cells from the root cap (Neimara *et al.,* 1996), but to date there is no experimental evidence to support this hypothesis. The non-host plant *Salsola kali* (Chenopodiaceae) blocked colonization by mixed *Glomus* species and *Gigaspora margarita* after penetration, with colonized roots autofluorescing, possibly due to the deposition of phenolic molecules in the cell walls, and with colonized root dying (Allen *et al*., 1989). This reaction is indicative of a plant defense response and occured after the formation of appressorium. However, in other reports Caryophyllales plants do not stimulate hyphal branching and appressorium formation of AM fungi, as if root exudates are lacking signal compounds necessary for fungal morphogenesis and penetration (Bécard and Piché, 1990; Giovannetti *et al*., 1993b; 1994; Schreiner and Koide, 1993c; Schwab *et al*., 1984). Also, AM-fungi pre-colonization stages may be inhibited by the extremely saline, xeric, or disturbed habitats where Caryophyllales plants often grow, and not mediated directly by plant-derived factors.

5. PHENOLIC COMPOUNDS AS SIGNAL MOLECULES FOR AM HOST RECOGNITION

Many non-host plants and even those that are host to mycorrhizae other than AM (such as ecto, arbutoid and ericoid mycorrhiza) fail to elicit functional appressorium formation in AM fungi (Giovannetti and Lioi, 1990; Giovannetti *et al*., 1993b; 1994). This evidence, in addition to others mentioned throughout this review, have led many researchers to speculate that the initiation of AM colonization is mediated by specific signal molecules produced by the host plant (Azcon and Ocampo, 1984; Becard and Piche, 1990; El-Atrach *et al*., 1989; Elias and Safir, 1987; Gianinazzi-Pearson *et al*., 1989; Giovannetti *et al*., 1985; 1993a, b; 1994; Glenn *et al*., 1988; Annapurna *et al*., 1996; Paula and Siqueira, 1990; Schreiner and Koide, 1993c). If certain compounds are necessary for host recognition, these compounds must be produced in a wide variety of plant species to account for the apparent non-specificity of AM fungi for plant hosts (Koide and Schreiner, 1992). Flavonoids, which are ubiquitous in plant roots (Wollenweber and Dietz, 1981), have a stimulatory effect on AM-fungal spore germination, hyphal growth, and colonization (Fig. 2) (Balaji *et al*., 1995; Bécard *et al*., 1992; Chabot *et al*., 1992; Gianinazzi-Pearson *et al*., 1989; Kape *et al.*, 1992; Nair *et al.*, 1991, Poulin *et al.*, 1993; Siqueira *et al*., 1991; Tsai and Phillips, 1991). In a couple of studies stimulatory flavonoids were those determined to actually be present in the host root exudates (Nair *et al*., 1991; Tsai and Phillips, 1991), and in one case stimulatory flavonoids were present in root exudates of P-stressed clover (Nair *et al*., 1991). Root flavonoids are found in greatest concentrations at the apical region of the root in the zone of elongation (Graham, 1991;

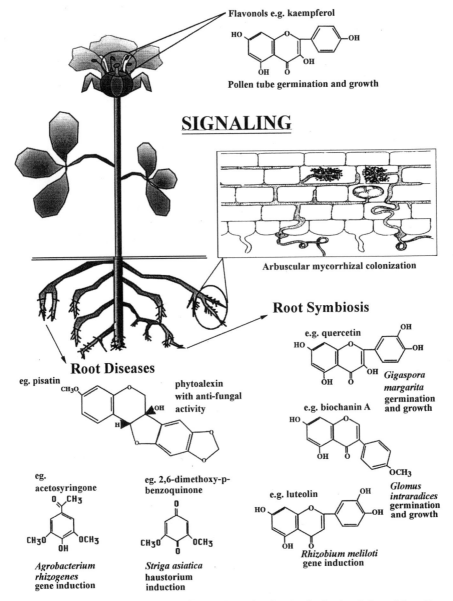

Flavonols e.g. kaempferol

Pollen tube germination and growth

SIGNALING

Arbuscular mycorrhizal colonization

Root Symbiosis

e.g. quercetin

Gigaspora margarita germination and growth

Root Diseases

eg. pisatin

phytoalexin with anti-fungal activity

e.g. biochanin A

Glomus intraradices germination and growth

eg. acetosyringone

eg. 2,6-dimethoxy-p-benzoquinone

e.g. luteolin

Agrobacterium rhizogenes gene induction

Striga asiatica haustorium induction

Rhizobium meliloti gene induction

Fig. 2. Plant phenolics active in cell signaling. Schematic of AM colonization (adapted from Brundrett *et al.,* 1994).

Peters and Long, 1988), the preferred sites for AM colonization (Smith and Walker, 1981) (Fig. 1). Also, root flavonoid concentrations and metabolism are altered by the presence of AM fungi (Harrison and Dixon, 1993; Kape *et al.*, 1992; Lambias and Mehdy, 1993; Mc Arthur and Knowles, 1992; Morandi, 1989; Morandi *et al.*, 1984; Volpin *et al.*, 1994; 1995); generally, however, these responses are small compared to those elicted by a fungal pathogen.

Flavonoids serve a variety of physiological roles in plants, including protection from UV stress (Li *et al.*, 1993), and as auxin transport inhibitors (Jacobs and Rubery, 1988). Flavonoids also function as plant phytoalexins, with some flavonoids exhibiting anti-fungal activity, particularly isoflavonoids (Fig. 2) (Dixon *et al.*, 1983; Smith and Banks, 1986). Generally, anti-fungal activity increases with increasing lipophilicity, possibly due to disruptions in the membranes of affected fungi, and inhibitory activity varies on different fungi (Dixon *et al.*, 1983; Smith and Banks, 1986). Flavonoids cause a variety of effects in mammalian systems, including direct interaction with known estrogen receptors (Miksicek, 1993). Flavonoids compounds are produced *via* the phenylpropanoid pathway, known to be correlated with the general induction of a defense response that includes increased cell wall deposition and lignin production (Koes *et al.*, 1994).

Flavonoids serve as signal molecules in the symbiosis between rhizobial bacteria and legumes that leads to the formation of nitrogen-fixing nodules (Fig. 2). Flavonoids exuded from host plant roots interact specifically with bacteiral *nod* gene products, stimulating transcription and the production of *nod* factors that begin nodule formation (Fischer and Long, 1992). Stimulatory flavonoids include flavones, flavanones, and chalcones for *Rhizobium* species and isoflavonoids for *Bradyrhizobium* species (Phillips, 1992). The host range of different rhizobium bacteria is determined by recognition of specific flavonoids exuded from the host plant species, mediated by interactions with *nod D* and other *nod* genes whose protein products regulate transcription of other *nod* genes (Phillips, 1992). Flavonoids also serve as the phytoalexins for legume plants. Legume plants are often used in mycorrhizal studies, because mycorrhizal mutants (myc-) that have been characterized are also nodulation mutants (Bradburg *et al.*, 1991; Duc *et al.*, 1989), and because of similarities in the interfaces that form between the plant and microorganism in both symbiotic interactions (Perrotto *et al.*, 1994).

Another example of flavonoids serving as molecular signals is the requirement of flavonols for pollen tube germination and growth, and thus for plant fertilization and reproduction (Fig. 2) (Ylstra *et al.*, 1994). The tip-directed growth pattern of pollen tubes is similar to that occurring in fungal hyphae; flavonoids may stimulate a general response in cell elongation (Bécard *et al.*, 1995). Response to flavonoids are also reported in fungal systems: there are several cases where plant pathogen spores are chemotropically attracted to host plant roots by flavonoids in the exudates (Tahara and Ibrahim, 1995), and for several pathogenic fungi flavonoids induce genes for phytoalexin detoxification (Van Etten *et al.*, 1989).

The results of studies comparing the *in vitro* effect of flavonoids on AM fungi are often conflicting. For example, in a study by Gianinazzi-Pearson *et al.*, (1989), two flavanones, naringenin and hesperitin, and a flavone, apigenin, added to solid media at 0.15 to 1.5 μM stimulated spore germination and growth of *Gigaspora margarita*: however, this study was not carried out under regulated CO_2 conditions (Gianinazzi-Pearson *et al.*, 1989). In two later studies by Becard *et al.*, (1992) and Chabot *et al.* (1992), a different isolate of the same species was used to monitor germination and growth effects of flavonoids added to solid media at 10 μM, with the experiments tested under controlled CO_2 conditions (Bécard *et al.*, 1992; Chabot *et al.*, 1992). In the latter studies, apigenin and hesperitin inhibited hyphal growth (Bécard *et al.*, 1992; Chabot *et al.*, 1992). These conflicting results may have been obtained because if flavonoids stimulated spore germination alone the germinating spore might evolve enough CO_2 to account for the subsequent stimulation of hyphal growth (Bécard *et al.*, 1992), or because the activity of flavonoids on AM-hyphal growth could be attenuated at higher concentrations. The authors of the two latter studies stressed the importance of maintaining controlled CO_2 conditions, and by comparing a wide variety of flavonoids to elucidate the structural specificity required for activity concluded that flavonol compounds caused the greatest growth stimulation (Bécard *et al.*, 1992; Chabot *et al.*, 1992). Discrepancies in future studies could be avoided by adopting standard procedures for methodology and analysis, comparing the effects of structurally similar compounds, and developing dose-responsiveness curves for active compounds. No studies on the *in vitro* effects of flavonoids on AM fungi other than *G. margarita* have been performed under controlled CO_2 conditions (Kape *et al.*, 1992; Nair *et al.*, 1991; Tsai and Phillips, 1991), nor have surveys of flavonoids to determine structure-activity relationships for other fungi been conducted. Even if standard conditions are used for future analysis, there may be different responses to flavonoids from different fungal species or even different isolates of the same species as was the case in the example mentioned above (Bécard *et al.*, 1992; Chabot *et al.*, 1992).

Tsai and Phillips (1991) found that some flavonoid constituents of alfalfa roots increased percentage germination in *Glomus etunicatum* spores, while another flavonoid constituent inhibited germination. Considering the variation in flavonoids that are produced by plants, it is likely that the total balance and quantity of flavonoids present in the plant root exudates alters AM growth in the field, as opposed to any one individual compound serving as a specific "signal". Obviously certain flavonoids stimulate the growth of AM fungi during the pre-colonization stages, but these compounds may only be active until a threshold value is obtained, or flavonoids may not be absolutely required for attachment to host plant root or subsequent colonization. For example, the addition of the highly stimulatory flavonol quercetin (Balaji *et al.*, 1995; Becard *et al.*, 1992) to the medium does not facilitate AM colonization of Ri-T-DNA transformed *myc*-pea roots by *Gigaspora margarita* (Balaji *et al.*, 1995). However, *myc*- mutant plants that have been characterized to date block AM fungus after appressorium formation, indicating that the fungus still recognizes the plant as host (Balaji *et al.*, 1995; Bradbury *et al.*, 1991; Duc *et al.*, 1989; Gollette *et al.*, 1993); in two reports the AM fungi formed more appressorium on *myc*-roots compared to myc+roots, perhaps due to repeated attempts at penetration (Balaji *et al.*, 1995; Bradbury *et al.*, 1991). Hence, myc-mutants are not necessarily the best system to use to elucidate the mechanism by which a fungus "recognizes" a host plant.

Another controversy about the role of flavonoids in AM initiation concerns the use of an Ri T-DNA-transformed root culture of carrot (Bécard *et al.*, 1995; Bel Rhlid *et al.*, 1993). While Bel Rhlid *et al.*, (1993) identified that this culture produced flavonoids, Becard *et al.* (1995), did not detect any flavonoids in a clone of the same culture. Bécard *et al.* (1995) determined that this root culture would become colonized by AM fungi *Gigaspora margarita*, even with the addition of the flavonoid-binding compound polyvinylpyrrolidone to the media, and concluded that flavonoids were not necessary for the colonization process. Experiments using transformed root cultures must be intepreted with caution because phenolic compounds fluctuate with changing environmental factors or root age. While a researcher can manipulate a root culture to produce greater quantities of exudate, this manipulattion results in a stressed condition and may change metabolic processes in the plant root. Again, the use of standard procedures for the study of AM interactions in transformed root systems would reduce the occurrence of such conflicting results, as would the use of reliable and accurate techniques for chemical analysis. Becard *et al.*, (1995) also found that maize mutants deficient in a gene crucial for flavonoid production were as heavily colonized by several AM fungi as were the wild-type plants, supporting the conclusion that flavonoids are not absolute requirements for the colonization process. These observations suggest that even if flavonoid synthesis is suppressed by unexpected genetic or biochemical events, there exists non-flavonoid inducers which are adaquate for AM formation. While flavonoids might not be absolutely required for AM colonization, they certainly accelerate the pre-colonization stages of fungal development through enhancement of hyphal growth. Hyphal growth of *Glomus intraradices* is stimulated by both the isoflavonoid compound biochanin A and the estrogen 17ß-estradiol, and stimulation by biochanin A is suppressed by the antiestrogen EM-652, indicating that flavonoids bind to specific receptors. Thus, flavonoids definitely have an effect on AM-fungal cellular processes, but the biochemical mechanisms that host plants use to regulate its fungsl symbiont is not clearly defined.

Because of the presence of rhizosphere microorganisms other than AM fungi on host plant roots, the activity of exogenously applied flavonoids stimulating root colonization in non-sterile systems (Nair *et al.*, 1991; Siqueira *et al.*, 1991) cannot be attributed to direct action on the AM fungus. This point is illustrated by studies examining the tripartite interaction that occurs between a host plant, AM mycorrhiza, and nodulating rhizobium bacteria. *Rhizobium* bacteria and mycorrhizal fungi act synergistically to enhance plant growth and reproduction, and in most reports both mediate increased colonization by the other symbiont (Xie *et al.*, 1995). The synergistic effect is probably at least in part due to the enhanced nutritional status of the host by both symbionts. Additionally, rhizobium bacteria metabolize flavonoids to different phenolic compounds (Rao and Cooper, 1995) and stimulate further flavonoid production from the host plant (Xie *et al.*, 1995), thus changing the composition of phenolics in the rhizosphere. Xie *et al.* (1995), suggested that the enhancemnt of mycorrhizal colonization before nodule formation and even in non-nodulating mutant soybean by rhizobium bacteria and isolated nod factors is mediated by the fluctuations of phenolic compounds in the rhizosphere.

Phenolic compounds other than flavonoids are known to be involved in cell signaling between host plants and parasitic endophytes. The parasitic plants *Striga* spp. and *Agalinis purpurea* (Scrophulariaceae) require host-derived signals for host recognition, germination, and haustorium induction; signal molecules consist of various classes of phenylpropanoid compounds, which become active after oxidation into quinones by enzymes from the parasite (Fig. 2) (Lynn and Chang, 1990). In *Striga* spp. interactions the chemical instability of the germination stimulants provides spatial information for the parasite, and a similar distance relationship may be important in the response of AM-fungi to host root exudates (Koide and Schreiner, 1992; Lynn, 1995). Also, phenolic exudates from wounded plants are required to induce gene expression and subsequent colonization by the pathogen *Agrobacterium tumifaciens,* which causes crown gall tumors on most dicotyledonous plants (Fig. 2) (Winans, 1992). *A. tumifaciens vir* genes are activated by acetosyringone, $\alpha-$ hydroxyacetosyringone, and a variety of other phenolic molecules including lignin precursors (Winans, 1992). Different strains of *A. tumefaciens* exhibit a different degree of induction respone to different phenolic molecules, which appear to interact directly with *vir* A proteins (Lee *et al.*, 1995). Additionally, plant phenolic molecules induce bacterial genes for catabolic pathways that degrade aromatic compounds (Peters and Verma, 1990).

In a greenhouse study, several phenolic acids reduced the growth of mycorrhizal asparagus (Pederson *et al.*, 1991). Recently, Douds *et al.*, (1996) examined the *in vitro* effect of several phenolic acids, which had been previously isolated from cell-wall extracts of Ri-T-DNA transformed roots of the host plant carrot and non-host plant sugar beet (Nagahashi *et al.*, 1996), on the germination and hyphal growth of *Gigaspora margarita* and *G. gigantea* (Douds *et al.*, 1996). Several of the individual henolic compounds induced different responses from the two fungi tested in this study (Douds *et al.*, 1996). Moreover, one compound found only in the host plant carrot roots, vanillic acid, was stimulatory to both fungi, but both carrot roots and non-host plant beet roots contained the generally inhibitory compound ferulic acid (Douds *et al.*, 1996). Once again, this study indicates that the balance of phenolics present in the roots probably produces a greater effect on phyphal growth than does the presence or absence of individual compounds. It was not determined in these studies whether the phenolic acids tested are present in the root exudates; more detailed investigation of the phenolic composition of host root exudates is required before it becomes clear how these compounds affect AM-fungal pre-colonization development. Also, the composition of other organic acids in root exudates and their effect on AM-fungi should be investigated for involvement in AM-initiation. A diterpene resin acid from pine roots, abietic acid, stimulates spore germination in the ectomycorrhizal fungi *Suillus* at concentrations as low as as 10^{-7}M (Fries *et al.*, 1987), and the pine root exudate constituent palmitic acid also stimulates growth of ectomycorrhizal fungi (Sun and Fries, 1992), but such compounds have yet to be investigated as possible signals for host recognition in AM interactions.

6.   FUTURE DIRECTIONS

Without question plant root volatiles and root exudates has an effect on the pre-colonization stages of AM-fungal development. However, without a uniformity in the methodology utilized in both the collection of root exudates and in examining the effects of exudate constituents on AM-fungal interactions, it is difficult to clearly define these effects. As individual AM-fungal species and isolates respond differently under experimental conditions, are present in a wide variety of environmental conditions, and are usually generalists for their plant host, it is possible that pre-colonization stages are mediated by numerous rhizosphere components that are not all required for continued development. Furthermore, although host root exudates and flavonoids stimulate the hyphal branching preceding attachment to the root (Balaji *et al.*, 1995; Gemma and Koske, 1988; Giovannetti *et al.*, 1993b; 1994; Graham, 1982; Mosse, 1988; Powell, 1976; Schreiner and Koide, 1993c; Tsai and Phillips, 1991), it is unclear if such morphogenesis are important for root inoculum (Powell, 1976). An AM fungus already supported by a functional symbiosis may not require the same inputs for the formation of further colonization units as do germ tubes supported only by spore reserves (Koide and Schreiner, 1992).

Nonetheless, it may be possible to identify root exudate constituents that have the greatest activity in AM-fungal interactions, and even potential signal molecules in the colonization process. From the available literature we can begin to see that plant root exudates alter the pre-colonization stages of AM interactions at several developmental points: during spore germination, chemotropic growth towards the plant root, hyphal branching in the vicinity of the root, and colonization. One step in the developmental process of AM-interactions that has been somewhat overlooked is the attachment of fungal hyphae to the plant root and appressorium formation. If specific signal molecules are required for host recognition, it is likely that these are only found in the immediate vicinity of the root, associated with root mucilage (Giovannetti *et al.*, 1994), remaining in the apoplast or only a short distance from the root, or only found in active form for a short time. For ericoid mycorrhiza host-specificity has been attributed to carbohydrate-carbohydrate interactions that occur at the cell surface (Bonfante-Fasolo,1988); the role of components of the root mucilage on AM-fungi has not been thoroughly examined. Likewise, the spatial and temporal fluctuation of exudate components in the vicinity of the root needs to be investigated.

Although it is tempting to correlate the phenomenon of non-host plants to a lack of recognition during pre-colonization stages, it is unclear whether root exudate components are the primary mediator of host recognition. There are numerous reports that during the early stages of AM colonization a plant defense response is transiently induced and then subsequently suppressed (Harrison and Dixon, 1993; Lambias and Mehdy, 1993; McArthur and Knowles, 1992; Volpin *et al.*, 1994; 1995), and that suppression is attenuated in plants grown in high-P conditions (Lambias and Mehdy, 1993). Inhibition of AM-fungal development in plants grown in high - P conditions has been attributed to inability of fungal suppressers to inactivate a defense response (McArthur and Knowles, 1992). If inadequate suppression of a defense response inhibits colonization of non-host plants such as *Salsola kali* (Chenopodiaceae) (Allen *et al.*, 1989), it is possible that host recognition is not required to initiation AM-colonization. Similarly, non-host plant might physically inhibit AM-development by virtue of cell wall composition (Tester *et al.*, 1987), rather than host recognition or the activation of a defense response.

Much of the work examining the pre-colonization stages of AM-development has focused on the effects of plant-derived factors on fungal development and not on changes produced in the plant. AM fungi also secrete exudates that alter the rhizosphere chemistry *via* nitrate depletion and pH changes (Bago *et al.*, 1996). Volpin *et al.* (1994), found that the flavonoid formononetin increased in alfalfa roots when *Glomus intraradices* was present in the rhizosphere, before hyphal contact. Thus, the effects of AM-fungi on the rhizosphere environment and plant root physiology, and how these effects facilitate AM development, requires further investigation. It is possible that a bi-directional exchange of chemical signals between plant and fungus is necessary for AM colonization to occur, such as that observed with the nodulation symbiosis. It is also possible that AM-colonization in the field is mediated in part by interactions with other rhizosphere microorganisms including bacterial inhabitants of the root mucigel. Obviously, the role of root exudates on AM-interactions is complex and difficult to define, but elucidation of the specific effects at particular stages in development would greatly benefit our understanding of the symbiosis.

ACKNOWLEDGMENTS

C.I. Mandelbaum would like to thank Espanta Seradge for her kind assistance in creating the figures, and John Clement, Prof. Ahmad S. Islam, Dr. T.J. Mabry, and Dr. R.P. Schreiner for their helpful comments reviewing the manuscript.

REFERENCES

Allen, M.F. 1993, Formation of vesicular-arbuscular mycorrhizae in *Atriplex gardneri* (Chenopodiaceae):seasonal response in a cold desert, *Mycologia* 75:733-776.

Allen, M.F., Allen, E.B. and Friese, C.F. 1989, Responses of the non-mycotrophic plant *Salsola kali* to invasion by vesicular-arbuscular mycorrhizal fungi, *New Phytol.* 111:4-49.

Amijee, F., Tinker, P.B. and Stribley, D.P. 1989, The development of endomycorrhizal root systems VII, A detailed study of the effects of soil phosphorus on colonization, *New Phytol.* 111:435-446.

Annapurna, K., Tilak, K.V.B.R. and Mukerji, K.G. 1996, Arbuscular mycorrhizal symbiosis - recognition and specificity. in : *Concepts in Mycorrhizal Research,* K.G. Mukerji, ed., Knwer Academic Publishers, Dordredit, The Netherlands, pp. 77-90.

Avio, L., Sbrana, C. and Giovannetti, M. 1990, The response of different species of *Lupinus* to VAM endophytes, *Symbiosis* 9:321-323.

Azcón, R. 1987, Germination and hyphal growth of *Glomus mosseae in vitro*: Effects of rhizosphere bacteria and cell-free culture media, *Soil Biol. Biochem.* 19:417-419.

Azcón, R. and Ocampo, J.A. 1981, Factors affecting the vesicular-arbuscular infection and mycorrhizal dependency of thirteen wheat cultivars, *New Phytol.* 87:677-685.

Azcón, R. and Ocampo, J.A. 1984, Effect of root exudation on VA mycorrhizal infection at early stages of plant growth, *Plant Soil* 82:133-138.

Azcón-Aguilar, C., Diaz-Rodriguez, R. and Barea, J.H. 1986, Effect of soil microorganisms on spore germination and growth of the vesicular-arbuscular mycorrhizal fungus *Glomus mosseae, Trans. Br. Mycol. Soc.* 86:337-340.

Bago, B., Vierheilig, H., Piché, Y. and Azcón-Aguilar, C.1996, Nitrate depletion and pH changes induced by extraradical mycelium of the arbuscular mycorrhizal fungus *Glomus intraradices* Smith & Schenk grown in monoaxenic culture, *New Phytol.* 133:273-280.

Balaji, B., Poulin, M.J., Vierhelig, H. and Piché, Y. 1995, Responses of an arbuscular mycorrhizal fungus, *Gigaspora margarita,* to exudates and volatiles from the Ri T-DNA-transformed roots of non-mycorrhizal and mycorrhizal mutants of *Pisum sativum* (L) Sparkle, *Ex. Mycol.* 19:275-283.

Bécard, G. and Piché, Y. 1989a, Fungal growth stimulation by CO_2 and roots exudates in vesicular-arbuscular mycorrhizal symbiosis, *App. Enviro. Microbiol.* 55:2320-2325.

Bécard, G. and Piché, Y. 1989b, New aspects on the aquisition of biotrophic status by a vesicular-arbuscular fungus, *Gigaspora margarita, New Phytol.* 112:77-83.

Bécard, G. and Piché, Y. 1990, Physiological factors determining vesicular-arbuscular mycorrhizal formation in host and nonhost Ri T-DNA transformed roots, *Can. J. Bot.* 68:1260-1264.

Bécard, G. and Piché, Y. 1992, Establishment of vesicular-arbuscular mycorrhiza in root organ culture: review and proposed methodology, *Methods in Microbiol.* 24:90-108.

Bécard, G., Douds, D.D. and Pfeffer, P.E. 1992, Extensive *in vitro* hyphal growth of vesicular-arbuscular mycorrhizal fungi in the presence of CO_2 and flavonols, *App. Enviro. Microbiol.* 58:821-825.

Bécard, G., Taylor, L.P., Douds, D.D., Pfeffer, P.E. and Doner, L.W. 1995, Flavonoids are not necessary plant signal compounds in arbuscular mycorrhizal symbiosis, *Mol. Plant-Microbe Intract.* 8:252-258.

Behnke, H.D. and Mabry, T.J. (eds) 1994, In : *Caryophyllales Evolution and Systematics*, Springer-Verlag, Heilderburg, Germany.

Bel Rhild, R., Chabot, S., Piche, Y. and Chenevert, R. 1993, Isolation and identification of flavonoids from Ri T-DNA-transformed roots (*Daucus carota*) and their significance in vesicular-arbuscular mycorrhiza, *Phytochem.* 33:1369-1371.

Black, R. and Tinker, P.B. 1979, The development of endomycorrhizal root systems, II, Effect of agromonic factors and soil conditions on the development of vesicular-arbuscular mycorrhizal infection in barley and on the endophyte spore density, *New Phytol.* 83:401-413.

Bolan, N.S., Robson, A.D. and Barrow, N.J. 1984, Increasing phosphorus supply can increase the infection of plant roots by vesicular-arbuscular mycorrhizal fungi, *Soil Biol. Biochem.* 16:419-420.

Bonfante, P. and Perotto, S. 1995, Strategies of arbuscular mycorrhizal fungi when infecting host plants, *New Phytol.* 130:3-21.

Bonfante-Fasolo, P.1988, The role of the cell wall as a signal in mycorrhizal associations, in: *Cell to Cell Signals in Plant, Animal and Microbial Symbiosis*, D. Scannerini, P. Smith, P. Bonfante-Fasolo and V. Gianinazzi-Pearson, eds., Springer-Verlag, Berlin, pp.219-235.

Bradbury, S.M., Peterson, R.L. and Bowley, S.R. 1991, Interactions between three alfalfa nodulation genotypes and two *Glomus* species, *New. Phytol.* 119:115-120.

Brundrett, M., Melville, L. and Peterson, R.L. 1994, *Practical Methods in Mycorrhiza Research*, Mycologue, Waterloo, Canada, pp. 1-161.

Carr, G.R., Hinkley, F., LeTacon,F., Hepper, C.M., Jones, M.G.K. and Thomas, E.1985, Improved hyphal growth of two species of vesicular-arbuscular mycorrhizal fungi in the presence of suspension-cultured plant cells, *New Phytol.* 101:417-426.

Chabot, S., Bel-Rhlid, R., Chênevert, R. and Piché, Y. 1992, Hyphal growth promotion *in vitro* of the VA mycorrhizal fungus *Gigaspora margarita* Becker and Hall, by the activity of structurally specific flavonoids under CO_2 enriched conditions, *New Phytol.* 122: 461-471.

Chaboud, A. and Rougier, M. 1981, Secretions racinaires mucilagineuses et role dans la rhizosphere, *Ann. Biol.* 20:313-326.

Daniels, B.A. and Trappe, J.M. 1980, Factors affecting spore germination of the vesicular-arbuscular mycorhizal fungus *Glomus epigaeus, Mycologia* 72:457-471.

Demars, B.G. and Boerner, R.E.J. 1994, Vesicular-arbuscular mycorrhizal fungi colonization in *Capsella bursa-pastoris* (Brassicaceae), *Am. Midl. Nat.* 132:377-380.

Demars, B.G. and Boerner, R.E.J. 1996, Vesicular arbuscular mycorrhizal development in the Brassicaceae in relation to plant life span, *Flora* 191:179-189.

Dixon, R.A., Dey, P.M. and Lamb, C.J. 1983, Phytoalexins:enzymology and molecular biology, *Adv. Enzymol. Related Areas of Mol. Biol.* 55:1-136.

Douds, D.D.Jr., Nagahashi, G. and Abney, G.D. 1996, The differential effects of cell wall-associated phenolics, cell walls, and cytosolic phenolics of host and non-host roots on the growth of two species of AM fungi, *New Phytol.* 133:289-294.

Drobnica, L., Zemanová, M., Nemec, P., Antos, K., Kristián, P., Stullerová, A., Knoppová V. and Nemec, P.J.R. 1967, Antifungal activity of isothiocyanates and related compounds, *App. Microbiol.* 15:701-709.

Duc, G., Trouvelot, A., Gianinazzi-Pearson, V. and Gianinazzi, S.1989, First report of non-mycorrhizal plant mutants (Myc-) obtained in pea (*Pisum sativum* L.) and fababean (*Vicia faba* L.), *Pl. Sci.* 60:215-222.

El-Atrach F., Vierheilig, H. and Ocampo, J.A. 1989, Influence of non-host plants on vesicular-arbuscular mycorrhizal infection of host plants and on spore germination, *Soil Biol. Biochem.* 21:161-163.

Elias, K.S. and Safir, G.R. 1987, Hyphal elongation of *Glomus fasciculatum* in response to root exudates, *App. Enviro. Microbiol.* 53:1928-1933.

Fischer, R.F. and Long, S.R. 1992, *Rhizobium*-plant signal exchange, *Nature* 357:655-660.

Fries, N., Serck-Hanssen, K., Dimberg, L.H. and Theander, O.1987, Abietic acid, an activator of basidiospore germination in ectomycorrhizal species of the genus *Suillus* (Boletaceae), *Ex. Mycol.* 11:360-363.

Gagnon, H., Tahara, S. and Inbrahim, R.K. 1995, Biosynthesis, accumulation, and secretion of isoflavonoids during germination and development of white lupin (*Lupinus albus* L.), *J. Ex. Bot.* 46:609-616.

Gemma, J.N. and Koske, R.E. 1988, Pre-infection interactions between roots and the mycorrhizal fungus *Gigaspora gigantea*: chemotropism of germ-tubes and root growth response, *Trans. Br. Mycol. Soc.* 91:123-132.

Gianinazzi-Pearson, V. and Gianinazzi, S. 1992, Influence of intergenic grafts between host and non-host legumes on formation of vesicular-arbuscular mycorrhiza, *New Phytol.* 120:505-508.

Gianinazzi-Pearson, V., Branzanti, B. and Gianinazzi, S. 1989, *In vitro* enhancement of spore germination and early hyphal growth of a vesicular-arbuscular mycorrhizal fungus by host root exudates and plant flavonoids, *Symbiosis* 7:243-255.

Giovannetti, M. and Lioi, L. 1990, The mycorrhizal status of *Arbutus unedo* in relation to compatible and incompatible fungi, *Can. J. Bot.* 68:1239-1244.

Giovannetti, M., Avio, L., Sbrana, C. and Citernesi, A.S. 1993a, Factors affecting appressorium development in the vesicular-arbuscular mycorrhizal fungus *Glomus mosseae* (Nicol. & Gerd.) Gerd. & Trappe, *New Phytol.* 123:115-122.

Giovannetti, M., Sbrana, C., Avio, L., Citernesi, A.S. and Logi, C. 1993b, Differential hyphal morphogenesis in arbuscular mycorrhizal fungi during pre-infection stages, *New Phytol.* 125:587-593.

Giovannetti, M., Sbrana, C. and Logi, C. 1994, Early processes involved in host recognition by arbuscular mycorrhizal fungi, *New Phytol.* 127:703-709.

Glenn, M.G., Chew, F.S. and Williams, P.H. 1985, Hyphal penetration of Brassica (Cruciferae) roots by vesicular-arbuscular mycorrhizal fungus, *New Phytol.* 99:463-472.

Glenn, M.G., Chew, F.S. and Williams, P.H. 1988, Influence of glucosinolate content of *Brassica* (Cruciferae) roots on growth of vesicular-arbuscular mycorrhizal fungus, *New Phytol.* 110:217-225.

Glinski, J. and Stepniewski, W. 1985, *Soil Aeration and its Role for Plants*, CRC Press Inc., Boca Raton, USA. pp. 91-104.

Gollette, A., Gianinazzi-Pearson,V., Giovannetti, M., Sbrana, C., Avio, L. and Gianinazzi, S. 1993, Cellular localization and cytochemical probing of resistance reactions to arbuscular mycorrhizal fungi in a `locus a' myc-mutant of *Pisum sativum* L., *Planta* 191:112-122.

Graham, J.H. 1982, Effect of citrus root exudates on germination of chlamydospores of the vesicular-arbuscular mycorrhizal fungus, *Glomus epigaeum, Mycologia* 74:831-835.

Graham, J.H., Leonard, R.T. and Menge, J.A. 1981, Membrane-mediated decrease in root exudation responsible for phosphorus inhibition of vesicular-arbuscular mycorrhiza formation, *Plant Physiol.* 68:548-552.

Graham, J.H., Leonard, R.T. and Menge, J.A. 1982, Interactions of light intensity and soil temperature with phosphorus inhibition of vesicular-arbuscular mycorrhiza formation, *New Phytol.* 91:683-690.

Graham, T.L. 1991, Flavonoid and isoflavonoid distribution in developing soybean seedling tissues and in seed and root exudates, *Plant Physiol.* 95:594-603.

Green, N.E., Graham, S.O. and Schenck, N.C. 1976, The influence of pH on the germination of vesicular-arbuscular mycorrhizal spores, *Mycologia* 69:929-934.

Hale, M.G., Foy, L.L. and Shay, F.J. 1971, Factor effecting root exudation, *Adv. Agron.* 2:89-109.

Harborne, J.B., Ingham, J.L., King, L. and Payne, M. 1976, The isopentenyl isoflavone, luteone as a pre-infection antifungal agent in the genus *Lupinus, Phytochem.* 15:1485-1487.

Harrison, M.J. and Dixon, R.A. 1993, Isoflavonoid accumulation and expression of defence gene transcripts during the establishment of vesicular-arbuscular mycorrhizal associations in roots of *Medicago truncatula, Mol. Plant-Microbe Intract.* 6:643-654.

Hayman, D.S., Johnson, A.M. and Ruddlesdin, I. 1975, The influence of phosphate and crop species on *Endogone* spores and vesicular-arbuscular mycorrhiza under field conditions, *Plant Soil* 43:489-495.

Hepper, C.M. and Smith, G.A. 1976, Observation's on the germination of *Endogone* spores, *Trans. Br. Mycol. Soc.* 66:189-194.

Hirrell, M.C., Mehravaran, H. and Gerdemann, J.W. 1978, Vesicular-arbuscular mycorrhizae in the Chenopodiaceae and Cruciferae:do they occur? *Can.J.Bot.* 56:2813-2817.

Ingham, J.L., Tahara, S. and Harborne, J.B. 1983, Fungitoxic isoflavones from *Lupinus albus* and other *Lupinus* species, *Z. Naturforsch.* 38:194-200.

Jacobs, M. and Rubery, P.H. 1988, Naturally occuring auxin transport regulators, *Science* 241:346-349.

Jasper, D.A., Robson, A.D. and Abbott, L.K. 1979, Phosphorus and the formation of vesicular-arbuscular mycorrhizas, *Soil Biol. Biochem.* 11:501-505.

Johnson, C.R., Graham, J.H., Leonard, R.T. and Menge, J.A. 1982a, Effect of flower bud development in *Chrysanthemum* on vesicular-arbuscular mycorrhiza formation, *New Phytol.* 90:671-675.

Johnson, C.R., Menge, J.A., Schwab, S. and Ting, I.P. 1982b, Interaction of photoperiod and vesicular-arbuscular mycorrhizae on growth and metabolism of sweet orange, *New Phytol.* 90:665-669.

Johnson-Green, P.C., Kenkel, N.C. and Booth, T. 1995, The distributon and phenology of arbuscular mycorrhizae along an inland salinity gradient, *Can. J. Bot.* 73:1318-1327.

Kape, R., Wex, K., Parniske, M., Göoge, E., Wetzel, A. and Werner, D. 1992, Legume root metabolites and VA - mycorrhiza development, *J. Plant Physiol.* 141:54-60.

Koes, R.E., Quattrocchio, F. and Mol, J.N.M. 1994, The flavonoid biosynthetic pathway in plants:function and evolution, *BioEssays* 16:123-132.

Koide, R.T. and Schreiner, R.P. 1992, Regulation of the vesicular-arbuscular mycorrhizal symbiosis, *Ann. Rev. Plant Physiol. Plant Mol. Biol.* 43:557-581.

Koske, R.E., 1982, Evidence for a volatile attractant from plant roots affecting germ tubes of a VA mycorrhizal fungus, *Trans. Br. Mycol. Soc.* 79:305-310.

Lambias, M.R. and Mehdy, M.C. 1993, Suppression of endochitinase, β-1, 3- endoglucanase, and chalcone isomerase expression in bean vesicular-arbuscular mycorrhizal roots under different soil phosphate conditions, *Mol. Plant-Microbe Intract.* 6:75-83.

Lee, Y.W., Jin, S., Sim, W.S. and Nester, E.W. 1995, Genetic evidence for direct sensing of phenolic compounds by the VirA protein of *Agrobacterium tumefaciens, Proc. Natl. Acad. Sci.* USA 92:12245-12249.

Li, J., Ou-Lee, T.M., Raba, R., Amundson, R.G. and Last, R.L. 1993, *Arabadopsis* flavonoid mutants are hypersensitive to UV-B irradiation, *Plant Cell* 5:171-179.

Lynn, D.G. 1995, Jan. 1. Letter to Dr. T.J. Mabry, University of Texas, Austin, TX.

Lynn, D.G. and Chang, M. 1990, Phenolic signals in cohabitation:implications for palnt development, *Ann. Rev. Plant Physiol. Plant Mol. Biol.* 41:497-526.

Manocha, M.S. and Chen, Y. 1990, Specificity of attachmant of fungal parasites to their hosts, *Can. J. Microbiol.* 36:69-76.

Marks, G.C. and Foster, R.C. 1973, Structure, morphogenesis, and ultrastructure of ectomycorrhizae, in : *Ectomycorrhizae:Their Ecology and Physiology*, G.C. Marks and T.T. Kozlowski, eds., Acedemic Press, New York, USA, pp. 1-41.

Marschner, H.1995, *Mineral Nutrition of Higher Plants,* Second edition, Acedemic Press, Cambridge, U.K.

McArthur, D.A.J. and Knowles, N.R. 1992, Resistance responses of potato to vesicular-arbuscular mycorrhizal fungi under varying abiotic phosphorus levels, *Plant Physiol.* 100:341-351.

Menge, J.A., Steirle, D., Bagyaraj, D.J., Johnson, E.L.V. and Leonard, R.T. 1978, Phosphorus concentrations in plants responsible for inhibition of mycorrhizal infection, *New Phytol.* 80:575-578.

Miksicek, R.J. 1993, Commonly occuring plant flavonoids have estrogenic activity, *Mol. Pharmacol.* 44:37-43.

Miller, R.M., Moorman, T.B. and Schmidt, S.K. 1983, Interspecific plant association effects on vesicular-arbuscular mycorrhiza occurrence in *Atriplex confertifolia*, *New Phytol.* 95:241-246.

Miranda, J.C.C. and Harris, P.J. 1994, Effects of soil phosphorus on spore germination and hyphal growth of arbuscular mycorrhizal fungi, *New Phytol.* 128:103-108.

Monz, C.A., Hunt, H.W., Reeves, F.B. and Elliot, E.T. 1994, The response of mycorrhizal colonization to elevated CO_2 and climate change in *Pascopyrum smithii* and *Bouteloua gracilis*, *Plant Soil* 165:75-80.

Morandi, D. 1989, Effect of xenobiotics on endomycorrhizal infection and isoflavonoid accumulation in soybean roots, *Pl. Physiol. Biochem.* 27:697-701.

Morandi, D., Bailey, J.A. and Gianinazzi-Pearson, V. 1984, Isoflavonoid accumulation in soybean roots infected with arbuscular-mycorrhizal fungi, *Physiol. Plant Path.* 24:357-364.

Morgan, J.A., Knight, W.G., Dudley, L.M. and Hunt, H.W. 1994, Enhanced root system C-sink activity, water relations and aspects of nutrient aquisition in mycotrophic *Bouteloua gracilis* subjected to CO_2 enrichment, *Plant Soil* 165:139-146.

Morley, C.D. and Mosse, B. 1976, Abnormal vesicular-arbuscular mycorrhizal infections in white clover induced by lupin, *Trans. Br. Mycol. Soc.* 67:510-513.

Mosse, B. 1959, The regular germination of resting spores and some observations on the growth requirements of an *Endogone* sp. causing vesicular-arbuscular mycorrhiza, *Trans. Br. Mycol. Soc.* 42:273-286.

Mosse, B.1988, Some studies relating to "independent" growth of vesicular-arbuscular endophytes, *Can. J. Bot.* 66:2533-2540.

Mosse, B. and Hepper, C. 1975, Vesicular-arbuscular mycorrhizal infections in root organ cultures, *Physiol. Plant Path.* 5:215-223.

Mukerji, K.G. and Kocher, B. 1983, Vesicular - arbuscular mycorrhiza in rape seed plant, *Proc. 6th Intern. Rape seed Conf., Paris*, Vol. II : 945-950.

Nagahashi, G., Abney, G.D. and Doner, L. 1996, A comparative study of phenolic acids associated with cell walls and cytoplasmic extracts of host and non-host roots for AM fungi, *New Phytol.* 133:281-288.

Nair, M.G., Safir, G.R. and Siqueira, J.O. 1991, Isolation and identification of vesicular-arbuscular mycorrhiza-stimulatory compounds from clover (*Trifolium repens*) roots, *App. Environ. Microbiol.* 57:434-439.

Newman, E.I. and Reddell, P. 1987, The distribution of mycorrhizas among families of vascular plants, *New Phytol.* 106:745-751.

Newsham, K.K., Fitter, A.H. and Watkinson, A.R. 1995, Multi-functionality and biodiversity in arbuscular mycorrhizas, *Trends Ecol. Evol.* 10:407-411.

Ocampo, J.A. 1980, Effect of crop rotations involving host and non-host plants on vesicular-arbuscular mycorrhizal infections of host plants, *Plant Soil* 56:283-291.

Ocampo, J.A. and Hayman, D.S. 1981, Influence of plant interactions on vesicular-arbuscular mycorrhizal infections, II, Crop rotations and residual effects of non-host plants, *New Phytol.* 87:333-343.

Ocampo, J.A., Martin, J. and Hayman, D.S. 1980, Influence of plant interactions on vescicular-arbuscular mycorrhizal infections, I, Host and non-host plants grown together, *New Phytol.* 84:27-35.

Paula, M.A. and Sequeira, J.O. 1990, Stimulation of hyphal growth of the VA mycorrhizal fungus *Gigaspora margarita* by suspension-cultured *Puerara phaseoloides* cells and cell products, *New Phytol.* 115:69-75.

Pederson, C.T., Safir, G.R. and Perent, S.1991, Effect of phenolic compounds on asparagus mycorrhiza, *Soil Biol. Biochem.* 23:491-494.

Perrotto, S., Brewin, N.J. and Bonfante, P. 1994, Colonization of pea roots by the mycorrhizal fungus *Glomus versiforme* and by *Rhizobium* bacteria: immunological comparison using monoclonal antibodies as probes for plant cell surface components, *Mol. Plant-Microbe Intrect.* 7:91-98.

Peters, N.K. and Long, S.R. 1988, Alfalfa root exudates and compounds which promote or inhibit induction of *Rhizobium meliloti* nodulation genes, *Plant Physiol.* 88:396-400.

Peters, N.K. and Verma, D.P.S. 1990, Phenolic compounds as regulators of gene expression in plant-microbe interactions. *Mol. Plant-Microbe Intract.* 3:4-8.

Phillips, D.1992, Flavonoids:Plant signals to soil microbes in: *Recent Advances in Phytochemistry* Vol. 26, *Phenolic Metabolism in Plants*, H.A. Stafford and R.K. Ibrahim, eds., Plenum Press, New York, USA, pp. 201-231.

Poulin, M.J., Bel-Rhlid, R., Piché, Y. and Chênevert, R. 1993, Flavonoids released by carrot (*Daucus carota*) seedlings stimulate hyphal development of vesicular-arbuscular mycorrhizal fungi in the presence of optimal CO_2 enrichment, *J.Chem. Ecol.* 19:2317-2327.

Powell, C.L. 1976, Development of mycorrhizal infections from *Endogone* spores and infected root segments, *Trans. Br. Mycol. Soc.* 66:439-445.

Powell, C.L. 1981, Inoculation of barley with efficient mycorrhizal fungus stimulates seed yield, *Plant Soil* 59:487-490.

Rao, J.R. and Cooper, J.E. 1995, Soybean nodulating rhizobia modify *nod* gene inducers daidzein and genstein to yield aromatic products that can influence gene-inducing activity, *Mol. Plant-Microbe Intract.* 6:855-862.

Ratnayake, M., Leonard, R.T. and Menge, J.A. 1978, Root exudation in relation to supply of phosphorus and its possible relevence to mycorrhizal formation, *New Phytol.* 81:543-552.

Rovira, A.D. 1969, Plant root exudates, *Bot. Rev.* 35:35-57.

Saif, S.R. 1984, The influence of soil aeration on the efficiency of vesicular-arbuscular mycorrhizas III, Soil carbon dioxide, growth and mineral uptake in mycorrhizal and non-mycorrhizal plants of *Eupatorium odoratum* L., *Guizotia abbyssinica* (L) Cass. and *Sorghum bicolor* (L.) Monech, *New Phytol.* 96 : 429-435.

Schreiner, R.P. and Koide, R.T. 1993a, Antifungal compounds from the roots of mycotrophic and non-mycotrophic plant species, *New Phytol.* 123: 99-105.

Schreiner, R.P. and Koide, R.T. 1993b, Mustards, mustard oils, and mycorrhizas, *New Phytol.* 123: 107-113.

Schreiner, R.P. and Koide, R.T. 1993c, Stimulation of vesicular-arbuscular mycorrhizal fungi by mycotrophic and non-mycotrophic plant roots systems, *App. Environ. Microbiol.* 59 : 2750-2752.

Schwab, S.M., Johnson, E.L.V. and Menge, J.A. 1982, Influence of simizine on formation of vesicular-arbuscular mycorrhizae in *Chenopodium quinona* Willd, *Plant Soil* 64 : 283-287.

Schwab, S.M., Leonard, R.T. and Menge, J.A. 1984, Quantitative and qualitative comparison of root exudates of mycorrhizal and nonmycorrhizal plant species, *Can. J. Bot.* 62 : 1227-1231.

Schwab, S.M., Menge, J.A. and Leonard, R.T. 1983, Quantitative and qualitative effects of phosphorus on extracts and exudates of sudangrass roots in relation to vesicular-arbuscular mycorrhiza formation, *Plant Physiol.* 73 : 761-765.

Schwab, S.M., Menge, J.A. and Tinker, P.B. 1991, Regulation of nutrient transfer between host and fungus in vesicular-arbuscular mycorrhizas, *New Phytol.* 117:387-398.

Simon, L., Bousquet, J., Lévesque, R.C. and Lalonde, M. 1993, Origin and diversification and endomycorrhizal fungi and coincidence with vascular land plants, *Nature* 363:67-69.

Siqueira, J.O., Safir, G.R. and Nair, M.G. 1991, Stimulation of vesicular-arbuscular mycorrhiza formation and growth of white clover by flavonoids, *New Phytol.* 118:87-93.

Smith, D.A. and Banks, S.W. 1986, Biosynthesis, elicitation and biological acivity of isoflavonoid phytoalexins, *Phytochem.* 25:979-995.

Smith, S.E. and Walker, N.A. 1981, A quantitative study of mycorrhizal infection in *Trifolium* : Separate determination of the rates of infection and of mycelial growth, *New Phytol.* 89:225-240.

St. John, T.V., Hays, R.I. and Reid, C.P.P. 1983, Influence of a volatile compound on the formation of vesicular-arbuscular mycorrhizas, *Trans. Br. Mycol. Soc.* 81 : 153-154.

Sun, Y.P. and Fries, N. 1992, The effect of tree-root exudates on the growth rate of ectomycorrhizal and saprotrophic fungi, *Mycorrhiza* 1:63-69.

Tahara, S. and Ibrahim, R.K. 1995, Prenylated isoflavonoids-an update, *Phytochem.* 38:1073-1094.

Tang, C. and Takenaka, T. 1983, Quantitation of a bioactive metabolite in undisturbed rhizosphere - benzyl isothiocyanate from *Carica papya* L., *J. Chem. Ecol.* 9:1247-1253.

Tawaraya, K., Watanabe, S. and Yoshida, E. 1996, Effect of onion (*Allium cepa*) root exudates on the hyphal growth of *Gigaspora margarita*, *Mycorrhiza* 6:57-59.

Tester, M., Smith, S.E. and Smith, F.A. 1987, The phenomenon of "nonmycorrhizal" plants, *Can. J. Bot.* 65 : 419-435.

Thomson, B.D., Robson, A.D. and Abbott, L.K. 1986, Effects of phosphorus on the formation of mycorrhizas by *Gigaspora calospora* and *Glomus fasciculatum* in relation to root carbohydrates, *New Phytol.* 103:751-765.

Tommerup, I.C. 1983, Temperature relations of spore germination and hyphal growth of vesicular-arbuscular mycorrhizal fungi in soil, *Trans. Br. Mycol. Soc.* 81: 381-387.

Tommerup, I.C. 1984, Development of infection by a vesicular-arbuscular mycorrhizal fungus in *Brassica napus* L. and *Trifolium subterraneum* L., *New Phytol.* 98 : 487-495.

Trinick, M.J. 1977, Vesicular-arbuscular infection and soil phosphorus utilization in *Lupinus* spp., *New Phytol.* 78 : 297-304.

Tsai, S.M. and Phillips, D.A. 1991, Flavonoids released naturally from alfalfa promote development of symbiotic *Glomus* spore *in vitro*, *App. Environ. Microbiol.* 57 : 1485-1488.

Van Etten, H.D., Mathews, D.E. and Mathews, P.S. 1989, Phytoalexin detoxification: Importance for pathogenicity and practical considerations, *Annu. Rev. Phytopath.* 27 : 143-164.

Vierheilig, H. and Ocampo, J.A. 1990a, Effect of isothiocyanates on germination of spores of *G. mosseae*, *Soil Biol. Biochem.* 22:1161-1162.

Vierheilig, H. and Ocampo, J.A. 1990b, Role of root extract and volatile substances of non-host plants on vesicular-arbuscular mycorrhizal spore germination, *Symbiosis* 9:199-202.

Volpin, H., Elkind, Y., Okon, Y. and Kapulnik, Y. 1994, A vesicular arbuscular mycorrhizal fungus (*Glomus intraradices*) induces a defense response in alfalfa roots, *Plant Physiol.* 104:683-689.

Volpin, H., Phillips, D.A., Okon, Y. and Kapulnik, Y. 1995, Suppression of an isoflavonoid phytoalexin defense response in mycorrhizal alfalfa roots, *Plant Physiol.* 108:1449-1454.

Walker, J.C., Morell, S. and Foster, H.H. 1937, Toxicity of mustard oils and related sulfur compounds to certain fungi, *Am. J. Bot.* 24:536-541.

Watrud, L.S., Heithaus, J.J. III and Jaworski, E.G. 1978, Geotropism in the endomycorrhizal fungus *Gigaspora margarita*, *Mycologia* 70: 449-452.

Winans, S.C. 1992, Two-way signaling in *Agrobacterium*-plant interactions, *Microbiol. Rev.* 56 : 12-31.

Wollenweber, E. and Dietz, V.H. 1981, Occurrence and distribution of free flavonoid aglycones in plants, *Phytochem.* 20 : 869-932.

Xie, Z., Staehelin, C., Vierheilig, H., Wiemken, A., Jabbouri, S., Broughton, W.J., Vögeli-Lange, R. and Boller, T. 1995, Rhizobial nodulation factors stimulate mycorrhizal colonization of nodulating and nonnodulating soybeans, *Plant Physiol.* 108:1519-1525.

Ylstra, B., Busscher, J., Fraken, J., Hollman, P.C.H., Mol, J.N.M. and van Tunen, A.J. 1994, Flavonols and fertilization in *Petunia hybrida:* localization and mode of action during pollen tube growth, *Plant J.* 6:201-212.

MYCORRHIZA IN CONTROL OF SOIL BORNE PATHOGENS

Reena Singh[1], Alok Adholeya[1], and K.G. Mukerji[2]

[1]Tata Energy Research Institute
Darbari Seth Block
Habitat Place, Lodhi Road
New Delhi - 110 003, INDIA

[2]C-29, Probyn Road
University of Delhi
Delhi-110 007, INDIA

1. INTRODUCTION

Speedy development of agriculture is vital to the progress of a country. For securing maximum crop production, the agricultural practices have evolved into highly technified and sophisticated plant production systems. Until recently, the management of plant diseases in these systems was done exclusively by the use of broad-spectrum chemical pesticides. However, there is reason for concern about the potential damage to ecosystems and pollution of groundwaters resulting from the widespread use of these chemicals on agricultural crops that led scientists to develop disease resistant transgenic plants as an alternative approach for controlling pathogens.

Breeding for plant resistance is, no doubt a most promising method for dealing with pests and diseases but it has its own limitations. The resistant variety to a particular pathogen has been shown to be highly susceptible to others and it has not been possible till now to breed for multiple resistant varieties to a wide range of races and pathogens. Moreover, it is not cost-effective; therefore, the use of pesticide has not been reduced with a genetic approach to disease control.

In order to address the problem of managing disease in an ecofriendly manner, the Common Agricultural Policy (CAP) of the European Union proposed the adequate management of natural, renewable resources and the reduction of chemical inputs. The replacement of the pesticides by the controlled use of microorganisms can be considered as a key component in the development of a sustainable agriculture (Barea *et al.*, 1996; Upadhyay *et al.*, 1997) to protect plants against pathogens (bacteria, fungi, nematodes) in an economically profitable manner with a minimized environmental pollution (Bethlenfalvay and Linderman, 1992).

Of the multitude of microorganisms that make up soil microbiota, mycorrhizal fungi are unique in their ability to link between plant and soil.Under natural conditions, plants strictly speaking do not have roots, they have mycorrhizas; the roots of most flowering plants form mutualistic symbiosis with soil fungi (Harley and Smith, 1983; Mukerji, 1996). These fungi have been shown to improve plant health against biotic and abiotic stresses and are now emerging as a potential biocontrol agents against soil-borne pathogens (Mukerji, 1999. The aim of this chapter is to understand the role of mycorrhiza in a biological control strategy and the underlying mechanisms involved.

2.   MYCORRHIZAL FUNGI AS DISEASE CONTROL AGENTS

The study of interactions between mycorrhizal fungi and plant pathogens began in the 1970s and these fungi have been reported to have inhibitory effect on plant pathogens and in reducing the disease severity in majority of the studies with some reports on their effect as neutral (Baath and Hayman, 1983; 1984) or occassional increase in disease severity (Davis *et al.*, 1979). Reduction of disease symptoms has been described for fungal pathogens such as *Phytophthora, Gaeumannomyces, Fusarium, Chalara (Thielaviopsis), Pythium, Rhizoctonia, Sclerotium, Verticillium, Aphanomyces, Pyrenochaeta, Ganoderma, Macrophomina, Bipolaris, Phoma, Urocystis, Microcyclus, Olpidium, Cylindrocarpon*, mycelium (Table 1), bacterial pathogens, *Erwinia* (Garcia-Garrido and Ocampo, 1989a), *Pseudomonas* (Garcia-Garrido and Ocampo, 1989b) for nematodes *Pratylenchus, Rotylenchus, Radopholus, Heterodera, Tylenchorynchus, Tylenchulus, Aphelenchus* and *Meloidogyne* (Table 2). However, vesicular-arbuscular mycorrhizal (VAM) fungi are reported to increase disease severity due to viruses (Jayaram and Kumar, 1995; Nemec and Myhre, 1984).

Table 1 : Influence of VAM on the control of soil-borne pathogens

Host	Disease	Pathogen	VAM fungi	Effect of VAM fungi on host	References
Allium cepa	Wilt	*Fusarium oxysporum* f. sp. *cepi*	-	Delayed disease development	Dehne, 1982
	-	*Phoma terrestris*	-	Reduced root infection, less stunting	Paget, 1975
	Wilt	*Pyrenochaeta terrestris*	-	Increased plant height	Becker, 1976
	White rot	*Sclerotium cepivorum*	*Glomus* sp.	Delayed disease epidemic and increased the yield by 22%	Torres-Barragain *et al.*, 1996
Arachis hypogaea	-	*Scherotium rolfsii*	*Glomus fasciculatum*	Reduction in the severty of disease	Krishna and Bagyaraj, 1983
Asparagus officinalis	Root rot	*Fusarium oxysporum*	*Glomus fasciculatum*	Lower incidence of disease	Wacker *et al.*, 1990
Brassica sp.	Damping off	*Rhizoctonia solani*	-	Survival percentage increased and plant growth increased	Iqbal and Nasim 1988, Iqbal *et al.*, 1998b

Cajanus cajan	Pigeon pea blight	*Phytophthora drechsleri* f. sp. *cajani*	*Gigaspora calospora*	Inhibitory effect on thedevelopment of diseae and improve plant growth	Bisht *et al.,* 1985
Capsicum frutescens	Wilt	*Fusarium oxysporum*	*Glomus mosseae*	No effect	Al-Momanty and Al-Raddad, 1988
Cassia tora	Wilt	*Fusarium oxysporum*	-	Reduction in the severity of disease and the population of pathogens plant growth	Chakrabarty and Mishra, 1986
Chamaecyparis lawsoniana	Root rot	*Phytophthora cinnamomi*	*Glomus mosseae*	Delayed onset of disease incidence and development and increased plant growth	Bartschi *et al.,* 1981
Cicer arietinum	Wilt	*Fusarium oxysporum*	*Glomus mosseae, Glomus constrictum Glomus monosporum*	No effect on wilt incidence	Reddy *et al.,* 1989
Citus	Wilt	*Phythophthora parasitica*	*Gigaspora margarita, Glomus constrictum, Glomus fasciculatum Glomus mosseae, Sclerocystis sinusa*	Significant increase in root dry weight & better growth of plant	Davis and Menge, 1981
Corchorus olitorius	-	*Fusarium solani*	*Glomus macrocarpum*	Reduction in disease severity and better growth in terms of height and dry weight	Bali and Mukerji, 1988
Cucumis melo	-	*Fusarium oxysporum*	-	Reduced infection and increased survival of seedlings and increase leaf area	Schonbeck, 1979
Cuminum cyminum	Wilt	*Fusarium oxysporum* f. sp. *cumini*	*Acaulospora laevis Gigaspora calospora, Glomus fasciculatum, Glomus mosseae*	Increased nutrient uptake and reduced disease severity	Champawat, 1991
Euphorbia pulcherima	-	*Pythium ultimum*	*Glomus fasciculatum*	Increased plant height and lower final pathongen population	Kaye *et al.,* 1994
Fragaria ananossa	-	*Phytophthora fragariae*	*Glomus etunicatum Glomus fasciculatum*	Reduced root necrosis	Norman *et al.,* 1996

175

Fragaria vesica var. *alpina*	Red core disease	*Phytophthora fragariae*	-	No effect	Baath and Hayman, 1984
Fragaria sp.	-	*Cylindrocarpon destructans*	-	Reduced root infection, less stunting	Paget, 1975
Glycine max	Root rot	*Fusarium solani* *Macrophomina phaseolina* *Rhizoctonia solani*	*Glomus mosseae*	Reduced disease symptoms	Zambolin and Schenck, 1983
	Root rot	*Phythophthora megasperma* var. *sojae*	*Glomus macrocarpum* var. *geosporum*	Increased severity of disease, internal stem discoloration	Ross, 1972
Gossypium sp.		*Fusarium oxysporum* f sp. *vesinfectum*		Reduced disease severity	Bali and Mukerji, 1988
	Seedling rot	*Thielaviopsis basicola*			Schonbeck and Dehne, 1977
Hevea brasiliensis	Leaf blight	*Microcyclus ulei*	*Glomus etunicatum*	Increased resistance to leaf blight; lesion size and production of spores of the pathogen significantly lower	Feldmann *et al.,* 1990
Hordeum vulgare	Root rot	*Bipolaris sorokiniana*	*Glomus* sp.	Severity of disease reduced	Boyetchko and Tewari, 1990
	Root rot	*Cochliobolus sativus*	-	Pathogen negated the effect of VAM	Grey *et al.,* 1989
Linum usitatissimum	Wilt	*Fusarium oxysporum, Oidium lini*	*Glomus intraradices*	VAM plants showed increased tolerance against *F. oxysporum* but become susceptible against *O. lini*	Duggasa *et al.,* 1996
Lucerne sp.	Wilt	*Fusarium oxysporum* f sp. *medicagnisis* *Verticilium aldoatrum*	*Glomus fasciculatum, Glomus mosseae*	Lower incidence of wilt and increase in plant growth	Hwang *et al.,* 1992
		Phytophthora cinnamonii	-	Increased infection	Davis *et al.,* 1978
Lycopersicon esculentum	Root rot	*Fusarium oxysporum* f. sp. *radices-lycopersici*	*Glomus intraradices*	Reduction in disease incidence and the population of pathogen and improvement in plant growth	Caron *et al.,* 1986b;c
	Wilt	*Fusarium* sp.	*Glomus mosseae*	Reduced wilt to 11% as against 45% in nonmycorrhizal plants and improvement in plant growth	Ramraj *et al.,* 1988

Lycopersicon esculentum	Corky root disease	*Pyrenochaeta lycipersici*	*Glomus caledonium*	Disease index lower and root growth better	Bochow and Abou-Shaar, 1990
	Root rot	*Phytophthora nicotiana* var. *parasitica*	*Glomus mosseae*	Decreased root necrosis	Cordier *et al.*, 1996, Trotta *et al.*, 1996
	Wilt	*Verticillium albo-atrum*	-	No effect	Baath and Hayman, 1983
Nicotiana tabacum	Root rot	*Pythium* sp.	-	-	Subhashini, 1990
	Root rot	*Thielaviopsis basicola*	*Glomus microcarpum*	Reduction in the number of pathogen propagules	Tosi *et al.*, 1988
	Root rot	*Thielaviopsis basicola*	*Glomus monosporum*	Better tolerance to pathogen, higher root and leaf dry weight	Giovanneti *et al.*, 1991
Oryza sativa	-	*Rhizoctonia solani*	*Glomus mosseae*	Increased root dry weight	Khadge *et al.*, 1990
	Stem rot	*Sclerotium oryzae*	-	-	Gangopadhyay and Das, 1987
Phaseolus aureus	Root rot	*Macrophomina phaseolina*	*Glomus mosseae*	Disease incidence reduced	Jalali *et al.*, 1990
Pisum sativum	Root rot	*Aphanomyces euteiches*	*Glomus fasciculatum*	Infection suppressed	Rosendahl, 1985
Populus sp.	-	*Cylindrocladium scoparium*	-	Increased plant growth	Barnard, 1977
Pyrus malus	Apple replant disease	-	*Glomos mosseae*	Increased growth	Utkhede *et al.*, 1992
Solanum melongena	Wilt	*Verticillium albo-atrum*	-	Increased plant growth	Melo *et al.*, 1985
Triticum sp.	Root rot	*Fusarium* sp., *Pythium* sp.	-	Increased plant growth	Kreischeheva and Millenina, 1978
	Take-all disease	*Gueumannomy-graminis*	-	Reduction of infection	Graham and Menge, 1982
	Foot root-rot	*Sclerotium (Corticium) rolfsii*	*Glomus fasciculatum*	Disease prevented	Harlapur *et al.*, 1988
Vicia faba	Wilt	*Fusarium oxysporum*	*Gigaspora calospora, Glomus fasciculatum, Glomus macrocarpum Glomus mosseae*	More incidence of disease when plants has more VAM infection	Sigh *et al.*, 1987
Vigna unguiculata	Root rot	*Macrophomina phaseolina*	*Glomus etunicatum*	Disease incidence reduced	Ramraj *et al.*, 1988

Table 2 : Influence of VAM fungi on the control of nematodes

Host	Nematode	VAM fungi	Effect of VAM fungi on host	References
Allium cepa	*Meloidogyne hapla* *Meloidogyne incognita*	*Glomus etunicatum*	Tolerance of plants against nematodes increased	Verdejo *et al.*, 1990
Avena sativa	*Meloidogyne incognita*	*Glomus mosseae*	Inhibitory effect on the disease incidence and development	Sikora and Schonbeck, 1975
Citrus jambhiri	*Tylenchulus semipenetrans*	-	No effect on plant growth	Baghel *et al.*, 1990
Citrus limon	*Radopholus citrophilus*	*Glomus intraradices*	Larger shoot & root weights, lower nematode population densities	Smith and Kaplan, 1988
Citrus sp.	*Radopholus citrophilus*	*Glomus etunicatum*	Increased host tolerance to nematode	O'Bannon and Nemec, 1979
	Radopholus citrophilis	*Glomus intraradices*	Increased host tolerance to nematode	Smith and Kaplan, 1988
	Tylenchulus semipenetrans	*Glomus fasciculatum*	Growth enhanced	O'Bannot *et al.*, 1979
	Tylenchulus semipenetrans	*Glomus mosseae*	Growth enhanced	O'Bannon and Nemec, 1978a
Cucumis melo	*Meloidogyne incognita*	-	Suppression of growth by nematode reduced to 21% as compared to 84% in non-mycorrhizal plants	Heald *et al.*, 1989
Cydonia oblonga	*Pratylenchus vulnus*	*Glomus intraradices*	Growth favoured and protection against nematodes	Calvers *et al.*, 1995
Cyphomendra betacea	*Meloidogyne incognita*	-	Inhibitory effect on disease development	Cooper and Grandison, 1987
Daucus carota	*Meloidogyne hapla*	*Glomus mosseae*	Inhibitory effect on disease development	Sikora and Schonbeck, 1975
Elettaria cardomomum	*Meloidogyne incognita*	*Gigaspora margarita, Glomus fasciculatum*	Improved plant growth & reduced nematode population	Thomas *et al.*, 1989
Glycine max	*Heterodera glycines*	*Gigaspora margarita*	Growth enhanced	Hussey and Roncadori, 1982

	Meloidogyne incognita	*Gigaspora heteogama*	Growth enhanced	Hussey and Roncadori, 1982
	Meloidogyne incognita	*Gigaspora margarita*	Growth enhanced	Schenk *et al.*, 1975
	Meloidogyne incognita	*Gigaspora margarita*	Growth enhanced	Hussey and Roncadori, 1982
Gossypium hirsutum	*Aphelenchus avenae*	*Gigaspora margarita*	Growth stimulated	Hussey and Roncadori, 1982
	Meloidogyne incognita	*Glomus fasciculatum*	Reduction of egg and nematode number/ gram of root	Saleh and Sikora, 1984
Ipomoea batatae	*Ciriconemella* sp., *Rotylenchus reniformis and Tylenchorhyus* sp.	-	No effect	Kassab and Taha, 1990
Lycopersicon esculentum	*Meloidogyne incognita*	*Glomus fasciculatum*	Reduction in number and size of root galls produced by nematodes	Bagyaraj *et al.*, 1979
Lycopersicon esculentum	*Meloidogyne incognita*	*Glomus mosseae*	-	Sikora and Schonbeck, 1975
	Meloidogyne incognita	*Gigaspora margarita*	Growth inhibited	Sikora and Schonbeck, 1975
	Meloidogyne javanica	*Glomus mosseae*	Suppression of gall index and the average number of galls per root system	Al-Raddad, 1995
	Rotylenchus reniformis	*Glomus fasciculatum*	Reduced juvenile penetration and development of nematode	Sitaramaiah and Sikora, 1982
Medicago sativa	*Meloidogyne hapla*	-	Increased tolerance and resistance of cultivars susceptible to *M. hapla* and improved resistance of resistant cultivars	Grandison and Copper, 1986
Musa acuminata	*Radopholus similis*	*Glomus fasciculatum*	Increased length, dry and fresh weight	Umesh *et al.*, 1988
Nicotiana tabacum	*Meloidogyne incognita*	*Glomus* mosseae	Improved plant growth of the plant with lower incidence of disease	Sikora and Schonbeck, 1975
	Meloidogyne incognita	*Glomus fasciculatum*	Improved growth of the plant with lower root-knot indices	Krishna Prasad, 1991

Phaseolus vulgaris	Meloidogyne incognita	Glomus sp.	Increase in fresh weight	Osman et al., 1990
	Meloidogyne javanica	Glomus etunicatus	Decrease in nematode egg population	Oleverira and Zambolin, 1988
Piper nigrum	Meloidogyne incognita	Glomus etunicatum, Glomus fasciculatum	Reduction of root-knot index and significantly increased growth	Sivaprasad et al., 1990
Prumus avium	Pratylenchus vulnus	Glomus intraradices	Increased host tolerance to nematode	Pinochet et al., 1995
Prunus domestica	Pratylenchus vulnus	Glomus intraradices	Increased host tolerance to nematode and improved plant growth	Camprubi et al., 1993
Prunus insititia	Pratylenchus vulnus	Glomus mosseae	Increased host tolerance to damaging nematode levels	Camprubi et al., 1993
Prunus persica	Pratylenchus vulnus	Glomus mosseae	None	Pinochet et al., 1995
Pyrus malus	Pratylenchus vulnus	Glomus mosseae	Increased host tolerance to nematode and improved plant growth	Pinochet et al., 1995
Trifolium alexandrium	Meloidogyne incognita	Gigaspora and Glomus sp.	Nullified the detrimental effects of post infection by nenatode; galls suppressed	Taha and Abdel-Kader, 1990
Trifolium alexandrium	Tylenchorynchus sp.	Glomus sp.	Increased plant growth and more number of nodules and inflorescence	Kassad and Taha, 1990b
Vigna unguiculata	Heterodera cajani	Glomus fasciculatum, Glomus epigaeus	Reduction in cyst formation	Jain and Sethi, 1988a,b
	Meloidogyne incognita	Glomus fasciculatum, Glomus epigaeus (versiforme)	Reduction in gall formation	Jain and Sethi 1988a,b
Vitis vinifera	Meloidogyne arenaria	Glomus fasciculatum	Growth stimulated	Atilano et al., 1976

3. FACTORS AFFECTING THE EFFICACY OF MYCORRHIZAL FUNGI AS DISEASE CONTROL AGENTS

Protective ability of mycorrhizal fungi against plant pathogens is often related to the nature of the host plant, mycosymbionts, plant pathogens, conditions of soil environment and soil microflora (Tello *et al.*, 1987).

3.1. Abiotic Factors

3.1.1. Temperature

Temperature exerts influence on the growth of both mycorrhizal fungi and root pathogens. As the optimum temperature needed for the growth may vary for both, it sometimes play decisive role in controlling the disease. In the early stages of damping off disease in cauliflower, more infection occured at 22-24°C by *Rhizoctonia solani* while the optimum temparature of VAM colonization was 20-22°C. At later stages, pathogen invasion was maximum at 22°C and VAM colonozation was high in pre-inoculated plants at 18-20°C compared with 20-22°C in mycorrhizal plants not exposed to the pathogen (Jain and Sethi, 1988).

3.1.2. Soil moisture

High soil moisture favours disease development but is not suitable for colonization of roots of mycorrhizal plants. In studies on *Brassica napus* plants grown at 18, 22, 26 and 38% moisture contents, mycorrhizal colonization was found negatively correlated with moisture percentage; being maximum at 18% and minimum at 38%. *Rhizoctonia solani* infection was higher at higher soil moisture contents. Damping off was maximum at 38% moisture while significantly lower at 18% moisture (Mahmood and Iqbal, 1982). It was found that pre-inoculated mycorrhizal seedlings of *Brassica napus* showed greater resistance to *Rhizoctonia solani* invasion at soil moisture contents favourable for VA mycorrhization (Iqbal and Mahmood, 1986). Higher moisture contents (25%) which are unfavourable for VAM colonization enhanced disease incidence by *Rhizoctonia solani* (Iqbal *et al.*, 1988). Studies on asparagus plants showed that *Fusarium oxysporum* inoculation of VAM infected plants reduced *Glomus fasciculatum* root colonization levels under well aerated (0 M Pa) but not under water stress conditions (-1.5 M Pa) (Wacker *et al.*, 1990).

3.1.3. Soil phosphorus content

Mycorrhizal colonization depends mainly on P concentration of soil. There is reduction in VAM colonization with increasing concentration of soluble phosphates (Mosse, 1973; Sparling and Tinker, 1978). Whether it has any role to play in deciding the efficacy of mycorrhizal fungi as disease control agents is not very clear. Phosphorus fertilization may exert positive or no effect on their efficacy as biocontrol agents.

In *Senecio vulgaris*, the effects of *Glomus* spp and the interaction between VAM fungus and foliar pathogens, *Puccinia lagenophorae* and *Albugo tragopogonis* was dependent on the level of P nutrition (West, 1995). It was found that rust infection could not establish on plants which were phosphorus limited. At medium and high phosphorus levels, however, total leaf number exhibiting infection was a function of leaf production. At high P level, leaf number of only mycorrhizal plants was reduced by rust infection while at medium P level, the rust was deleterious to both mycorrhizal as well as non-mycorrhizal plants; effects being less marked in mycorrhizal plants. At high P level, percentage germination of seeds from mycorrhizal plants was reduced while it enhanced in seeds from rusted mycorrhizal plants. Mycorrhizal colonization of mother plants grown with medium P supply benefitted the offspring in increasing leaf & bud production but rusted mother plants at same P level produced offsprings with reduced leaf number and potential fecundity (West, 1995). On the other hand, in *Tagetes patula* P fertilization played no role in biocontrol ability of the VAM fungi. Although the number of *Pythium ultimum* propagules in soil were decreased ten times in mycorrhizal plant (inoculated by *Glomus intraradices*) than in controls, this was not affected by P concentration (St-Arnaud *et al.*, 1994). Similar results revealed that an increase in available P in the substrate as well as P contents of roots and leaves had no effect on the population of *Fusarium oxysporum* f sp. *radices lycopersici*. Only the presence of *Glomus intraradices* resulted in a significant decrease in the population of *F. oxysporum* and root necrosis but had no significant effect on mean dryness of plant at all P concentrations (Caron *et al.*, 1986b). Protection of tomato plants against *Erwinia carotovora* and *Pseudomonas syringae* resulted in reduction of the number of colony forming units of the pathogen in the rhizosphere of mycorrhizal plants (inoculated with *Glomus mosseae*) were also shown to be independent of the concentration of P in the plants (Garcia-Garrido and Ocampo, 1989 a,b).

Phosphorus fertilization in plants may check nematode attack directly or through augmentation by mycorrhizal colonization in plant roots. Rough lemon seedlings when inoculated with different dosages of nematode, *Radopholus citrophilus* and *Glomus intraradices* in P amended soil (25 & 300 g P/Kg soil) showed lower nematode population densities as compared to low P and non-mycorrhizal plants (Smith and Kaplan, 1988). However, in some studies P fertilization had no effect on the nematode control ability of VAM fungi. In *Cucumis melo*, *Meloidogyne incognita* was shown to suppress the growth of non-mycorrhizal plants by 84% as compared to 21% suppression in mycorrhizal plants amended with 50 mg/ g P and similar trend was observed in soil with 100 mg/g P (Heald *et al.*, 1989).

3.2. Biotic Factors

3.2.1. Host genotype

Host genotype controls the rate and extent of mycorrhiza formation. The magnitude of mycorrhizal dependency varies both between and within species (Menge *et al.*, 1978; Plenchette *et al.*, 1983). And there are some studies which indicate that a particular mycorrhizal fungus against a particular pathogen may be effective on one host and less effective on the other host. When the clones of *Fragaria vesica* were colonized with VA mycorrhizal fungus (*Glomus fistulosum*), the susceptible clones became resistant to *Phythophthora fragariae* whereas the others remained highly disease susceptible (Mark and Cassels, 1996). The resistant and partially resistant clones were less affected. Disease suppression by EM fungus, *Paxillus involutus* was most effective on *Pinus resinosa* while lower level of protection was observed on *Pinus strobus*, *P. banksiana* and *P. sylvestris* all grown in test tubes and inoculated with *P. involutus* and *F. oxysporum* f sp. *pini* (Duchesne *et al.*, 1989a).

3.2.2. Mycorrhizal fungus

Mycorrhizal fungus may differ in their efficiency to control root pathogens. Particular fungus may be effective against one pathogen and ineffective against other pathogens. *Fusarium oxysporum* blight in cotton plants inoculated with *Glomus intraradices* was as severe as in non-mycorrhizal plants but it was less severe when cotton plants were inoculated with *Glomus mosseae* (Hu and Gui, 1991). There was no recovery of *Phytophthora cinnamomi* from ectomycorrhizas formed by white type basidiomycete even after the 24th day of inoculation. However, ectomycorrhizae formed by the ascomycete, *Cenococcum geophilum* and the basidiomycete *Hysterangium inflatum* failed to limit root infection in soil and the pathogen was recovered from both types after 55 days of incubation (Malajczuk, 1988). *Laccaria laccata* showed an inhibitory effect on *in vitro* growth of the damping off fungus, *Fusarium oxysporum* when grown in paired cultures while other mycorrhizal fungi had no such inhibitory effect (Chakravarty and Hwang, 1991).

In contrast to their beneficial role, some VAM fungi are reported to cause diseases in plants. Tobacco plants when inoculated with sievings (containing endogonaceous spores) from soils suspected of containing the tobacco stunt pathogen and a single spore of *Glomus macrocarpum* developed symptoms identical to those of naturally stunted plants. Stunting was correlated with colonization of roots by arbuscules and external hyphae (Modjo and Hendrix, 1986).

3.2.2.1. Time of mycorrrhizal inoculation

Time of inoculation of plants with mycorrhizal fungi, i.e. before, simultaneously or after inoculation with the pathogens greatly influences the efficacy of mycorrhizal fungi in the control of pathogens. Infection of peas with *Aphanomyces euteiches* in irradiated soil was suppressed by *Glomus fasciculatum* when plants were challenge inoculated after two weeks. No reduction of the pathogen was, however, detected when plants were inoculated with both fungi at the same time (Rosendahl, 1985). Root colonization by *Glomus intraradices* was significantly increased when *Fusarium oxysporum* was inoculated simultaneously with or four weeks before the VAM fungus. Also, compensation for dry mass due to *F. oxysporum* occured when the VAM fungus was inoculated after *Fusarium* (Caron *et al.*, 1986a). Pre-inoculation of cocoa seedling with VAM fungi at three weeks followed by inoculation with *Ganoderma pseudoferreum* at eleven weeks significantly reduced pathogenic infection of the roots. Biological control by EM fungi was observed when inoculations of seedlings with an EM fungus was performed either simultaneously (Chakravarty and Unestam, 1986b) or prior to as long as 11 weeks (Stack and Sinclair, 1975) infection with a pathogen. No instance is known where EM fungi protected seedlings after the onset of pathogenesis.

Dual inoculation of *Cyphomandra betacea* (Tamarillo) with *Meloidogyne incognita* and VAM fungi improved plant growth and suppressed nematode reproduction and development in roots, but nematode infection and development was less in plants pre-inoculated with VAM fungi than in plants inoculated simultaneously with both organisms (Cooper and Grandison, 1987). In *Elattaria cardamomum*, VAM improved plant growth and reduced nematode population both when inoculation was made simultaneously or after nematode infection (Thomas *et al.*, 1989). Studies on *Phaseolus vulgaris* showed a significant increase in fresh weight when plants were inoculated with *Glomus* spp. 15 or 30 days before *M. incognita* gall index and nematode population significantly increased with simultaneous inoculation with VAM and nematode but there was a significant decrease when nematodes were inoculated 15 or 30 days after VAM fungi (Suresh and Bagyaraj, 1984). In *Musa acuminata*, inoculation with burrowing nematode, *Radopholus similis* retarded root growth but the effect was offset by mycorrhizal inoculation and the nematode number both in roots and soil became significantly lowered when VAM was supplied simultaneously or 7 days prior to nematode inoculation (Umesh *et al.*, 1988). Pre-inoculation of *Trifolium alexandrium* with VAM fungi (mainly *Glomus* and *Gigaspora* sp.) significantly nullified the detrimental effects of host infection by *M. incognita* on plant growth, though the nematode population was unaffected (Taha and Abdel-Kader, 1990). In *Vigna unguiculata*, inoculation of seedlings with *Glomus fasciculatum / G. epigaeus (G. versiforme)*, 15 days prior to inoculation with *M. incognita* or *Heterodera cajani* alleviated the deleterious effect of both nematodes to a considerable extent. Prior inoculation by *G. fasciculatum* reduced gall formation by *M. incognita* and cyst formation by *H. cajani*. Prior inoculation by *G. versiforme* resulted in increased number of cysts of *H.*

cajini. Pre-inoculation with mycorrhizal fungi (*G. mosseae*) significantly reduced root infection with *M. javanica* in tomato plants (St-Arnold *et al.*, 1995).

3.2.2.2. Levels of mycorrhizal inoculum

Levels of inoculum also affects the efficacy of mycorrhizal fungi. Inoculation with 30 spores (*G. fasciculatum*) per *Gossypium hirsutum* plant resulted in 38% root colonization and this had no effect on nematode, *M. incognita*. At densities of 60, 120 and 240 spores/plant, mycorrhizal root colonization increased to 55-60% and nematode egg per plant was significantly reduced. At concentration of 480 spores /plant and root colonization level of 87%, total egg and nematode densities were greatly suppressed (Saleh and Sikora, 1984).

3.2.3. Virulence and inoculum potential of pathogen (s)

The potential effectiveness of mycorrhizal fungi also depends on the virulence and inoculum potential of the pathogen(s) present in the soil. A high pathogen inoculum and pathogen density in the rhizosphere may render ineffective the biocontrol ability of mycorrhizal fungi (Azcon-Aguilar and Barea, 1996). The presence of VAM fungi had no effect on *Fusarium* wilt of pigeon pea when high levels of *F. udam* were present in the soil (Reddy *et al.*, 1989).

3.2.4. Soil microflora

Various members of soil microflora exert direct or indirect influence on both root pathogens and mycorrhizal fungi. *Gliocladium virens*, which reduces population densities of and damping off by *Pythium ultimum* in cucumber, when added to soil artificially infested with *P. ultimum* increased mycorrhizal colonization by *G. etunicatum* and is compatible if applied as a dual inoculum (Paulitz and Linderman, 1991). Similarly, combination of VAM inoculation with additional N, K, and soil inoculation by *Bacillus subtilis* str. T99 resulted in the lowering of disease index and highly increased plant growth and yield (Bochow and Abou-Shaar, 1990).

4. MECHANISMS OF DISEASE CONTROL BY MYCORRHIZAL ASSOCIATION

Mechanisms that could be accounted for the disease control ability of mycorrhizal fungi include competition for infection site and host photosynthates, improvement of host nutrition, root damage compensation, morphological changes in the host root, changes in microbial communities in the mycorrhizosphere, antibiosis and the activation of defence mechanisms.

4.1. Competition for Host Photosynthates

Both mycorrhizal fungus and pathogen depends on host photosynthates for their growth. The major substrate for microbial activity in the rhizosphere or on the rhizoplane is organic C released by plant roots (Azaizeh *et al.*, 1995). Mycorrhizal fungi and root pathogens compete for the carbon compounds reaching the root (Linderman, 1994; Smith, 1987). When mycorrhizal fungus has primary access to photosynthates, the higher C demand may inhibit pathogen growth (Azcon-Aguilar and Barea, 1996).

4.2. Competition for Infection Sites

The pathogens and the VAM fungi, although colonizing the same root system, usually develop in different cortical cells indicating some sort of competition for space (Azcon-Aguilar and Barea, 1996). If VAM fungus colonizes the root then this would limit the colonization of the pathogenic fungi to areas of the root which had not been colonized, thus affording protection (Goncalves *et al.*, 1991). Cordier *et al.* (1996) showed that *Phytophthora* does not penetrate arbuscule-containing cells and its development is also reduced in adjacent uncolonized regions.

4.3. Increased Nutrient Status of the Host Plant

Mycorrhizal symbiosis increases the nutrient uptake and results in more vigorous plants, the plant itself may thus become more resistant/ tolerant to pathogen attack (Azcon Aguilar and Barea, 1996). Although most of the studies indicate that enhanced P nutrition play an important role in the higher tolerance of mycorrhizal plants to pathogens (Dehne, 1982; Hussey and Roncadori, 1982) , there are reports that such impoved P nutrition is associated with an increase in disease severity (Davis *et al.*, 1979; West, 1995) or has no effect on bioprotection (Caron *et al.*, 1986 b). Benefits achieved by mycorrhizal inoculation against pathogens could not be duplicated by adding P fertilizer (Cooper and Grandison, 1987) and this suggests that it is not due merely to improved P nutrition of the host but other mechanisms are also involved.

4.4. Root Damage Compensation

Mycorrhizal fungi compensate for the root disruption caused by the pathogen , allowing the plant to continue functioning inspite of damaging levels of the fungi (Cordier *et al.*, 1996) and the nematodes (Calves *et al.*, 1995; Pinochet *et al.*, 1996).

4.5. Morphological Changes in Host Root

Plants colonized by VAM fungi develop a dense root system, with a higher number of shorter, more branched adventitious roots of greater diameter (Berta *et al.*, 1993) leading to root system comprising of relatively larger proportion of higher older roots in the root system (Hooker *et al.*, 1994). The role of VAM alterations to root architecture in plant protection has not been understood fully. Observations by Hickman (1970) suggests that a more branched root system would lead to more disease. However, Norman *et al.* (1996) found negative correlation between disease incidence and branching of root system in VAM fungal-colonized plants which suggests that VAM alterations to root architecture are not directly involved in inducing resistance but that factors correlated with these changes may be responsible.

4.6 Physiological Changes in Host Root

Modification of the basic physiology of plant roots may contibute to disease suppression. There is promotion of water and nutrient uptake by mycorrhizal fungi which affects the course of pathogenesis.VAM roots have higher respiratory activity than non-VAM roots (Dehne, 1987; Harley and Smith, 1983; Snellgrove *et al.*, 1982). Part of the increase is likely due to VAM fungal respiration itself (Baas *et al.*, 1989). Increased respiration rate of VAM roots is an indication for the higher metabolic activity which might enable plants to react more rapidly and more effectively against root pathogens (Dugassa *et al.*, 1996). Furthermore, VAM roots showed increased ethylene production and DNA methylation (Dugassa *et al.*, 1996). Colonization with VAM fungi also results in increased amino acid production especially arginine which inhibits pathogen spore formation (Baltuschat and Schonkbeck, 1975). Exudation pattern of root also changes both qualitatively as well as quantitatively (Graham *et al.*, 1981). EM fungi are known to synthesize plant growth regulators such as auxins and ethylene (Harley and Smith, 1983; Nylund, 1988; Rupp *et al.*, 1989) that may alter the susceptibility of tree roots to pathogens.

4.7. Changes in Microbial Community in the Mycorrhizosphere

There are evidences that microbial shifts occur in the mycorrhizosphere (Bansal and Mukeji, 1994; Garbaye, 1991; Jalali and Jalali, 1991) and that the resulting microbial equilibria could influence the growth and health of plants. The possible mechanisms which lead to these shifts include changes in plant-derived organic factors in the soil induced by VAM (Ames *et al.*, 1984; Bagyaraj and Menge, 1978; Christensen and Jakobsen, 1993), qualitative and quantitative changes in root exudation caused by VAM colonization (Bansal and Mukerji, 1994) or a direct interaction between the hyphal network of VAM fungi and rhizospheric microorganisms (St-Arnold *et al.*, 1995). Changes in soil microorganisms community induced by VAM formation may lead to stimulation of those microorganisms which may be antagonistic to root pathogens (Azcon-Aguilar and Barea, 1996).More pathogen-antagonistic actinomycetes were isolated from the rhizosphere of VAM plants than from non-mycorrhizal controls (Secilia and Bagyaraj, 1987). Caron (1989) reported a reduction in *Fusarium* populations in the soil surrounding mycorrhizal tomato roots as compared with the soil of non-mycorrhizal controls.

4.8. Antibiosis

The production of antibiotics, although, not observed in VAM fungi, has often been put forward as a reason for the protective effect that EM fungi can confer on their plant roots against plant pathogens (Duchesne *et al.*, 1973; Zak, 1964). Over 100 species of EM fungi have been reported to produce antibiosis in pure culture, including antibacterial, antifungal, or antiviral compounds (Marx, 1973).

4.9. Activation of Defence Mechanisms

Plants possess a broad spectrum of basic defence mechanisms, pre-established or induced, which restricts the ingress of potential pathogens. Inducible defence involves a complex series of plant processes which are triggerred as a general response to attack by a pathogen and renders the plant to be resistant. VAM fungi can activate certain plant defence genes. However, the induction of plant defence genes is transient, uncoordinated, weak and/ or occurs in localized fashion (Gianinazzi-Pearson *et al.*, 1996) in contrast to more extensive tissue response to root pathogens even during susceptible infections (Tahiri-Alaoui *et al.*, 1993 a,b). In VAM, the defence-related responses are based on the elicitation of host responses to invasion by the fungi which are then somehow suppressed or maintained at a low level compatible with symbiotic interactions by specifically activated plant and/or fungal machanisms (Gianinazzi-Pearson *et al.*, 1996). Whether this low priming of defence genes by the symbiotic infections can pre-dispose roots to respond more rapidly to a secondary infection, and hence contribute to their enhanced resistance to soil-borne pathogens, as suggested by Gianinazzi (1991), needs to be tested.

In most cases, plant resistance is associated with a multiple defence response which include (Kalpulnik *et al.*, 1996):

a) a hypersensitive response (HR) which is characterizd by rapid, localized chemical defence, and death of plant cells surrounding the infection site (Meier *et al.*, 1993; Uknes *et al.*, 1996).

b)   the accumulation of secondary metabolites, such as antimicrobial phytoalexins (Dixon
      et al., 1993) and phenolics (Dehne and Schonbeck, 1979).
c)   structural defensive barriers such as lignin, callose and hydroxyproline-rich proteins
      (Dixon and Harrison, 1990).
d)   the production of several new enzymes (van Loon, 1985), some with antifungal
      activities (Mauch *et al.*, 1988 a,b).

Among the compounds involved in plant defence (Bowles, 1990), the compounds studied in relationship to VAM formation are phytoalexins, phenolics, lignins, callose, hydroxyproline rich glycoproteins (HRGP), chitinases, β-1,3-glucanases and peroxidases. The compounds studied in relation to EM fungi include chitinases and peroxidases.

4.9.1. Producion of secondary metabolites

4.9.1.1. Phytoalexins

These are low-molecular weight isoflavonoid compounds derived by phenylpropanoid pathway. These are the toxic compounds released at the site of infection. The isoflavonoids, glyceollin which is fungitoxic (Kaplan *et al.*, 1980) and coumestrol which is nematostatic, together with the coumestan isojagol, showed greater amounts of accumulation in mycorrhizal roots than in non-mycorrhizal roots (Morandi and Le Querre, 1991). Enhanced levels of glyceollin have only been detected in later stages of symbiosis and not during early stages of root colonization. The level of phytoalexin medicarpin, medicarpin-malonyl glucoside, daidzein, coumestral, formononetin, formononetin-malonylglucoside and 4',7-dihydroxy flavone showed transient increase in roots of *Medicago truncatula* 7-40 days after *Glomus* inoculation (Harrison and Dixon, 1993) showing that phytoalexins and their precursors are activated by VAM inoculation.

Increase in the activities and transcripts of two enzymes: phenylalanine ammonia-lyase (PAL), the first in the phenylpropanoid pathway, and chalcone isomerase (CHI), specific to the isoflavonoid/flavonoid biosynthesis, and followed by decrease in both to levels below those in *Medicago sativa* (Volpin *et al.*,1994; 1995). Increase in transcripts of chalcone synthase (CHS), another enzyme specific to isoflavonoid/flavonoid synthesis has also been found in *M. truncatula* mycorrhizal roots. However, no such changes has been observed in mycorrhizal interactions in bean (Lambais and Mehdy, 1993) and parsley (Franken and Gnadinger, 1994). Furthermore, transcripts of isoflavone reductase (IFR), a late enzyme in medicarpin biosynthesis did not increase but gradually decreased in mycorrhizal *M. truncatula* and *M. sativa* (Harrison and Dixon, 1993; Volpin *et al.*, 1995).

4.9.1.2. Phenolics

These are aromatic compounds having various attached substituent groups, such as hydroxyl, carboxyl and methoxyl groups. Many of these compounds appear to be simply by-products of metabolism; but this probably reflects in part of our ignorance of allelochemical interactions. Phenolics may have a role in protecting VAM roots against pathogenic fungi (Grandmaison *et al.*, 1993). Both oxidation and polmerization of tomato root phenols have been reported to increase following VAM association (Dehne and Schonbeck, 1979). A continuous increase of total soluble phenols has been found in VAM roots of *Arachis hypogaea* (Krishna and Bagyaraj,1984). Grandmaison *et al.* (1993) found no qualitative difference in the soluble and bound phenolics isolated from non-mycorrhizal *Allium cepa* roots with those from mycorrhizal (*Glomus intraradices*) roots but the mycorrhizal roots showed higher concentration of wall bound phenolic compounds. Binding of phenolic compounds to cell walls could be responsible for the resistance of VAM roots to pathogenic fungi, as it results in increased resistance by the cell wall to the action of digestive enzymes (Grandmaison *et al.*, 1993). Under *in vitro* conditions, the phenolic compounds were observed as electron-dense deposits in mycorrhizal roots (*G. intraradices*) at sites of fungus penetration as well as at strategic sites of fungal spread, such as intracellular spaces (Benhamou *et al.*, 1994). These depositions were not seen in *Fusarium*-infected carrot roots that were not mycorrhizal. These substances are reported to cause strong morphological and cytochemical deformations of the pathogen (Benhamou *et al.*, 1994).

4.9.2. Structural defence barriers

4.9.2.1. Lignins

Lignins are polymers of aromatic alcohols. VAM fungus stimulates lignification of endoderm cell walls and vascular tissues (Becker, 1976; Dehne and Schonbeck, 1979). Development of collosites or lignitubers has also been reported in the host cell (Schonbeck, 1979). As lignins provide strengthening to cell wall, it gives protection against attack by pathogens.

4.9.2.2. Callose

It is a polysaccharide composed of β-D-glucopyranose (a glucan) residues. β-1,3 glucans has been detected within the structural host wall material around the point of penetration of VAM hyphae into the plant cell which later on disappear along with the wall material as the fungus branches to form an arbuscule (Gianinazzi-Pearson *et al.*, 1996). Synthesis of β-1,3-glucans has been found to occur in mycorrhizas of pea, tobacco and leak (Gianinazzi-Pearson *et al.*, 1995; Gollotte *et al.*, 1995) but has not been found during interactions in maize (Balestrini *et al.*, 1994). This suggests that the synthesis of β-1,3 glucans is not a constant feature of the wall reaction in all plants (Gianinazzi-Pearson *et al.*, 1996).

4.9.2.3. Hydroxyproline rich glycoproteins (HRGP)

HRGPs are structural cell wall glycoproteins which accumulate in intercellular spaces or as wall appositions and papillae (Benhamou *et al.*, 1990; O' Connel *et al.*, 1990). Significant amounts of HRGP-encoding mRNA has also been found to accumulate in mycorrhizal parsley roots (Franken and Gnadinger,1994). In maize roots, corresponding glycoprotein has been immunolocated around arbuscule hyphae, both in the structured wall material at the point of cell penetration and in the interfacial contact zone (Belestrini *et al.*, 1994; Bonfante *et al.*, 1991).

4.9.3. Production of enzymes

4.9.3.1. Pathogenesis-related (PR) proteins

PR proteins are induced in plants during resistance to pathogen infection (Carr and Klessig, 1989; Stintzi *et al.*, 1993). These PR proteins have been grouped into eleven families, designated PR-1 to PR-11 (van Loon *et al.*, 1994). PR-2, PR-3 and PR-11 proteins are hydrolytic enzymes with β-1,3-glucanase or chitinase activities (Gianinazzi-Pearson *et al.*, 1996; Stintzi *et al.*, 1993; van Loon *et al.*, 1994).

In most of the research concerning plant defence responses to VAM fungi, the studies have focussed on the induction of these two hydrolytic enzymes as they can degrade chitin (chitinases) and β-1,3-glucans (β-1,3-glucanases), the major structural components of the cell walls of many fungi (Batricki-Garcia, 1987).

4.9.3.2. Chitinases

Plant chitinases are potential anti-fungal hydrolases, which have also been investigated in various host-pathogen interactions (Mauch *et al.*, 1988a,b; Pegg and Young, 1982). An increased chitinase activity has been detected during early interactions between roots of *Allium porrum* (Spanu *et al.*, 1989), bean (Lambais and Mehdy, 1993) and alfalfa (Volpin *et al.*, 1994) with VAM fungi in the first stages of root colonization which later on repressed to levels that are below those in non-mycorrhizal roots during the establishment of mycorrhiza, although higher levels can persist in some plant-fungus combinations (Dehne and Schonbeck, 1978). Genes encoding acid and basic chitinases are also differentially expressed during mycorrhiza formation (Gianinazzi-Pearson *et al.*, 1996). New chitinase isomers have been reported by Dumas-Gaudot *et al.* (1992,a,b) that were specifically induced in several VAM associations. Extra chitinase activity was also induced in *Eucalyptus* roots following ectomycorrhizal inoculation (*Pisolithus* sp.) and root colonization (Albrecht *et al.*, 1994). However, chitinolytic activity in roots of *E. pilularis* Sm and *Pinus sylvestris* challenged with the EM fungus, *Pisolithus tinctorious* (Pers) Coker and Couch, was found to be same as those of unchallenged roots (Hodge *et al.*, 1995).

4.9.3.3. β-1,3 glucanases

β-1,3 glucanase activity showed transient increase during initial phases of root invasion. However, decreased activities has been observed during phases of rapid colonization in bean roots, but to a much lesser extent in tomato (Lambais and Mehdy, 1993; Vierheilig *et al.*, 1994). β-1,3 glucanase transcripts also showed a similar behaviour with increased level as plant/fungal interactions initiate, followed by their strong repression (Lambais and Mehdy, 1993).

4.9.3.2 Peroxidases

Peroxidases are involved in cell wall reinforcement during plant reactions to pathogens (Collinge *et al.*, 1994; Dixon and Harrison, 1990). Peroxidase activity associated with epidermal and hypodermal cells also increased in mycorrhizal roots (Gianinazzi and Gianinazzi-Pearson, 1992). Induction of peroxidases by EM fungi, *Pisolithus* sp. has also been reported (Arbrecht *et al.*, 1994).

5.   CONCLUSION

Valuable data generated on induced suppression of soil-borne pathogens by mycorrhizae has no doubt proved their potentiality in controlling plant pathogens. But, so far, examples of successful practical applications are scarce (Hooker *et al.*, 1994; Linderman, 1994) because of the complexity of this tripartite association (mycorrhiza-soil-plant) and the direct influence of prevailing environmental conditions. Experiments have only usually tested single mycorrhizal fungus and a single host genotype, diversity within mycorrhizal fungi for biocontrol of plant pathogens is unknown. There are certain pathogenic antagonists , *Trichoderma, Gliocladium, Pseudomonas, Bacillus* and PGPR which cooperate with mycorrhizal fungi for biocontrol and the phytosanitary role of mycorrhizal fungi can be made more effective when integrated with other plant protection measures. Thus, for exploiting the prophylactic activity of mycorrhizal fungi in a best way, the right combinations of factors should be found out, the most important being the selection of appropriate/efficient mycorrhizal fungi.

REFERENCES

Albrecht, C. , Burgess, T. , Dell , B. and Lapeyrie, F. 1994, Chitinase and peroxidase activities are induced in eucalyptus roots according to aggessiveness of Australian ectomycorrhizal strains of *Pisolithus* sp., *New Phytol.* 127: 217-222.

Al-Momany, A. and Al-Raddad, A. 1988, Effect of vesicular-arbuscular mycorrhizae on *Fusarium* wilt of tomato and pepper, *Alexandria J. Agri. Res.* 33: 249-261.

Al-Raddad, A. M. 1995, Interaction of *Glomus mosseae* and *Paecilomyces lilacinus* on *Meloidogyne javanica* of tomato, *Mycorrhiza* 5: 233-236.

Ames, R. N. , Reid, C. P. P. and Ingham, E. R. 1984, Rhizosphere bacterial population responses to root colonization by a vesicular-arbuscular mycorrhizal fungus, *New Phytol.* 96: 555-563.

Atilano,R. , Menge, J. and Van Gundy, S. 1982. Interaction between *Meloidogyne arenaria* and *Glomus fasciculatum* in grapes, *J. Nematol.* 13: 52-57.

Azaizeh, A. , Marschner, H. , Romheld, V. and Wittenmayer, L.1995, Effects of a vesicular-arbuscular mycorrhizal fungus and other soil microorganisms on growth, mineral nutrient acquisition and root exudation of soil-grown maize plants, *Mycorrhiza* 5: 321-327.

Azcon-Aguilar, C. and Barea, J. M. 1996, Arbuscular mycorrhizas and biological control of soil-borne plant pathogens- an overview of mechanisms involved, *Mycorrhiza* 6: 457-464.

Baas, R., Van der Werf, A. and Lambers, H. 1989, Root respiration and growth in *Plantago major* as affected by vasicular arbuscular mycorrhizal infection, *Plant Physiol.* 91: 227-232.

Baath, E. and Hayman, D. S. 1983, Plant growth responses to vesicular- arbuscular mycorrhiza, XIV, Interactions with *Verticillium* wilt on tomato plants, *New Phytol.* 95: 419-426.

Baath , E. and Hayman, D. S. 1984, No effect of vesicular-arbuscular mycorrhiza on red core disease of strawberry, *Trans. Br. Mycol. Soc.* 82: 532-536.

Baghel, P. P. S., Bhatti, D. S. and Jalali, B. L. 1990, Interaction of VA mycorrhizal fungus and *Tylenchulus semipenetrans* on citrus, in : *Current Trends in Mycorrhizal Research*, B.L. Jalali and H. Chand, eds., Proc. Natl. Conf. on Mycorrhiza, Haryana Agric. Univ., Hisar, India, 14-16 Feb., 1990, TERI New Delhi, pp. 118-119.

Bagyaraj, D. G. and Menge, J. A. 1978, Interaction between a VA mycorrhiza and *Azotobacter* and their effects on rhizosphere, *New Phytol.* 80: 567-573.

Bagyaraj, D. J., Manjunath, A. and Reddy, D. D. R. 1979, Interaction of vesicular-arbuscular mycorrhiza with root knot nematode in tomato, *Plant Soil* 51: 397-403.

Balestrini, R., Romera, C., Puigdomenac, P. and Bonfante, P. 1994, Location of a cell wall hydroxyproline-rich glycoprotein, cellulose and β-1,3 glucans in apiçal and differentiated regions of maize mycorrhizal roots, *Planta* 195: 201-209.

Bali, M. and Mukerji, K. G. 1988, Effect of VAM fungi on *Fusarium* wilt of cotton and jute, Mycorrhiza for green Asia: Proc. of the first Asian Conference on Mycorrhizae, January, 1988, pp. 233-234.

Baltruschat, H. and Schonbeck, F. 1975, The influence of endotrophic mycorrhiza on the infestation of tobacco by *Thielaviopsis basicola*. *Phytopathol. Z.* 84: 172-188.

Bansal , M. and Mukerji, K. G. 1994, Positive correlation between VAM-induces changes in root exudation and mycorrrhizosphere mycoflora, *Mycorrhiza* 5: 39-44.

Barnard, E. L. 1977, Ph. D. Diss. Duke Univ. Darham, N. C. pp.147.

Bartnicki-Garcia, S. 1987. The cell wall : a crucial structure in fungal evolution, in: *Evolutionary Biology of the Fungi*, A. D. M. Rayner, C. M. Brasio and D. Moore, eds., Cambridge University Press, Cambridge, pp. 389-403.

Bartschi, H. , Gianinazzi-Pearson, V. and Vegh, I. 1981. Vesicular arbuscular mycorrhizae formation and root rot disease (*Phytophthora cinnamomi*) development in *Chamaecyparis lawsaniana*, *Phytopathol. Z.* 102: 212-218.

Becker, W. N. 1976, Quantification of onion vesicular-arbuscular mycorrhizae and their resistance to *Pyrenochaeta terrestris,* Ph. D. Dissertation, University of Illinois, Urbana, pp 72.

Benhamou, N., Mazau, D., Esquerre-Tugaye, M. T. and Asselin, A. 1990, Immunogold localization of hydroxyproline-rich glycoproteins in necrotic tissue of *Nicotiana tabacum* L. cv. *xanthinc* infected by tobacco mosaic virus, *Physiol. Mol. Plant Pathol.* 36: 129-145.

Benhamou, N., Fortin, J. A., Hamel, C., St-Arnaud, M. and Shatilla, A. 1994, Resistance responses of mycorrhizal Ri T-DNA- transformed carrot roots to infection by *Fusarium oxysporum* f. sp. *chrysanthemi*, *Phytopathol.* 84: 958-968.

Berta, G. , Fusconi, A. and Trotta, A. 1993, VA mycorrhizal infection and the morphology and function of root systems, *Environ. Expt. Bot.* 33: 159-173.

Bethlenfalvay, G. J. and Linderman, R. G. (eds) 1992, *Mycorrhizae in Sustainable Agriculture*, ASA Special Publication No. 54, Madison, Wis.

Bisht, V. S. , Krishna, K. R. and Nene, Y. L. 1985, Interaction between vesicular-arbuscular mycorrhiza of three cultivars of potato (*Solanum tuberosum* L.), *Plant Soil* 79: 299-303.

Bochow, H. and Abou-Shaar, M. 1990, On the phytosanitary effect of mycorrhiza in tomatoes to the corky-root disease, *Zentralbratt fur mikrobiologie* 145: 171-176.

Bonfante, P. , Tamagrove, L., Peretto, R., Esquerre-Tugaye, M.T., Mazau, D., Mosinik, M. and Vian, B. 1991, Immunocytochemical location of hydroxyproline- rich glycoproteins at the interface between a mycorrhizal fungus and its host plant, *Protoplasma* 165: 127-138.

Bowles, D. J. 1990, Defence-related proteins in higher plants, *Ann. Rev. Biochem.* 59: 873-907.

Boyetchko, S. M. and Tewari, J. P. 1990, Effect of phosphorus and VA mycorrhizal fungi on common root rot of barley, 8th NACOM, Innovation and Hierarchical Integration Jackson, Wyoming, 5-8 September 1990, 33pp.

Calvet, C., Pinochet, J. , Camprubi, A. and Fernandez, C. 1995, Increased tolerance to the root-lesion nematode *Pratylenchus vulnus* in mycorrhizal micropropagated BA-29 quince rootstock, *Mycorrhiza* 5: 253-258.

Camprubi, A. , Pinochet, J. , Calvet, C. and Estaum, V. 1993, Effects of the root-lesion namatode *Pratylenchus vulnus* and the vesicular arbuscular mycorrhizal fungus *Glomus mosseae* on the growth of three plum root stocks, *Plant Soil* 153: 223-229.

Caron, M. 1989, Potential use of mycorrhiza in control of soil-borne diseases, *Can. J. Plant Pathol.* 11: 177-179.

Caron, M., Fortin, J. A. and Richard, C. 1986a, Effect of inoculation sequence on the interaction between *Glomus intraradices* and *Fusarium oxysporum* f. sp. *radices-lycopersici* in tomatoes, *Can. J. Plant Pathol.* 8: 12-16.

Caron, M., Fortin, J. A. and Richard, C. 1986b, Effect of phosphorus concentration and *Glomus intraradices* on fusarium crown and root rot of tomatoes, *Phytopathol.* 76: 942-946.

Caron, M., Richard, C. and Fortin, J. A. 1986, Effect of preinfestation of the soil by a vesicular-arbuscular mycorrhizal fungus, *Glomus intraradices*, on *Fusarium* crown and root rot of tomatoes, *Phytoprotec.* 67: 15-19.

Carr, J. P. and Klessig, D. F. 1989c, The pathogenesis-related proteins in plants, in: *Genetic Engineering : Principles and Methods*, Plenum Press, New York, pp. 65-109

Chakravarty, P. and Hwang, S. F. 1991, Effect of an ectomycorrhizal fungus, *Laccaria laccata*, on *Fusarium* damping-off in *Pinus banksiana* seedlings, *Eur. J. For. Pathol.* 21: 97-106.

Chakravarty, P. and Mishra, P. R. 1986a, Influence of endotrotrophic mycorrhizae on the Fusarial wilt of *Cassia tora, Phytopathol.* 115: 130-133.

Chakravarty, P. and Unestam, T. 1986b, Role of mycorrhizal fungi in protecting damping-off of *Pinus sylvestris* seedlings, in: *Proc. Ist Eur. Sym. Mycorrhiza*, 1985, V. Gianinazzi-Pearson and S. Gianinazzi, eds., Dijon, France.

Champawat, R. S. 1991, Interaction between vesicular arbuscular fungi and *Fusarium oxysporum* f. sp. *cumini* - their effects on cumin, *Proc. Indian Natl. Sci. Aca. (Part B), Biol. Sci.* 57: 59-62.

Christensen, H. and Jakobsen, I. 1993, Reduction of bacterial growth by a vesicular arbuscular mycorrhizal fungus in the rhizosphere of cucumber (*Cucumis sativa L.*), *Biol. Fertil. Soil* 15: 253-258.

Collinge, D.B., Gregersen, P. L. and Thordal-Christensen, H. 1994, The induction of gene expression in response to pathogenic microbes, in: *Mechanisms of Plant Growth and Improved Productivity:Modern Approaches and Perspectives*, A.S. Basea, ed., Marcel Dekker, New York, pp. 391-433.

Cooper, K. M. and Grandison, G. S. 1987, Effects of vesicular-arbuscular mycorrhizal fungi on infection of tomarillo (*Cyphomandra betacea*) by *Meloidogyne incognita* in fumigated soil, *Plant Dis.* 71: 1101-1106.

Cordier, C., Gianinazzi, S. and Gianinazzi-Pearson, V. 1996, Colonization patterns of root tissues by *Phytophthora nicotianae* var. *parasitica* related to reduced disease in mycorrhizal tomato, *Plant Soil* 185: 223-232.

Davis, R. M., Menge, J. A. and Zentmyer, G. A. 1978, Influence of vesicular arbuscular mycorrhizae on *Phytophthora* root rot of three crop plants, *Phytopathol.* 68: 1614-1617.

Davis, R. M., Menge, J. A. and Erwin, D. C. 1979, Influence of *Glomus fasciculatum* and soil phosphorus on *Verticilium* wilt of cotton, *Phytopathol.* 9: 453-456.

Davis, R. M. and Menge, J. A. 1981, *Phytophthora parasitica* inoculation technique and intensity of vesicular arbuscular mycorrhiae in citrus, *New Phytol.* 87: 705-715.

Dehne, H. W. 1982, Interaction between vesicular-arbuscular mycorrhizal fungi and plant pathogens, *Phytopathol.* 72: 1115-1119.

Dehne, H. W. 1987, Zur Bedentung der vesikular arbuskularen (VA) mykorrhiza fur die Pflanzengesundheit, Habil. Univ. Hannover, Germany.

Dehne, H. W. and Schonbeck, F. 1978, The influence of endotrophic mycorrhiza on plant diseases, 3, Chitinase-activity and ornithine-cycle, *J. Plant Dis.Protec.* 85: 666-678.

Dehne, H.W. and Schonbeck, F. 1979, The influence of endotrophic mycorrhiza on *Fusarium* wilt of tomato, *Z. Pflkrankh. Schutz.* 82: 630-632.

Dixon R. A. and Harrison, M. J. 1990, Activation, structure and organization of genes involved in microbial defence in plants, *Adv. Genet.* 28: 165-234.

Dixon, R. A., Dey, P. M. and Lamb, C. J. 1993, Phytoalexin : Enzymology and molecular biology, *Adv. Enzymol.* 55: 1-136

Duchesne, L. C., Ellis, B. E. and Peterson, R. L. 1989, Disease suppression by the ectomycorrhizal fungus *Paxillus involutus* : contributations of oxalic acid, *Can. J. Bot.* 67: 2726-2723.

Dugassa, G. D., von Alten, H. and Schonbeck, F. 1996, Effects of arbuscular mycorrhiza (AM) on health of *Linum usitatissimum* L. infected by fungal pathogens, *Plant Soil* 185: 173-182.

Dumas-Gaudat, E., Furlan, V., Grenics, J. and Asselin, A. 1992, New acidic chitinase isoforms induced in tobacco roots by vesicular-arbuscular mycorrhizal fungi, *Mycorrhiza* 1: 133-136.

Feldmann, F., Junqueira, N. T. V. and Liebei, R. 1990, Utilization of VA mycorrhiza as a factor in integrated plant protection, *Agric. Eco. Environ.* 29: 131-136.

Franken, P. and Gnadinger, F. 1994, Analysis of parsley arbuscular endomycorrhiza:infection development and mRNA levels of defence related genes, *Mol Plant-Microbe Interac.* 7: 612-620.

Gangopadhyay, S. and Das, K. M. 1987, Control of soil borne disease of rice through vesicular arbuscular mycorrhiza, in: *Mycorrhiza Round Table Proc. of a Natl. Workshop*, Jawaharlal Nehru Univ., New Delhi, pp. 560-580.

Garbaye, J. 1991, Biological interactions in the mycorrhizosphere, *Experientia* 47:370-375.

Garcia-Garrido, J. M. and Ocampo, J. A. 1989a, Interaction between *Glomus mosseae* and *Erwinia carotovora* and its effect on the growth of tomato plants, *New Phytol.* 110: 551-555.

Garcia-Garrido, J. M. and Ocampo, J. A. 1989b, Effect of VA mycorrhiza infection of tomato on damage caused by *Pseudomonas syringae, Soil Biol. Biochem.* 121: 165-167.

Gianinazzi, S. 1991, Vesicular-arbuscular (endo) mycorrhizas : cellular, biochemical and genetic aspects, *Agric. Eco. Environ.* 35: 105-119.

Gianinazzi, S. and Gianinazzi-Pearson, V. 1992, Cytological, histochemistry and immunocytochemistry as tools for studying structure and function in endomycorrhiza, in: *Techniques for the Study of Mycorrhiza : Methods in Microbiology*, Vol.24, J. P. Norris, D. J. Read and A.K. Varma, eds., Academic Press, London, pp. 109-139.

Gianinazzi-Pearson, V., Dumas-Gaudat, E., Gollote, A., Tahiri-Alaoui, A. and Gianinazzi, S. 1996, Cellular and molecular defence-related root responses to invasion by arbuscular mycorrhizal fungi, *New Phytol.* 133: 45-57.

Giovanneti, M., Tosiu, L., Torre, G. D. and Zazzerini, A. 1991, Histological, physiological and biochemical interactions between vesicular-arbuscular mycorrhizae and *Thielaviopsis basicola* in tobacco plants, *J. Phytopathol.* 131: 265-274.

Gollotte, A., Gianinazzi-Pearson, V. and Gianinazzi, S. 1995, Etude immunocytochimique des interfaces plante-champignon endomycorrhizen a arbuscules chez des pois iso-geniques myc+ ou resistant a endomycorrization (myc-), *Acta Bot. Gallica* 141: 449-454.

Goncalves, E. J., Muchovej, J. J. and Muchovej, R. M. C. 1991, Effect of kind and method of fungicidal treatment of bean seed on infection by the VA mycorrhizal fungus *Glomus macrocarpum* and by the pathogenic fungus *Fusarium solani*, I, Fungal and plant parameters, *Plant Soil* 132: 41-46.

Graham, J. H. and Menge, J. A. 1982, Influence of vesicular arbuscular mycorrhizae and soil phosphorus on take all disease of wheat, *Phytopathol.* 72: 95-98.

Graham, J. H., Leonard, R. T. and Menge, J.A. 1981, Membrane-mediated decrease in root exudation responsible for phosphorus inhibition of vesicular arbuscular mycorrhizae formation, *Plant Physiol.* 68: 548-552.

Grandmaison, J., Olah, G. M., van Calsteren, M.R. and Furlan, V. 1993, Characterization and localization of plant phenolics likely involved in the pathogen resistance expressed by endomycorrhizal roots, *Mycorrhiza* 3: 155-164.

Grandison, G. S. and Cooper, K. M. 1986, Interaction of vesicular-arbuscular mycorrhizae and cultivar of alfalfa susceptible and resistant to *Meloidogyne hapla, J. Nematol.* 18: 141-149.

Grey, W. E., Leur, J. A. G., van Kashour, G. and El-Naimi, M. 1989, The intercation of vesicular-arbuscular mycorrhizae and common root rot (*Cochliobolus sativus*) in barley, *Rachis* 8: 18-20.

Harlapur, S. I., Kulkarni, S. and Hegre, R. K. 1988, Studies on some aspects of root rot of wheat caused by *Sclerotium rolfsii* Sacc, *Plant Pathol. Newsletter* 6: 49.

Harley, J. L. and Smith, S. E. 1983, *Mycorrhiza Symbiosis*, Academic Press, London, UK, pp. 483.

Harrison, M. J. and Dixon, R.A. 1993, Isoflavonoid accumulation and expression of defence gene transcripts during the establishment of vesicular-arbuscular mycorrhizal associations in roots of *Medicago trunculata, Mole. Plant Microbe Interac.* 6: 643-654.

Heald, C. M., Bruton, B. D. and Davis, R. M. 1989, Influence of *Glomus intraradices* and soil phosphorus on *Meloidogyne incognita* infecting *Cucumis melo, J. Nematol.* 21: 69-73.

Hickman, C.J. 1970, Biology of *Phytophthora* zoospores, *Phytopathol.* 60: 1128-1135.

Hodge, A., Alexander, I. J. and Gooday, G. W. 1995, Chitinolytic activities of *Eucalyptus pisularis* and *Pinus sylvestris* root systems challenged with mycorrhizal and pathogenic fungi, *New Phytol.* 131: 255-261.

Hooker, J. E., Jaizme-Vega, M. and Atkinson, D. 1994, Biocontrol of plant pathogens using arbuscular mycorrhizal fungi, in: *Impact of Arbuscular Mycorrhizas on Sustainable Agriculture and Natural Ecosystems*, S. Gianinazzi, H. Schuepp., eds., Biekhauser, Basel, pp. 191-200.

Hussey, R. S. and Roncadori, R. W. 1982, Vesicular-arbuscular mycorrhiza may limit nematode activity and improve plant growth, *Plant Dis.* 66: 9-14.

Hu, Z. J. and Gui, X. D. 1991, Pretransplant inoculation with VA mycorrhizal fungi and *Fusarium* blight of cotton, *Soil Biol. Biochem.* 23: 201-203.

Hwang, S. F., Chang, K. F. and Chakravarty, P. 1992, Effects of vesicular-arbuscular mycorrhizal fungi on the development of *Verticilium* and *Fusarium* wilts of alfalfa, *Plant Dis.* 76: 239-243.

Iqbal, S. H. and Mahmood, T. 1986, Vesicular-arbuscular mycorrhizae as a deterrant to damping off caused by *Rhizoctonia solani* in *Brassica napus*, *Biologia* 32: 193-200.

Iqbal, S. H., Nasim, G. and Niaz, M. 1988a, II Role of vesicular-arbuscular mycorrhiza as a deterrant to damping off caused by *Rhizoctonia solani* in *Brassica oleracea*, *Biologia Pakistan* 34: 79-84.

Iqbal, S. H. and Nasim, G. 1988b, IV VA mycorrhiza as a deterrent to damping off caused by *Rhizoctonia solani* at different temperature regimes, *Biologia Pakistan* 34: 215-221.

Jain, R. K. and Sethi, C. L. 1988a, Influence of endomycorrhizal fungi *Glomus fasciculatum* and *G. epigaeus* on penetration and development of *Heterodera cajani* on cowpea, *Indian J. Nematol.* 18: 89-93.

Jain, R. K. and Sethi, C. L. 1988b, Interaction between vesicular-arbuscular mycorrhiza, *Meloidogyne incognita* and *Heterodera cajani* on cowpea as influenced by time of inoculation, *Indian J. Nematol.* 18: 263-268.

Jalali, B. L. and Jalali, I. 1991, Mycorrhiza in plant disease control, in : *Handbook of Applied Mycology*, Vol. 1; Soil and Plants, D.K. Arora, B. Rai, K.G. Mukerji and G.R. Knudsen, eds., Maxcel Dekker Inc., New York, pp. 131-154.

Jalali, B. L., Chhabra, M. L. and Singh, R. P. 1990, Interaction between vesicular-arbuscular mycorrhizal endophyte and *Macrophomina phaseolina* in mungbean, *Indian Phytopathol.* 43: 527-530.

Jayaram, J. and Kumar, D. 1995, Influence of mungbean yellow mosaic virus on mycorrhizal fungi associated with *Vigna radiata* var. PS 16, *Indian Phytopathol.* 48: 108-110.

Kaplan, D. R., Keen, N. T. and Thomason, I. J. 1980, Study of the mode of action of glyceollin in soybean, Incompatibility to the root knot nematode *Meloidogyne incognita*, *Physiol. Plant Pathol.* 16: 319-325.

Kapulnik, Y., Volpin, H., Itzhaki, H., Ganon, D., Galili, S., David, R., Shaul, O., Elad, Y., Chet, I. and Okon, Y. 1996, Suppresion of defence responses in mycorrhizal alfalfa and tobacco roots, *New Phytol.* 133 : 59-64.

Kassab, A. S., and Taha, A. H. Y. 1990a, Aspects of the host-parasite relationships of nematodes and sweet potato 1, Population dynamics and interaction of *Criconemella* spp., *Rotylenchus reniformis, Tylenchorhynchus* spp. and endomycorrhiza, *Ann. Agri. Sci.* 35: 497-508.

Kassab, A. S. and Taha, A. H. Y. 1990b, Interaction between plant-parasitic nematode, vesicular-arbuscular mycorrhiza, rhizobia, and Egyptian clover, *Ann. Agric. Sci.* (Cairo) 35: 509-520.

Kaye, J. W., Pflegger, F. L. and Stewart, E. L. 1984, Interaction of *Glomus fasciculatum* and *Phythium ultimum* on greenhouse grown Poinsettia, *Can. J. Bot.* 62: 1575-1579.

Khadge, B. R., Ilag, L. L. and Mew, T. W. 1990, Interaction study of *Glomus mosseae* and *Rhizoctonia solani*, in: *Current Trends in Mycorrhizal Research*, Proc. Natl. conference on Mycorrhiza, pp. 94-95.

Khruscheva, E. P. and Milenina, N. P. 1978, Mycorrhiza and root rot of spring wheat, *Mycologiya, Fitopatologia* 12: 43-45.

Krishna Prasad, K. S. 1991, Influence of a vesicular-arbuscular mycorrhiza on the development and reproduction of root-rot nematode affecting flue cured tobacco, *Afro-Asian J. Nematol.* 1: 130-134.

Krishna,K. R. and Bagraraj, D. J. 1983, Interaction between *Glomus fasciculatum* and *Sclerotium rolfsii* in Peanut, *Arachis hypogaea*, *Can. J. Bot.* 61: 2349-2351.

Krishna, K.R. and Bagyaraj, D. G. 1984, Phenols in mycorrhizal roots of *Arachis hypogaea*, *Experientia* 40: 85-86.

Lambais, M. R. and Mehdy, M. C. 1993, Suppression of endochitinase, β-1,3-endoglunase and chalcone isomerase expression in bean vesicular-arbuscular mycorrhizal roots under different soil phosphate conditions, *Mole. Plant-Microbe Interac.* 6: 75-83.

Linderman, R. G. 1994, Role of VAM fungi in biocontrol, in: *Mycorrhiza and Plant Health*, F.L. Pleger and R.G. Linderman, eds., APS, St. Paul, pp 1-26.

Mahmood, T. and Iqbal, S. H. 1982, Influence of soil moisture contents on VA mycorrhizae and pathogenic infection by *Rhizoctonia solani* in *Brassica napus*, *Pak. J. Agri. Res.* 3: 45-49.

Malajczuk, N. 1988, Interaction between *Phytophthora cinnamomi* zoospores and microorganisms on non-mycorrhizal and ectomycorrhizal roots of *Eucalyptus marginata*, *Trans. Brit. Mycol. Soc.* 90: 375-382.

Mark, G. L. and Cassels, A. C. 1996, Genotype dependence in the interaction between *Glomus fistulosum, Phytophthora fragariae* and the wild strawberry (*Fragaria vesica*), *Plant Soil* 185: 233-239.

Marx, D. H. 1973, Mycorhizae and feeder root diseases, in: *Ectomycorrhizae: Their Ecology and Physiology*, G. C. Marks and T. T. Kozlowski, eds., Academic Press, New York.

Mauch, F., Mauch-Mani, B. and Boller, T. 1988a, Antifungal hydrolases in pea tissue, I, Purification and characterization of two chitinases and two β-1,3-glucanases differrentiallyβ regulated during development in response to fungal infection, *Plant Physiol.* 87: 325-333.

Mauch, F., Mauch-Mani, B. and Boller, T. 1988b, Antifungal hydrolases in pea tissues, II, Inhibition of fungal growth by the combinations of chitinase and β-1,3-glunase, *Plant Physiol.* 88: 936-942.

Meier, B. M., Shaw, N. and Slusarenko, A. J. 1993, Spatial and temporal accumulation of defence gene transcripts in bean (*Phaseolus vulgaris*) leaves in relation to bacteria-induced hypersensitive cell death, *Mole. Plant-Microbe Interac.* 6 : 453-466.

Melo, I. S., Costa, C. P. and Silveria, A. P. D. 1985, Efect of vesicular arbuscular mycorrhizae on aubergine wilt caused by *Verticilium alboatrum* Reinke and Berth. Summ, *Phytopathol.* 11: 173-179.

Menge, J. A., Munnecke, D. E., Johnson, E. L. V. and Carnas, D. W. 1978, Dosage response of the vesicular-arbuscular mycorrhizal fungi *Glomus fasciculatum* and *G. constrictum* to methyl bromide, *Phytopathol.* 68: 1368-1372.

Modjo,H. S. and Hendrix, J. W. 1986, The mycorrhizal fungus *Glomus macrocarpum* as a cause of tobacco stunt disease, *Phytopathol.* 76: 688-691.

Morandi, D. and Le Querre, J. L. 1991, Influence of nitrogen on accumulation of isosojagol (a newly detected coumestan in soybean) and associated isoflavonoids in roots and nodules of mycorrhizal and non-mycorrhizal soybean, *New Phytol.* 117: 75-79.

Mosse, B. 1973, Advances in the study of vesicular arbuscular mycorrhiza, *Ann Rev. Phytopathol.* 11: 171-196.

Mukerji, K.G. 1996, Taxonomy of endomycorrhizal fungi, in : *Advance in Botany*, K.G. Mukerji, B. Mathur, B.P. Chamola and P. Chitralekha, eds., APH Publishing Corporation, New Delhi, India, pp. 213-221.

Mukerji, K.G. 1999, Mycorrhiza in control of plant pathgoens : molecular approaches, in : *Biotechnological Approaches in Biocontrol of Plant Pathogens*, K.G. Mukerji, B.P. Chamola and R.K. Upadhyay,eds., Kluwer Academic/Plenum Press, USA, pp. 135-155.

Nemac, S. and Myhre, D. 1984, Virus-*Glomus etunicatum* interactions in *Citrus* rootstocks, *Plant Dis.* 68: 311-314.

Norman, J. R., Atkinson, D. and Hooker, J. E. 1996, Arbuscular mycorrhzal fungal-induced alteraction to root architecture in strawberry and induced resistance to the root pathogen *Phytophthora fragariae*, *Plant Soil* 185: 191-198.

Nylud, J. A. 1988, The regulation of mycorrhiza formation: carbohydrate and hormone theories reviewed, *Scand. J. For. Res.* 3: 465-479.

O' Bannon, J. H. and Nemec, S. 1979, The response of *Citrus limon* seedlings to a symbiont, *Glomus etunicatum* and a pathogen, *Radopholus similes, J. Nematol.* 11: 270-275.

O' Connel, R. J., Brown, I. R., Mansfield, J. W., Bailey, J. A., Mazau, D., Rumeau, D. and Esquerr-Tugaye, M. Y. 1990, Immunological localisation of a hydroxyproline-rich glycoproteins accumulating in melon and bean at sites of resistnce to bacteria and fungi, *Mole. Plant-Microbe Interac.* 3: 33-40.

Oliveira, A. A. R. and Zambolin, L. 1987, Interaction between the endomycorrhizal fungus *Glomus etunicatum* and the gall nematode *M. javanica* on beans with split roots, *Fitopatologia Brasilera* 12: 222-225.

Osman, H. A., Korayem, A. M., Ameen, H. H. and Badr-Eldin, S. M. S. 1990, Interaction of root-knot nematode and mycorrhizal fungi on common bean *Phaseolus vulgaris* L., *Anzeiger fur Schadlingskunde, Pflanzenschutz, Umweltschutz* 63: 129-131.

Paget, D. K. 1975, In : *Endomycorrhizas*, F. E. Sanders, B. Mosseae and P. B. Tinker, eds., Academic Press, London, pp. 593-606.

Paulitz, T. C. and Linderman, R. G. 1991, Lack of antagonism between the biocontrol agent *Gliocladium virens* and vesicular arbuscular mycorrhizal fungi, *New Phytol.* 117: 303-308.

Pinochet, J., Calvet, C., Camprubi, A. and Fernandez C. 1996, Interactions between migratory endoparasitic nematodes and arbuscular mycorrhizal fungi in perrenial crops : A review, *Plant Soil* 185: 183-190.

Plenchette, C., Furlan, V. and Fortin, J. A. 1983, Responses of endomycorrhizal plants grown in a calcined montmorillonite clay to different levels of soluble phosphorus, I, Effect of growth and mycorrhizal development, *Can. J. Bot.* 61: 1377-1383.

Ramraj, B., Shanmugam, N. and Reddy, D. A. 1988, Biocontrol of *Macrophomina* root rot of cowpea and *Fusarium* wilt of tomato by using VAM fungi, Mycorrhizae for Green Asia, Proc. first Asian Conf. on Mycorrhza, January, 29-31, pp. 250-251.

Reddy, M. V., Rao, J. N. and Krishna, K. R. 1989, Influence of vesicular-arbuscular mycorrhizae on *Fusarium* wilt of pigeonpea, *Internatl. Pigeonpea Newsleter* 9: 23.

Rosendahl, S. 1995, Interactions between the vesicular-arbuscular mycorrhizal fungus *Glomus fasciculatum* and *Aphanomyces euteiches* root rot of peas *(Pisum sativum), Phytopathol.* 114: 31-40.

Ross, J. P. 1972, Influence of *Endogone* mycorrhiza on *Phytophthora* rot of soybean, *Phytopathol.* 62: 876-897.

Rupp, L. A., Mudge, K. W. and Negm, F. B. 1989, Involvement of ethylene in ectomycorrhiza formation and dichotomous branching of roots of mugo pine seedlings, *Can. J. Bot.* 67: 477-482.

Saleh, H. and Sikora, R. A. 1984, Relatoinshiop betwween *Glomus fasciculatum* root colonozation of cotton (*Gossypium hirsutum* cultivar coker 201) and its effect on *Meloidogyne incognita, Nematologica* 30: 230-237.

Schenck, N. C., Kinoch, R. A. and Dickson, D. W. 1975, Interaction and endomycorrhizal fungi and root knot nematode on soybean, in: *Endomycorrhizas*, F. F. Sanders, B. Mosse and P. B. Tinker, eds., Academic, London, pp. 605-617.

Schonbeck, F. 1978, IIIrd Int. Plant Pathol. Cong. (Abstr.) Munchen, pp. 173.

Schonbeck, F. 1979, Endomycorrhizas in relation to plant diseases, in: *Soil-Borne Plant Pathogens*, B. Schippers, and W. Gama, eds., Academic Press, New York, pp. 271-280.

Schonbeck, F. and Dehne, H. W. 1977, Damage to mycorrthizal cotton seedlings by *Theilaviopsis basicola, Plant Dis. Repr.* 61: 266-268.

Secilia, J. and Bagyaraj, D. G. 1987, Bacteria and actinomycetes associated with pot cultures of vesicular-arbuscular mycorrhizas, *Can. J. Microbiol.* 33: 1069-1073.

Sikora, R. A. and Schonbeck, F. 1975, Efffect of vesicular mycorrhiza on the population dynamics of the root knot nematode, *8th Int. Cong. Plant Prot.* 5: 158-166.

Singh, K., Varma, A. K. and Mukerji, K. G. 1987, Vesicular arbuscular mycorrhizal fungi in diseased and healthy plants of *Vicia faba, Acta Bot. Ind.* 15: 304-310.

Sitamaiah, K. and Sikora, R. A. 1982, Effect of the mycorrhizal fungus, *Glomus fasciculatum* on the host parasite relationship of *Rotylenchus reniformis* in tomato, *Nematologica* 28: 412-419.

Sivaprasad, P., Jacob, A., Nair, S. K. and George B. 1990, Influence of VA mycorrhizal colonization on root-knot nematode infestation in *Piper nigrum* L., in: *Current Trends in Mycorrhizal Research*, B. L. Jalali and H. Chand, eds., Pro. Natl. Con. on Mycorrhiza, Haryana Agri. Univ., Hisar, India, (14-16 Feb. 1990) and TERI, New Delhi.

Smith, G. S. 1987, Interactions of nematodes with mycorrhizal fungi, in: *Vistas on Nematology*, J. A. Veech and D. W. Dickon, eds., *Soc. Namatol.*, Hyattsville, Md., pp. 292-300.

Smith, G. S. and Kaplan, D.T. 1982, Influence of mycorrhizal fungus phosphorus and burrowing nematode interactions on growth of rough lemon seedlings, *J. Nematol.* 20: 539-544.

Snellgrove, R. C., Splittstosser, W. E., Stribley, D. P. and Tinker, P. B. 1982, The distribution of carbon and the demand of the fungal symbiont in leek plants with vesicular arbuscular mycorrhizas, *New Phytol.* 92: 75-85.

Spanu, P., Boller, T., Ludwig, A., Wiemkin, A., Faccio, A. and Banfante-Fasolo, P. 1989, Chitinase in roots of mycorrhizal *Allium porrum* : regulation and localisation, *Planta* 177: 447-455.

Sparling, G. P. and Tinker, P. G. 1978, Mycorrhizal infection in Pennine Grassland, I, Levels of infection in the field, *J. App. Ecol.* 15: 943-950.

Stack, R. W. and Sinclair, W. A. 1975, Protection of Douglas-fir seedlings against *Fusrium* root rot by a mycorrhizal fungus in absence of mycorrhizal formation, *Phytopathol.* 65: 468-472.

Subhashini, D. V. 1990, The role of VAM in controlling certain root diseases of tobacco, in: *Current Trends in Mycorrhizal Research*, B. L. Jalali and H. Chand, eds., Proc. Natl. Conf. on Mycorrhiza Haryana Agril. Univ., Hisar, India.

Suresh, C. K. and Bagyaraj, D. J. 1984, Interaction between a vesicular arbuscular mycorrhiza and a root knot nematode and its effect on growth and chemical composition of tomato, *Nematologia Mediterranea* 12: 31-39.

St-Arnaud, M., Hamel, C., Caron, M. and Fortin, J. A. 1994, Inhibition of *Pythium ultimum* in roots and growth substrate of mycorrhizal *Tagetes patula* colonized with *Glomus intraradices, Can. J. Plant Pathol.* 16: 187-194.

St-Arnaud, M., Hamel, C., Vimard, B., Caron, M. and Fortin, J. A. 1995, Altered growth of *Fusarium oxysporum* f. sp. *chrysanthemi* in an *in vitro* dual culture system with the vesicular arbuscular mycorrhizal fungus *Glomus intraradices* growing on *Daucus carota* transformed roots, *Mycorrhiza* 5: 431-438.

Stinzi, A., Heitz, T., Prasad, V., Wiedemann-Meidinoglu, S., Kaufmann, S., Geoffroy, P., Legrand, M. and Fritig, B. 1993, Plant pathogenesis-releateed proteins and their role in defence against pathogens, *Biochimie* 75: 687-706.

Strobel, N. E. and Sinclair, W.A. 1991, Role of flavanolic wall infusions in the resistance induced by *Laccaria laccata* to *Fusarium oxysporum* in primary roots of Doulas-fir, *Phytopathol.* 81: 420-425.

Taha, A. H. Y. and Abdel-Kader, K. M. 1990, The reciprocal effects of prior invasion by root-knot nematode or by endomycorrhiza on certain morphological and chemical characteristics of Egyptian clover plants, *Ann. Agri. Sci.* (Cairo) 35: 521-532.

Tahiri-Alaoui, A., Dumas-Gaudot, E. and Gianinazzi, S. 1993a, Immunocytochemical localisation of pathogenesis-related PR-1 proteins in tobacco root tissues infected *in vitro* by the black rot fungus *Chalara elegans, Physiol. Mole. Plant Pathol.* 42: 69-82.

Tahiri-Alaoui, A., Dumas-Gaudot, E., Gianinazzi, S. and Antoniw, J. F. 1993b, Expression of the PR-1 gene in roots of two *Nicotiana* species and their amphidiploid hybrid infected with virulent and avirulent races of *Chalara elegans, Plant Pathol.* 42: 728-736.

Tello, J. C., Vares, F., Notorio, A. and Lacasa, A. 1987, Mycorrhizae: a potential against plant diseases, *ITEA* 18: 40-64.

Thomas, G. V., Sundaraju, P., Ali, S. S. and Ghai, S. K. 1989, Individual and interactive effects of VA mycorrhizal fungi and root-knot nematode, *Meloidogyne incognita*, on cardomum, *Tropical Agric.* 66: 21-24.

Torres- Barragan, A., Zavalta-Mejia, E., Gonzalez-Chavez and Ferrera-Cerrato, R. 1996, The use of arbuscular mycorrhizae to control onion white rot (*Sclerotium cepivorum*) under field conditions, *Mycorrhiza* 6: 253-257.

Tosi, L., Giovanneti, M., Zazzerini, A. and Torre, G. 1988, Influence of mycorrhizal tobacco roots incorporated into the soil on the development of *Theilaviopsis basicola, J. Phytopathol.* 122: 186-189.

Trotta, A., Varese, G. C., Gnavi, E., Fusconi, A., Sampo, S. and Berta, G. 1996, Interactions between the soil borne root pathogen *Phytophthora nicotianae* var. *parasitica* and the arbuscular mycorrhizal fungus *Glomus mosseae* in tomato plants, *Plant Soil* 185: 199-209.

Uknes, S., Vernooj, B., Morris, S., Chandler, D., Specker, N., Hunt, M., Neuenschwander, U., Lawton, K., Starret, M., Friedrich, L., Weymann, K., Negrotto, D., Gorlach, J., Lanahan, M., Salmeron, J., Ward, E., Kessmann, H. and Ryals, J. 1996, Reduction of risk for growers: methods for the development of disease-resistant crops, *New Phytol.* 133: 3-10.

Umesh, K. C., Krishnappa, K. and Bagyaraj, D. J. 1988, Interaction of burrowing nematode, *Radophilus similis* and VA mycorrhiza, *Glomus fasciculatum* (Thaxt). Gerd and Trappe in banana (*Musa acuminata* Colla.), *Indian J. Nematol.* 18: 6-11.

Utkhede, R. S., Li, T. S. C. and Smith, E. M. 1992, The effect of *Glomus mosseae* and *Enterobacter aerogenes* on apple seedlings grown in apple replant disease soil, *J. Phytopathol.* 135: 281-288.

Van Loon, L. C. 1985, Pathogenesis-related proteins, *Plant Mole. Biol.* 4: 111-116.

Van Loon, L. C., Pierpoint, W. S., Boller, T. and Conejero, V. 1994, Recommendations for naming plant pathogenesis-related proteins, *Plant Mole. Boil. Repr.* 12: 245-264.

Verdejo, S., Calvet, C. and Pinochet, J. 1990, Effect of mycorrhiza on kiwi infected by the nematodes *Meloidogyne hapla* and *M. javanica, Buletin de Sanidad Vegetal, Plagas* 16: 619-624.

Volpin, H., Elkind, Y., Okon, Y. and Kapulnik, Y. 1994, A vesicular arbuscular mycorrhizal fungus *Glomus intraradices* induces a defence response in alfalfa roots, *Plant Physiol.* 104: 683-689.

Volpin, H., Phillips, D. A., Oken, Y. and Kapulnik, Y. 1995, Suppression of an isoflavonoid phytoalexin defence response in mycorrhizal alfalfa roots, *Plant Physiol.* 108: 1449-1454.

Vierheilig, H., Alt, M., Mohr, V., Boller, T. and Wiemkin, A. 1994, Ethylene biosynthesis and activities in the roots of host and non-host plants of vesicular-arbuscular mycorrhizal fungi after inoculation with *Glomus mosseae, J. Plant Physiol.* 143: 337-343.

Wacker, T. L., Safir, G. R. and Stephens, C. T. 1990, Effects of *Glomus fasciculatum* on the growth of *Asparagus* and the incidence of *Fusarium* root rot, *J. Am. Soc. Hort. Sci.* 115: 550-554.

West, H. M. 1995, Soil phosphate status modifies response of mycorrhizal and non-mycorrhizal *Senecio vulgaris* L. to infection by the rust, *Puccinia lagenophorae* Cooke, *New Phytol.* 129: 107-116.

Zak, B. 1964, Role of mycorrhizae in root disease, *Ann. Rev. Phytopathol.* 2: 377-392.

Zambolin, L. and Schenck, N. C. 1983, Reduction of the effect of pathogenic root infecting fungi in soybean by the mycorrhizal fungus *G. mosseae, Phytopathol.* 73: 1402-1405.

MASS PRODUCTION OF VAM FUNGUS BIOFERTILIZER*

David D. Douds, Jr.[1], Vijay Gadkar[2] and Alok Adholeya[2]

[1]USDA-ARS ERRC
600 E. Mermaid Lane
Wyndmoor, PA, 19038
USA

[2]Tata Energy Research Institute
Habitat Place, Lodi Road
New Delhi 110003, INDIA

1.   INTRODUCTION

Vesicular-arbuscular mycorrhizal (VAM) fungi are symbiotic soil fungi which colonize the roots of approximately 80% of plant families (Harley and Harley, 1987). They impart to their hosts a variety of benefits which include increased growth and yield due to enhanced nutrient acquisition (Diederichs and Moawad, 1993; Mosse, 1973), water relations (Davies *et al.*, 1993; Subramanian *et al.*, 1995), pH tolerance (Clark and Zeto, 1996; Maddox and Soileau, 1991), and disease and pest resistance (Lopez *et al.*, 1997; Mark and Cassells, 1996; Newsham *et al.*, 1995; Trotta *et al.*, 1996). The most common beneficial effect of mycorrhizae is increased uptake of immobile nutrients, notably P, from soil (Bolan, 1991). The extraradical mycelium of the mycorrhizal fungus acts in effect as an extension of the root system, more thoroughly exploring the soil volume. The P depletion zone around a non-mycorrhizal root extends to only 1-2 mm, approximately the length of a root hair (Li *et al.*, 1991) whereas extra radical hyphae of VAM fungi extend 8 cm or more beyond the root making the P in this greater volume of soil available to the host (Rhodes and Gerdemann, 1975).

An example of the potential economic benefit of the enhancement of P uptake through VAM fungi is found in the work of Plenchette and Morel (1996). *Glycine max* colonized by the VAM fungus *Glomus intraradices* produced 80% of maximum growth at a soil solution P concentration of 0.110 mg/ml. Non-mycorrhizal plants required 0.148 mg/ml for the same rate of growth. This calculated to a savings to the farmer of 222 kg P_2O_5 ha⁻¹.

---

*Mention of a brand or firm name does not constitute an endorsement by the US Department of Agriculture or the Tata Energy Research Institute over others not mentioned.

The symbiotic association between VAM fungi and plant roots goes back 400 million years (Taylor *et al.*, 1995), coincident with the first colonization of land by the plant kingdom. Studies have confirmed that almost all the agroclimatic conditions on this earth carry natural population of VAM (Bhatia *et al.*, 1996) Metabolic regulation, loss of genetic information, regulation of gene expression, or other events are thought to have occurred over these years to result in the obligately symbiotic nature of these fungi. These fungi require fixed carbon, or other as yet unknown compounds, from their host plant and as a result, have not been cultured axenically to date.

Given that the most effective potential method of production of VAM fungus inoculum, namely axenic culture in a sterile vessel such as a fermenter, is not possible at present, dual culture with a plant host or root organ culture is the current alternative. We describe in this review the methods for production of VAM fungus inoculum that are available at present available and those under development, and biological and practical considerations necessary for the production and utilization of these fungi.

1.1. Situations where Inoculation would be Beneficial

Vesicular-arbuscular mycorrhizal fungi are native and ubiquitous in most areas and therefore most agronomic soils need not be inoculated to produce mycorrhiza on crop roots. However, native isolates of species may not be as effective in growth promotion as potentially introduced isolates. One reason for this could be the past management history of the soil. There is evidence that high chemical input agriculture selects for less beneficial isolates of VAM fungi (Johnson, 1993). A shift to low-input, sustainable agriculture on such sites could benefit from large scale inoculation with effective isolates (Bethlenfalvay and Schuëpp, 1994; Hooker and Black, 1995). Intensive agricultural management practices can also supresspress populations of VAM fungi in soils (Douds *et al.*, 1993; Kurle and Pfleger, 1994). Phosphorus fertilization (Vivekanandan and Fixen, 1991), tillage (Galvez *et al.*, 1995; McGonigle and Miller, 1993;), non-mycorrhizal plants in a rotation (Baltruschat and Dehne, 1988), pesticide application (Schreiner and Bethlenfalvay, 1996; Timmer and Leyden, 1978;1980,), and long fallows (Thompson, 1987) have a negative effect on VAM fungi.

The use of VAM fungus inoculum in other plant production systems is readily justified. These include systems in which the soil has been severely disturbed or where plants are grown in sterilized potting mixes. Severely disturbed sites in which mycorrhizal inoculum would be beneficial include mine land (Gould *et al.*, 1996; Jasper *et al.*, 1987; Lumini *et al.*, 1994), severely eroded areas (Aziz and Habte, 1989; Sharma *et al.*, 1996) and flooded fields (Wetterauer and Killorn, 1996). Inoculation with VAM fungi has long been known to increase growth and relieve nutrient deficiency in citrus grown in nursery soils routinely fumigated to control pathogens (Timmer and Leyden, 1978,1980). Inoculation of horticultural crops with VAM fungi has been shown to increase the growth of pot-grown plants (Yeager *et al.*, 1990) and micropropagated plantlets (Elmeskaoui *et al.*, 1995) and cut production costs (Johnson and Menge, 1982).

1.2. Biological and Practical considerations in Inoculum Production

Approximately 155 species of VAM fungi have been described so far (Morton and Bentivenga, 1994). Individual isolates of a species may have different growth requirements and abilities to enhance plant growth. Some VAM fungi are broadly effective (Sylvia *et al.*, 1993) while others, though they colonize many host plant species, may promote the growth of only some (Hung *et al.*, 1990; Monzon and Azcon., 1996). Others, such as members of the *Gigasporineae*, may not promote plant growth directly but have beneficial effects upon soil structure by stabilizing soil aggregates (Miller and Jastrow, 1992; Schreiner and Bethlenfalvay., 1995; Tisdall and Oades., 1979). Therefore, the first consideration in inoculum production is the choice of the fungal isolate that meets the needs of a given plant production system. This will require preliminary trials with a variety of isolates to find the most effective one.

A second consideration is the persistence of the inoculated isolate in the field. Is it sufficiently competitive in relation to the native population and adaptable to that site so as to be present in significant numbers in the years to come? Yearly reinoculation can be expensive. Molecular methods are being utilized to study the persistence of introduced isolates of VAM fungi. Some molecular probes allow for the identification of specific isolates of VAM fungi in the presence of other isolates of the same species (Lloyd-MacGilp et al., 1996). Some evidence suggests that introduced isolates may not persist (Weinbaum et al., 1996).

Another biological consideration in the production of inoculum is the host plant upon which the fungus will grow.The degree of importance of this consideration depends in part upon the production system to be used. Some inoculum production systems also allow for the colonization of the media by other microorganisms, including potential pathogens. Therefore, one should choose a host plant species for inoculum production which has a suite of pathogens that are not harmful to the eventual target plant species in the field (Menge, 1984). In addition, sporulation by a VAM fungus will vary with host plant (Bever et al., 1996 Hetrick and Bloom, 1986). Choice of the host plant for inoculum production for maximization of spore production becomes less of a factor if colonized roots can be utilized as inoculum. This allowance can be made for members of the *Glomineae*, but not for those of *Gigasporineae*. The latter do not produce vesicles in roots and therefore colonized roots are poor inocula.

1.3. Pest Control During Culture

Control of pests during inoculum production is also an issue. One does not want pests in the inoculum to adversely affect either the production of inoculum or the eventual target host plant in the field. Menge (1984) outlines six rules to ensure the production of pest-free inocula of VAM fungi in a greenhouse:

i)   sterilize pots and growth media.
ii)   begin with pest-free inoculum.
iii)   monitor cultures and discard those that are contaminated.
iv)   alternate host species, especially among those that do not share pests with the crop for which the inoculum is intended.
v)   utilize selected pesticides with minimal side effects upon the mycorrhizae or the host and
vi)   practice sanitary procedures in the greenhouse.

The effects of some common fungicides and fumigants in common use upon VAM fungi have been reviewed (Menge, 1982). Recent research has focused primarily upon their utilization to inhibit colonization of roots in the field to produce non-mycorrhizal plants as experimental controls (West et al., 1993). One fungicide to avoid in inoculum production due to its deleterious effect upon mycorrhizal fungi is Benlate (benomyl) (Perrin and Plenchette, 1993; Sukarno et al., 1993). Fungicide coatings on seeds have no effect upon the formation of mycorrhizae on the resultant seedlings (Spokes et al., 1989). The anti-oomycete (e.g *Pythium* spp) fungicide metalaxyl did not adversely affect mycorrhizal development and has been recommended by Seymour et al. (1994) for routine application in VAM fungus pot cultures.

The response of VAM fungi to selected fungicides, and their effectiveness *vis-a-vis* pathogens, varies with soil type (Perrin and Plenchette, 1993), and a preliminary experimentation under local conditions may be necessary.

1.4. Delivery to the Field

Besides the reliable production of large quantities of pest-free inoculum, the other factor currently inhibiting the application of VAM fungus inoculum in large-scale operations is economical delivery to the field. This is a primary flaw in soil-based pot culture systems. Utilization of these inocula would require delivery and application of tons of a non-uniform mixture to the site.

The problem of application to the field has been addressed in a variety of ways. One is to use of expanded clay particles, sized to fit pre-existing farm implements designed to deliver seeds or fertilizer pellets to the field (Baltruschat, 1987), as a support matrix in the hydroponic culture method (see below). Another method is to encapsulate the VAM fungus propagules in beads of a gelatin such as carrageenan. This method has been used successfully with inocula grown in the aeroponic method of Hung and Sylvia (1988) which was sheared (Sylvia and Jarstfer, 1992). Sheared root inocula of *Glomus* sp. or *Glomus etunicatum* were suspended in autoclaved and cooled 2.5% [w/v] Kappa carrageenan (Sylvia and Jarstfer, 1992). Pellets were formed by injecting drops of the suspension into 0.3 M KCl to solidify the gel. Inocula from the nutrient film technique (see below) should also be amenable to encapsulation. Another method of encapsulation is to use 2% [w/v] alginate solidified with 0.1 M $CaCl_2$ has been used (Strullu *et al.*, 1991). Test plants (*Allium porrum*) receiving as few as 300.5 mg capsules containing root inoculum had 40% of their root length colonized after six weeks. The infectivity of alginate bead inocula was maintained after one month in storage at 4°C. Spores of *Glomus versiforme* produced *in vitro* culture with Ri-T-DNA transformed roots of *Daucus carota* were successfully encapsulated in alginate beads (Declerck *et al.*, 1996a). The gelatin and alginate bead methods offer an advantage in that the salt solution used to solidify the beads can contain other inorganic nutrients and as such, can promote root growth in the area of inoculation. They may, however, serve as nutrient-rich areas which stimulate the growth of aggressive saprophytic fungi. Purified gellan gum has proved to be a potential agent for the production of beads to encapsulate inocula (Doner and Douds, 1995). Inocula have also been mixed in 2.5% [w/v] hydroxyethylcellulose gel carrier which acts as a sticking agent (Hung *et al.*, 1991; Sylvia and Jarstfer, 1992).

Another encapsulation process with potential application to VAM fungus delivery to the field is granulation with semolina (coarse durum wheat flour) and kaolin clay ("Pesta"). This process has been used successfully to encapsulate biocontrol agents such as fungi (Connick *et al.*, 1991) and nematodes (Connick *et al.*, 1994). Semolina (80 gm), kaolin (20 gm) and fungal culture (52 ml) are mixed, kneaded, and rolled into a sheet 1.0 to 1.1 mm thick and allowed to air dry to approximately 10% water content. The resulting dry sheet is ground and sieved to collect the proper size fractions for application via conventional farm machinery. These preparations have a satisfactory shelf life. Preparations containing inocula of *Colletotrichum truncatum*, a mycoherbicide for *Sesbania exalta*, remained viable after 24 weeks storage at 25 °C at water activities of 0.33 or less (Connick *et al.*, 1996).

Another way to transfer a VAM fungus isolate to the field is to transplant precolonized seedlings to the field. This is applicable to vegetable production in which seedlings are typically raised in a greenhouse prior to transplanting. Seedlings may be grown in preinoculated potting mix (Yeager *et al.*, 1990; Niemi and Vestberg, 1992) or micropropagated plantlets may be inoculated *in vitro* (Elmeskaoui *et al.*, 1995).

2. LABORATORY AND GREENHOUSE METHODS OF INOCULUM PRODUCTION

2.1. Greenhouse Pot Culture

The most widely used method of inoculum production for research purposes today is greenhouse pot culture. Starter inoculum is added to the potting medium in which is placed a seedling of a suitable host plant. The variables to be optimized in this production system include potting mix recipe, disinfestation method, and nutrient regime under which to grow the plants.

2.2. Potting Mix

The concept of substrate receptiveness was introduced by Perrin *et al,* (1988). A receptive host plant, such as *Allium porrum*, is planted into a variety of potting media or field soils preinoculated with a range of concentration of propagules of the VAM fungus isolate of interest. Colonization of plants is measured 14-24 days later and the relative promotion of mycorrhizal formation among the media types is noted. Clay loam soils were more conducive to mycorrhizal development than calcareous soils and calcined clay and calcined clay/ vermiculite mixes were more conducive than peat based mixes in these tests. A variety of peat-based potting mixes of interest to horticulturists were examined by Caron and Parent (1988) with respect to compatibility with five isolates of VAM fungi. They found that peat-based mixes supported mycorrhizal colonization levels of 25-44% of root length when P was kept at minimal levels to allow good plant growth, determined in this study as 8.6 mg P per each 6 litre pot every four weeks. Final P concentrations of P in the media at the end of the 14 week experiment ranged for 2 to 10 mg/kg.

2.3. Media Disinfestation

Soil-based potting media require treatment before they can be used to culture of VAM fungi for two reasons. First, it is necessary to kill the indigineous VAM fungi present in the soil which may compete with the introduced isolate and contaminate the culture. Second, it is necessary to kill organisms potentially harmful to the VAM fungi, the pot culture host plant, or the eventual target plant at the site of application.

A variety of methods are available for disinfesting soil prior to establishing pot cultures: sterilization via steam, fumigation, or g irradiation; pasteurization with dry or moist heat; or microwave treatment. Irradiation may be impractical for large-scale operations and a common fumigant, methyl bromide, will soon be unavailable in the US. Heat treatment appear to be the most useful. A major disadvantage of heat treatment is the potential for alteration of chemical properties of the soil (Jakobsen and Andersen, 1982; Sonneveld, 1979). In particular, heating may alter the availability of nutrients in the soil, whether due to the release of N and P from organisms killed by the treatment (Speir *et al.,* 1986) or affecting the chemical form of the nutrients in the soil itself (Payne and Rechcigl, 1989; Wolf *et al.,* 1986). Heating the soil may also cause the release of toxic compounds (Cawse, 1975; Rovira and Bowen, 1966). These factors may affect the ability of the introduced VAM fungus to grow in the soil and the eventual level of colonization of host roots (Jakobsen and Andersen, 1982). In addition, total sterilization of the soil removes all competition for saprophytic or pathogenic organisms that may colonize the media during culture. The more mild dry or moist heat pasturization is preferred since alteration of the chemical properties of the media is minimized and portions of the populations of nonpathogenic organisms in the soil remain active (Sylvia, 1984). A good routine procedure to kill indigineous VAM fungi as well as fungal pathogens is heating to 80 °C for 30 minutes with aerated steam (Sylvia and Schenck, 1984).

The following experiment will serve as an example of the effects of disinfestation treatment upon media chemistry and potential complications in VAM fungus culture. Half liter volumes of moist media (1:1:1:0.75 [v/v] field soil: sand: vermiculite: calcined clay) containing equal parts of pot culture inocula of the VAM fungi *Gigaspora margarita* (DAOM 194757), *Glomus intraradices* (DAOM 197198), and *Glomus mosseae* (INVAM 336) were microwaved up to 16 minutes. Treated and control media then were placed in pots with a seedling of *Paspalum notatum* Flugge for 12 weeks of growth in a greenhouse while a portion of the media was analyzed for available P (Bray I).

Microwave treatment longer than 2 minutes was necessary to kill the VAM fungi present in the potting media in this experiment (Fig. 1a). Plants grew best both when colonized by mycorrhizal fungi and in media microwaved for 16 minutes (Fig. 1b). Shoot P concentrations and contents exhibited the same response as plant growth (Fig. 1c), suggesting that the availability of P was a determining factor in plant growth. Finally, soil P analyses showed relatively constant level of P through 8 minutes of microwaving, but a 50% increase with the 16 minute treatment (Fig. 1d). These results indicate an optimal microwave treatment of 4 minutes to disinfest the soil of indigeneous VAM fungi. Longer treatments may increase soil P levels to the point where colonization of the pot culture host may be inhibited. Efficiency of disinfestation and effect upon nutrient availability will change with sample size and soil type (Ferriss, 1984), and some experimentation is necessary before this method is used.

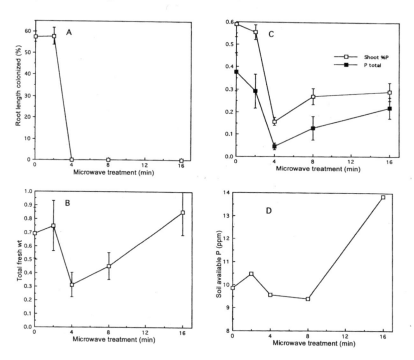

Fig. 1. Effect of microwave radiation treatment upon availability of P, growth, and colonization of *Paspalum notatum* by a mixture of *Glomus intraradices*, *Glomus mosseae*, and *Gigaspora margarita*. One half L batches of media (soil : sand : vermiculite : calcined clay; 1:1:1:0.75 [v/v]) were microwaved for 0 to 16 minutes. Plants were grown in a greenhouse in conical, 160 cm³ plastic pots for 12 weeks. Means of three observations ± SEM, except soil analyses which were conducted on pooled samples. A) percentage root length colonized by AM fungi; B) total fresh weight of *P. notatum*; C) concentration and total P in shoots; and D) available P in soil (Bray I).

2.4. Nutrient Regime

There are two important phenomena to consider when developing a nutrient regime to maximixe the colonization of pot culture host plants and sporulation by VAM fungi. First is the negative effect of high levels of nutrients, notably P, upon hyphal growth, root exudation, and colonization of roots (Allen *et al.*, 1981; Graham *et al.*, 1981; Nagahashi *et al.*, 1996). If addition of P is necessary, insoluble forms such as rock phosphate may be a wiser choice than soluble P fertilizer (Graham and Timmer, 1985). Secondly, one must bear in mind that the carbon compounds necessary for extraradical hyphal growth and spore production are transported to the fungus from the host plant and are ultimately products of photosynthesis (Bevenge *et al.*, 1975; Koch and Johnson, 1984), so the vigor of the plant is important. Therefore, a nutrient regime which provides conditions conducive to the bilateral transfer of nutrients in the mycorrhiza, i.e. P from the fungus to the host and C from the host to the fungus, should result in increased sporulation and colonization. Indeed, Douds and Schenck, (1990) found that nutrient solution without P increased colonization of roots and spore production in the soil by a variety of VAM fungi relative to cultures receiving only water or solutions in which P was balanced or overabundant relative to N levels (Hoagland and Arnon, 1938).

3. SOILESS CULTURE METHODS

3.1. Sand Culture

A soilless culture method one step removed from routine greenhouse pot culture is sand culture. This method provides the solid substrate necessary for hyphal growth for spread of colonization, yet allows for more control over nutrient levels than culture media containing soil or organic materials.

One method of sand culture is the "Beltsville Method" described by Millner and Kitt, (1992). An automated drip irrigation system is used to apply buffered nutrient solution. Irrigation frequency, nutrient concentration, plant host, and sand particle size are factors to be optimized in this method. The authors achieved their best results using half strength nutrient solution (Hoagland and Arnon, 1938) modified with $PO_4 = 20$ mM, with 0.5 mM MES buffer applied 5 times a day, 60 ml per application, to each of the 15 cm diameter plastic pot. Addition of $CaCO_3$ to the sand allowed for the culture of the basophyllic species *Glomus mosseae*. Sporulation by the fungi grown in this method was up to 3.6 fold times higher than that in soil-based potting media.

One advantage of this method is that host plants do not need to be precolonized and transplanted into the pot. The culture is initiated by inoculating directly in the sand medium. In addition, the automated system decreased labor and the resulting inocula (colonized roots and spores) are more easily isolated from the sand than from soil or other organic-based media. Further, this system is the only one besides soil-based culture and *in vitro* methods, that has been shown to satisfactorily produce inocula of *Gigaspora* species. A disadvantage is that, unlike aeroponic or the nutrient film techniques, the inocula must be isolated from the media.

3.2. Expanded Clay

Another variation of hydroponic culture for VAM fungus inoculum production utilizes expanded clay (Dehne and Backhaus, 1986). Precolonized plants, e.g, maize colonized by *G. etunicatum*, are transplanted into the hydroponic system with expanded clay as structural support media instead of sand. Composition of the nutrient solution is carefully regulated as in other culture methods. Extraradical hyphae proliferate in the media as the culture matures. Hyphae are found tightly appressed to the clay particles, growing along and within microscopic fissures (Grunewalt-Stšcker and Dehne, 1989). Intact spores were associated with only approximately 15% of the particles, whereas hyphae were present on 94-99% of the particles.

The primary advantage of this method is that the inoculum comes "ready-made", i.e. already affixed to a carrier clay particle for delivery to the field. Expanded clay particles can be cracked and sorted by size prior to initiating the culture so that they are compatible with the farm machinery designed to drill seeds or apply fertilizer to the soil (Baltruschat, 1987). Expanded clay particles colonized by G. etunicatum injected into the seed row next to the seeding or 5 cm to the side of the seed row with a sidedress fertilizer apparatus enhanced colonization of maize in the field (Baltruschat, 1987). Another advantage is the inoculum is readily stored. Air dried expanded clay particles with mycorrhizal fungus hyphae stored at room temperature remained infective for at least five years (Grunewalt-Stšcker and Dehne, 1989). A disadvantage of this procedure is that it may not work for members of the *Gigasporineae*. Hyphae of these species are typically ineffective inocula and are usually propagated using spores.

3.3.   Other Techniques

Another inert material used as a substrate/matrix in the hydroponic type culture system is rock wool (Heinzemann and Werlitz, 1990). *Glomus manihotis* and *G. etunicatum* sporulated abundantly and colonized 70% of the root length of the culture host *Cucumis sativus* in 70 days.

Vestberg and Uosukainen, (1992) utilized ploymeric hydrogel crystals, first allowed to hydrate and swell to their final size, mixed with vermiculite. Strawberry plantlets were transplanted into this mix, inoculated with *Glomus intraradices*, and fertilized with bone meal. The fungus sporulated profusely within the hydrogel matrix after only seven weeks. Further work is necessary both to investigate the ease of handling and storage of this type of inoculum and to determine if this production system works for VAM fungi other than those which sporulate within roots, such as *G. intraradices*.

4.   AEROPONIC CULTURE

Growing mycorrhizal roots in the absence of a substrate further enhances the ease of recovery and purity of VAM fungus inoculum. This is the philosophy of the developers of the aeroponic culture method (Hung and Sylvia, 1988; Jarstfer and Sylvia, 1995). Culture host plants are precolonized by surface disinfested spores of the VAM fungus isolate of interest by growing them for a short time in vermiculite. These plants are then transfered to the aeroponic culture chamber. The chamber is basically a waterproof box. The root system is suspended into the chamber through a hole in the top. A nutrient solution, present in a reservoir in the bottom of the chamber, is applied to the roots as a mist or fog by one of the several pump/spray methods.

The variables to be adjusted in this system are much the same as these for routine greenhouse pot culture. These include host plant, concentration of the nutrient solution, light, temperature, and water quality (Jarstfer and Sylvia, 1995). The authors have used *P. notatum* and cuttings of *Ipomea batatas* (L.) Lam as host plants. Concentration of nutrients in the solution is more critical than in soil culture because the solution mist is always in direct contact with the roots and hyphae. Phosphorus concentrations of 0.3 mM and monitoring the pH to keep it between 5 and 7.5 are recommended.

There are several important advantages of this method over the routine pot culture method of inoculum production. First, one has exacting control over the cultural conditions. Secondly, the presence of plant pests and contaminating microbes is more readily controled and monitored. One merely has to lift the lid of the chamber periodically and remove a sample for analysis. Thirdly, under proper conditions, with optimal host - fungus (notably *Glomus* species) combinations, large quantities of inocula are produced. Sporulation and colonization may be greater than that in soil based pot culture (Sylvia and Hubbell, 1986). Further, this inoculum does not need to be isolated from a substrate such as sand or soil. The inoculum may be sheared in a food processor/blender to maximize propagule numbers (Sylvia and Jarstfer,1992).Sheared aeroponically produced roots of *I. batatas* colonized by *Glomus* sp. (INVAM 925) produced an average of 1.35 x10^5 propagules g-1 dry root. Optimal size of a sheared root piece was 63ìm in this study. If 20 propagules per plant produces an acceptable colonization level, this inocula would cost only $0.05 per 100 plants (Sylvia and Jarstfer, 1992). Results would vary based upon the percentage root length colonized, percentage of viable hyphae in the roots, and, in particular, percentage of root length with vesicles.

5.   NUTRIENT FILM TECHNIQUE

The nutrient film technique (NFT) was adapted for VAM fungus inoculum production by Mosse and Thompson (1984). Culture host plants are placed on an inclined tray over which flows a layer of a nutrient solution. As in the aeroponic method, seedlings must be precolonized in another media due to the lack of a solid substrate. Nutrient solution pH can be adjusted for the requirements of individual isolates. If legumes are used, nutrient solutions without N, e.g. 0.05 to 0.10 strength Hoagland's solution (Hoagland and Arnon, 1938) can be used to ensure against the negative effects of N form upon mycorrizae and solution pH changes due to N uptake. Better colonization of *Phaseolus vulgaris* L. was found when insoluble forms of P such as rock phosphate or bonemeal were utilized to yield final concentrations of P in solution from 11 to 65 ìM. The N source for nonlegumes can be manipulated to help keep the pH within the range that maintains low levels of P from rock phosphate in solution (Elmes and Mosse, 1984). For example, utilizing $(NH_4)_2SO_4$ as the N source for maize colonized by *G. mosseae,* solution pH dropped to 6.1 and P concentration of roots rose to 0.60% dry weight, resulting in negligible infection. Supplying N as $Ca(NO_3)_2 : (NH_4)_2SO_4$ (19:1), solution pH became 6.9. P concentration of roots was 0.23% dry weight and roots were well colonized. This method produced abundant spores and extraradical mycelium of the *Glomus* species tested. Further, the colonized roots are likely to be conducive to the shearing process of Sylvia and Jarstfer, (1992) to increase the efficiency utilization as inocula. However, not all host fungus combinations may grow in this culture system, or may require further refinement of the composition of the nutrient solution for the success (Elmes and Mosse, 1984).

6.   DUAL *IN VITRO* CULTURE OF VAM FUNGI AND ROOTS

The technology available today that is closest to axenic culture of VAM fungi is dual, *in vitro* culture with plant roots. The great advantage of this method of production, i.e. final inocula free of contaminating microorganisms, is matched by the prolific sporulation of some species of VAM fungi under these conditions.
Vesicular-arbuscular mycorrhiza formation under *in vitro* conditions was first reported in the early 1960s (Mosse, 1962) when the presence of *Pseudomonas* sp was considered

necessary to ensure colonization. A bacteria-free symbiosis was achieved by the early 1970s. Mosse and Hepper (1975) reported the use of root organ culture to obtain typical infections with *G. mosseae*. Miller-Wideman and Watrud, (1984) provided new scientific insight by successfully inducing *Gigaspora margarita* to sporulate *in vitro* using tomato roots as host. The root organ culture technique now has greater potential with the utilization of roots genetically transformed by the Ri plasmid of *Agrobacterium rhizogenes* (Mugnier and Mosse, 1987; Bècard and Piche, 1992), i.e. "hairy roots" Mugnier and Mosse (1987) inoculated Ri-T-DNA transformed roots of *Convolvulus sepium* with *G. mosseae* and obtained successful colonization. This was the first utilization of Ri T-DNA transformed roots, but unfortunately, their method required peat in the media.

Bècard and Fortin, (1988) presented an in-depth evaluation of the root organ culture technique and reported basic improvements necessary for VAM fungus colonization of roots. They identified the rate-limiting factors in the colonization of transformed roots of carrot (*Daucus carota* L.) by *G. margarita* Becker and Hall. Subsequent studies identifying factors influencing spore germination, nutitional quality of the medium, nature of inoculum, etc. further helped in fine-tuning the technique and thus improve the reliability and reproducibility of the technique (Bècard and Piche,1989,1990; Chabot *et al.,* 1992b; Diop *et al.,* 1992). Among the species of VAM fungi that have been cultured with Ri-T-DNA transformed roots, in addition to *G. margarita* and *G. mosseae*, are: *Gigaspora gigantea* (Douds and Bècard,1993), *G. etunicatum* (Schreiner and Koide, 1993), *G. intraradices* (Chabot *et al.*, 1992a; St. Arnaud *et al.*, 1996), and *Glomus versiforme* (Declerck *et al.*, 1996b).

Cultures are initiated by the transfer of pregerminated, surface sterilized spores or surface sterilized, colonized root pieces onto petri plates of minimal media (Bècard and Fortin, 1988; Diop *et al.*, 1994) or modified Strullu-Romand medium (Declerck *et al.*, 1996b). Decontamination of highly ornamented spores of those with particularly recalcitrant associated bacteria, is aided by short periods of sonication. Germination and growth of hyphae are enhanced by elevated CO_2 levels (2%) and a temperature of 32 °C. Cultures are typically incubated at 24 °C after inoculation.

Both resulting colonized root tissue and spores may be used as inoculum, though roots colonized by *G. versiforme* were found to be a better inoculum than spores produced *in vitro* (Declerck *et al.*, 1996b). Sporulation differs among VAM fungus species and ranges from 450 per plate after one year of culture for *G. margarita* (Diop *et al.*, 1992) to 9,500 after five months of culture for *G. versiforme* (Declerck *et al.*, 1996b). A recent improvement in the *in vitro* method in which hyphae of *G. intraradices* are allowed to grow into media in a root-free compartment, free of accumulated byproducts of metabolism as the roots age, resulted in as many as 34,000 spores per Petri plate (St Arnaud *et al.*, 1996).

Studies currently underway at Tata Energy Research Institute [TERI] for the mass production of selected VAM fungi have yeilded tangible results. The aim of the study was for the production of VAM inoculum which has the added benifits of the material produced under *in vitro* conditions while maintaining the efficacy. (Fig 2.)

7.   ON-FARM PRODUCTION

The greatest value and potential of VAM fungi lie in the ability of these fungi to improve the efficiency and sustainability of crop production in vast areas of low fertility, especially in developing countries (Jefferies and Dodd, 1991). The most suitable form of inoculum should therefore be produced on-farm, ie. directly on the site of its application using local resources.This approach was used in an extensive project conducted in Columbia

Multiplication in cell culture plates

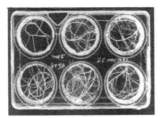

Plate based production

Mass Production in jar

Fig. 2. Stratagies used for the production and bulking of VAM inoculum under *in vitro* conditions.

(Sierverding, 1991) which produced a commercial VAM fungus inoculant named "Manihotina" which contained *Glomus manihotis*.The field production of VAM fungi basically involves the adoption of different farming practices which are important for successful multiplication of VAM fungi.

The choice of VAM fungus for mass production, and the host plant(s) to use, are factors to consider here as in the smaller-scale methods discussed above. Plants such as maize, bahiagrass (*P. notatum*), and sorghum (*Sorghum bicolor* (L.) Moench.) have been suggested as suitable hosts for this type of inoculum production. However, general recommendations cannot be made since soil and climate vary regionally. Thus, locally available materials for inoculum producation must be tested. Therefore, it is necessary to know how to multiply, evaluate, and mass-produce different isolates of VAM fungi under a variety of conditions.

The purpose of on-farm or field production of inocula is to make VAM inocula available in high volume at low cost. If the volume of soil inocula needed to successfully inoculate one ha of cassava, beans, or other agronomic crop can be assumed to be 3000 to 5000 L, the following are needed for the production of this volume of soil inocula:

25 m2 land area
1.25-1.5 kg of fumigant
plastic sheeting
2 kg fertilizer
3.0 - 6.25 kg starter inocula of selected VAM isolate, and
seeds of host plant appropriate to local conditions, that do not share pathogens with the eventual target host crop.

7.1   Procedure of On-farm Production

STEP 1. Preparation of the area: Near or on the field to be inoculated is selected. Clear the vegetation and till the soil over 25m2. Do not apply any fertilizer at this time, and the site is bordered off to prevent trespassing

STEP 2. Soil disinfestation: Use appropriate soil fumigant and apply to the prepared area by broadcasting. The fumigant is incorporated to a depth of 25 cm. Irrigate with a minimum of 20 L per m2 to ensure the fumigant dissolves. Cover the area with plastic sheeting and seal the edges to prevent escape of the fumigant. Uncover the area after 10-14 days and incorporate (without stepping on the soil) inorganic fertilizer to supply the equivalent of 50-50-50-10 kg per ha of N-P-K-Mg. Loosen the soil after 5 days to improve aeration and escape of the residual fumigant.

STEP 3. Inoculation: Inoculae only after the fumigant has completely evaporated (up to 10 days after uncovering). Make small, 3-5 cm deep holes in the soil along a 20 × 20 cm grid and apply 5-10 g of the mycorrhizal fungus starter inoculum to each hole. Sow the seds of a host plant, chosen for its ability to germinate quickly, directly on the inoculum.

STEP 4. Multiplication: Irrigate if the inoculum production process is conducted during the dry season. Clip the flowers to avoid contaminating the inocula with seeds. Apply suitable pesticide if necessary.

STEP 5. Harvest of inoculum: Concentrations of infective propagules will be the greatest after 4-6 months. Cut the host plant at ground level and harvest the soil substrate, including roots, to a depth of 20 cm. This exercise yields 5000 L of soil inocula from the 25 m2 area. Reclaim the production site with compost.

7.2. On-farm Production at TERI: Pilot Project

A long term experiment was conducted at Tata Energy Research Institute (TERI) to develop an on-farm system for local edapho-climatic conditions. Multiplication rates of various indigenous and exotic isolates of VAM fungi were compared, and suitable hosts were tested for multiplication of the VAM fungi in diferent seasons. Indigenous and exotic VAM fungus isolates were maintained in the nursery from March 1993 through February 1996 with three 4-month multiplication cycles each year. The same host plants were used for all VAM fungus isolates in a single multiplication cycle. The hosts used in the three cycles were, sudangrass (*Sorghum sudanese* (Piper) Stapf.), maize and carrot in the first year; maize, sudangrass and onion (*Allium cepa* L.) the second year; and sudangrass, maize and oats (*Avena sativa* L.) in the third year.

A nursery area of 3 × 9 m was selected and prepared as outlined above. Twenty seven raised beds (60 × 60 × 16 cm) were constructed. The experiment was a completely randomized design with 9 VAM fungus isolates and three replicate beds per isolate. The VAM fungi used in the experiment included seven exotics: *Glomus etunicatum* (8969 from Native Plant Industries, Salt Lake City, UT, USA), *Glomus intraradices* (8975 from Native Plant Industries), *G. intraradices* (DAOM 197198 from Laval University, Quebec, Canada), *Glomus fasciculatum* (from Gainesville, FL, USA), *Gigaspora margarita* (INVAM 185), and *Glomus mosseae* (INVAM 156), *G. mosseae* (INVAM 336) and indigenous inocula of *Glomus caledonium* and a mixed indigenous consortium containing *Glomus*, *Gigaspora*, and *Scutellospora* species. These fungi were initially cultured for three months in earthen pots (20 x 24 cm) containing sterilized nursery soil. Inoculum for the pilot study therefore contained spores, hyphae, and infected root pieces from the starter cultures.

The inoculum of the VAM fungus isolate assigned to a bed was placed in furrows in the bed. Surface sterilized seeds of the host species were sown directly on the layer of inoculum and covered with nursery soil. The beds were subsequently irrigated at regular intervals. After the three month production cycle, irrigation was stopped and the crop was allowed to dry. After one week, the shoots were removed and the roots were chopped into 1 cm pieces and remixed in the soil. The soil was then again made into beds for the next multiplication cycle.

A variety of patterns of mycorrhization of the host plants was seen. Colonization of roots by *G. intraradices* 8975 increased through the three years from 27% root length of sorghum in the first cycle to 68% of the root length of oats in the final cycle. Other patterns reflected the greenhouse observations that there are optimal host x fungus combinations for colonization levels (Pope *et al.*, 1983). Overall colonization of roots was greatest with the indigenous consortium, 68% root colonization on carrot in year one, 86% colonization of onion in the second year two, and 87% colonization of sorghum in the final year.

Inoculum production in nursery beds was achieved (Table 1). Interestingly, the two indigenous inocula, the consortium and *G. caledonium*, produced the greatest number of propagules indicating their adaptation to the site. Variability in the number of infectious propagules from cycle to cycle could be attributed to environmental conditions, host specificity (Hetrick and Bloom,1986; Simpson and Daft, 1990; Struble and Skipper., 1988) and seasonal fluctuations (Dodd and Jeffries, 1986; Jacobsen and Nielsen., 1983; Sylvia., 1986). Physiological condition of the inocula (e.g., dormancy) may also cause problems in assessment of the numbers of propagules.

Table1: Production of VAM fungi at the end of a three year pilot study of on-farm inoculum production at TERI, New Delhi, India

AM fungus	Total infectious propagules $(\times 10^5)$	Increase from first to third year (fold)
Indigenous mix	25	33
Glomus caledonium	22	24
G. intraradices (DAOM 197198)	16	47
Gig. margarita (INVAM 185)	13	35
Glomus fasciculatum	10	40
G. mosseae (Invam 336)	9	15
G. mosseae (Invam 156)	9	23
G. etunicatum (NPI 8969)	8	21
G. intraradices (NPI 8975)	6	34

8. CONCLUSION

Vesicular-arbuscular mycorrhizal fungi are symbiotic soil fungi which have great potential utility in plant production. Their obligate symbiotic nature currently prohibits axenic culture. Inoculum production technologies available at present include soil-based methods, variations of hydroponic culture, and dual *in vitro* culture with Ri-T-DNA transformed roots. Each method has advantages and limitations which must be weighed, as well as production variables to be optimized, before one attempts mass production of VAM fungus inocula.

REFRENCES

Allen, M.F., Sexton, J.C., Moore, T.S. and Christensen, M. 1981, Influence of phosphate source on vesicular-arbuscular mycorrhizae of *Bouteloua gracilis*, *New Phytol.* 87:687-694.

Aziz, T. and Habte, M. 1989, The sensitivity of three vesicular-arbuscular mycorrhizal species to simulated erosion, *J. Plant Nutri.* 12: 859-870.

Baltruschat, H. 1987, Evaluation of the suitability of expanded clay as a carrier material for VA mycorrhiza spores in field inoculation of maize, *Ange. Bota.* 61: 163-169.

Baltruschat, H. and Dehne, H.W. 1988, The occurrence of vesicular-arbuscular mycorrhiza in agro-ecosystem, I, Influence of nitrogen fertilization and green manure in continuous monoculture and in crop rotation on the inoculum potential of winter wheat, *Plant Soil* 107: 279-284.

Bècard, G., Douds, D.D. and Pfeffer, P.E. 1992a, Extensive *in vitro* hyphal growth of vesicular-srbuscular mycorrhizal fungi in the presence of CO_2 and flavonols, *App. Environl. Microb.* 58: 821-825.

Bècard, G. and Piché, Y. 1992b. Establishment of vesicular-arbuscular mycorrhiza in root organ culture: review and proposed methodology, in: *Techniques for the Study of Mycorrhiza,* J. Norris, D. Read and A. Varma, eds., Academic Press, New York. pp.89-108.

Bècard, G. and Piché, Y. 1989, New aspects on the acquition of biotrophic status by a vesicular-arbuscular mycorrhizal fungus, *Gigaspora margarita*, *New Phytol.* 112: 77-83.

Bècard, G. and Fortin, J.A. 1988, Fungal growth stimulation by CO_2 and root exudates in vesicular-arbuscular mycorrhizal symbiosis, *Appl. Environ. Microb*, 55: 2320-2325.

Bethlenfalvay, G.J. and Schuepp, H. 1994, Arbuscular mycorrhizas and agrosystem stability, in: *Impact of Arbuscular Mycorrhizas on Sustainable Agriculture and Natural Ecosystems,* S. Gianinazzi and H. Schuepp, eds., BirkhŠuser Verlag, Basel pp.117-131.

Bevenge, D.I., Bowen, G.D. and Skinner, M.F. 1975, In: *Comparative Carbohydrate Physiology of Ecto- and Endomycorrhizas,* F.E. Sanders, B. Mosse and P.B. Tinker, eds., Academic Press, New York. pp.149-174.

Bever, J.D., Morton, J.B., Antonovics, J. and Schultz, P.A. 1996, Host-dependent sporulation and species diversity of arbuscular mycorrhizal fungi in a mown grassland, *J. Ecol.* 84: 71-82.

Bhatia, N. P., Sundari, K. and Adholeya, A. 1996, Diversity and selective dominance of vesicular-arbuscular fungi, in: *Concepts in Mycorrhizal Research*, K G Mukerji, ed., Kluwer Academic Publishers, Netherlands pp. 133-178.

Bolan, N.S. 1991, A critical review on the role of mycorrhizal fungi in the uptake of phosphorus by plants, *Plant Soil* 134: 189-208.

Caron, M. and Parent, S. 1988, Definition of a peat-based medium for the use of vesicular-arbuscular mycorrhizae (VAM) in horticulture, *Acta Horticul.* 221: 289-294.

Cawse, P.A. 1975, Microbiology and biochemistry of irradiated soils, in: *Soil Biochemistry*, Vol 3, E.A. Paul and A.D. McLaren, eds., Marcel Dekker, Inc. New York, pp. 213-267.

Chabot, S., Bècard, G. and Pichè, Y. 1992a, Life cycle of *Glomus intraradices* in root organ culture, *Mycologia* 84: 315-321.

Chabot, S., Bel-Rhlid, R., Chenevert, R. and Pichè, Y. 1992b, Hyphal growth promotion *in vitro* of the VA mycorrhizal fungus, *Gigaspora margarita* Becker and Hall, by the activity of structurally specific flavonoid compounds under CO_2-enriched conditions, *New Phytol.* 122: 461-467.

Clark, R.B. and Zeto, S.K. 1996, Growth and root colonization of mycorrhizal maize grown on acid and alkaline soil, *Soil Biol. Biochem.* 28: 1505-1511.

Connick, W.J., Daigle, D.J., Boyette, C.D., Williams, K.S., Vinyard, B.T. and Quimby, P.C. 1996, Water activity and other factors that affect the viability of *Colletotrichum truncatum* conidia in wheat flour-kaolin granules ('Pesta'), *Biocon. Sci. Biotech.* 6: 277-284.

Connick, W.J., Boyette, C.D. and McAlpine, J.R. 1991, Formulation of mycoherbicides using a pasta-like process, *Biol Cont.* 1: 281-287.

Connick, W.J., Nickle, W.R., Williams, K.S. and Vinyard, B.T. 1994, Granular formulations of *Steinernema carpocapsae* (strain A11) (Nematoda: Rhabditida) with improved shelf-life, *J. Nematol.* 26: 352-359.

Davies, F.T., Potter, J.R. and Linderman, R.G. 1993, Drought resistance of mycorrhizal pepper plants- independent of leaf phosphorus concentration resonse in gas exchange and water relations, *Physiol Plantarum* 87: 45-53.

Declerck, S., Strullu, D.C., Plenchette, C. and Guillemette, T. 1996a, Entrapment of *in vitro* produced spores of *Glomus versiforme* in alginate beads: *in vitro* and *in vivo* inoculum potentials, *J. Biotechnol.* 48: 51-57.

Declerck, S., Strullu, D.C and Plenchette, C. 1996b, *In vitro* mass production of arbuscular mycorrhizal fungus, *Glomus versiforme* with Ri-T-DNA transformed carrot roots agriculture and natural ecosystems associated, *Mycol. Res.* 100: 1237-1242.

Dehne, H. W. and Backhaus, G.F. 1986, The use of vesicular-arbuscular mycorrhizal fungi in plant production, I, Inoculum production. *Z. fr Pflanzenkrankheiten und Pflanzenschutz.* 93: 415-424.

Diedrichs, C. and Moawad, A.M. 1993, The potential of VA mycorrhizae for plant nutrition in the tropics, *Angew Bot.* 67: 91-96.

Diop, T. A., Bècard. G. and Pichè, Y. 1992, Long-term *in vitro* culture of an endomycorrhizal fungus, *Gigaspora margarita*, on Ri-T-DNA transformed roots of carrot, *Symbiosis* 12: 249-259.

Diop, T. A., Plenchette, C. and Strullu, D. G. 1994, Dual axenic culture of sheared-root inocula of vesicular-arbuscular mycorrhizal fungi associated with tomato roots, *Mycorrhiza* 5: 17-22

Dodd, J.C. and Jeffries, P. 1986, Early development of vesicular-arbuscular mycorrhizas in autumn-sown cereals, *Soil Biol. Biochem.* 18: 149-154

Doner, L.W. and Douds, D.D. 1995, Purification of commercial gellan to monovalent cation salts in acute modification of solution and gel forming properties, *Carbohydrate Res.* 273: 225-233.

Douds, D.D., Janke, R.R. and Peters, S.E. 1993, VAM fungus spore populations and colonization of roots of maize and soybean under conventional and low-input sustainable agriculture, *Agric. Ecosyst. Environ.* 43: 235-335.

Douds, D.D. and Schenck, N.C. 1990, Increased sporulation of vesicular-arbuscular mycorrhizal fungi by manipulation of nutrient regimens, *App. Environ. Microbiol.* 56: 413-418

Douds, D.D. and Bècard, G. 1993, Competitive interactions between *Gigaspora margarita* and *Gigaspora gigantea in vitro*, in : *Proceedings of the Ninth North American Symposium on Mycorrhizae*, Guelph, ONT, Canada, August 8-12, 1993

Elmes, R.P. and Mosse, B. 1984, Vesicular-arbuscular endomycorrhizal inoculum production, II, Experiments with maize (*Zea mays*) and other hosts in nutrient flow culture, *Can. J. Bot.* 62: 1531-1536.

Elmeskaoui, A., Damont, J.P., Puolin, M.J., Piché, Y. and DesJardins, Y. 1995, A tripartite culture system for endomycorrhizal inoculation of micropropagated strawberry plantlets, *in vitro*, *Mycorrhiza* 5: 313-319.

Ferriss, R.S. 1984, Effects of microwave oven treatment on microorganisms in soil, *Phytopathol.* 74: 121-126.

Galvez, L., Douds, D.D., Wagoner, P., Longnecker, L.R., Drinkwater, L.E. and Janke, R.R. 1995, An overwintering cover crop increases inoculum of VAM fungi in agricultural soil, *Am. J. Alter. Agricul.* 10: 152-156.

Gould, A.B., Hendrix, J.W. and Ferriss, R.S. 1996, Relationship of mycorrhizal activity to time following reclamation of surface mine land in western Kentucky, I, Propagule and spore population densities, *Can. J. Bot.* 74: 247-261.

Graham, J.H. and Timmer, L.W. 1985, Rock phosphate as a source of phosphorus for vesicular-arbuscular mycorrhizal development and growth of citrus in a soilless medium, *J. Am. Soc. Horticul. Sci.* 110: 489-492.

Graham, J.H., Leonard, R.T. and Menge, J.A. 1981, Membrane-mediated decrease in root exudation responsible for phosphorus inhibition of vesicular-arbuscular mycorrhiza formation, *Plant Physiol.* 68: 548-552.

Grunewaldt-Stöcker, G. and Dehne, H.W. 1989, The use of vesicular-arbuscular mycorrhizal fungi in plant production, II, Characterization of inocula on inorganic carrier material, *Z. Pflanzenkrankheiten und Pflanzenschutz* 96: 615-626.

Harley, J.L. and Harley, E.L. 1987, A check-list of mycorrhiza in the British flora, *New Phytol.* 105 (suppl.): 1-102.

Heinzemann, J. and Weritz, J. 1990, Rockwool: and new carrier system for mass multiplication of vesicular-arbuscular mycorrhizal fungi, *Angew. Bot.* 64: 271-274.

Hetrick, B.A.D. and Bloom, J. 1986, The influence of host plant on production and colonization ability of vesicular-arbuscuar mycorrhizal spores, *Mycologia* 78: 32-36.

Hoagland, D.R. and Arnon, D.I. 1938, The water-culture method for growing plants without soil, Univ. of California, College of Agriculture, Agriculture Experimental Station Circular 347. Berkley, CA.

Hooker, J.E. and Black, K.E. 1995, Arbuscular mycorrhizal fungi as components of sustainable soil-plant systems, *Cri. Rev. Biotech.* 15: 201-212.

Hung, L.L., O'Keefe, D.M. and Sylvia, D.M. 1991, Use of Hydrogel as a sticking agent and carrier for vesicular-arbuscular mycorrhizal fungi, *Mycol. Res.* 95: 427-429.

Hung, L.L, Sylvia, D.M. and O'Keefe, D.M. 1990, Isolate selection and phosphorus interaction of vesicular-arbuscular mycorrhizal fungi in biomass crops, *Soil Sci. Soc. Am. J.* 54: 762-768.

Hung, L.L, and Sylvia, D.M. 1988, Inoculum production of vesicular-arbuscular mycorrhizal fungi in aeroponic culture, *Appl. Environ. Microbiol.* 54: 353-357.

Jakobsen, I. and Nielsen, N.E. 1983, Vesicular-arbuscular mycorrhiza in field grown crops-I, Mycorrhizal infection in cereals and peas at various times and soil depths, *New Phytol.* 93: 401-413.

Jakobsen, I. and Andersen, A.J. 1982, Vesicular-arbuscular mycorrhiza and growth in barley: effects of irridiation and heating of soil, *Soil Biol. Biochem.* 14: 171-178.

Jarstfer, A.G. and Sylvia, D.M. 1995, Aeroponic culture of VAM fungi, in: *Mycorrhiza: Structure, Function, Molecular Biology and Biotechnology*, A. Varma and B. Hock, eds., Springer-Verlag, Berlin pp. 427-441.

Jasper, D.A., Robson, A.D. and Abbott, L.K. 1987, The effect of surface mining on the infectivity of vesicular-arbuscuar mycorrhizal fungi, *Aust. J. Bot.* 6: 641-652.

Jeffries, P. and Dodd, J.C. 1991, The use of mycorrhizal inoculants in forestry and agriculture, in: *Handbook of Applied Mycology,* Vol. 1, Soil and plants, D.K Arora, B. Rai, K.G. Mukerji and G.R. Knudsen, eds., Marcel Dekker Inc, New York, pp. 155-186.

Johnson, N.C. 1993, Can fertilization of soil select less mutualistic mycorhizae ? *Ecol. Appl.* 3: 749-757.

Johnson, C.R. and Menge, J.A. 1982, Mycorrhizae may save fertilizer dollars, *Am. Nurseryman* 155: 79-86.

Koch, K.E. and Johnson, C.R. 1984, Photosynthate partitioning in split-root citrus seedlings with mycorrhizal and non-mycorrhizal root systems, *Plant Physiol.* 75: 26-30.

Kurle, J.E. and Pfleger, F.L. 1994, Arbuscular mycorrhizal fungus spore populations respond to conversions between low-input and conventional management practices in a corn-soybean rotation, *Agronomy J.* 86: 467-475.

Li, X.L., George, E. and Marschner, H. 1991, Extension of the phosphorus depletion zone in VA-mycorrhizal white clover in a calcareous soil, *Plant Soil* 136: 41-48.

Lloyd-MacGilp, S.A., Chambers, S.M., Dodd, J.C., Fitter, A.H., Walker, C. and Young, J.P.W. 1996, Diversity of the ribosomal transcribed spacers within and among isolates of *Glomus mosseae* and related mycorrhizal fungi, *New Phytol.* 133: 103-111.

Lopez, A., Pinochet, J., Fernandez, C., Calvert, C. and Camprubi, A. 1997, Growth response of OHF-333 pear rootstock to arbuscular mycorrhizal fungi, phosphorus nutrition and *Pratylenchus vulnus* infection. *Fundamen, App. Nematol.* 20: 87-93.

Lumini, E., Bosco, M., Puppi, G., Isopi, R., Frattegiani, M., Buresti, E. and Favilli, F. 1994, Field performance of *Alnus cordata* Loisel (Italian alder) inoculated with *Frankia* and VA-mycorrhizal strains in minespoil afforestation plots, *Soil Biol. Biochem.* 26: 659-661.

Maddox, J.J. and Soileau, J.M. 1991, Effects of phosphate fertilization, lime ammendments and inoculation with VA-mycorrhizal fungi on soybeans in an acid soil, *Plant Soil* 134: 83-93.

Mark, G.L. and Cassells, A.C. 1996, Genotype-dependence of the interaction between *Glomus fistulosum*, *Phytophthora fragariae* and the wild strawberry (*Fragaria vesca*), *Plant Soil* 185: 233-239.

McGonigle, T.P. and Miller, M.H. 1993, Mycorrhizal development and phosphorus absorption in maize under conventional and reduced tillage, *Soil Sci. Soc. Am. J.* 57: 1002-1006.

Menge, J.A. 1982, Effect of soil fumigants and fungicides on vesicular-arbuscular fungi, *Phytopatho.* 72: 1125-1132.

Menge, J.A. 1984, Inoculum production, in: *VA Mycorrhiza*, C.L. Powell and D.J. Bagyaraj, eds., CRC Press, Boca Raton, FL. pp. 187-203.

Millar-Wideman, M. A. and Watrud, 1984, Sporulation of *Gigaspora margarita* in root cultures of tomato, *Can. J. Bot.* 30: 642-646.

Miller, R.M. and Jastrow, J.D. 1992, The role of mycorrhizal fungi in soil conservation, in: *Mycorrhizae in Sustainable Agriculture*, G.J. Bethlenfalvay and R.G. Linderman, eds., Agronomy Society of America Special Publication No. 54, Madison, WI. pp 29-44.

Millner, P.D. and Kitt, D.G. 1992, The Beltsville method for soilless production of vesicular-arbuscular mycorrhizal fungi, *Mycorrhiza* 2: 9-15.

Monzon, A. and Azcon, R. 1996, Relevence of mycorrhizal fungal origin and host plant genotype to inducing growth and nutrient uptake in *Medicago* species, *Agricult. Ecosyst. Environ.* 60: 9-15.

Morton, J.B. and Bentivenga, S.P. 1994, Levels of diversity in endomycorrhizal fungi (Glomales, Zygomycetes) and their role in defining taxonomic and non-taxonomic groups, *Plant Soil* 159: 47-59.

Mosse, B 1962, The establishment of mycorrhizal infection under asceptic conditions, Rothamsted Experimental Station report for 1961, p 80.

Mosse, B. and Hepper, C.M. 1975, Vesicular arbuscular infection in root organ cultures, *Physiol. Plant Pathol.* 5:215-223.

Mosse, B. and Thompson, J.P. 1984, Vesicular-arbuscular endomycorrhizal inoculum production, I, Exploratory experiments with beans (*Phaseolus vulgaris*) in nutrient flow culture, *Can. J. Bot.* 62:1523-1530.

Mosse, B. 1973, Advances in the study of vesicular-arbuscular mycorrhiza, *Ann. Rev. Phytopathol.* 11: 171-196.

Mugnier, J. and Mosse, B. 1987, Vesicular-arbuscular mycorrhizal infections in transformed Ri-T-DNA roots grown axenically, *Phytopathol.* 77: 1045-1050.

Nagahashi, G., Douds, D.D. and Abney, G.D. 1996, Phosphorus amendment inhibits hyphal branching of the VAM fungus *Gigaspora margarita* directly and indirectly through its effect on root exudation, *Mycorrhiza* 6: 403-408.

Newsham, K.K., Fitter, A.H. and Watkinson, A.R. 1995, Arbuscular mycorrhiza protect an annual grass from root pathogenic fungi in the field, *J. Ecol.* 83: 991-1000.

Niemi, M. and Vestberg, M. 1992, Inoculation of commercially grown strawberry with VA mycorrhizal fungi, *Plant Soil* 144: 133-142

Payne, G.G. and Reichcigl, J.E. 1989, Influence of various drying techniques on the extractability of plant nutrients from selected Florida USA soils, *Soil Science* 148: 275-283.

Perrin, R., Duvert, P. and Plenchette, C. 1988, Substrate receptiveness to mycorrhizal association: concepts, methods and application, *Acta Horticult.* 221: 223-228.

Perrin, R. and Plenchette, C. 1993, Effect of some fungicides applied as soil drenches on the mycorrhizal activity of two cultivated soils and their receptiveness to *Glomus intraradices*, *Crop Protec.* 12: 127-133.

Plenchette, C. and Morel, C. 1996, External phosphorus requirement of mycorrhizal and non-mycorrhizal barley and soybean plants, *Biol. Fert. Soils* 21: 303-308.

Pope, P.E., Chaney, W.R., Woodhead, S. and Rhodes, J.D. 1983, Vesicular-arbuscular mycorrhizal species influence the mycorrhizal dependency of four hardwoood tree species, *Can. J. Bot.* 61:412-417

Rhodes, L.H. and Gerdemann, J.W. 1975, Phosphate uptake zones of mycorrhizal and non-mycorrhizal onions, *New Phytol.* 75: 555-561.

Rovira, A.D. and Bowen, G.D. 1966, The effects of microorganisms upon plant growth, II, Detoxification of heat sterilized soils by fungi and bacteria, *Plant Soil* 25: 129-142.

Schreiner, R.P. and Bethlenfalvay, G.J. 1995, Mycorrhizal interactions in sustainable agriculture, *Crit. Rev. Biotech.* 15: 271-285.

Schreiner, R.P and Bethlenfalvay, G.J. 1996, Mycorrhizae, biocides, and biocontrol, Response of a mixed culture of arbuscular mycorrhizal fungi and host plant to three fungicides, *Biol. Fert. Soils* 23: 189-195.

Seymour, N.P., Thompson, J.P. and Fiske, M.L. 1994, Phytotoxicity of Fosetyl Al and phosphonic acid to maize during production of vesicular-arbuscular mycorrhizal inoculum, *Plant Disease* 78: 441-446.

Sharma, M.P., Gaur A., Bhatia, N.P. and Adholeya, A. 1996, Growth response and dependance of *Acacia nilotica* var. *cupriciformis* on the indigenous arbuscular mycorrhizal consortium of a marginal wasteland soil, *Mycorrhiza* 6: 441-446.

Sieverding, E. 1991, Vesicular-arbuscular mycorrhiza management in tropical agrosystems, Deutshe GTZ Eshborn, Germany, p. 371

Simpson, D. and Daft, M.J. 1990, Spore production and mycorrhizal development in various tropical crop hosts infected with *Glomus clarum, Plant Soil* 121: 171-178.

St-Arnaud, M., Hamel, C., Vimard, B., Caron, M. and Fortin, J.A., 1996, Enhanced hyphal growth and spore production of arbuscular mycorrhizal fungus *Glomus intraradices* in an *in vitro* system in the absence of host roots, *Mycol. Res.*100: 328-332.

Struble, J. and Skipper, H. 1988, Vesicular-arbuscular mycorrhizal fungal spore production as influenced by plant species, *Plant Soil* 109: 277-280.

Sylvia, D.M. 1986, Spatial and temporal distribution of vesicular-arbuscular mycorrhizal fungi associated with *Uniola paniculata* in Florida foredune, *Mycologia* 78: 728-734.

Sonneveld, C. 1979, Changes in chemical properties of soil caused by steam sterilization, in: *Soil Disinfestation*, D. Mulder, ed., Elsevier,NY. pp. 39-50.

Speir, T.W., Cowling, J.C., Sparling, G.P., West, A.W. and Corderoy, D.M. 1986, Effects of microwave radiation on the microbial biomass, phosphatase activity, and levels of extractable nitrogen and phosphorus in a low fertility soil under pasture, *Soil Biol. Biochem.* 18: 377-382.

Spokes, J.R., Hayman, D.S. and Kandasamy, D. 1989, The effects of fungicide-coated seeds on the establishment of VA mycorrhizal infection, *Ann. App. Bio.* 115: 237-241.

Strullu, D.G., Romand, C. and Plenchette, C. 1991, Axenic culture and encapsulation of the intraradical forms of *Glomus* spp., *World J. Microbiol. Biotechnol.* 7: 292-297.

Subramanian, K.S., Charest, C., Dwyer, L.M. and Hamilton, R.I. 1995, Arbuscular mycorrhizal and water relations in maize under drought stress at tasselling, *New Phytol.* 129: 643-650.

Sukarno, N., Smith, S.E. and Scott, E.S. 1993, The effect of fungicides on vesicular-arbuscular mycorrhizal symbiosis, I, The effects on vesicular-arbuscular mycorrhizal fungi and plant growth, *New Phytol.* 25: 139-147.

Sylvia, D.M. 1984, Production of inocula of VA mycorrhizal fungi, in: *Applications of Mycorrhizal Fungi in Crop Production,* J.J. Ferguson, ed., University of Florida, Gainesville, FL. p 8-16.

Sylvia, D.M. and Schenck, N.C. 1984, Aerated-steam treatment to eliminate VA mycorrhizal fungi from soil, *Soil Biol. Biochem.* 16: 675-676.

Sylvia, D.M. and Jarstfer, A.G. 1992, Sheared-root inocula of vesicular-arbuscular mycorrhizal fungi, *App. Environ. Microbiol.* 58: 229-232.

Sylvia, D.M., Wilson, D.O., Graham, J.H., Maddox, J.J., Millner, P., Morton, J.B., Skipper, H.D., Wright, S.F. and Jarstfer, A.G. 1993, Evaluation of vesicular-arbuscular mycorrhizal fungi in diverse plants and soils, *Soil Biol. Biochem.* 25: 705-713.

Taylor, T.N., Remy, W., Hass, H. and Kerp, H. 1995, Fossil arbuscular mycorrhizae from the Early Devonian, *Mycologia* 87: 560-573.

Thompson, J.P. 1987, Decline of vesicular-arbuscular mycorrhizae in long fallow disorder of field crops and its expression in phosphorus deficiency of sunflower, *Austr. J. Agricul. Res.* 38: 847-867.

Timmer, L.W. and Leyden, R.F. 1978, Stunting of citrus seedlings in fumigated soils in Texas and its correction by phosphorus fertilization and inoculation with mycorrhizal fungi, *J. Am. Soc. Horticult. Sci.* 103: 533-537.

Timmer, L.W. and Leyden, R.F. 1980, The relationship of mycorrhizal infection to phosphorus-induced copper deficiency in sour orange seedlings, *New Phytol.* 85: 15-32.

Tisdall, J.M. and Oades, J.M. 1979, Stabilization of soil aggregates by the root systems of ryegrass, *Austr. J. Soil. Res.* 17: 429-441.

Trotta, A., Varese, G.C., Gnavi, E., Fusconi, A., Sampo, S. and Berta, G. 1996, Interactions between the soilborne root pathogen *Phytophthora nicotianae* var. *parasitica* and the arbuscular mycorrhizal fungus *Glomus mosseae* in tomato plants, *Plant Soil* 185: 199-209.

Vestberg, M. and Uosukainen, M. 1992, A new method for producing VA-mycorrhiza inoculum in a soil-free substrate, *Mycologist* 6:38.

Vivekanandan, M. and Fixen, P.E. 1991, Cropping systems effects on mycorrhizal colonization, early growth, and phosphorus uptake of corn, *Soil Sci. Soc. Am. J.* 55: 136-140.

Weinbaum, B.S., Allen, M.F. and Allen, E.B. 1996, Survival of arbuscular mycorrhizal fungi following reciprocal transplanting across the Great Basin, USA, *Ecolo. Appli.* 6: 1365-1372

West, H.M., Fitter, A.H. and Watkinson, A.R. 1993, Response of *Vulpia ciliata* ssp. *ambigua* to removal of mycorrhizal infection and to phosphate application under natural conditions, *J. Ecol.* 81: 351-358.

Wetterauer, D.G. and Killorn, R.J. 1996, Fallow- and flooded-soil syndromes: effects on crop production, *J. Prod. Agric.* 9: 39-41.

Wolf, D.C., Dao, T.H., Scott, H.D. and Lavy, T.L. 1989, Influence of sterilization methods on selected microbiological, physical and chemical properties, *J. Environ. Quality* 18: 39-44.

Yeager, T.H., Johnson, C.R. and Schenck, N.C. 1990, Growth response of *Podocarpus* and *Ligustrum* to VA mycorrhizae and fertilization rate, *J. Environ. Horticul.* 3: 128-132.

Pope, P.E., Chaney, W.R, Rhodes J.D. and Woodland, S.H. 1983, The mycorrhizal dependency of four hardwood tree species, *Can. J. Bot.* 412-417.

MYCORRHIZAL TECHNOLOGY IN PLANT MICROPROPAGATION SYSTEM

Nikhat Sarwar Naqvi and K.G.Mukerji

Applied Mycology Laboratory
Department of Botany
University of Delhi
Delhi 110 007, INDIA

1.   INTRODUCTION

The present growing concern about the negative effects of various chemicals which are used in plant production on environment is promoting the development of sustainable management practices. The approach for sustainable plant production systems to ensure increased productivity is through the development of alternative strategies by decreasing the chemical inputs to economic but non- polluting levels. One such strategy is the better exploitation or management of biological components i.e. microbes present in soil which contribute to soil fertility (Mulongoy *et al.*, 1992). The microorganisms present in the rhizosphere viz. several bacteria, actinomycetes and fungi, play an important role in increasing plant growth. Certain soil microbes colonize and form symbiotic associations with roots of the plants like root nodule bacteria and mycorrhiza. Mycorrhizae are mutualistic associations occurring between roots of most plant species and certain groups of fungi. The major beneficial effects associated with the use of mycorrhiza include : better absorption of moisture and nutrients from soil, reduction of pathogenic root infections and improvement of soil structure. These effects, in turn, result in increased survival rates at outplanting as well as enhancing growth of plants with reduced fertilizer inputs. Thus, mycorrhizal fungi are biological alternative to expensive chemical fertilizers and pesticides used in improving productivity of plants (Mukerji *et al.*, 1991; Mukerji and Sharma, 1996; Srivastava *et al.*, 1996).

Over the past few years the use of *in vitro* micropropagation technique for growth and multiplication of wide range of plants - Plantation crops, ornamentals, fruits, vegetables and forest trees, has increased rapidly (Paranjothy *et al.*, 1990). The tissue culture technique is not only being used as research tool in plant improvement programmes but has also been widely adopted commercially for micropropagation and mass multiplication of varieties of plants. One of the most common problems associated with *in vitro* micropropagated plants, is the low survival rate and poor growth while shifting these plants to field conditions. The low survival results in high losses to the industry and efforts to increase survival and growth require increased fertilizer and pesticides inputs. The role of mycorrhizae in improving the

Mycorrhizal Biology, edited by Mukerji *et al.*
Kluwer Academic/Plenum Publishers, 2000

growth and productivity of seed grown plants, as a natural symbiont has been well established (Srivastava *et al.,* 1996). During last few years, the mycorrhizal technology has been used in a number of micropropagated horticultural, plantation, forest trees and in many other crops in order to improve their survival and growth and to minimize the fertilizer and pesticides inputs in micropropagation system (Puthur *et al.,* 1998; Subhan *et al.,* 1998; Varma and Schuepp, 1994)) (Table 1).

2.   MYCORRHIZAL SYMBIOSIS - GENERAL CONCEPT AND IMPORTANCE

Mycorrhizae are mutualistic associations occurring between roots of most plant species and certain groups of fungi. They are of world wide distribution in the plant kingdom and are reported from Bryophytes, Pteridophytes, many Gymnosperms and most Angiosperms. They are one of the most ancient associations having very long evolutionary history (Fitter, 1991; Marks, 1991; Varma and Shanker, 1994). Morphologically 7 different types of mycorrhizae are classified on the basis of nature of fungal hyphae, extent of penetration of roots, production of external fungal sheath and various intercellular and intracellular structures produced by the mycobiont upon association with root (Harley and Smith, 1983). Among the seven types of mycorrhizae, vesicular arbuscular mycorrhiza (VAM) is the most important ecologically and economically. It is widespread in distribution and occur in plants growing in arctic, temperate, subtropical and tropical regions (Read, 1991). Because most agricultural, horticultural and forest tree species are colonized by them, they are attracting much attention (Raman and Mahadevan, 1996). The fungi responsible for the formation of VAM have been placed in the order Glomales of division Zygomycotina (Marton and Benny, 1990; Morton and Bentivenga, 1994; Mukerji, 1995). Lot of work has been done in the past few decades to show that this association is beneficial to plants in many ways. Increase in uptake of soil phosphorus by mycorrhizal roots is well documented (Bolan, 1991; Marshner and Dell, 1994). Not only P, but uptake of several other poorly mobile micronutrients like Cu, Zn, Fe and Mn are increased in presence of VAM fungi (Hussey and Roncadori, 1982; Mittal *et al.,* 1987). Suppressed uptake of Mn has also been reported thereby protecting the plant from toxic effect of high concentration of Mn (Arines *et al.,* 1989; Posta *et al.,* 1994). It also improves the plant water relations and provide protection to plant against soil toxins (Filter, 1991; Jeffries and Dodd, 1991). The extramatrical hyphae of VAM fungi can improve soil structure by binding the soil particles into more stable aggregates (Tisdall, 1994). VAM fungi have a potential to biologically suppress the root pathogens (Bansal and Mukerji, 1994; Fitter and Garbage, 1994; Hooker *et al.,* 1994a). The plants earlier inoculated with VAM fungi have been shown to exhibit increased resistance to several root pathogens like *Phytophthora* spp. *Macrophomina* spp., *Aphanomyces* spp., *Fusarium* spp., *Verticillium* spp., and *Rhizoctonia* spp. (Jalali and Jalali, 1991). In addition these fungi also provide resistance to plants against nematodes (Berkert and Robsen, 1994; Krishna and Bagyaraj, 1984; Li *et al.,* 1991; Thompson, 1994). VAM fungi are also reported to have an important implication in reforestation of wastelands. Potential for increasing the growth of various reforestation crops by mycorrhizal inoculation and management in semiarid and tropical zones is well recognized (Jagpal and Mukerji, 1991; Mukerji and Dixon, 1992; Sarwar and Mukerji, 1995; 1998). The potential of VAM in increasing crop yield may also be an alternative to rising fertilizer costs (Menge, 1983). VAM fungi are also known to interact synergistically with certain soil micro-organisms like phosphate-solubilizing bacteria and N_2 fixing micro-organisms (Azcon *et al.,* 1976; Raj *et al.,* 1981). There is enough literature available on the effect of VAM on nodulation, N_2 fixation and growth of legumes (Azcon *et al.,* 1991; Subbu Rao and Krishna, 1990).

3.   MICROPROPAGATION SYSTEM

Among the techniques of biotechnology, micropropagation is the most successful example, has established itself as multi million dollar industry. This technique is used widely and successfully as a research tool as well as by private companies on commercial scale for the mass production of wide range of plant species which include horticultural, medicinal, ornamental, cereals, tubers, spices, forest and foliage, etc. (Bajaj, 1986). In Europe alone 175 million plants belonging to 30 genera were raised during 1992 from this technique. There was 3 - 4 fold increase in number of commercial and official micropropagation laboratories between 1982 and 1992 (O' Riordain, 1992). In India, the plant tissue culture industry has recorded a tremendous pace of growth and about 60 laboratories are involved in this business (Varma and Schuepp, 1996). There are various reasons for utilizing micropropagation techniques in plant production. One major benefit is the capacity to implement the breeding programmes rapidly with large number of cloned micropropagated plants being produced in relatively short time. Secondly, micropropagated plantlets are usually free from diseases due to their initiation from meristems and maintenance in axenic conditions. Thus for most plant species, technique of micropropagation offers advantage of rapid multiplication as well as production of high quality, uniform plants free from the diseases. However, one of the greatest challenges for micropropagation technique is to identify and overcome the factors affecting the survival and establishment of micropropagated plants after their transfer from culture conditions to the greenhouse or field conditions. During acclimatization stage, plants are subjected to severe environmental stresses due to poor root, shoot and cuticular development. Although some plant species can withstand these stresses, due to the rapid onset of root and shoot growth but during weaning about 10 - 40 % of plantlets either die or show stunted growth which incur high losses at the commercial level. Such plants are also easily susceptible to attack by various fungal pathogens. This results in large increase in inputs of fertilizers and pesticides. The synthetic media used for micropropagation of plants do not contain VAM fungi - an important symbiont and the plants obtained are non-mycorrhizal. Such micropropagated plants establish mycorrhizal association only after their transplant in field soil. However, an early inoculation of these plants with selected / specific VAM fungi bring modifications in root morphology, improve plant performance and thereby provide same benefits to micropropagated hosts as in the case of field grown plants (Lovato *et al.*, 1995).

4.   POTENTIAL BENEFITS OF VAM FUNGI IN MICROPROPAGATED SYSTEMS

During last one decade, lot of information have been generated on mycorrhizal inoculation in micropropagated plants especially in fruit crops and some other crops for their better survival and improved growth. A large number of micropropagated plants have shown positive response when inoculated with VAM fungi (Table 1). In micropropagated strawberry, besides improved vegetative growth, mycorrhizal inoculation also increased runner development (Verma and Adholeya, 1996; Vestbery, 1992a). Significant increase in leaf area has been reported in three micropropagated pineapple varieties inoculated with five different VAM fungi (Guillemin *et al.*, 1992).Blocking of shoot apical growth after transplanting could be prevented by VAM fungal inoculation (Berta *et al.*, 1990; Fortuna *et al.*, 1992; Naqvi and Mukerji, 1998). Mycorrhizal fungal inoculation has also been found to influence root morphology and increase in the production of lateral roots in micropropagated grapevine (Schellenbaum *et al.*, 1991). Inoculation with *Glomus mosseae* induced more branched root system in micropropagated plum rootstock (Giovannethi *et al.*, 1996).

Table 1: Some examples of application of VAM fungi to micropropagated plants

Plant	VAM fungi	References
I. Horticultural crops :		
1. *Annona cherimola*	*Glomus* sp.	Azcon-Aguilar *et al.* (1994).
2. Apple	*Glomus fasciculatum,*	Branzanti *et al.* (1992);
	G.mosseae, G.intraradices,	Fortuna
	G.viscosum	*et al.* (1996).
3. Avocado	*Glomus deserticola, G.mosseae*	Azcon-Aguilar
	G. fasciculatum	*et al.,* 1992;
		Vidal *et al.* (1992).
4. Banana	*Glomus* spp.	Lin and Chang, (1987);
	G. fasciculatum	Alonso Reyes *et al.*(1995).
5. Cherry	*Gigaspora margarita*	Pons *et al.* (1983)
	G. intraradices, G. deserticola	Lovato *et al.* (1994);
	G.mosseae, G. geosporum	Declercket *al.*(1994).
6. *Gerbera jamesonii*	*Glomus etunicatum*	Chi Luan *et al.* (1994)
	G.fasciculatum	
	G.fasciculatum	
7. Grapevine	*Gigaspora margarita,*	Ravolanirina *et al.* (1989);
	Glomus mosseae, G.caledonium	Schellenbaum
	G.fasciculatum,	*et al.* (1991);
	Commercial inoculant	Lovato *et al.* (1992)
8. Hortensia	*Glomus* intraradices,	Varma and Schuepp, (1994)
9. Jackfruit	*Acaulospora morroweae*	Sivaprasad *et al.*
	Glomus fasciculatum,	(1995)
	G.constrictum, G. mosseae,	
	G. etunicatum	
10. Kiwifruit	*Glomus* sp. strain E3	Schubert *et al.* (1992)
11. Oilpalm	*Glomus* spp.	Blal *et al.*(1990)
12. Pineapple	*G.clarum, Scutellospora*	Guillemin *et al.*
	pellucida, Glomus sp.	(1992;1995); Noval
	Gigaspora margarita	*et al.* (1995);
	Acaulospora sp.	Matos and Silva (1996).
13. Plum	*Glomus mosseae, G.caledonium*	Fortuna *et al.* (1992)
	G.coronatum, Glomus strain A6,	Giovannetti *et al.* (1996);
	G. intraradices, G.visosum	Fortuna *et al.*(1996).
14. Strawberry	*Glomus* sp., *G.geosporum*	Williams *et al.*(1992);
	G.mosseae; G.etunicatum	Vosatka *et al.*(1992);
	G.macrocarpum V3,	Vestberg, (1992a,b);
	G.versiforme, G.etunicatum	Taube-Baab and
		Balhruschat,(1990)
II. Forest trees :		
15. *Populus deltoides*	*Glomus* sp. *Acaulospora* sp.	Adholeya and Cheema,
		(1990)
16. *Anthyllis cystisoides*	*Glomus fasciculatum*	Salamanca *et al.* (1992).
17. *Spartium junceum*	*Glomus fasciculatum*	Salamanca *et al.* (1992)
18. *Fraxinus excelsior*	*Glomus intraradices*	Lovato *et al.*(1994)
19. *Leucaena leucocephala*	*Glomus fasciculatum*	Naqvi and Mukerji, (1997).

Endomycorrhizal inoculation is also reported to have a positive influence on the homogeneity of micropropagated apple (Branzanti *et al.,* 1992). Uosukainen and Vestberg

(1994) found that in micropropagated apple, the mycorrhizal plants were more uniform in size and also culture time was a quarter to one third shorter than with uninoculated plants. The application of VAM fungi also offers an opportunity to reduce fertilizer inputs as well as to limit use of pesticides (Hooker *et al.*, 1994b). Williams *et al.* (1992) found that fertilizer inputs can be reduced in mycorrhizal micropropagated strawberry to levels considerably lower than those used in commercial practice and yet plant development can be maintained that is equivalent to that of non-mycorrhizal plants receiving full fertilizer inputs.

Increase in P concentration and content has been recorded in micropropagated *Anthyllis* after mycorrhizal inoculation (Salamanea *et al.*, 1992).In *Gerbera jamesonii* plantlets, mycorrhizal inoculation with *Glomus etunicatum* increased P, K, Zn and Cu uptake besides increasing the dry weight of plants. VAM inoculation also advanced flowering by 16 days as compared with the control (Chiluan *et al.*, 1994). Improvement in survival rate - the main constraint in micropropagated plants, has also been observed in micropropagated avocado (Azcon-Agricular *et al.*, 1994), *Anthyllis* (Salamanea *et al.*, 1992), jackfruit (Sivaprasad *et al.*, 1995) and in *Leucaena leucocephala* (Naqvi and Mukerji, 1998).

VAM inoculation also reduces the acclimatization period by about 8 weeks in *Anthyllis cystisoides* (Salamanea *et al.*, 1992). Since *Anthyllis* is used in revegetation programme for desertified Mediterranean ecosystem, shorter propagation cycle is of high value. There are also reports of the effect of VAM fungal inoculation on the plant tolerance to fungal pathogens like *Phytophthora* (Guillemin *et al.*, 1994; Taube-Baab and Balhruschat, 1996; Vestberg *et al.*, 1994). The protective effect against fungal pathogens seem to be of even greater importance in micropropagated systems where plants are at greater risks to pathogen attack due to poor cuticular and root development. Lin *et al.* (1987) conducted a survey regarding inoculation at the commercial scale of 7 micropropagated plants viz. strawberry, apple, rose, pineapple, potato, blackberry and ginger with appropriate VAM fungal strains. Out of the above plants, blackberry and apple were found to be most promising.

5.   MYCORRHIZAL INOCULATION TECHNIQUE

Micropropagation involves two main stages - an *in vitro* and an *ex vitro* and mycorrhizal inoculation can be done at either of these two stages. (i) *In vitro* - rooting stage, and (ii) *Ex vitro* - a). immediately after the rooting stage at the beginning of acclimatization period or b). after the acclimatization period, before starting the post acclimatization period under greenhouse condition.

5.1.   Inoculation at *in vitro* Stage

The rooting of micropropagated plants is commonly done in agar medium supplemented with auxins. Extensive studies have been done on the germination of VAM fungal spores and early stages of growth of mycelium. Results indicate that most VAM fungal spores require low nutrient content media for rapid germination. High concentration of nutrients particularly phosphorus has a inhibitory effect on spore germination and growth (Green *et al.*, 1976; Koske 1981; Moske, 1959). VAM fungal spore inoculation *in vitro* at the root initiation stage to establish vesicular-arbuscular mycorrhizal symbiosis has been successfully obtained (Becard and Fortin, 1988; Diop *et al.*, 1992; Verma and Adholeya, 1995; 1996). At this stage, for each plant and fungus combination, appropriate medium for the growth of both the micropropagated plant and VAM fungus is required. Pons *et al.* (1983) achieved *in vitro* inoculation of *Prunus avium* using *Gigaspora margarita*. Ravolanirina *et al.*(1989) also reported a functional mycorrhizal symbiosis in *Vitis vinifera* in the rooting medium with surface sterilized spores of *Gigaspora margarita, Glomus caledonium* and *G. mosseae.* Elmeskaovi *et al.* (1995) obtained a *in vitro* system for culturing *Glomus intraradices* with 4 Ri T-DNA-transformed carrot roots or no-transformed tomato roots which was used as a potential active source of inoculum for colonization of micropropagated strawberry plantlets after root induction in growth chamber. Colonized plantlets were reported to have more extensive root system and better shoot growth than control plants. *In vitro* inoculation of micropropagated plants with VAM fungal spores is possible but it is a cumbersome process which involves technical expertise. The technique , therefore, is not practically feasible on commercial scale and is limited to research studies only.

5.2.   Inoculation at *ex vitro* Stage

This can be achieved either at the beginning of acclimatization stage or before the start of hardening phase after the acclimatization stage. A number of references are available on *ex vitro* inoculation studies of micropropagated horticultural plants like grapevine (Ravolanirina *et al.,* 1989; Schubert *et al.,* 1990), oilpalm (Vestberg, 1972a;c), strawberry (Blal *et al.,* 1990), Boston fern (Porton *et al.,* 1990), pineapple (Guillemin *et al.,* 1992; 1994; 1995), avocado (Azcon-Aguilar *et al.,* 1992; Vidal *et al.,* 1992), kiwifruit (Schubert *et al.,* 1992), banana (Naqvi and Mukerji, 1998), apple (Branzanta *et al.,* 1992; Vosukainen, 1992), jackfruit (Sivaprasad *et al.,* 1995), plum (Fortuna *et al.,* 1992). VAM fungi have also been used to inoculate micropropagated woody legumes like *Anthyllis* and *Spartium* sp. used in revegetation and reclamation programme in the Mediterranean region (Salmanca *et al.,* 1992) and *L. leucocephala* (Naqvi and Mukerji 1998).

The VAM fungal inoculum can be used in different forms like mixture of soil-based inoculum and mycorrhizal roots (Naqvi and Mukerji, 1998; Schubert *et al.,* 1990), mycorrhizal root only or surface sterilized mycorrhizal roots (Ponton *et al.,* 1990). Different workers have inoculated micropropagated plants at various stages. In *Pistacia integenima*, VAM inoculation was done after the acclimatization stage (Schubert and Marthnelli, 1988). Vidal *et al.* (1992) inoculated avocado both at the onset of acclimatization stage and at the beginning of hardening phase. Their findings revealed that although VAM symbiosis could be established at both the stages but the inoculation at the beginning of hardening stage showed better results. In *Annona cherimola*, mycorrhizal inoculation was assayed at two different stages of the micropropagation process. First, immediately after the *in vitro* phase, before starting the acclimatization period and second, after the acclimatization phase before starting the post-acclimatization period under greenhouse conditions. The greatest effect of VAM fungi on plant growth was observed when they were inoculated after the acclimatization period (Azcon Aguilar *et al.,* 1994). The different responses can be due to the differences in the root growth and development rate of the plants. Thus for each plant species, an inoculating protocol should be worked out taking account of the growth and development of each plant. Ravolanirina *et al.* (1989) studied the impact on plant growth of both *in vitro* and *ex vitro* inoculation. The later was more effective for increasing growth of micropropagated grapevine rootstocks in potting mixture. Hence *ex vitro* inoculation is easy as compared to *in vitro* and gives better results.

6.   EFFECT OF GROWTH SUBSTRATES

The substrate used for growth of micropropagated plants are important for growth of plants and development of VAM symbiosis. The most common substrates used in plant tissue culture are peat based without soil and synthetic substrates like perlite or vermiculite. Studies have revealed that different growth substrates used have varying effect on development of VAM fungi. In micropropagated avocado plants soil-sand mix was found to be better for

VAM fungal colonization as compared to peat-sand mix (Vidal *et al.*, 1992). In three different micropropagated plants, peat based substrates were found to enhance establishment of VAM symbiosis and also growth of the plants (Wany *et al.*, 1993). In another study, certain types of peat and composted substrates have been found to have negative effect on mycorrhizal symbiosis establishment, although there was no effect on germination and early mycelial growth of VAM fungi (Calvet *et al.*, 1992). In micropropagated grapevine, VAM colonized the roots in all peat-based media used, but significant growth response was noted in plants when soil was added to the substrate (Schubest *et al.*, 1990). Vestberg (1992b) found that in micropropagated strawberry, sand fertilized with bone meal was better than peat based substrates in rapid VAM colonization and sporulation. VAM inoculation greatly increased the percentage survival of *Rubus arcticus* on sand substrate (Vestberg, 1992b). Ponton *et al.* (1990) showed that only one out of three peat based substrates was effective in enhancing growth of Boston fern. Substrate of soil and zeolite (3:1) resulted in highest colonization as compared to other substrate combinations in pineapple (Noval *et al.*, 1995). Thus growth media used seem to be major factor in establishment of functional mycorrhizal symbiosis in micropropagated plants.

7.   VAM SPECIFICITY IN RELATION TO HOST

VAM fungi are known to colonize roots of wide variety of host plants belonging to different families. In general the VAM fungi are not host specific but for inoculation of micropropagated plants, selection of an appropriate VAM fungal partner is essential to derive maximum benefit of mycorrhizal symbiosis. Difference in response to mycorrhizal inoculation is also observed among cultivars. Studies carried out on 3 cultivars of micropropagated pineapple and 5 VAM fungi indicated some specificity of fungi for promoting growth of different pineapple varieties (Guillemin *et al.*, 1992). Vestberg (1992b,c) found that only 3 out of 6 fungal strains tested with 10 strawberry cultivars were highly efficient for significant growth responses. He also found that early strawberry cultivars showed higher root colonization and sporulation than late cultivars. Varma and Schuepp (1994) also reported that *Glomus intraradices* did not colonize roots of different hosts viz. strawberry, raspberry and hortensia to some extent and considerable differences were observed between the varieties of strawberry and raspberry. Their studies confirmed that endomycorrhizal root colonization is affected by the host-fungus combination. The effects ranged from mutualistic (hortensia), through neutral (strawberry var. Avanta and raspberry var. Zeva 1) to negative (raspberry var. Himboqueen and strawberry var. Elsanta). Differences in growth responses by use of different VAM fungi have also been observed in wild cherry (Lovato *et al.*, 1994), grapevine (Schubert *et al.*, 1990). The question of using VAM fungal inocula containing single or several fungal species has been discussed by Sieverding (1989). Responses to mycorrhizal inoculation are dependent both on the species of micropropagated host (Salamanca *et al.*, 1992) and VAM fungal species (Williams *et al.*, 1992). Above studies coerce the importance of screening of different species/ isolates/ strains of VAM fungi for a particular micropropagated host and the one which gives optimum benefits under given set of environmental conditions should be selected for commercial use.

8.   EFFECT OF FERTILIZER APPLICATION

The role of VAM fungi in uptake of nutrients particularly P is well documented (Bolan, 1991; Koida, 1991; Marschner and Dell, 1994; Smith *et al.*, 1994). Mycorrhizal inoculation economizes the use of P fertilizer in micropropagated plants like grapevine (Ravolanirina *et al.*, 1989), apple (Branzanti *et al.*, 1992; Fortuna *et al.*, 1996). Mycorrhizal inoculated plants of apple grew at par with non-mycorrhizal plants supplied with a nutrient solution containing 40 ppm P (Branzanti *et al.*, 1992). In micropropagated strawberry, the mycorrhizal plants receiving 25 % of the minimum recommended commercial rate of osmocote (commercial controlled-release fertilizer) had a similar yield to non-mycorrhizal plants which received the full recommended rate of this fertilizer (Williams *et al.*, 1992). In micropropagated kiwifruit (*Actinidia deliciosa*) VAM inoculation induced higher growth with intermediate fertilizer doses as compared to non-mycorrhizal plants with highest fertilizer application (Schubest *et al.*, 1992). The addition of zeolite which can absorb some nutrients and allows their slow release has been shown to enhance the growth of micropropagated strawberry (Vasatka *et al.*, 1992). In mycorrhizal micropropagated oilpalm, the coefficient of fertilizer utilization has been found to increase from almost 5 times for superphosphate to 4 time for rock phosphate (Blal *et al.*, 1990).

9.   PRODUCTION OF VAM INOCULUM

The most important and practical approach pertaining to the utilization of VAM fungi in micropropagated systems is the readily available supply of inoculum. Since these fungi cannot be grown on synthetic media, VA mycorrhizal inoculum has to be prepared by multiplication of the selected fungi in roots of susceptible host plants growing in sterilized soil or substrates such as perlite, vermiculite, peat, sand or a mixture of these. Methods for the production and use of VAM inocula have been reviewed from time to time (Sylvia and Jarstfer, 1994; Verma and Adholeya, 1996; Walker and Vestberg, 1994). Ri-plasmid transformed root culture have also been used to obtain colonized root culture (Beeard and Fortin, 1988; Verma and Adholeya, 1995). Surface sterilized spores can be used to establish *in vitro* infections in whole plants or excised roots growing on synthetic media (Verma and Adholeya, 1995). Commercial inoculant are now available in North America and Europe (Mulongoy *et al.*, 1992). Lovato *et al.* (1992) tested two commercially available granular inoculant from Agricultural Genetics Company (AGC, UK) and from Phyotec (Belgium) with micropropagated grapevine and pineapple. Although there were effects of soil pH on the effectiveness of the inocula but their findings confirmed that utilization of commercial inoculant is technically feasible. Thus there is a need to carefully match inoculum type to the micropropagated plant system supported by relevant research regarding optimal dose to be used, efficacy under different soil conditions and host specificity in response to the endomycorrhizal fungi used.

10.   MYCORRHIZATION OF MICROPROPAGATED *LEUCAENA LEUCOCE-PHALA* (LAM.) DE WIT.

L. leucocephala is a multipurpose leguminous tree of high commercial value as forage and firewood and usually employed in reforestation programmes in developing countries.

Cotyledonary nodes excised from 15 days old seedlings were cultured on modified B5 medium supplemented with cytokinin BAP at 5 x 10^{-6} M concentration to obtain multiple shoots. Rooting of individual shoot was induced in half strength of B5 medium supplemented with IBA at 5 x 10^{-6} M concentration. Cultures were maintained at 28 ± 2°C with 12 hr./12 hr light and dark cycle. The light supplied from fluorescent tubes at 200 μE/m2/s-1 intensity. The plantlets obtained were transferred to sterilized soil rite (synthetic substrate containing perlite and vermiculite) for acclimatization and given one fourth strength of liquid B5 medium.

After 3 weeks plants were further transferred to polybags containing sterilized soil and covered with polythene to maintain high humidity for a week. The plants were given one tenth strength of liquid B5 medium at the time of transplant in soil. During this transfer, *Glomus fasciculatum* inoculum (10 g) containing mycorrhizal roots with external mycelium and soil containing spores, was added to each plantlet. *Rhizobium* inoculum (strain PRGL 001) was also added at two stages - first when plants were transferred to autoclaved soilrite and second at the time of transfer to fumigated soil in polybags. Four different treatments were given to these plants (i). *Rhizobium* inoculum only (ii). *Glomus fasciculatum* inoculum only (iii). *Rhizobium* + *G. fasciculatum* and (iv). Control (no inoculum). Plants were kept in polyhouse at 30 ± 2°C with 55- 60 % RH. No fertilizer or nutrient solution was supplied to plants during the observation period except irrigation with tap water (Naqvi and Mukerji, 1998).

Mycorrhizal inoculation had a significant effect on root and shoot length, root and shoot dry weight, number of leaves and number of nodules of mycorrhizal plants. The maximum growth response was observed in plants with dual inoculation as compared to single treatment of either *Rhizobium* or *G.fasciculatum* (Fig.1). There was no plant mortality in dual inoculated plants after 150 days whereas plant survival was 65. 21 % in *Rhizobium* inoculated plants and 91. 66 % in *G.fasciculatum* inoculated plants (Fig.2). All the uninoculated plants died after 150 days (Naqvi and Mukerji, 1998). The plants inoculated had good colonization arbuscular mycorrhiza with external hyphae, internal hyphae with arbuscules and vesicles (Fig.3).

11. CONCLUSION

Micropropagation is a technique of increasing importance for the mass multiplication of quality and disease free plants of many commercial fruit crops and forest trees. The technique of *in vitro* mass multiplication of commercial crops face the major problem of low survival rate of these plants with poor growth performance. Introduction of VAM fungi in micropropagated system has been proved to be important biotechnology to over come these problems up to great extent. Mandatory introduction of VAM fungi in micropropagated system of plants would be inexpensive, eco-friendly and feasible approach with a number of naturally available benefits to plants system. A large number of plant genera have developed an intimate association with VAM fungi probably from the course of their evolution in the nature. Like other important synthetic chemicals essential for micropropagation of plants, the use of VAM fungi as bio-regulator, bio-fertilizer, biocontrol agents will be the reality in near future on commercial scale. The use of VAM in plant micropropagation system will also reduce pesticides and fertilizer inputs and make this system more cost effective, eco-friendly and sustainable. Better insight of different factors affecting mycorrhizal dependency viz. host plant, fungal isolate and its compatibility with the micropropagated plant, soil conditions is needed to extract the maximum benefit from mycorrhizal association in micropropagated system. There is a need to further refine and improve the techniques for inoculum production for research purpose and on commercial scale without compromising with the quality.

Fig. 1. Growth response of 90 days out micropropagated *Leucaena leucocephala* given different treatments

C	=	control
R	=	+ *Rhizobium*
R	=	+ *Glomus fasciculatum*
R+F	=	+ *Rhizobium* + *G. fasciculatum*

226

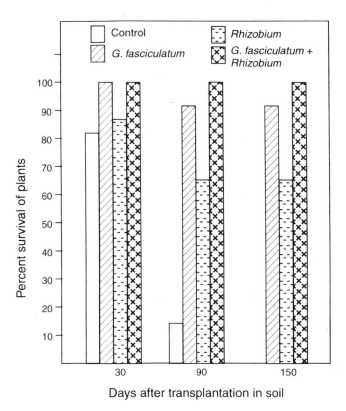

Fig. 2. Percent surrival of micropropagated *Leucaena leucocephala* in different treatments.

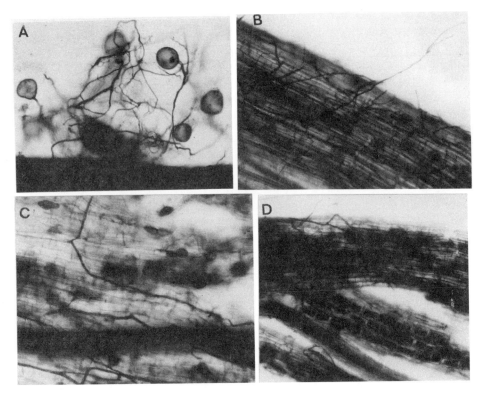

Fig. 3. VAM colonization in roots of micropropagated *Leucaena leucocephala* at different stages of growth
    A. Germinating chlamydospore (c) of VAM fungi with hypha entering the root.
    B. Root epidermis penetrated by external hypha (EM), Root cortex is filled with internal hyphae (IH) and Vesicles (V).
    C. Internal hyphae (IH) and Vesicles (V) of various shapes inside the root.
    D. Root segment heavily colonized by VAM fungi showing darkly stained asbuscules (Ar).

ACKNOWLEDGEMENTS

Senior author is thankful to the Council of Scientific and Industrial Research, New Delhi, India for financial support during the study.

REFERENCES

Adholeya, A. and Cheema,G.S. 1990, Evaluation of VA mycorrhizal inoculation in micropropagated *Populus deltoides* Marsh clones, *Curr Scie.* 59 (23): 1245-1247.

Alonso Reyes, R., Gongalez, P.M., Exposito Gracia, L., Crukelo Rodriguez, R. and Roque Martinez,L. 1995, The influence of mycorrhizae and phosphate solubilizing bacteria on the growth and development of banana vitroplants, *Infomusa*, 4(2): 9-10.

Arines, J., Vilarino, A. and Sainz, M. 1989, Effect of vesicular-arbuscular mycorrhizal fungi on Mn uptake by red clover, *Agric. Ecosys. Environ.* 29: 1-4.

Azcon, R., Barea, J.M. and Hayman, D.S. 1976, Utilization of rock phosphate in alkaline soils by plants inoculated with mycorrhizal fungi and phosphate solubilizing bacteria, *Soil Biol. Biochem.* 8: 135-138.

Azcon, R., Rubio, R. and Barea, J.M. 1991, Selective interactions between different species of mycorrhizal fungi and *Rhizobium meliloti* strains, and their effects on growth, N_2-fixation ^{15}N and nutrition of *Medicago sativa*, *New Phytol.* 117 : 399-404.

Azcon-Aguilar, C., Barcelo, A., Vidal, M.T. and De la Vina, G. 1992, Further studies on the influence of mycorrhizae on growth and development of micropropagated avocado plants, *Agronomie* 12: 837-840.

Azcon-Aguilar,C., Encina, C.L., Azcon, R. and Barea, J.M. 1994, Effect of arbuscular mycorrhiza on the growth and development of micropropagated *Annona cherimola* plants, *Agric. Scie. Finland* 3(3): 281-288.

Bajaj, Y.P.S. (eds). 1986, *Biotechnology in Agriculture and Forestry* Vol. I, Springer-Verlag, Berlin, Heidelberg, New York, Tokyo.

Bansal, M. and Mukerji, K.G. 1994, Positive correlation between VAM- induced changes in root exudation and mycorrhizosphere mycoflora, *Mycorrhiza* 5: 39-44.

Becard, G. and Fortin, J.A. 1988, Early events of vesicular-arbuscular mycorrhiza formation on Ri T-DNA transformed roots, *New Phytol.* 108: 211-218.

Berta,G., Fusconi,A., Trotta,A. and Scannerini, S. 1990, Morphogenetic modifications induced by the mycorrhizal fungus *Glomus* strain E_3 in the root system of *Allium porrum* L., *New Phytol.* 114: 207-215.

Blal, B., Morel, C., Gianinazzi-Pearson, V., Fardeau, J.C. and Gianinazzi, S. 1990, Influence of vesicular-arbuscular mycorrhizae on phosphate fertilizer efficacy in two tropical acid soils planted with micropropagated oilpalm (*Elaeis guinensis* Jacq.), *Biol. Fert. Soils* 9: 43-48.

Bolan, N.S. 1991, A critical review on the role of mycorrhizal fungi in the uptake of phosphorus by plants, *Plant Soil* 134: 189-207.

Branzanti, B., Gianinazzi-Pearson,V. and Gianinazzi, S. 1992, Influence of phosphate fertilization on the growth and nutrient status of micropropagated apple infected with endomycorrhizal fungi during the weaning stage, *Agronomie* 12: 841-845.

Burkert, B. and Robson, A.1994, ^{65}Zn uptake in subterranean clover (*Tifolium subterraneum* L.) by three vesicular-arbuscular mycorrhizal fungi in a root-free sandy soil, *Soil Biol. Biochem.* 26: 117-1124.

Calvet, C., Estaun, V. and Camprubi, A. 1992, Germination, early mycelial growth and infectivity of a vesicular-arbuscular mycorrhizal fungus in organic substrates, *Symbiosis* 14 : 405-411.

Chi-Luan,W., Chi-Ning,C., Chang,D.C.N. and Wen,C.L. 1994, Effects of *Glomus* sp. on the growth of micropropagated *Gerbera jamesonii* plantlets, Memoirs of the College of Agriculture, National Taiwan University, 34(2): 97-110.

Declerck,S., Devos, B., Delvaux, B. and Plenchette, C. 1994, Growth response of micropropagated banana plants to VAM inoculation, *Fruits* 49(2):103-109.

Diop, T.A., Becard, G. and Piche, Y. 1992, Long term *in vitro* culture of endomycorrhizal fungus, *Gigaspora margarita* on Ri-TDNA transformed roots of carrot, *Symbiosis* 12 : 149-259.

Elmeskaovi, A., Demont, J.P., Poulin,M.J.,Piche,Y.and Desjardins,Y. 1995, A tripartite culture system for endomycorrhizal inoculation of micropropagated strawberry plantlets *in vitro, Mycorrhiza* 5(5): 313-319.

Estaun, V., Calvet,C. and Hayman, D.S. 1987, Influence of plant genotype on mycorrhizal infection; Response of three pea cultivars, *Plant Soil.* 103 :295-298.

Fitter, A.H. 1991, Costs and benefits of mycorrhizas: Implications for functioning under natural conditions, *Experientia* 47: 350-355.

Fitter, A.H. and Garbaye, J. 1994, Interactions between mycorrhizal fungi and other soil organisms, *Plant Soil* 159:123-132.

Fortuna,P., Citernesi, S., Morini, S., Giovannetti, M. and Loveli, F. 1992, Infectivity and effectiveness of different species of arbuscular mycorrhizal fungi in micropropagated plants of Mr S2/5 plum rootstock, *Agronomie* 12: 825-829.

Fortuna, P., Citernesi, A.S., Morini, S., Vitangliano, C. and Giovannetti, M. 1996, Influence of arbuscular mycorrhizae and phosphate fertilization on shoot apical growth of micropropagated apple and plum rootstocks, *Tree Physiol.* 16(9):757-763.

Giovannetti, M., Fortuna, P., Loreti, F. and Morini, S. 1996, Effects of arbuscular mycorrhizal inoculation on *in vivo* root induction and development in shoots of Mr S2/5 plum rootstock grown *in vitro, Acta Horticul.* 374 : 99-102.

Green, N.E., Graham,S.O. and Schenck, N.C. 1976, The influence of pH on the germination of vesicular-arbuscular mycorrhizal spores, *Mycologia* 68 : 929-934.

Guillemin,J.P., Gianinazzi,S. and Trouvelot, A. 1992, Screening of arbuscular endomycorrhizal fungi for establishment of micropropagated pineapple plants, *Agronomie* 12: 831-836.

Guillemin,J.P., Gianinazzi,S.,Gianinazzi-Pearson,V. and Marchal, J. 1994, Contribution of endomycorrhizas to biological protection of micropropagated pineaple [*Annanas comosus* (L.)Merr.] against *Phytophthora cinnamomi* Rands, *Agricu. Sci. Finland* 3: 241-252.

Guillemin, J.P., Gianinazzi, S., Gianinazzi-Pearson, V. and Marchal, J. 1995, Effect of arbuscular endomycorrhizae on growth and mineral nutrition of pineapple *vitro*plants in a soil with high salinity, *Fruits* 50(5):333-334.

Harley,J.L. and Smith, S.E. 1983, *Mycorrhizal Symbiosis*, Academic Press, New York.

Hooker, J.E., Jaizme-vega,M. and Atkinson, D. 1994a, Biocontrol of plant pathogens using arbuscular mycorrhizal fungi, in: *Impact of Arbuscular Mycorrhizas on Sustainable Agricultural and Natural Ecosystems*, Birkhauser, Switzerland.

Hooker, J.E., Gianinazzi, S., Vestberg, M., Borea,J.M. and Atkinson,D. 1994b, The application of arbuscular mycorrhizal fungi to micropropagation systems: an opportunity to reduce chemical inputs, *Agric. Sci. Finland* 3: 227-232.

Hussey, R.S. and Roncadori, R.W. 1982, Vesicular-arbuscular mycorrhizae may limit nematodes activity and improve growth, *Plant Disease* 66: 9.

Jagpal,R. and Mukerji,K.G. 1991, VAM fungi in reforestation, in: *Plant Roots and their Environment,* B.L. McMichael and H. Persson, eds., London, New York, Tokyo, pp. 309-313.

Jalali, B.L. and Jalali, I. 1991, Mycorrhiza in plant disease control, in *Handbook of Applied Mycology* Vol.1., Soil and Plants, D.K. Arora, B. Rai, K.G. Mukerji and G.R. Knudsen, eds., Marcel Dekker Inc., New York, USA,pp. 131-154.

Jeffries, P. and Dodd, J.C. 1991, The use of mycorrhizal inoculations in forestry and agriculture, in : *Handbook of Applied Mycology*, Vol. 1, Soil and Plants, D.K. Arora, B. Rai, K. G. Mukerji and G.R. Knudsen, eds., Marcel Dekker Inc., New York, USA,pp. 155-185.

Koide, R.T. 1991, Nutrient supply, nutrient demand and plant response to mycorrhizal infection, *New Phytol.* 117: 35-386.

Koske,R.E. 1981, *Gigaspora gigantea* : Observation on spore germination of a VA mycorrhizal fungus, *Mycologia* 73: 288-300.

Krishna, K.R. and Bagyaraj, D.J. 1984, Growth and nutrient uptake of peanut inoculated with the mycorrhizal fungus, *Glomus fasciculatum* compared with non-inoculated ones, *Plant Soil* 77: 404-408.

Li, X.L., Marschner, H. and George, E. 1991, Acquisition of phosphorus and copper by VA-mycorrhizal hyphae and root to shoot transport in white clover, *Plant Soil* 136: 49-57.

Lin,C.H. and Chang,D.C. 1987, Effect of three *Glomus* endomycorrhizal fungi on the growth of micropropagated banana plantlets, *Trans. Mycol Soc. R.O.C.* 2(1): 37-45.

Lin, M.T., Lucena,F.B., Matos, M.A.M., Paiva, M., Assis, M. and Caldos, L.S.1987, Greenhouse production of mycorrhizal plants of nine transplanted crops, in: *Mycorrhizae in the Next Decade : Practical Applications and Research Priorities*, D.M. Sylvia, L. L. Hun and J. H. Graham, eds., 7th NACOM. Gainesville, Fl. pp. 281.

Lovoto,P.,Giullemin, J.P. and Gianinazzi, S. 1992, Application of commercial arbuscular endomycorrhizal fungal inoculant to the establishment of micropropagated grapevine rootstock and pineapple plants, *Agronomie* 12: 873-880.

Lovato, P., Hammatt, N., Gianinazzi-Pearson,V. and Gianinazzi, S. 1994, Mycorrhization of micropropagated mature wild cherry (*Prunus avium* L.) and common ash (*Fraxinus excelsior* L.), *Agric. Scie. Finland* 3: 297-302.

Lovato, P., Schuepp,H., Trouvelot, A. and Gianinazzi, S. 1995, Application of arbuscular mycorrhizal fungi (AMF) in orchard and ornamental plants, in: *Mycorrhiza*, A. Varma and B. Hock, eds., Springer-Verlag, Berlin, Heidelberg,pp. 443-467.

Marks, G.C. 1991, Causal morphology and evolution of mycorrhizas, *Agric. Ecosy. Environ.* 35: 89-104.

Marschner, H. and Dell, B. 1994, Nutrient uptake in mycorrhizal symbiosis, *Plant Soil* 159: 89-102.

Matos, R.M.B.and Silva, E.D. 1996, Effect of inoculation by arbuscular mycorrhizal fungi on the growth of micropropagated pineapple plants, *Fruits* 51(2): 115-119.

Menge, J.A. 1983, Utilization of vesicular-arbuscular mycorrhizal fungi in agriculture, *Can. J. Bot.* 61: 1015-1024.

Mittal, N., Saxena,G., Jagpal, R. and Mukerji,K.G. 1987, Interaction between nematophagous fungi and vesicular-arbuscular mycorrhizal fungi, *Plant Cell Incomp. Newsletter* 19: 54-59.

Morton, J.B. and Benny, G.L. 1990, Revised classification of arbuscular mycorrhizal fungi (Zygomycetes) : A new order, Glomales, two new suborders Glomineae and Gigasporineae and two new families, Acaulosporaceae and Gigasporaceae with an emendation of Glomaceae, *Mycotaxon* 37: 471-491.

Morton, J.B. and Bentivenga, S.P. 1994, Levels of diversity in endomycorrhizal fungi (Glomales, Zygomycetes) and their role in defining taxonomic and non-taxonomic groups, *Plant Soil* 159: 47-60.

Mosse, B. 1959, The regular germination of resting spores and some observations on the growth requirements of an *Endogone* sp. causing vesicular-arbuscular mycorrhiza, *Trans. Brit. Mycol. Soc.* 42: 273-286.

Mukerji, K.G. 1996, Taxonomy of endomycorrhizal fungi, in: *Advances in Botany,* K.G. Mukerji, B. Mathur, B.P. Chamola and P. Chitralekha, eds., APH Publishing Corporation, New Delhi, pp. 212-219.

Mukerji, K.G. and Dixon, R.K. 1992, Mycorrhizae in reforestation, in: *Proc. Int.Symp. Rehabilitation of Tropical Rainforest Ecosystems : Research and Development Priorities,* N.M. Majid, I.A.A. Malik, M.Z., Hamzah and K. Jusoff, eds., Kuching, Sarawak, Malaysia, pp.66-82.

Mukerji, K.G. and Sharma, M. 1996, Mycorrhizal relationships in forest ecosystes, in: *Forest-A Global Perspective,* S.K. Mazumdar, E.W. Miller and F.J. Brenner, eds., The Penn. Acad. Sci., USA, pp. 95-125.

Mulongoy, K., Gianinazzi,S., Roger,P.A. and Dommergues, Y. 1992, Biofertilizers : Agronomic and environmental impacts and economics, in: *Microbial Technology : Economic and Social Aspects,* E.D. Silva, A. Sasson and C. Rotledge, eds., Cambridge University Press, Cambridge, pp. 59-69.

Naqvi, N.S. and Mukerji,K.G. 1998, Mycorrhization of micropropagated *Leucaena leucocephala* (Lam.)de Wit., *Symbiosis* 24: 103-114.

Noval, B.D.L., Fernandez,F.and Jerrera, R.1995, Effect of the use of arbuscular mycorrhiza and substrate combinations on the growth and development of *in vitro* raised pineapple plants, *Cultivos Tropicales* 16(1): 19-22.

O'Riordain, F. 1992, The European plant tissue culture industry-1990, *Agronomie*12: 743-746.

Paranjothy, K., Saxena, S., Banerjee, M., Jaiswal, V.S. and Bhojwani, S.S. 1990, Clonal multiplication of woody perennials, in: *Plant Tissue Culture: Applications and Limitations, Development in Crop Science,* Bhojwani, S.S. ed., Elsevier Publisher, Amsterdam, pp. 1-33.

Pons, F., Gianinazzi-Pearson,V., Gianinazzi,S. and Navatel, J.C. 1983, Studies of VA mycorrhiza *in vitro*: Mycorrhizal synthesis of axenically propagated wild cherry (*Prunus avium* L.) plants, *Plant Soil* 71: 217-221.

Ponton,F., Piche,Y., Parent,S. and Caron,M. 1990, Use of vesicular-arbuscular mycorrhizae in Boston fern production II, Evaluation of four inocula, *Horticul. Sci.* 25: 416-419.

Posta, K., Marschner, H. and Romheld,V. 1994, Manganese reduction in the rhizosphere of mycorrhizal and non mycorrhizal maize, *Mycorrhiza* 5: 119-124.

Puthur, J. T., Prasad, K.V.S.K., Sharimla, P. and Pardha Saradhi, P. 1998, Vesicular arbuscular mycorrhizal fungi improve establishment of micropagated *Leueaena leucocephala* plant, *Cell Tissue Org. Cult.* 53: 41-47.

Raj,J., Bagyaraj, D.J. and Manjunath,A. 1981, Influence of soil inoculation with vesicular-arbuscular mycorrhiza and a phosphate dissolving bacterium on plant growth and ^{32}P uptake, *Soil Biol. Biochem.* 13: 105-108.

Raman, N. and Mahadevan, N. 1996, Mycorrhizal research - a priority in agriculture, in: *Concepts in Mycorrhizal Research,* K.G. Mukerji, ed., Kluwer Academic Publishers, Netherlands, pp. 41-75.

Ravolanirina, F., Gianinazzi,S., Trouvelot,A. and Carre, M. 1989, Production of endomycorrhizal explants of micropropagated grapevine rootstocks, *Agric. Ecosy. Environ.* 29: 323-327.

Read,D.J. 1991, Mycorrhizae in ecosystems, *Experientia* 47: 376-290.

Salamanca, C.P., Herrera, M.A. and Barea, J.M. 1992, Mycorrhizal inoculation of micropropagated woody legumes used in re-vegetation programmes for desertified Mediterranean ecosystems, *Agronomie* 12: 869-972.

Sarwar, N. and Mukerji, K.G. 1995, Mycorrhizal technology in forestation, in: *Mycorrhizae : Biofertilizers for the Future,* A. Adholeya and S. Singh, eds., TERI, New Delhi, pp. 342-344.

Sarwar, N. and Mukerji, K.G. 1998, Mycorrhiza in forestation in arid, semi-arid and wastelands, in: *Ecological Techniques and Approaches to Vulnerable Environment, Hydrosphere-Geosphere Interaction,* R.B. Singh, ed., Oxford IBH Publishing Co. Ltd., New Delhi, pp.263-271.

Schellenbaum, L., Berta, G., Ravolanirina, F., Tissevant, B., Gianinazzi, S. and Fitter, A.H. 1991, Influence of endomycorrhizal infection on root morphology in a micropropagated woody plant species (*Vitis vinifera* L.), *Ann. Bot.* 68: 135-141.

Schubert, A., Bodrino, C. and Gribaudo, I. 1992, Vesicular-arbuscular mycorrhizal inoculation of kiwifruit (*Actinidia deliciosa*) micropropagated plants, *Agronomie* 12:847-850.

Schubert, A. and Martinelli, A. 1988, Effects of vesicular-arbuscular mycorrhizae on growth of *in vitro* propagated *Pistacia integerrima*, *Acta Hort.* 441-443.

Schubert, A., Mazzitelli, M., Ariusso, O. and Eynard, I. 1990, Effects of vesicular-arbuscular mycorrhizal fungi on micropropagated grapevines : Influence of endophyte strain, P fertilization and growth medium, *Vitis* 29: 5-13.

Sieverding, E. 1989, Should VAM inocula contain single or several fungal species? *Agric. Ecosy. Environ.* 29: 391-396.

Sivaprasad, P., Ramesh, B., Mohanakumaran, N., Rajmohan, K. and Joseph, P.J. 1995, Vesicular-arbuscular mycorrhizae for the *ex vitro* establishment of tissue culture plantlets, in: *Mycorrhizae: Biofertilizers for the Future*, A. Adholeya and S. Singh, eds., TERI, New Delhi, pp. 281-283.

Smith, S.E., Gianinazzi-Pearson, V., Koide, R. and Cairney, J.W.G. 1994, Nutrient transport in mycorrhizas : Structure, physiology and consequences for efficiency of the symbiosis, *Plant Soil* 159: 103-114.

Srivastava, D., Kapoor, R., Srivastava, S.K. and Mukerji, K.G. 1996, Vesicular-arbuscular mycorrhiza - an overview, in: *Concepts in Mycorrhizal Research*, K.G. Mukerji, ed., Kluwer Academic Publishers, Netherlands, pp. 1-39.

Subba Rao, N.S. and Krishna, K.R. 1990, Interactions between vesicular-arbuscular mycorrhiza and nitrogen-fixing microorganisms and their influence on plant growth and nutrition, in: *New Trends in Biological Nitrogen Fixation*, N.S. Subba Rao, ed., Oxford IBH Publishing Co. Ltd., New Delhi, pp. 2-30.

Subhan, S., Sharimla, P. and Pardha Saradhi, P. 1998, *Glomus fasciculatum* alleviates transplanttion shock of micropropagated *Sesbania sesban*, *Pl. Cell Rep.* 17: 268-272.

Sylvia, D.M. and Jarstfer, A.G. 1994, Production of inoculum and inoculation with arbuscular-mycorrhizal fungi, in: *Management of Mycorrhiza in Agriculture, Horticulture and Forestry*, A.D. Robson, L.K. Abbott and N. Malajczuk, eds., Kluwer Academic Publishers, Netherlands, pp. 231-238.

Taube-Baab, H. and Balhruschat, H. 1996, Productivity of strawberry (*Fragaria* x *Ananassa* Duch.) plants following inoculation with VA mycorrhizal fungi, *Erwerbsobstbau* 38(5): 144-147.

Thompson, J.P. 1994, Inoculation with vesicular-arbuscular mycorrhizal fungi from cropped soil overcomes long fallow disorder of linseed (*Linum usitatissimum* L.) by improving P and Zn uptake, *Soil Biol. Biochem.* 26: 1133-1143.

Tisdall, J.M. 1994, Possible role of soil microorganisms in aggregation in soils, *Plant Soil* 159: 115-122.

Uosukainen, M. 1992, Rooting and weaning of apple rootstock, *Agronomie* 12: 803-806.

Uosukainen, M. and Vestberg, M. 1994, Effect of inoculation with arbuscular mycorrhizas on rooting, weaning and subsequent growth of micropropagated *Malus* (L.) Moench, *Agric. Sci. Finland* 3: 269-280.

Varma, A. and Schuepp, H. 1994, Infectivity and effectiveness of *Glomus intraradices* on micropropagated plants, *Mycorrhiza* 5: 29-38

Varma, A. and Schuepp, H. 1996, Influence of mycorrhization on the growth of micropropagated plants, in: *Concepts in Mycorrhizal Research*, K.G. Mukerji, ed., Kluwer Academic Publishers, Netherlands, pp. 113-132.

Verma, A. and Shankar 1994, Mycorrhizae, in: *History and Progress of Botany in India - Modern Period*, B.M. Johri, ed., Oxford and IBH Publishing Co., New Delhi, pp.345-360.

Verma, A. and Adholeya, A. 1995, Selection of suitable host for *in vitro* inoculation by vesicular-arbuscular mycorrhizal fungi *Gigaspora margarita*, in: *Mycorrhizae : Biofertilizers for the Future*, A. Adholeya and S. Singh, eds., TERI, New Delhi, pp. 482-484.

Verma, A. and Adholeya, A. 1996, Cost-economics of existing methodologies for inoculum production of vesicular-arbuscular mycorrhizal fungi, in; *Concepts in Mycorrhizal Research*, K.G. Mukerji, ed., Kluwer Academic Publishers, Netherlands, pp. 177-192.

Vestberg, M. 1992a, Arbuscular mycorrhizal inoculation of micropropagated strawberry and field observation in Finland, *Agronomie* 12: 865-867.

Vestberg, M. 1992b, The effect of growth substrate and fertilizer on the growth and vesicular-arbuscular mycorrhizal infection of three hosts, *Agric. Sci. Finland* 1:95-105.

Vestberg, M. 1992c, The effect of vesicular-arbuscular mycorrhizal inoculation on the growth and root colonization of ten strawberry cultivars, *Agric. Sci. Finland* 1: 527-535.

Vestberg, M., Palmujoki, H., Parikka, P. and Uosukainen, M. 1994, Effect of arbuscular mycorrhizas on crown rot (*Phytophthora cactorum*) in micropropagated strawberry plants, *Agric. Sci. Finland* 3: 289 - 296.

Vidal, M.T., Azcon-Aguilar, C., Barea, J.M. and Pliego-Alfaro, F. 1992, Mycorrhizal inoculation enhances growth and development of micropropagated plants of avocado, *Hort. Sci.* 27: 785-787.

Vosatka, M., Gryndler,M. and Prikryl, Z. 1992, Effect of the rhizosphere bacterium *Pseudomonas putida*, arbuscular mycorrhizal fungi and substrate composition on the growth of strawberry, *Agronomie* 12: 859-863.

Walker,C. 1992, Systematics and taxonomy of the arbuscular endomycorrhizal fungi (Glomales)- a possible way forward, *Agronomie* 12:887-897.

Walker,C. and Vestberg, M. 1994, A simple and inexpensive method for producing and maintaining closed pot cultures of arbuscular mycorrhizal fungi, *Agric. Sci. Finland* 3: 233-240.

Wang, H., Parent,S., Gosselin,A. and Desjardins,Y. 1993, Vesicular-arbuscular mycorrhizal peat based substrates enhance symbiosis establishment and growth of three micropropagated species, *Ame. Soc. Hort. Sci.* 118: 896-901.

Williams, S.C.K., Vestberg, M., Uosukainen, M., Dodd, J.C. and Jeffries, P. 1992, Effects of fertilizers and arbuscular mycorrhizal fungi on the *post-vitro* growth of micropropagated strawberry, *Agronomie* 12: 851-857.

VESICULAR ARBUSCULAR MYCORRHIZAL FUNGI IMPROVES ESTABLISHMENT OF MICROPROPAGATED PLANTS

P. Sharmila, Jos T. Puthur, and P. Pardha Saradhi

Plant Physiology and Biotechnology Laboratory
Department of Biosciences
Jamia Millia Islamia
New Delhi - 110 025, INDIA

1.   INTRODUCTION

Plant tissue culture has become an important and advantageous tool for rapid propagation of several plant species. Although this technique has got several successful applications, there are still some hurdles which limits its widespread use. For instance the transplantation of plantlets developed *in vitro* to soil claims very little success in many cases (Pierik, 1988; Puthur *et al.,* 1998; Subhan *et al.,* 1998). Weak root system is one of the major hindrances in the successful establishment of the micropropagated plantlets in the field conditions. In general, mycorrhizal fungi helps in the development of a stronger root system (Ponton *et al.* 1990). Moreover, the conditions to which micropropagated plantlets are transferred from *in vitro* conditions to which they are accustomed to distinct *in vivo* conditions would be a kind of stress (popularly referred to as transplantation shock) to them. Some plants exhibit considerable dependence on mycorrhizae to thrive in stressed situations (Barea *et al.,* 1993; Bethlenfalvay *et al.,* 1987). These potentials of symbiotic association between VAM fungal species and plant roots strengthens the belief of its significance in averting the transplantation shock brought about by unfavorable environmental conditions (such as alteration in humidity and nutritional conditions). In this review efforts are made to briefly highlight on some salient features related to the increased interests in using VAM fungi in successful establishment of micropropagated plants in the soil/field conditions.

2.   RAPID MULTIPLICATION THROUGH TISSUE CULTURE

Micropropagation has become an important tool in speeding up propagation of several plant species of commercial value. Axillary bud sprouting is the most important means to produce large number of uniform plants that are genotypically and phenotypically similar to the agricultural or forestry plant of ones choice. Following flow chart gives an idea regarding the rapidity with which the plants of our choice can be multiplied through axillary bud sprouting:

Mycorrhizal Biology, edited by Mukerji *et al.*
Kluwer Academic/Plenum Publishers, 2000

	28 days	
Single node from selected plants	---------->	10 shoots each with ~4 nodes (total nodes 10x4=40)
	28 days	
~400 shoots (each with ~4 nodes)	<-----------	culture individual nodes after microcuttings
	28 days	
culture individual nodes	---------->	~16, 000 shoots (each with ~4 nodes)
	28 days	
6,40, 000 shoots or plants	<-----------	culture individual nodes

Often there had also been interests in raising and multiplying somaclonal variants with certain superior characters, haploids, triploids/polyploids, etc. also through tissue culture. In all these cases the plantlets are developed under specific *in vitro* conditions in various types of tissue culture vessels. Plantlets grown in such small culture containers are exposed to high levels of inorganic and organic nutrients, high relative humidity, elevated carbohydrate and growth regulator levels, low irradiance and limited CO_2 and O_2 gas exchange. These factors often induce physiological, anatomical and morphological abnormalities which interfere with the acclimatization subsequent to transplantation resulting in low survival rates *ex vitro* (Puthur *et al.*, 1998; Subhan *et al.*, 1998; Ziv, 1995).

3. HURDLES BEHIND THE SUCCESSFUL ESTABLISHMENT OF MICROPROPAGATED PLANTLETS IN THE FIELD

Weak root system is one of the major hindrances in the successful establishment of micropropagated plantlets in the field conditions (Pierik, 1988; Ziv, 1995). Rooting the shoots on semi solid medium results in the formation of adventitious roots with several anomalous features such as a) poorly developed vascular connections, b) little or no secondary thickening, c) loose cortical cell arrangement and d) pigmented cells, that interfere with the *ex vitro* acclimatization which results in poor survival percentage (McClelland *et al.,*1990; Smith and McClelland,1991). Many a times there is also an inferior vascular connection between the root and the shoot.

The stems developing *in vitro* are hypolignified, cell walls are thin with large intercellular air spaces with a limited development of vascular tissue. Micropropagated plants often show considerably less supportive tissues (such as sclerenchyma and collenchyma) than *ex vitro* plants (Donnelly *et al.*, 1985). Some times the vascular connections within the root and the stem are incomplete (Ziv, 1995).

The cuticular waxy layer is under developed which results in excessive cuticular transpiration *in vivo* (Fuchigami *et al.*, 1981; Pierik, 1988). The cuticular layer in many species was reported to be structurally and chemically different from field grown plants (Cappelades *et al.*, 1990).

Due to failure of the stomata to respond to stimuli that normally induce closure, water is lost rapidly from the leaves (Santamaria *et al.*, 1993). Transplanted plantlets cannot compensate the excessive water loss due to nonfunctional stomata (Lee and Wetzstein, 1982) and excessive cuticular transpiration. Therefore, the plantlets raised in an environment of 100% humidity under *in vitro* conditions fail to survive under *in vivo* conditions wherein RH is generally very low if appropriate precautionary measures are not taken.

Hyperhydricity is one of the major anomally associated with majority of *in vitro* raised shoots/plants. Unusual cellulose and lignin depositions cause reduced cell wall pressure which increases water uptake by the cells and thus results in glassy turgescence/ hyperhydricirty of leaves and stems. This phenomenon is also termed vitrification (Roberts *et al.*, 1990; Ziv, 1995). This condition exhibits a glassy and water logged tissue appearance besides distorted growth. This ultimately influences photosynthesis, transpiration and CO_2/O_2 gas exchange (Preece and Sutter, 1991).

The leaves formed *in vitro* : a) are thin with underdeveloped leaf mesophyll and b) have little chlorophyll content, poor chloroplast organisation and low ribulose biphosphate carboxylase activity. These factors make the plantlets photosynthetically less efficient or even inactive (Desjardins *et al.*, 1994; Ziv , 1995).

Micropropagated plantlets are grown in a completely sterile environment. In nature leguminous plants are known to remain in symbiotic association with *Rhizobium*. Several plant species including leguminous ones are known to have symbiotic association with VAM under natural conditions (Mukerji *et al.*, 1996; Puthur *et al.*, 1998; Subhan *et al.*, 1998). Therefore, micropropagated plantlets of certain plant species which otherwise require the symbiotic association with microorganisms in nature may die or fail to grow luxuriantly upon transplantation *in vivo*. Hence, it becomes necessary to develop such an association for their survival (Pierik, 1988).

4.   SUCCESSFUL ESTABLISHMENT OF MICROPROPAGATED PLANTLETS THROUGH MYCORRHIZAL ASSOCIATION

In order to overcome the problems associated with the establishment of micropropagated plants and to achieve cent per cent success in their survival under field conditions several scientists and biotechnological companies (that are involved in large scale multiplication of certain important and/or rare plants that are of commercial value) have been trying to evolve suitable methods. To begin with several groups have made efforts to improve the establishment of micropropagated plantlets by trying to acclimatize them for a period of two to six weeks by a) transferring the plantlets from tissue culture medium to vessels with potting mixture consisting of one or few of commercially available materials such as vermiculite, perlite, peat, sawdust, washed sand, etc.; b) altering the water potential of medium that is used for watering; c) using growth retardants such as paclobutrazol and ancymidol which are believed to strengthen roots and shoots (Chin, 1982; Khunachak, *et al.*, 1987; Roberts *et al.*, 1990, 1992; Smith *et al.*, 1990, 1992; Ziv, 1989, 1990); d) using specially designed trays/apparatus for planting the plantlets; e) lowering the production of ethylene by the plantlets (Kevers and Gasper, 1985); f) acclimatization in microcomputer controlled chambers (Fujiwara *et al.*, 1988; Hayashi *et al.*, 1988), etc. in addition to the regulation of humidity. However, the establishment of micropropagated plantlets in field conditions with such means has been realized to be laborious, time consuming, uneconomical besides being unsuccessful in achieving cent per cent success. There had also been efforts to strengthen plantlets for better acclimatization under field conditions by exposing them to certain special treatments such as : a) significant alteration in the composition of the medium including supplementing liquid medium as a second phase over the existing semi solid medium (Aitken-Christie and Jones, 1987; Maene and Debergh, 1985); b) use of culture vessels with a permeable membrane which lowers relative humidity in the vessel due to free diffusion of water vapor (Roberts *et al.*, 1992); c) use of desiccants like polyethylene glycol, silica gel, saturated salt solutions or a lanolin layer over the semisolid medium or agar to reduce the RH (Whish *et al.*,

1992; Ziv *et al.*, 1983). Again such protocols that are practiced prior to the transfer of plantlets to the field are uneconomical and many a times lead to several other complications in the physiology of the plant system.

In nature several species of angiosperms as well as gymnosperms have symbiotic association with certain mycorrhizal fungal species (Mukerji *et al.*, 1996). Amongst mycorrhizal fungi, vesicular arbuscular mycorrhizal fungal species constitute an important group which form symbiotic association with the underground roots of various plant species. In mycorrhizal symbiosis both root of the plant and fungus sacrifice their independence and form a sort of unity to help in each others metabolic events for better survival (Pankow *et al.*, 1991). In fact, mycorrhizal association helps certain plants to survive well even under stressed environmental conditions such as high temperature, drought, salinity and mineral deficiency (Caldwell and Virginia, 1989; Harley and Smith, 1983).

As transplantation shock is a kind of stress to which plantlets are forced to face it has been realized that establishing a mycorrhizal association can help in alleviating the transplantation shock. Initial studies in this direction had been demonstrated in the afforestation and reforestation programmes with non-micropropagated i.e. nursery plantlets/seedlings. Seedlings of *Acacia auriculiformis, A. catechu, Albizia lebeck, Gliricidia sepium, Leucaena leucocephala, Prosopis juliflora*, etc. with mycorrhizal association could successfully be transplanted and established into barren lands or under stressed conditions (Mukerji *et al.*, 1996; Osonubi, *et al.*, 1991; Pardha Saradhi, 1995; Pardha Saradhi *et al.*, 1994). In general it had been shown that VAM association improves the strength of the root system as well as in increasing the extent of nodulation in leguminous tree species.

5. ESTABLISHMENT OF MYCORRHIZAL ASSOCIATION

5.1. Physiological Studies

In general, the percentage of survival of well rooted micropropagated shoots i.e. plantlets was shown to be low in many cases for instance : a) *Sesbania sesban* showed 30% survival (Pardha Saradhi *et al.* 1995; Shahnaz *et al.*, 1998; Pant *et al.*, unpublished), b*) Leuceana leucocephala* showed 20-40% survival, (Dhawan and Bhojwani, 1985; Pardha Saradhi and Alia, 1995); c) *Meconopsis simplicifolia* showed 5-30% survival depending on temperature (Sulaiman and Babu, 1993) and d) *Quercus acutissima* showed 34% survival (Kim *et al.*, 1994). In contrast, all the plantlets of *Sesbania sesban* and *Leucaena leucocephala* inoculated with *Glomus fasciculatum* or *G. macrocarpum* survived (Puthur *et al.*, 1998; Subhan *et al.*, 1998). The low percentage survival in non-mycorrhizal micropropagated plantlets could be due to a weak root system besides one or few of other factors.

At any given time, the growth (in terms of shoot height, number of leaves, fresh weight and dry weight of shoots and roots) of mycorrhizal plants will be superior to that of non-mycorrhizal plants. For instance in *Sesbania sesban*, 16 weeks after transplantation, the height, fresh weight and dry weight of shoots of mycorrhizal plantlets were 37, 109 and 52% higher than that of non-mycorrhizal plantlets, respectively (Table 1). Within 4 weeks after transplanting to garden soil, pale yellow to brownish nodules measuring 1-3 mm in diameter developed on the roots of both mycorrhizal and non-mycorrhizal micropropagated plantlets of both *Sesbania sesban* and *Leucaena leucocephala* inoculated with required specific strains of *Rhizobium* at the time of transplantation. However, micropropagated plants with mycorrhizal association showed better nodulation than non-mycorrhizal. Therefore, the roots of mycorrhizal plants were found to be healthier/stronger than that of non-mycorrhizal plants. Sixteen weeks after transplantation, the fresh weight and dry weight of roots of mycorrhizal plantlets were found to be 50 and 38% higher than that of non-mycorrhizal plantlets (Table 1). Superiority of

VAM fungal species inoculated plantlets due to mycorrhizal association to that of uninoculated plantlets can be seen also in Fig. 1 and 2. The development of superior root system in the plants inoculated with VAM fungi such as *Glomus fasciculatum* and *G. macrocarpum*, indicated that endomycorrhizal association help in the establishment of micropropagated plantlets under field conditions (Pardha Saradhi, 1995; Pardha Saradhi *et al.*, 1994).

Table 1 : Growth characteristics of mycorrhizal (M) and nonmycorrhizal (NM) micropropagated *Sesbania sesban* plantlets 16 weeks after transplantation to garden soil

Parameter	NM	M
Height (cm)	21.4±0.3[a]	29.3±0.4[b]
Fresh weight of shoot (mg)	618.2±71.3[a]	1291.8±194.3[b]
Dry weight of shoot (mg)	55.71±5.8[a]	84.7±9.8[b]
Fresh weight of root (mg)	193.1±12.1[a]	290.2±24.8[b]
Dry weight of root (mg)	24.9±3.3[a]	34.3±1.8[b]
Degree of nodulation	++	+++
Number of leaves	8.5±0.6[a]	10.0±1.4[a]

Data represent mean±standard error from three independent experiments (each with at least 12 replicates). Values for each parameter within rows followed by similar letter are not significantly different from each other (P≤0.05; Student's 't' test). ++ means moderate and +++ means profuse nodulation.

Fig. 1. Micropropagated *Sesbania sesban* plants inoculated with (right) and without (left) *Glomus fasciculatum*, 16 weeks after transplantation.

Fig. 2. Micropropagated *Leucaena leucocephala* plants inoculated with (right) and without (left) *Glomus fasciculatum*, 12 weeks after transplantation.

Increase in leaf area, fresh weight of the shoots and the roots in mycorrhizal fungus inoculated micropropagated plantlets in comparison to uninoculated controls was also observed in *Avocado* (Schubert *et al.*, 1992). Sbrana *et al.* (1994) demonstrated that the inoculation of the micropropagated plantlets of horticultural tree species viz. apple, peach and plum results in improving the overall growth besides increasing the survival percentage. Similarly Ravolanirina *et al.* (1989) observed improved growth and survival of the micropropagated oilpalm and grapevine upon inoculation with VAM fungi. There is mounting evidence that mycorrhizal fungi increase the survival rate of micropropagated plantlets by helping in the development of stronger root system and enhancement in root functions (Puthur *et al.*, 1998).

The benefits of VAM fungi had also been demonstrated in some micropropagated horticultural plants. In strawberry Elmeskaoui *et al.* (1995) observed more extensive root systems and better shoot growth in the mycorrhizal than nonmycorrhizal micropropagated plantlets. These investigators showed that VAM fungi reduces the plantlets osmotic potential. The increase in root/shoot length ratio is taken for significant preadaptation of micropropagated plantlets to acclimatization. Elmeskaoui *et al.* (1995) and coworkers attempted for the first time to coculture mycorrhizal fungus with micropropagated plantlets under *in vitro* conditions. They attempted this cumbersome procedure based on the reports that inoculation with VAM fungi do not significantly improve the growth of micropropagated plantlets during the acclimatization stage (Wang *et al.*, 1993). Inoculating roots of micropropagated *Anthyllis cytisoides* and *Spartium junceum* with VAM fungi during an early *post vitro* weaning stage shortened the acclimatization process by 8 weeks. Moreover, overall growth and nutrient status of these micropropagated plantlets was found to be improved to great extent due to mycorrhizal association (Salamanca *et al.*, 1992). Mycorrhizal formation ensures satisfactory *ex vitro* development of micropropagated plants of avocado (Azcon-Aguillar *et al.*, 1992). Wang *et al.* (1993) showed that inoculation of micropropagated plantlets of *Gerbera jamesonii*, *Nephrolepis exaltata* and *Syngonium podophyllum* with VAM fungal species *Glomus intraradices* or *G. vesiculiferum* resulted in early flowering besides reducing their mortality.

Wang *et al.* (1993) also reported that mycorrhizal micropropagated plants bear higher number and larger sized flowers.

5.2. Histochemical Studies

The roots of micropropagated plantlets inoculated with the spores or crude inoculum (consisting spores, hyphae and other components of the VAM fungal species and soil particles along with root segments of host plant such as sudan grass or *Trigonella* which are popularly used for the rapid multiplication of VAM fungi) show establishment of a good mycorrhizal association. Figs. 3 to 5 show the roots of *Sesbania sesban* and *Leucaena leucocephala* plantlets (cleared with 10% KOH, stained in 0.1% trypan blue in lactophenol and mounted in 1:1 of lactic acid:glycerol as per the protocol of Phillips and Hayman, 1970) demonstrating the establishment of a good mycorrhizal association. These root preparations show intracellular hyphae with well developed fine network of intercellular hyphae, extrametrical hyphae, arbuscules, vesicles associated with hyphae and spores. A large number of intercellular vesicles develop within 5-6 weeks, and in some cases endospores were seen in vesicles after an interval of 10 weeks (Fig. 4). Invariably, there is a significant increase in the surface area for absorption of water and various essential mineral nutrients in the mycorrhizal plants in comparison to nonmycorrhizal plants. Fine and ramifying net work of intercellular hyphae, extrametrical hyphae and cells filled with arbuscules is a clear indication of several fold enhancement in the surface area for absorption. The extent of the formation of vesicles and arbuscules is known to influence nutrient exchange processes (Hayman, 1983). This could be one of the major reasons behind the observed enhancement in the growth of mycorrhizal plants as compared to the nonmycorrhizal plants.

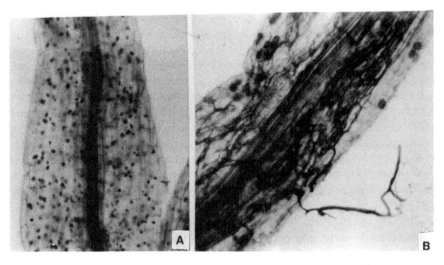

Fig. 3. Squash preparation of roots of micropropagated *Sesbania sesban* without *Glomus fasciculatum* (A) and inoculated with *Glomus fasciculatum* (B) (6 weeks after transplantation to garden soil). Note : the extensive ramification of intercellular hyphae in the roots of micropropagated *Sesbania sesban* inoculated with *Glomus fasciculatum* and the same devoid in roots of micropropagated *Sesbania sesban* non-inoculated *Glomus fasciculatum.*

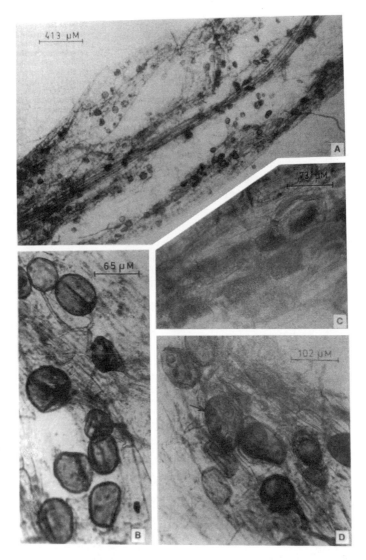

Fig. 4. A. Squash preparation of mycorrhizal roots of micropropagated *Sesbania sesban* inoculated with *Glomus fasciculatum*, 6 weeks after transplantation to garden soil. Note : ramifying intercellular hyphae with large number of vesicles and also certain deeply stained cells with arbuscules; B. An enlarged portion of A showing well developed vesicles; C. An enlarged portion of A showing some cells with deeply stained arbuscules; D. Enlarged portion from squash preparation of root from micropropagated *Sesbania sesban* inoculated with *Glomus fasciculatum*, showing some mature vesicles with endospores (arrow), 10 weeks after transplantation.

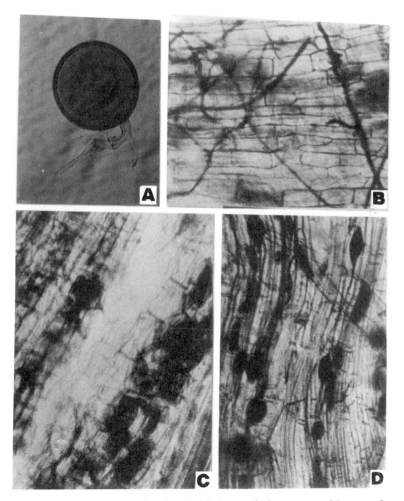

Fig. 5. Portions from the squash preparation of mycorrhizal roots of micropropagated *Leucaena leucocephala* (after 12 weeks of transplantation) cleared with KOH and stained with trypan blue, showing germinating spore (A), ramifying intercellular hypae (B), deeply stained arbuscules (C) and well developed intercellular vesicles (D).

6. SPECIFICITY OF FUNGAL STRAIN FOR POSITIVE ASSOCIATION

As in several other symbiotic associations there seems to be a sort of specificity existing between certain genera/species/strains of mycorrhizal fungi and the host plant for good mycorrhizal association. Earlier certain, physiological specificity was observed between the different *Rhododendron* cultivars and the mycorrhizal fungal isolates (Lemoine, 1992). In fact, some mycorrhizal fungal isolates had been reported to have negative effects on growth and homogeneity of certain *Rhododendron* clones. This stresses the point that it is essential to screen various strains of mycorrhizal fungi for recognizing the beneficial ones (Lemoine, 1992). Wang *et al.* (1993) observed variation in survival of the transplanted micropropagated plantlets depending on the isolate of mycorrhizal fungus used. Azcon-Aguilar *et al.* (1992) also stressed on the need to define the endomycorrhizal fungus to be used for obtaining maximum benefits out of the association.

7. SIGNIFICANCE OF VAM ASSOCIATION FOR THE ESTABLISHMENT OF MICROPROPAGATED PLANTLETS

a) Development of a superior root system

Mycorrhizal inoculation helps in the development of a stronger root system (Ponton *et al.*, 1990). Mycorrhizal association increases rooting intensity and surface area of root (Mukerji and Dixon, 1992; Puthur *et al.*, 1998). Mycorrhizal inoculation is reported to increase the formation of lateral roots in mycorrhizal, micropropagated grapevine (Schellenbaum *et al.*, 1991).

b) Increase water conducting capacity

Daft and Okusamya (1973) reported an increase in number of vascular bundles in maize following infection with mycorrhizal fungus. This reduces the resistance to water transport. The increased water conducting capacity is due to increased hydraulic conductivity which ultimately influences the water potential, transpiration rate and leaf resistance (Nelson and Safir, 1982). Graham and Syvertsen (1984) attributed the increase in water conducting capacity to the greater transpiration rate. Moreover, mycorrhizae form the primary absorptive organs of most vascular plants.

c) Increase phosphorous uptake

VAM are known to increase P uptake of many plant species even in soils with low levels P (Herrera *et al.*, 1993; Osonubi, 1991). VAM fungus mediated P uptake occurs in three stages viz. uptake by the endophyte hyphae in the soil, translocation to the hyphae inside the root cortex via entry points and release to the plant (Hayman, 1983). VAM association has been demonstrated to improve crop production (Bagyaraj, 1992; Sieverdig, 1989) especially that of legumes by increasing P uptake. In general many tropical legumes are poor scavengers of P as they have root system with sparse root hairs, this draw back is taken care by VAM association (Habte and Aziz, 1985).

d) Enhance nitrogen levels

Under low levels of nutrient solution particularly N, it was seen that plants inoculated with VAM would use nearly all the available N sources. Although the predominant form of assimilable N is the readily mobile NO_3^-, VAM also makes NH_4^+ available to the plant in higher proportion (Ponton *et al.*, 1990).

e) Increase photosynthetic efficiency

A 33% reduction in stomatal resistance and 67% reduction in mesophyll resistance to carbon dioxide uptake was observed in mycorrhizal plants in comparison to nonmycorrhizal plants. Mycorrhizal association has been shown to increase chlorophyll content and CO_2 fixation (Allen *et al.*, 1981; Hayman, 1983; Levy and Krikun, 1980).

Alberte *et al.* (1975) reported an increase of photosynthetic units as a result of mycorrhizal infection. Hayman (1983) also reported that VAM association reduces the leakiness of membranes which store the energy for photophosphorylation. Romano *et al.* (1996) had reported that mycorrhization increases the quantum yield of photochemistry as well as the electron transport and the phtosynthetic metabolism. They further reported that the density of photosynthetic units per leaf area increases in mycorrhizal plantlets. VAM association has been shown to increase the level of cytokinin (Allen *et al.*, 1980) which is also known to increase photosynthetic activity (Herold 1980).

f) Averts the possible attack by harmful soil borne pathogens

Micropropagated plants are highly susceptible to soil pathogens such as *Fusarium* (which cause wilt), *Phytophthora*, various nematodes, etc. VAM fungal association has been shown to avert the attack by such soil borne pathogens (Dugassa *et al.*, 1996; Gianinazzi *et al.*, 1982; Norman *et al.*, 1996).

g) Enhance uptake of micro and immobile nutrients

VAM forms a bridge between roots and nutrient sites in soil through hyphal connections and facilitate efficient uptake of immobile nutrients by host plants particularly those with restricted/weak root system (Aziz and Habte, 1989). Mycorrhizal infection could take care of the total Zn requirement of the higher symbiont (Swaminathan and Verma, 1979). The extraradical mycelium of the mycosymbiont links the root and the soil environment and helps the plant to use various soil nutrients (in particular trace elements which diffuse slowly towards the root surface) more effeciently (Herrera *et al.*, 1993). Symbiotic association of mycorrhizal fungal species with roots is in particular highly beneficial for plants in the areas/soils with high resource limitations as mycorrhizal roots improve the nutrient and water acquisition capacity of plants (Russo *et al.* 1993; Sbrana *et al.* 1994).

h) Helps in osmotic adjustments within the leaf

Mycorrhizae enabled plants to maintain leaf turgor and helps in conductance even at greater tissue water deficits and lower soil water potentials (Auge *et al.*, 1986). In general, mycorhizal plants have higher leaf water potentials (Nelsen and Safir, 1982). The mechanism of this sort of adaptation is still unclear but may include increased root hydraulic activity, osmotic adjustment, cell wall elasticity and stomatal conductance (Allen *et al.*, 1982; Auge *et al.*, 1986; 1987)

i) Enhance nodulation

VAM association enhances nodulation of the legumes (Bethelenfalvay *et al.*, 1987).N_2-fixing legumes have a increased requirement of phosphorous and at the early stages of transplantation of micropropagated plantlets the uptake of P from the soil will be poor due to restricted root system and sparse root hairs. Availability of P is one of the most limiting factors for nodule function in legumes. Moreover, in legumes nitrogen status is enhanced by VAM by enhancing nitrogen fixation (Munns and Mosse, 1980; Russo *et al.*, 1993). VAM forms symbiotic associations with roots of most leguminous trees (Mukerji *et al.*, 1996).

j) Alleviate environmental stress

Mycorrhizal association has been shown to increase the tolerance of plants to various environmental stresses such as drought, toxic metals, saline soil, root pathogens, high soil temperatures and adverse pH (Allen and Boosalis, 1983; Caldwell and Virginia, 1989; Harley and Smith, 1983). In addtion symbiotic association between VAM fungus and roots of micropropagated plantlets enhances the ability of later to establish and cope with stress situations arising as a result of sudden change in environmental conditions encountered as a result of their shift from *in vitro* to *in vivo* conditions.

8.   METHODS USED FOR ESTABLISHING VAM ASSOCIATION

In general subsequent to the isolation and identification of spores of various VAM fungal species, they can be individually multiplied in large scale by inoculating the roots of seedlings of certain plant species such as sudan grass (*Sorghum halepense*), *Trifolium pratense* and *Fragaria vesca* in flat earthen pots. Two to three month later, the mycorrhizal roots of these plants show profusely developed hyphae, arbuscules, vesicles as well as spores. Fig. 6 show flat earthen pots with sudan grass used for multiplication of certain VAM fungal species.

a)   The successful colonization by VAM fungi in micropropagated plants was achieved previously by using peat based substrate during transplantation (Ponton *et al.,* 1990; Sbrana *et al.,* 1994; Schubert *et al.,* 1990; Wang *et al.* 1993). However, peat had been shown to retard the growth of VAM fungi as well as the development of mycorrhizal association with the roots of host plants (Biermann and Linderman, 1983).

b)   Successful colonization of micropropagated plantlets with VAM fungi had been achieved under *in vitro* conditions (Elmeskaoui *et al.,* 1995; Pons *et al.,* 1983; Ravolanirina *et al.,* 1988). Under *in vivo* conditions the majority of the roots developed *in vitro* are replaced by new ones (Conner and Thomas, 1981) and as a result most of the mycorrhizal roots will be lost at this stage (Schubert *et al.,* 1990). Moreover, the *in vitro* methodology is lengthy and involves cumbersome practices requiring isolation and sterilization of fungal spores (Schubert *et al.,* 1990).

c)   Inoculation of VAM along with the somatic embryos which is later encased in a nutritive coating to form the artificial seed (Strullu *et al.,* 1989). Although mycorrhizal association could successfully be achieved through this method, the procedure involved is laborious, time consuming and expensive.

Fig. 6. Plants of sudan grass, fodder sorghum (*Sorghum halepense*) growing in flat earthen pots used for multiplication and maintenance of VAM fungi.

d)   A single step process, involving simple laying of crude inoculum consisting of spores and root fragments with hyphae, vesicles, arbuscules besides soil particles around the roots of micropropagated plantlets at the time of their transplantation to garden soil seems to be most suitable. This method is easier, cost effective and most importantly less time consuming. Hence, this method is suitable for commercial application. Moreover, the formation of mycorrhizal association in this method is more rapid as the crude inoculum has root fragments with various fungal propagules. This could probably be due to the presence of exudates of root fragments in the crude inoculum which are known to stimulate the growth of the VAM fungi and their penetration in to fresh roots (Azcon and Ocampo, 1984). In addition, VAM fungal association with roots of micropropagated plants is better in soil (Schubert *et al.,* 1990).

ACKNOWLEDGEMENTS:

Part of the work presented here was sponsored by New Energy and Industrial Technology Development Organization/Research Institute of Innovative Technology For the Earth (Japan).

REFERENCES

Aitken-Christie, J. and Jones, C. 1987, Towards automation: radiata pine and root hedges *in vitro, Plant Cell Tissue Organ Cult.* 8: 185-196.

Alberte, R. S., Fiscus, E. L. and Naylor, A. W. 1975,. Effects of water stress on the development of the photosynthetic apparatus in greening leaves, *Plant Physiol.* 5: 317-321.

Allen, M. F., Moore, T. S. Jr. and Christinsen, M. 1982, Phytohormone changes in *Bouteloua gracilis* infected by vesicular-arbuscular mycorrhizae, II, Altered levels of gibberellin-like substances and abscisic acid in the host plant, *Can. J. Bot.* 60: 468-471.

Allen, M. F. and Boosalis, M. G. 1983, Effects of two species of vesicular – arbuscular mycorrhizal fungi on drought tolerance of winter wheat, *New Phytol.* 93: 67-76.

Auge, R. M., Schekel, K. A. and Wample, R. L. 1986, Osmotic adjustment in leaves of VA Mycorrhizal and non-mycorrhizal rose plants in response to drought stress, *Plant Physiol.* 82: 765-770.

Auge, R. M., Schekel, K. A. and Wample, R. L. 1987, Rose leaf elasticity changes in response to mycorrhizal colonization and drought acclimation, *Plant Physiol.* 70: 175-182.

Azcon-Aguilar, C., Barcelo, A., Vidal, M. T. and Vina, G. de la. 1992, Further studies on the influence of mycorrhizae on growth and development of micropropagated avocado plants, *Agronomie* 12: 837-840.

Azcon, R. and Ocampo, J. A. 1984, Effect of root exudation on VA mycorrhizal infection at early stages of plant growth, *Plant Soil* 82: 133-138.

Aziz, T. and Habte, M. 1989, Interaction of *Glomus* species and *Vigna unguculata* in an oxisol subjected to stimulated erosion, *New Phytol.* 87, 63-67.

Bagyaraj, D. J. 1992, Vesicular arbuscular mycorrhiza: Application in agriculture, *Methods Microbiol.* 24: 360-373.

Barea, J. M., Salamanca, C. P., and Herrera, M. A. 1993, Inoculation of woody legumes with selected arbuscular mycorrhizal fungi and rhizobia to recover desertified mediterranean ecosystem, *Appl. Environ. Microbiol.* 59: 129-133.

Bethlenfalvay, G. J., Brown, M. S., Mihara, K. L. and Stafford, A. E. 1987, *Glycine- Glomus-Rhizobium* Symbiosis V, Effects of mycorrhiza on nodule activity and transpiration in soyabeans under drought stress, *Plant Physiol.* 85: 115-119.

Biermann, B. J. and Linderman, R. G. 1983, Increased *Geranium* growth using pretransplant inoculation with a mycorrhizal fungus, *J. Amer. Soc. Hort. Sci.* 108: 972-976 .

Caldwell, M. M. and Virginia, R. A. 1989, Root systems, in : *Plant Physiological Ecology-Field Methods and Instrumentation*, R.W. Pearcy, J. A. Ehleringer, H. A. Mooney and P. W. Rundel, eds., Chapman and Hall, London, pp. 367-398.

Cappelades, M., Fontarnau, R., Carulla, C. and Debergh, P. 1990, Environment influences anatomy of stomata and epidermal cells in tissue cultured *Rosa multiflora*, *J. Amer. Soc. Hort.. Sci.* 115: 141-145.

Chin, C. K. 1982, Promotion of shoot and root formation in asparagus *in vitro* by ancymidol, *Hort. Sci.* 17: 590-591.

Cooner, A. J. and Thomas, M. B. 1981, Re-estabilishing plantlets from tissue culture: a review, *Plant Propagators Soc. Proc.* 31: 342-357.

Daft, M. J. and Okusanya, B. O. 1973, Effect of *Endogone* mycorrhiza on plant growth, V, Influence of infection on the anatomy and reproductive development in four hosts, *New Phtol.* 72:1333-1339.

Desjardins, Y., Haider, C. and Riek, J. de 1994, Carbon nutrition *in vitro*, Regulation and manipulation of carbon assimilation in micropropagated systems, in : *Automation and Environmental Control in Plant Tissue Culture*, J. A. Christie, T. Kozai and A. L. S. Mary, eds., Kluwer Academic Publishers, Dordrecht, The Netherlands, pp. 493-516.

Dhawan, V. and Bhojwani, S. S. 1985, *In vitro* vegetative propagation of *Leucaena leucocephala* (Lam.) de Wit., *Plant Cell Rep.* 4: 315-318.

Donnelly, D. J., Vidaver, W. E. and Lee, K. Y. 1985, The anatomy of tissue cultured red rasperry prior to and after transfer to soil, *Plant Cell Tissue Organ Cult.* 4: 43-50.

Dugassa, G. D., Alten, H. von. and Schonbeck, F. 1996, Effects of arbuscular mycorrhiza (AM) on health of *Linum usitatissimum* L. infected by fungal pathogens. *Plant Soil* 185: 173-182.

Elmeskaoui, A., Damont J. P., Poulin M. J., Piche, Y. and Desjardins, Y . 1995, A tripartite culture system for endomycorrhizal inoculation of micropropagated strawberry plantlets *in vitro*, *Mycorrhiza* 5: 313-319.

Fuchigami, L. H., Cheng, T. Y. and Soeldner, A. 1981, Abaxial transpiration and water loss in aseptically cultured plum, *J. Amer. Soc. Hort. Sci.* 106: 519-522.

Fujiwara, L., Kozai, T. and Watanabe, I. 1988, Development of a photoautotrophic tissue culture system for shoot and/or plantlets at rooting and acclimatization stages, *Acta Hort.* 230: 153-158.

Gianinazzi, S., Gianinazzi-Pearson, V. and Trouvelot, A. 1982, Les Mycorrhizes, Partie Integrante de la plante biologie et perspectives d'Utilisation, *Coll INRA* 13, INRA, Paris.

Graham, J. H. and Syvertsen, J. P. 1984, Influence of vesicular-arbuscular mycorrhiza on the hydraulic conductivity of roots of two citrus rootstocks, *New Phtol.* 97: 277-284.

Habte, M. and Aziz, T. 1985, Response of *Sesbania grandiflora* to inoculation of soil with vesicular arbuscular mycorrhizal fungi, *Appl. Env. Microbiol.* 50(3): 701-703.

Harley, J. L. and Smith, S. E. 1983, *Mycorrhizal Symbiosis*, Academic Press, New York.

Hayashi, M., Nakayama, T. and Kozai, T. 1988, An application of the acclimatization unit for growth of carnation explants,and for rooting and acclimatization of the plantlets, *Acta Hort.* 230: 189-194.

Hayman, D. S. 1983, The physiology of vesicular-arbuscular endomycorrhizal symbiosis, *Can. J. Bot.* 61: 944-963.

Herrera, M. A., Salamanca, C. P. and Barea, J. M. 1993, Inoculation of woody legumes with selected arbuscular mycorrhizal fungi and rhizobia to recover desertified mediterranean ecosystems, *Appl. Environ. Micro. Bio.* 59: 129-133.

Kevers, C. and Gaspar, T.H. 1985, Vitrification if carnation *in vitro*: changes in ethylene production, ACC level and capacity to convert ACC to ethylene, *Plant Cell Tissue Org. Cult.* 4: 215-223.

Kim, Y. W., Lee, B. C., Lee, S. K. and Jang, S. S. 1994, Somatic embryogenesis and plant regeneration in *Quercus acutissima*, *Plant Cell Rep.* 13: 315-318.

Khunachak, A., Chin, C. K., Le, T. and Gianfagna, T. 1987, Promotion of asparagus shoot and root growth by growth retardants, *Plant Cell Tissue Organ Cult.* 11: 97-110.

Lee, N. and Wetzstein, H. Y. 1988, Quantum flux density effects on the anatomy and surface morphology of *in vitro*- and *in vivo*-developed sweetgum leaves, *J. Amer. Soc. Hort. Sci.* 113: 167-171

Levy, Y. and Krikun, J. 1980, Effect of vesicular-arbuscular mycorrhiza on *Citrus jambhiri* water relations, *New Phytol.* 85: 25-36.

Lemoine, M. C., Gianinazzi, S. and Gianinazzi-Pearson, V. 1992, Application of endomycorrhizae to commercial production of *Rhododendron* microplants, *Agronomie* 12: 881-885.

Maene, L. and Debergh, P. 1985, Liquid medium additions to established tissue cultures to improve elongation and rooting *in vivo*, *Plant Cell Tissue Organ Cult.* 5: 23-24.

McClelland, M. T., Smith, M. A. L. and Carothers, Z. 1990, The effect of *in vitro* and *ex vitro* root initiation on subsequent microcutting root quality, *Plant Cell Tissue Organ Cult.* 23: 11-123.

Mukerji, K. G. and Dixon, R. K. 1992, Mycorrhizae in Forestation. *Proc. International Symp. Rehabilitation Trop. Rainforest Ecosystem*, N. M. Majid, I. A. A. Malek, M. Z. Hamzah, and K. Jusoff, eds., Univ. Pert. Malaysia, pp 66-82.

Mukerji, K. G., Chamola, B. P., Kaushik, A. Sarwar, N. and Dixon, R. K. 1996, Vesicular Arbuscular Mycorrhiza: Potential biofertilizer for nursery raised multipurpose tree species in tropical soils, *Ann. For.* 4(1): 12-20.

Munns, D. N. and Mosse, B. 1980, Mineral nutrition of legume crops, in : *Advances in Legume Science*, R. J. Summerfield and A. H. Bunting, eds., HMSO London, pp. 115-125.

Nelson, C. E. and Safir, G. R. 1982, The water relations of well watered, mycorrhizal and non-mycorrhizal onion plants, *J. Amer. Soc. Hort. Sci.* 107(2): 271-274.

Norman, J. R., Atkinson, D. and Hooker, J. E. 1996, Arbuscular mycorrhizal fungal-induced alteration to root arcitecture in strawberry and induced resistance to the root pathogen *Phytophthora fragariae*, *Plant Soil*, 185: 191-198.

Osonubi, O., Mulongoy, K., Awotoye, O. O., Atayese, O. O. and Okali, D. U. U. 1991, Effects of ectomycorrhizal and vesicular-arbuscular mycorrhizal fungi on drought of four leguminous woody seedlings, *Plant Soil* 136: 131-143.

Pankow, W., Boller, T. and Wiemken, A. 1991, Structure function and ecology of the mycorrhizal symbiosis, *Experentia* 47: 391-394.

Pardha Saradhi, P., Mukerji, K. G. and Alia 1994, Micropropagation of fast growing tree species with high CO_2 assimilating potential and their establishment through mycorrhizal association in barren lands, *Abstracts of RITE Workshop*, October 30-November 1, 1994, Research Institute of Innovative Technology for the Earth, Kyoto, Japan, pp. 131 - 135.

Pardha Saradhi, P. 1995, Micropropagation of fast growing tree species with high CO_2 assimilating potential and their establishment through mycorrhizal association in barren lands, *RITE NOW* 16: 15 (Research Institute of Innovative Technology for the Earth, Kyoto, Japan).

Pardha Saradhi, P. and Alia 1995, Production and selection of somaclonal variants of *Leucaena leucocephala* with high carbon dioxide assimilating potential, *Energy Convers. Mgmt.* 36: 759-762.

Phillips, J. M. and Hayman, D. S. 1970, Improved procedures for clearing roots and staining parasitic and vesicular arbuscular mycorrhizal fungi for rapid assessment of infection, *Trans. Brit. Mycol. Soc.* 55: 158-162.

Pierik, R. L. M. 1988, *In vitro* culture of higher plants as a tool in the propagation of horticultural crops, *Acta Hort.* 226: 25-40.

Pons, F., Gianinazzi-Pearson, V., Gianinazzi, S. and Navatel, J. C. 1983, Studies of VA mycorrhizae *in vitro*: mycorrhizal synthesis of axenically propagated wild cherry (*Prunus avium* L.) plants, *Plant Soil* 71: 217-221.

Ponton, F., Piche, Y., Parent, S. and Caron, M. 1990, The use of vesicular-arbuscular mycorrhizae in boston fern production, I, Effect of peat-based mixes, *Hort. Science* 25: 183-189.

Preece, J. E. and Sutter, E. 1991, Acclimatization of micrpropagted plants to the greenhouse and field, in: *Micropropagation: Technology and Application*, P. C. Debergh and Zimmermann, eds., Kluwer Academic Publishers, Dordrecht pp. 71-91.

Puthur, J. T., Prasad, K. V. S. K., Sharmila, P. and Pardha Saradhi, P. 1998, Vesicular arbuscular mycorrhizal fungi improves establishment of micropropagated *Leucaena leucocephala* plantlets, *Plant Cell Tissue Org. Cult.* 53: 41-47.

Ravolanirina, F., Blal, B., Gianinazzi, S. and Gianinazzi Pearson, V. 1989, Mise au point d'une methode rapide d'endomycorhization de vitro plant, *Fruits* 44: 165-170.

Roberts, A. V., Smith, E. F. and Mottley, J. 1990, The preperation of micropropagated plantlets for transfer to soil without acclimatization, in : *Methods in Molecular Biology : Plant Cell and Tissue Culture*, J. W. Pollard, and J. M. Walker, eds., Humana Press, Clifton, New Jersey, pp. 220-227.

Roberts, A. V., Walker, S., Horan, I., Smith, E. F. and Mottley, J. 1992, The effect of growth retardants, humidity and lightining at stage III on stage IV of micropropagation in chrysanthemum and rose, *Acta Hort.* 319: 153-158.

Romano, A., Strasser, R. J., Eggenberg, P. and Martins-Loucao, M. A. 1996, Proceedings EPFL-IBIOS GS (Ecublens), CH-1015 Lasanne.

Russo, R. O., Gordon, J. C. and Berlyn, G. P. 1993, Evaluating alder-endophyte (*Alnus acuminata-Frankia-mycorrihzae*) interaction: Growth response of *Alnus acuminata* seedlings to inoculation with *Frankia* strain Ar13 and *Glomus intraradices*, under three phosphorus levels, *J. Sustainable Forest.* 1: 93-109.

Salamanca, C. P.; Herrera, M. A. and Barea, J. M. 1992, Mycorrhizal inoculation of micropropagated woody legumes used in revegetation programmes for desertified mediterranean ecosystems, *Agronomie* 12: 869-872.

Santamaria, J. M., Dan W. J. and Atkinsen, C. J. 1993, Stomata of micropropagated *Delphinium* plants respond to ABA, CO_2, light and water potential but fail to close fully, *J. Exp. Bot.* 44: 99-107.

Sbrana, C., Giovannetti, M. and Vitagliano 1994, The effect of mycorrhizal infection on survival and growth renewal of micropropagated fruit rootstocks, *Mycorrhiza* 5: 153-156 .

Schellenbaum, L., Berta, G., Ravolanirina, F., Tisserat, B., Gianinazzi, S. and Fitter, A. H. 1991, Influence of endomycorrhizal infection on root morphology in a micropropagated woody plant species (*Vitis vinifera* L.). *Ann. Bot.* 698: 135-141.

Schubert, A., Mazzitelli, M., Ariusso, O. and Eynard, I. 1990, Effects of vesicular- arbuscular mycorrhizal fungi on micropropagated grapevines: Influence of endophyte strain, P fertilization and growth medium, *Vitis* 29: 5-13.

Schubert, A., Bodrino, C. and Gribaudo, I. 1992, Vesicular- arbuscular mycorrhizal inoculation of kiwifruit (*Actinidia deliciosa*) micropropagated plants, *Agronomie* 12: 847-850.

Sieverding, E. 1989, Ecology of VAM fungi in tropical agrosystems, *Agric. Ecosyst. Environ.* 29: 369-390.

Smith, E. F., Roberts, A. V. and Mottley, J. 1990, The preperation *in vitro* of chrysanthemum for transplantation to soil, 2, Improved resistance to desiccation conferred by paclobutrazol, *Plant Cell Tissue Organ Cult.* 21: 133-140.

Smith, M. A. L. and McClelland, M. T. 1991, Gauging the quality and performance of woody plants produced *in vitro*, *In vitro Cell Devel. Biol.* 27P: 52-56.

Smith, M. A. L., Eichorst, S. M. and Rogers, R. B. 1992, Rhizogenesis pretreatments and effects on microcuttings during transition, *Acta Hort.* 319: 77-82.

Strullu, D. G., Romand, C., Callac, P., Teoule, E. and Demarly, Y. 1989, Mycorrhizal synthesis *in vitro* between *Glomus* spp. and artificial seeds of alfalfa, *New Phtol.* 113: 545-548.

Sulaiman, I. M. and Babu, C. R. 1993, *In vitro* regeneration through organogenesis of *Meconopsis simplicifolia* – An endagered ornamental species, *Plant Cell Tissue Organ Cult.* 34: 295-298.

Subhan, S., Sharmila, P. and Pardha Saradhi, P. 1998, *Glomus fasciculatum* alleviates transplantation shock of micropropagated *Sesbania sesban*, *Plant Cell Rep.* 17: 268-272.

Swaminathan, K. and Verma, B. C. 1979, Responses of three crop species to vesicular- arbuscular mycorrhizal infection on Zinc-deficient Indian soils, *New Phytol.* 82: 481-487.

Wang, H., Parent, S., Gosselin, A. and Desjardins, Y. 1993, Vesicular- arbuscular mycorrhizal peat-based substrates enhance symbiosis establishment and growth of three micropropagated species, *J. Amer. Soc. Hort. Sci.* 118: 896-901.

Wetzstein, H .Y. and Sommer, H. Y. 1982, Leaf anatomy of tissue-cultured *Liquidambar -styraciflua* (hamamelidaceae) during acclimatization, *Amer. J. Bot.* 69: 1579-1586.

Whish, J. P. M., Williams, R. R. and Taji, A. M. 1992, Acclimatization-effects of reduced humidity *in vitro*, *Acta Hort.* 319: 231-236.

Ziv, M., Meir, G. and Halevy, A. H. 1983, Factors influencing the production of hardened glaucous carnation plantlets *in vitro*, *Plant Cell Tissue Organ Cult.* 2: 55-60.

Ziv, M.1989, Enhanced shoot and cormlet proliferation in liquid cultured gladiolus buds by growth retardants, *Plant Cell Tissue Organ Cult.* 17: 101-110.

Ziv, M. 1990, The effect of growth retardants on shoot proliferation and morphogenesis in liquid cultured Gladiolous plants, *Acta Hort.* 280: 207-214.

Ziv, M. 1995, *In vitro* acclimatization, in : *Automation and Environmental Control in Plant Tissue Culture*, J. A. Christie, T. Kozai and A. L. S. Mary, eds., Kluwer Academic Publishers, Dordrecht, The Netherlands, pp. 493-516.

VAM ASSOCIATION IN WEEDS : ITS SIGNIFICANCE

Rajni Gupta and K.G.Mukerji

Applied Mycology Laboratory
Department of Botany
University of Delhi
Delhi-110007, INDIA

1. INTRODUCTION

Plant roots provide an ecological niche for many of the microorganisms that abound in soil. German Botanist Albert Bernand Frank in 1885 introduced the Greek world "mycorrhiza", which literally means "fungus root". In natural ecosystems much of the root system can be colonized by mycorrhizal fungi. Colonization is restricted to the root cortex and does not enter the vascular cylinder. Two major types of mycorrhizal associations, vesicular arbuscular mycorrhiza (VAM) and ectomycorrhiza (ECM) occupy the roots of the majority of plants in natural ecosystems throughout the world (Brundrett, 1991; Sharma and Mukerji, 1996). These associations play a valuable role in plant nutrient uptake in nature. The nature and abundance of propagules of mycorrhizal fungi determine their persistence in soil during periods of inactivity, their response to disturbance, their resistance to predation by other soil organisms and their capacity for dispersal to new locations (Brundrett, 1991; Brundrett and Abbott, 1994). Propagules of vesicular arbuscular mycorrhizal fungi are thought to include spores, dead root fragments and other colonized organic material as well as networks of hyphae in soil. Hyphal networks are considered to be especially important in many soils, where they can survive drought conditions, but are more sensitive than other propagules to soil disturbance (Jasper *et al.*, 1989). Propagules of ECM fungi include networks of mycelial strands, old mycorrhizal roots, sclerotia and basidiocarps (Ba *et al.*, 1991).

Two hypotheses concerning the distribution of mycorrhizal and non-mycorrhizal species have been proposed. First, non-mycorrhizal plant species are taxonomically related. Early workers noted that non-mycorrhizal species are confined to certain family groups. A list of families which tend to be non-mycorrhizal was compiled by Gerdemann (1968). Since that time, other families have been examined which appear to be non-mycorrhizal (Khan, 1974, 1978). Although the division of plant families into mycorrhizal and non-mycorrhizal groups is useful as a general rule, Gerdemann (1975) noted that the number of exceptions will likely increase as more species are examined. Second, the distribution of mycorrhizal and non-mycorrhizal plants may be related to successional stages within the community. Reeves *et al.* (1979) suggested that under arid conditions, colonizers of disturbed soils tend to be non-mycorrhizal. Disturbance of soils reduces the number of mycorrhizal propagules (Moorman and Reeves, 1979). Reeves *et al.* (1979) hypothesized that under conditions of low soil inoculum, the adaptive superiority of the invader plants is the ability to live without mycorrhizal fungi. Non-mycorrhizal species found on the disturbed sites were members of the Chenopodiaceae, Brassicaceae, Polygonaceae and Amaranthaceae. These families, traditionally considered non-mycorrhizal (Gerdemann, 1968), contain a high percentage of weedy species. Miller (1979) suggested that the weedy growth habit or ruderal strategy is related to the plant's ability to do well without mycorrhizae.

Weeds cause enormous losses in crop yields both in quantity and quality besides they create health and aesthetic problems. The losses are of subtle nature and are not recognized immediately, but if not controlled timely and properly, the yield losses in different cereals, pulses, fibre and oilseed crops have been reported varying from 16 to 58% under different agroclimatic conditions. Besides the losses in field crops, the weeds also cause considerable losses in plantation, horticultural and vegetable crops and interferes in production of fauna and flora in acquatic systems and impede water flow in irrigation canals. They also cause problems in highways, rail tracks and industrial sites. Nor much research work has been done in India on different aspects of crop-weed association, losses in crop yield due to weeds, weed biology and ecology and weed management in different crops and cropping system upto the first quarter of this century except the extensive identification of flora of India inclusive of Pakistan, Ceylon and Burma. 'Weed is a plant growing where it is not desired' was the first definition given by Jethro Tull (1931). The definition is simple as one may note, it may be a crop plant but is not desired where it is growing. Though *Cynodon dactylon* is among the world"s most difficult weeds but when put in lawns, it is not a weed. *Imperata cylindrica* is a tough and hard to control weeds of barren area and plantation crops but when used as a soil binder, it may not be called a weed. Weeds in general are characterized as a plant having better competitive ability and grow profusely under marginal conditions and escapes adverse conditions by shortening their life cycles (Wilson and Hartnett, 1998).

2.   STATUS OF VAM IN WEEDS

In India, the total losses in plant yield have been reported to be around Rs.500 crores during 1973-74, of which 33% was contributed by weeds alone. The amount lost due to pests, other than weeds were : disease 26%, insect 20%, rodents 6% and storage 6-8%. At ICRISAT (International Crop Research Institute for Semi Arid Tropics), Hyderabad, the percent yield reduction due to weed competition in sorghum was found upto 70%, pearl millet 60% and chickpea 40.1%. The data of the numerous field experiments conducted during 10 years, i.e., 1978 to 1987, on different aspects of weed management both under AICRP - weed control and else were computed to estimate losses in yield in cereal, pulse, oilseed, fibre and other commercial crops. Due to its wide diversity in climate, edaphic factors and system of cultivation, the weed flora also widely differs from one region to another due to different seasons, cropping patterns adopted and association with the crops. Various weeds, i.e., *Cyperus compressus, Cyperus bulbosus, Stelaria glauca, Eclipta alba, Phalaris minor, Paspalum disticum, Leptochola penicea, Phyllanthus niruri, Cynodon dactylon, Dactylolectinium aegypticum, Echinochola colonum, Eragrostis japonica, Elusine indica, Commelina benghalensis, Trianthema monogyna, Panicum colonum, Cleome viscosa, Oplismenus burmanii, Euphorbia hirta, Sida parviflora, Tribulus terestris, Corchorus tridens, Convolvulus arvensis, Amaranthus virdis, Boerhavia diffusa, Xanthium strumarium, Hibiscus lobatus, Rumex dentatus, Acalypha indica, Trigonella polycerata, Medicago indica, Bidens, Tridax, Chenopodium album,* etc. are most common weeds of various crops i.e. wheat, bajra, rice, sorghum, pearl millet, cotton and gram. Besides the crop loss aspect there is another aspect of mycorrhizal association in weeds. Research on floristics and endomycorrhizal occurrence in natural ecosystems in especially limited in the coastal regions of tropical countries (Mahadevan *et al.,* 1989). Investigations of this type are not only of great botanical interest but are also of importance in social forestry and land reclamation (Raghupathy and Mahadevan, 1991). In the present article, we are dealing with the significance of mycorrhizal association in weeds.

A number of studies have shown that vesicular arbuscular (VA) mycorrhizal colonization can significantly improve the phosphorus nutrition and yield of plants grown in soils of low fertility (Mosse, 1973). Crop plants have been used in most of these studies and relatively little information has accumulated about the significance of VA mycorrhizal colonization in natural and seminatural plant communities. This type of association occurs in many types of vegetation including tropical rain forest (Mukerji and Mandeep, 1998; Redhead, 1968), semi arid grassland (Jagpal and Mukerji, 1988; Khan, 1974), sand dune systems (Nicolson, 1955) and temperate deciduous woodland (Mosse and Hayman 1973). The presence of the endophyte has been demonstrated in many species assessment of the biological significance of infection depends upon quantitative measurement of the extent of colonization both within the community as a whole and within individual plants. Apart from one investigation in upland grassland (Sparling and Tinker, 1975) such quantitative studies have been lacking.

Mukerji and Dixon (1992) reported that VAM influences soil fertility and thus the growth and development of plants, such investigations are not only of botanical interest but are also of importance in social forestry and land reclamation. The first report of true VA mycorrhizal association in rape seed and the prospects of improving growth and yield of rape on large scale through inoculation with VAM fungi was observed by Mukerji and Kochar (1983). Recently, Gupta and Mukerji (1996) characterized and quantified the VAM association in roots of grass weeds growing in Delhi. In *Eragrostis ciliaris* and *Oplisemenus* 100% colonisation was observed. External hyphae were observed forming distinct appresoria, at the entry points in the root tissues. The spores of *Glomus fasiculatum* were present mainly in the cells of mycorrhizal roots of *Chloris*. Out of 14 grass species examined, only in 2 species viz., *Bracharia reptans* and *Setaria tomentosa*, VAM colonization and morphology was observed in six plant species over a period of two years. These plants were *Festuca rubra* L., *Holcus lanatus* L., *Lathyrus pratensis* L., *Plantago lanceolata* L., *Rumex acetosa* L. and *Trifolium pratense*. This comprises two grasses, two legumes and two dicots and includes spores which are known normally to be heavily infected by VAM (Table 1).

Table 1 : VAMycorrhizal colonization in grasses

Host	Percent of root segment colonized	No. of vesicles in per one cm root bits	No. of spores per 10g rhizosphere soil
Arachne racemosa	40	2	11
Botrichola pertusa	70	10	21
Bracharia reptans	0	0	10
Chloris dolichostachya	25	0	20
Cynodon dactylon	50	10	0
Dactylolectinium aegyptium	60	10	3
Elusine indica	75	12	0
Eragrostis ciliaris	100	15	25
Hemarthria compressa	20	7	24
Lectochola panicea	20	2	0
Oplismenus compositus	100	8	14
Phalaris minor	100	30	15
Setaria glauca	20	4	5
S. tomentosa	0	0	0

The measurement of mycorrhizal infection indicated that in some species the extent of fungal infection varied greatly both within one year and in successive years, but the causes of this variation are unknown. Mycorrhizal fungi form hyphal connections between plants of the same and different species and nutrients can be transferred from one plant to another along these connections. If these connections are abundant in the field then this could have important implications for mycorrhizal functioning and plant performance. Thus, when considering both plant nutritional benefits of VAM and the cost of the plant in terms of photosynthate, overall levels of mycorrhizal fungi in the roots of a group co-existing species may be as important as individual levels of infection within plant species. *Rumex acetosa* has previously been reported as being non-mycorrhizal (Harley and Harley, 1987) but Sanders and Fitter (1992) reported it to be VA mycorrhizal. In a Danish grassland community, Rosendahl *et al.* (1990) observed great differences in the level of morphologically distinct VAM infections in co-existing plants, although the species involved in VAM formation was not identified. The appearance of one fungal species can be distinctly different in two host plant species (Gerdemann, 1965) and that this difference may be result of variation in root tissue structure, producing distinct constraints of the fungus (Brundrett and Kendrick, 1990).

Plants with mycorrhizal associations are known to predominate in most natural ecosystems but little is known about phenological variations in mycorrhizal activity in many habitats (Brundrett, 1991). Seasonal fluctuations in numbers of mycorrhizal roots and spores have been examined in deciduous forests (Mayer and Godoy, 1989), grasslands (Sanders and Fitter, 1992), salt marsh (Van Duin *et al.,* 1989), sand dune, tropical forests and arid communities (Allen, 1983; Louls and Lim, 1987). Seasonal fluctuations in numbers of spores of VAM fungi have been observed and there may also be seasonal variations in their capacity to form mycorrhizas because newly formed spores require periods of dormancy (Gemma and Koske, 1988). Brundrett *et al.* (1994) studied the seasonal variations in the capacity of VAM and ECM fungal propagules in Jarrah forest soil to colonize host roots by using bioassays where bait plants were grown in intact soil cores. Mycorrhizal activity in Jarrah forest soil have been expected to be greatest in spring and autumn, when warm soil temperature coincide with adequate soil moisture levels. These two periods have been found to favour root growth in Jarrah forest trees. However, mycorrhizal fungal activity was not significantly greater during these times. Jarrah forest soil remains dry for extended periods in the summer (Gentilli, 1989) and mycorrhizal fungi would be expected to remain inactive at these times.

Vesicular arbuscular mycorrhizae are abundant in temperate ecosystems (Harley and Harley, 1987). This prevalence of VAM in the field has been one of the main aspect of evidence in the argument that they are generally beneficial to plants (Fitter, 1990). If mycorrhizal infection is high in a number of individual plants in a community, then potentially a significant part of the photosynthate produced by those individuals could be directed to the fungal symbionts i.e. the carbon cost to the hosts would be high. McGonigle (1987) found little temporal variation in the total fractional infection of species, in a species rich grassland, although large differences in colonization between species were observed. VAM was also found to be most abundant in roots at a depth of 3-6 cm in the soil. The tropical forests of India have received almost no attention from mycorrhizologists, although all microorganisms including mycorrhizal fungi in tropical forests most profoundly influence soil fertility and thus the growth and development of plant (Mahadevan *et al.,* 1989). Shamsuddin (1979) reported mycorrhizal association in 99 out of 200 species of Malaysian forest. The occurrence of mycorrhizal fungi in the tropical forest trees of Tamil Nadu was reported by Mohan Kumar and Mahadevan (1987). They found mycorrhizal association in 131 out of 178 species. Kannan and Lakshminarasimhan (1989) reported that 48 plant species belonging to 38 families differed in their mycorrhizal associations. Raghupathy and Mahadevan (1993) made a detailed floristic and endomycorrhizal survey in Thanjavur district, Tamil Nadu. A variety of vegetation and soil types exist in Thanjavur district. Depending upon the vegetation cover, moisture, salinity, colour and nature of the soil, nine distinct soil types were recognized, sandy, red, cultivated, wet, grassland, forest, mangrove, brackish and acquatic sediments (Nair *et al.,* 1980). Of the six genera from the families of Acaulosporaceae, Gigasporaceae and Glomaceae including 136 species, 40 species were isolated from rhizosphere soils of plants growing in Thanjavur district. There was little variation between coastal sandy soil and river sandy soil. The principal VAM species in both soils are *Gigaspora albida, G. gigantea, Glomus aggregatum, G. ambisporum, G. fasciculatum, G.heterosporum, Scutellospora heterogama, S. nigra* and *Sclerocystis pachycaulis.*

A total of 737 plant species from 121 families of angiosperms and four species of pteridophytes were examined for mycorrhizal colonization. Mycorrhizal root colonization was recorded in 372 species. The quantum of colonization was in the range of 10-90%. Of the four pteridophytes, only *Isoetes coromandelina* displayed mycorrhizal association. Out of 549 dicotyledonous species surveyed, about 301 displayed mycorrhizal association, whereas in monocotyledons, 71 species out of 188 were mycorrhizal. Of the 733 angiosperm species examined, roots of 371 species were mycorrhizal. In Fabaceae, the largest family, represented in the district, 94 species were examined of which 62 species were mycorrhizal. Of the 85 species of Poaceae, 39 were mycorrhizal. About 51 species of Cyperaceae were screened for VAM colonization but only 12 species were mycorrhizal. 50% members of the Asteraceae, Euphorbiaceae, Convolvulaceae and Rubiaceae showed mycorrhizal association. Other families such as Cyperaceae, Asteraceae, Scrophulariaceae, Amaranthaceae and Poaceae displayed less than 40% mycorrhizal colonization. Meney *et al.* (1993) reported the presence of VAM fungi in Cyperaceae and Rostionaceae. VAM fungi were detected in roots of two species of Cyperaceae (*Lepidosperma gracile* and *Tetraria capillaris*) and two species of Rostionaceae (*Alexgeorgea nitens* and *Lyginia barbata*), all representing the first record of VAM fungi were prominent from late autumn to early winter and in upto 30% of the young, new season's root as they penetrated the upper 10 cm region of the soil profile. Mycorrhizal infection was not evident during the dry summer months. Harley and Harley (1987) compiled the mycorrhizal association of plants in the British Isles and concluded that most plants in Britain were mycorrhizal. Mycorrhizal colonization was more frequent in forest areas, where there was a very higher diversity of plant species, than in the cultivated fields. It is likely that fertilizer application to cultivated fields reduces VAM species (Mosse and Hayman, 1980). A similar observation was made by Grime *et al.* (1987), who worked on the mechanism of floristic diversity with reference to mycorrhizal. Species of the families Nelumbonaceae, Nymphacaceae, Brassicaceae, Tamaricaceae, Erythroxylaceae, Oxilidaceae, Balsaminaceae, Simaroubaceae, Rhizophoraceae, Sphenocleaceae, Goddeniaceae, Myrsinaceae, Salvadoraceae, Avicenniaceae, Chenopodiaceae, Polygonaceae, Aristolochiaceae, Urticaceae, Ulmaceae, Salicaceae, Ceratophyllaceae, Arecaceae, Typhaceae and Potamogetonaceae were non-mycorrhizal. However, VAM spores were collected from the rhizosphere soils of some of the species. Although VAM colonization was studied in certain pteriodophytes (Mishra *et al.,* 1980; 1992), the incidence of VAM spores in their rhizosphere has not been investigated. Chaubal *et al.* (1982) reported the rare occurrence of VAM fungi in acquatic plants. Raghupathy *et al.* (1993) reported that 47% of the aquatic plants were mycorrhizal. They also reported percentage variation of VAM in plants of various families (Table 2). Changes in edaphic factors greatly influences mycorrhizal associations. The number of VAM spores increased in summer (Mason, 1964). Hayman (1970) suggested that the increased number of mycorrhizal spores in wheat fields during summer was related to the seasonal changes, particularly temperature and moisture.

Table 2 : Ditribution of Mycorrhizae in some major families

Family	Species examined	No. of mycorrhizal species	% infection
Leguminosae	94	62	23-90
Poaceae	85	39	12-82
Cyperaceae	51	12	13-40
Euphorbiaceae	35	188	12-72
Asteraceae	24	11	13-90
Acanthaceae	22	9	40-60
Rubiaceae	22	14	12-60
Scrophulariaceae	21	5	33-86
Convolvulaceae	19	11	28-80
Amaranthaceae	17	6	30-68

Source : From Raghupathy and Mahadevan, 1993.

Gupta and Mukerji (1996b) also characterized and quantified the VAM associations in roots of Compositae weeds growing in Delhi. Most of the plants belonging to the family Compositae are vesicular arbuscular mycorrhizal. The amount of vesicles and arbuscules formed in roots varied in all plants (Table 3). The shape of vesicles were oval to rounded. In plants of *Galinsoga parviflora*, arbuscules were densely present. They were dichotomously branched and were of dominant type. Percent formation of arbuscules was 40%, number of vesicles per one cm in root bits approximately was 5. Dominant VAM fungal spores in the rhizosphere soil consisted of *Glomus macrocarpum* and *Glomus fasciculatum*. A comparative study of diseased and healthy plants of *Sonchus arvensis* was made. Severally infected *Sonchus arvensis* plants with downy mildew *Bremia lactuca* showed presence of dense extramatrical hyphae and appresoria formation was quite distinct, while in the healthy plants number of vesicles and arbuscules per one cm was quite high. Therefore, it can be concluded that VAM increased the foliar infection in this plant (Zaidi and Mukerji, 1983).

Table 3 : VA Mycorrhizal colonization in Compositae

Host	Percent of root segment colonized	No. of vesicles in per one cm roots bits	No. of spores per gm Rhizosphere soil
Ageratum conyzoides	40	2	4
Bidens pilosa	70	20	20
Blumea bifoliata	-	-	-
Eclipta alba	70	10	12
Galinsoga parviflora	40	5	11
Gnaphalium purpureum	-	-	-
Launaea nudicaulis	60	2	6
Sonchus arvensis (Diseased)	80	20	28
S. arvensis (Healthy)	70	14	24
Tridax procumbens	50	12	12
Vernonia cinerea	80	8	20

Since 1959, the possible importance of arbuscular mycorrhizal fungi (AMF) in sand dune succession has been recognized and numerous surveys have shown AMF to be common inhabitants of temperate and tropical dune sites in coastal and inland dunes throughout the world (Koske and Gemma, 1992; 1996). In these primary successional sites, AMF arrive early with the first plant colonizers (Gemma and Koske, 1988). The development of the belowground AMF community is closely linked to above ground vegetational changes during colonization and succession. In a variety of nautral sites, AMF have been found to play a vital role in plant succession, affecting the success not only of obligately mycotrophic species, but also of species that are facultatively mycotrophic and even non-mycotrophic. Depending on the level of disturbance and existing plant population in a site, the ability of plant species to become established after dispersal site is determined in part by the population of AMF in the soil (Francis and Read, 1995). Vegetation is an effective means of slowing sand movements in dunes (Woodhouse, 1982). Koske and Gamma (1996) studied the species of AMF present in the dunes and identified changes occurring in the AMF community as well as the inoculum potential of the dune soil during the early stages of primary succession of the plant community.

AMF are common members of the biota of sand dune systems throughout the world, where they appear to play an important role in the development of plant community structure (Koske and Gamma, 1996). The simultaneous arrival of AMF and plant colonists in primary successional sites, represented by the planted areas in this study, allows the fungi to interact with the deveoping plant community from the earliest sereal stage (Nicolson, 1960). Most of the species of AMF occurring in the Province Lands have been previously reported from sand dunes of the mid-Atlantic coast of the U.S. (Koske, 1987). The formation of AMF community is closely linked to changes in the aboveground vegetation. In a variety of natural sites, AMF have been found to play a vital role in plant succession, affecting the succession not only of obligately mycotrophic species, but also of those that are facultatively mycotrophic plant species. These are prevented from colonizing a site if the VAM fungi are absent or if their abundance is below a critical level (Gange *et al.,* 1993). It is the hyphal network of AMF that first exerts these beneficial or antagonistic effects on the seedlings of potential invaders. Hyphal network is responsible for the preponderance of initial contacts with roots of seedlings (Evans and Miller, 1988).

Many researchers report no vesicular arbuscular mycorrhizal fungal colonization in aquatic macrophytes, particularly in species of the families Cyperaceae and Juncaceae (Allen, 1991). However, other investigators report VAM fungal colonization in temperate wetland plants (Read *et al.,* 1976), in rice (Dhillon and Ampornapn, 1992), in halophytic wetland grasses, and in aquatic plants (Tanner and Clayton, 1995). The presence of mycorrhizae in wet lands suggest that they are ecologically significant, but their function is not well understood (Mukerji and Mandeep, 1998). VAM fungi were also found to increase plant water uptake and reduce wilting due to water stress (Harley and Smith, 1983) or salinity (Rozema *et al.,* 1986). Salinity also increases during drought periods in the saline wet lands of North Dakota (Swanson *et al.,* 1988). Drawdowns caused by periodic droughts can significantly reduce the growth of wet land species. Lodge (1989) found that VAM colonization was greater in moist soil than in very dry or flooded soils in a green house moisture gradient experiment. VAM fungi may be advantageous to wet land plants during periods of low soil moisture associated with annual and interannual drawdowns, in wet lands with low soil phosphorus, and in wetlands with high salinity. Netzel and Vander Valk (1995) reported the VAM colonization in plants growing in the wettest zone. Percent VAM fungal colonization of plants found in other vegetational zones varied greatly, but species growing in North Dakotan wetlands generally had higher VAM fungal colonization.

Members of the Poaceae generally had the highest levels of colonization, and species in the Cyperaceae and Juncaceae had lower rates of colonization, this trend is consistent with previous reports, i.e. *Carex stricta* was reported as having no colonization by Anderson *et al.* (1984) compared to 6-22% in the present study. Rickert *et al.* (1994) reported no colonization in *Carex atherodes.* Anderson *et al.* (1984) found fungal spores of the genus *Gigaspora* at wet sites in Illinois. Spores of *Glomus manihotis* were found only in North Dakotan wet lands. The significance of different fungal floras in different wet lands is unknown and needs further study. Soil moisture and the resulting reduction - oxidation levels is the single most important environmental gradient in a wet land vegetation composition and zonation. Many of the early season grasses and sedges (*Poa pratensis, Calamagrostis canadensis, Distichlis stricta, Hordeum jubatum, Carex stricta, Carex vesicaria* and *Carex atherodes*) had a higher level of VAM fungal colonization in June than in August. A pattern of increasing VAM fungal colonization of prairie grasses during the growing season, with a decline after plant fruiting, was observed by Hetrick *et al.* (1989). The percentage of VAM fungal colonization varied significantly with host plant species in wetland ecosystems (Mukerji and Mandeep, 1998).

3. SIGNIFICANCE OF VAM IN RELATION TO CROP YIELD, PATHOGEN AND WEED

3.1. Plant-Fungal Symbiosis

Mycorrhizae, as a major interface or connection between soil and plant, are a keystone in integrated systems. In spite of the substantial progress made in recent years, we still know relatively little about the full range of functions of mycorrhizal symbiosis. The concept of sustainability in agriculture aims to conserve the productive capacity of the land, minimizing energy and resource use, and optimizing the rate turnover and recycling of matter and nutrients. Sustainable agriculture can be considered as the maintenance of soil fertility and structure over a long period of time, such that the economic yields from crop plants can be achieved through the minimum input of fertilizer necessary to reach such yields. It is difficult to develop any form of agriculture that could be truly sustainable. Instead, in sustainable systems we should be seeking a modification of existing strategies, such that fertilizer inputs are reduced but not eliminated, and that maximum use is made of the soil microbiota, in efficient nutrient capture and in cycling nutrients to the plants root system (Varma, 1995). Symbiosis causes a significant increase in plant nutrition, water, control of root pathogens and pests, the overall health of the plant, maintenance of soil fertility, sustainable agriculture, arboriculture, viticulture and flori-horticulture (Bagyaraj and Varma, 1995; Mukerji and Sharma, 1996).

The presence of fungi in organic matter zones may provide a very efficient interface to microsites, "mycorrhizosphere" where active mineralization by other microbes is taking place (Barea and Jeffries, 1995; Mukerji, et al., 1998). Storage of phosphorus as polyphosphates in the arbuscules also occurs and may be as great as 40% of the total P in the fungus. Advantage of this capacitance to accumulate nutrients when they are in the soil environment and then continue to provide nutrients to the host plant during times when the soil supply is reduced. The external hyphae bind the soil near the root and prevent the development of gaps between soil and root as the root expands and contracts due to hydrostatic tension. Also, the soil shrinks and swells with change in water content, thereby maintaining liquid continuity across the root soil interface. Hyphae, coupled with mucilaginous material (polysaccharides) produced in the mycorrhizosphere should provide an effective binding matrix in the soil. The ability of mycorrhizal hyphae to contribute to the aggregation of sand particles and stabilization and soil provides support of this suggestion especially in low input disturbed ecosystems (Varma, 1995). When fungal hyphae or germ tube contact roots, the communication between the symbionts become more refined. In earlier stages of contact, water soluble and volatile exudates function as chemical messengers to the fungi perhaps directing them to particular regions of the root where penetration is possible. Penetration of roots occurs in young areas near the root tip, where extensive suberization has not occurred (Bonfante and Bianciotto, 1995). Hydrolytic enzymes are important factors in determining strategies of colonization by fungi (Perotto et al., 1994). AM fungi, which are obligate biotrophs, produce limited and regulated quantities of cell wall dissolving enzymes. Roots release diffusible water soluble, diffusible volatile and non-diffusible compounds (Koske and Gemma, 1994). Water soluble compounds include mono and disaccharides, numerous amino acids, organic acids, flavonones and nucleotides, enzymes and other substances. Volatile exudates are composed of organic acids, alcohols, aldehydes, ketones, phenols, esters, terpenoids and miscellaneous compounds. Fungus, host, soil climate and microorganisms interact in an immensely complex fashion and the contribution of individual components may be difficult to assess. Chemical exudates act as messengers in the events of AMF development. They may elicit directional hyphal growth by affecting the molecular mechanisms by which members of the symbiosis recognize each other and permit the morphological development necessary to allow synthesis of the mycorrhiza are unknown (Smith and Gianinazzi-Pearson, 1988) (Fig. 1). Two mechanisms are established by which AMF may increase their ability to respond to roots of host plants.

First mechanism involves the stimulation of root branching after the roots have been contacted by germ tubes and development of the primary infection has occurred (Berta et al., 1990). This stimulation in rooting may result from plant hormones produced or induced by the fungus. Then it is linked to surface bound recognition factors at the host cell wall or to transfer of a plasmid from fungus to host (Annapuran, et al., 1996; Gemma and Koske, 1988). Second mechanism that brings the partner close together is the attraction of root tips to spores, whose germ tubes have not successfully located roots. High levels of exudation have been correlated with extensive fungal development in the rhizosphere and rhizoplane (Simon et al., 1993).

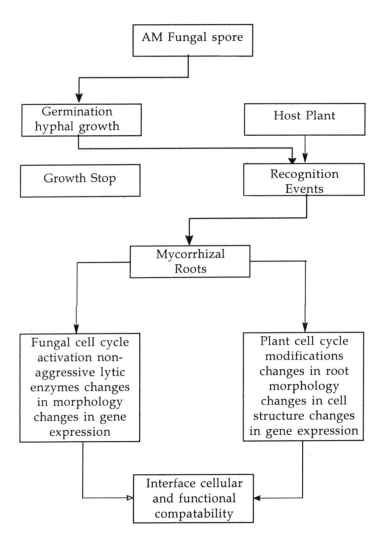

Fig. 1 : Flow chart summarizes some of the check points which control the establishment of the AM symbiosis.

Schwab and Reeves (1981) also indicated that in addition to temperature and light, mycorrhizal infection in semi arid region is greatly influenced by the season. Variation in VAM infection between plant species and seasonal changes in infection under field conditions have been ascribed to a variety of causes (Fitter, 1990). The interpretation of patterns of infection is complicated by the difficulty of distinguishing different VAM fungal species and lack of knowledge of their ecology. VAM fungi may infect plant species or even parts of the root system differentially. Spatial differentiation of different fungal types within individual plant root systems could lead to multiple infection. The spatial distribution of spores of different mycorrhiza forming species within the soil is extremely variable (Walker *et al.,* 1982) and if spores are a common form of VAM propagation,new roots exploting new areas of soil would be exposed to a different relative abundance of VAM spores in different areas. As a consequence, the potential for different parts of a root system to become infected by different species exists. Once such patterns were established, they could be maintained by re-infection from roots.

The presence or absence of mycorrhizal fungi in plant species strictly followed taxonomic divisions. It appears that a particular family determine the mycorrhizal status of a species. Stebbins (1965) speaking of the relationship between polyploidy and weediness remarked, "where there is a polyploidy there is a correlation between polyploidy and weediness, but where there isn't polyploidy you can still get weeds". The same might be said for plant groups which are non-mycorrhizal. Weedy species are common in non-mycorrhizal families such as the Amaranthaceae, Brassicaceae and Chenopodiaceae, but species from mycorrhizal families may also become weeds once established facultative species may function as hosts for wind blown spores of VA fungi and allow the inoculum to build. The mycorrhizal status of plant species which colonize a disturbed site depends on many factors. Colonizers of severely disturbed habitats have been characterised as predominantly mycorrhizal (Daft and Nicolson, 1974) or predominantly non-mycorrhizal (Miller, 1979). Lack of dependence of mycorrhizal fungi may be advantageous to colonizers under certain circumstances, but little is known about conditions for which, this is true. Water availability seemed extremely important to these sites in determining the number and proportion of mycorrhizal colonizers. Pendleton and Smith (1983) reported that certain families possess some characteristic which inhibits the formation of mycorrhizal associations. A weedy growth habit is not strongly linked to the non-mycorrhizal state.

4.   CONCLUSION

However, the rhizospheres of the same species growing on adjacent dry soils contaminated many spores and displayed VAM infection in the roots. Clearly soil moisture profoundly affects spore number. Mosse and Hepper (1975) proposed that variation in the effectiveness of colonization may be due to physiological differences between species. One further possible reason for low infection or lack of infection is the inability of the fungi to compete with other soil microorganisms. Thus unsuitable characteristics of soil and host are responsible for the near absence of VAM infection. Sanders and Tinker (1971) suggested that VA mycorrhiza simply increases the surface area for absorption of available phosphate ions and such a view is supported by the field observations. Since marked seasonal fluctuations of available P can occur in grassland (Gupta and Rorison, 1975), it is possible that mycorrhizal infection increases the capacity of plants to accumulate phosphate as it is released by other microorganisms. It is well known that mycorrhizal infection is heaviest on infertile soils. Low levels of nitrogen or phosphorus can be themselves lead to an increased intensity of infection (Mosse and Philips, 1971).

In general investigation of the distribution of VA mycorrhizal spores in relation to floristics has been undertaken only by a few workers (Grime *et al.,* 1987). Endomycorrhizal study of the colonization of mycorrhizal in the flora of different districts will give a clear picture of the distribution of VA mycorrhizal in different classes, families, genera and species.

REFERENCES

Allen, M.F. 1983, Formation of vesicular arbuscular mycorrhizal in *Atriples gardnert* (Chenopodiaceae) seasonal response in cool desert, *Mycologia* 75 : 773-776.
Allen, M.F. 1991, *The Ecology of Mycorrhiza*, Cambridge University Press, Cambridge, England.
Ammani, K., Venkateswarlu,K. and Rao, A.S. 1994, Vesicular arbuscular mycorrhizal in grasses : their occurrence, identify and development, *Phytomorph.* 44 : 159-168.
Anderson, R.C., Liberta, A.E. and Dickman, L.A. 1984, Interaction of vascular plants and vesicular arbuscular mycorrhizal fungi across a soil moisture nutrient gradient, *Oecologia* 64 : 111-117.

Annapurna, K., Tilak, K.V.B.R. and Mukerji, K.G. 1996, Arbuscular mycorrhizal symbiosis recognition and specificity, in : *Concepts in Mycorrhizal Research*, K.G. Mukerji, ed., Kluwer Academic Publishers, Netherlands, pp. 77-90.

Ba, M.A., Garbaye, J. and Dexheimer, J. 1991, Influence of fungal propagules during the early stage of the time sequence of ectomycorrhizal colonization on *Afzelia africana* seedlings, *Can. J. Bot.* 69 : 2442-2447.

Barera, J.M. and Jeffries, P. 1995, Arbuscular mycorrhizas in sustainable soil-plant systems, in : *Mycorrhiza : Structure, Function, Molecular Biology and Biotechnology*, A. Varma and B. Hock, eds., Springer verlag, Germany, pp.521-560.

Berta, G., Fusconi, A., Trotta, A. and Scannerini, S. 1970, Morphogenetic modifications induced by the mycorrhizal fungus *Glomus* strain E_3 on the root system of *Allium porrum*. L., *New Phytol.* 114 : 207-215.

Bonfante, P. and Bianciotto, V. 1995, Pre-sysmbiotic versus symbiotic phase in arbuscular endomycorrhizal fungi morphology and cytology, in : *Mycorrhiza : Structure Function, Molecular Biology and Biotechnology*, A. Varma, and B. Hock, eds., Springer-verlag, Germany, 229-250.

Brundrett, M.C. 1991, Mycorrhizas in natural ecosystem, in : *Advances in Ecological Research*, Vol. 21, A.Macfayden, M. Begon and A.H. Fitter, eds., Academic Press, London, pp. 171-313.

Brundrett,M.C. and Abbott, L.K. 1994, Mycorrhizal fungus propagules in the Jarrah forest, I, Seasonal study of inoculum levels, *New Phytol.* 127 : 539-546.

Brundrett, M.C., Murase, G. and Kendrick, B. 1990, Comparative anatomy of roots and mycorrhizal of common Ontario trees, *Can. J. Bot.* 68 : 551-578.

Chaubal, R., Sharma, G.D. and Mishra, R.R. 1982, Vesicular arbuscular mycorrhizal in subtropical aquatic and marsh plant communities, *Proc. Indian Acad. Sci.* 91B : 69-77.

Daft, M.J. and Nicolson, J.H. 1974, Arbuscular mycorrhizas in plants colonizing coal wastes inScotland, *New Phytol.* 73 : 1129-1138.

Dhillon, S.S. and Ampornpan, L. 1992, The influence of inorganic nutrient fertilization on the growth, nutrient composition and vesicular-arbuscular mycorrhizal colonization of pretransplant rice (*Oryza sativa*) plants, *Biol. Fertil. Soils* 13 : 85-91.

Evans, D.G. and Miller, M.H. 1988, Vesicular arbuscular mycorrhizal and the soil disturbance induced reduction of nutrient absorption in maize I, causal relations, *New Phytol.* 110 : 67-74.

Fittter, A.H. 1990, The role and ecological significance of vesicular arbuscular mycorrhizas in temperate ecosystems, *Agri. Ecosyst. Environ.* 29 : 137-151.

Francis, R. and Read, D.J. 1995, Mutalism and antagonism in the mycorrhizal symbioses, with special reference to impacts on plant community structure, *Can. J. Bot.* 73 (Suppl.): S1301-S1309.

Gange, A.C., Brown, V.K. and Sinclair, G.S. 1993, Vesicular arbuscular mycorrhizal fungi : a determinant of plant community structure in early succession, *Func. Eco.* 7 : 616-622.

Gemma, J.N. and Koske, R.E. 1988, Seasonal variation in spore abundance and dormancy of *Gigaspora gigantea* and in mycorrhizal inoculum potential of a dune soil, *Mycologia* 80 : 211-216.

Gentilli, J. 1989, Climate of the Jarrah forest, in : *The Jarrah forest*, B. Dell, J.J. Harvel and N. Malajuczuk, eds., Kluwer Adacemic Publishers, Netherlands, pp.23-40.

Gerdemann, J.W. 1965, Vesicular arbuscular mycorrhizal of maize and tulip tree by *Endogone fasiculata, Mycologia* 57 : 562-575.

Gerdemann, J.W. 1968, Vesicular arbuscular mycorrhizae and plant growth, *Ann. Rev. Phytopathol.* 6 : 397-418.

Gerdemann, J.W. 1975, Vesicular arbuscular mycorrhizal, in : *The Development and Function of Roots*, J.R. Torrey and D.T. Clarkson, eds., Academic Press, London, pp.575-591.

Grime, J.P., Mackey, M.L., Miller, S.H. and Read, D.J. 1987, Mechanisms of floristic diversity : a key role for mycorrhiza, in : *Mycorrhizal in the Next Decade : Practical Applications and Research Priorities*, D.M. Sylvia, L.L. Hung and J.H. Graham, eds., North American Conference on Mycorrhizal, Gainesville, Fla. pp.151.

Gupta, P.L. and Rorison, I.H. 1975, Seasonal differences in the availability of nutrients down a podsolic profile, *J. Ecol.* 63 : 521.

Gupta, Rajni and Mukerji, K.G. 1996a, Studies on the vesicular arbuscular mycorrhizal association in the root of common Delhi grasses, *Phytomorphology* 46 (2) : 117-121.

Gupta, Rajni and Mukerji, K.G. 1996b, Vesicular arbuscular mycorrhizal in plants in compositae: Occurrence identify and significance, *Comp. News Letter* 29 : 40-46.

Harley, J.L. and Harley, E.L. 1987, A checklist of mycorrhiza in the British Flora, *New Phytol.* (Supplement) 105 : 1-102.

Harley, J.L. and Smith, S.E. 1983, *Mycorrhizal symbiosis*, Academic Press, London.

Hayman, D.S. 1970, *Endogone* spore numbers in soil and VA mycorrhiza in wheat as influenced by seasons and soil treatments, *Trans. Br. Mycol. Soc.* 54 : 53-63.

Hetrick, B.A.D., Wilson, G.W.T. and Hartnett, D.C. 1989, Relationship between mycorrhizal dependence and competitive ability of two tall prairie grasses, *Can. J. Bot.* 67 : 2608-2615.

261

Jagpal, R. and Mukerji, K.G. 1988, VAM fungi - Tool for reforestation, *Phytol. Res.* 1:31-35.

Jasper, D.A., Robson, A.D. and Abbott, L.K. 1989, Soil disturbance reduces the infectivity of external hyphae in VA mycorrhizal fungus maintain infectivity in dry soil, except when the soil is disturbed, *New Phytol.* 112 :101-107.

Kannan, K. and Lakshminarasimhan, C. 1989, Survey of VAM maritime strand plants of Point Calimere, in : *Mycorrhizae for Green Asia,* A. Mahadevan, N. Raman and K. Wata Rafan, eds., University of Madras, Madras.

Khan, A.G. 1974, The occurrence of mycorrhizal in halophytes, hydrophytes and xerophytes and of *Endogone* spores in adjacent soils, *J. Gen. Microbiol.* 81 : 9-14.

Khan, A.G. 1978, Vesicular arbuscular mycorrhizae in plants colonizing black wastes from bituminoous coal mining in the Illawara region of New South Wales, *New Phytol.* 81 : 57-63.

Koske, R.E. 1987, Distribution of VA mycorrhizal fungi along a latitudinal temperature gradient, *Mycologia* 79 : 55-68.

Koske, R.E. and Gemma, J.N. 1992, Restoration of early and late successional dune communities at Province Lands, Cape Cod National Seashore, Tech. Rep. NPS/NARURI/NRTR-92/03, Coop. NPS Studies Unit, University of Rhode Island, Narraganselt Bay Campus, Narragasett, RI.

Koske, R.E. and Gemma,J.N. 1994, Fungal interactions to plant prior to mycorrhizal formation, in : *Mycorrhizal Functioning : An Integrative Plant-Fungal Process*, M. Allen, ed., Chapman and Hall, New York, pp.3-36.

Koske, R.E. and Gemma, J.N. 1996, Arbuscular mycorrhizal fungi in Hawaiian sand dunes : Island of Kanai, *Pacific Science* 50 : 36-45.

Lodge, D.J. 1989, The influence of soil moisture and flooding on formation of VA endo and ectomycorrhizal in *Populas* and *Salix, Plant Soil* 117 : 243-253.

Louis, I. and Lim, G. 1987, Spore density and root colonization of vesicular arbuscular mycorrhizal in tropical soil, *Trans. Brit. Mycol. Soc.* 88 : 207-212.

Mahadevan, A., Raman, N. and Natarajan, K. 1989, *Mycorrhiza for Green Asia*, University of Madras, Madras.

Mason, D.T. 1964, A survey of the number of *Endogone* spores in soil cropped with barley, raspberry and strawberry, *Hostic. Res.* 4 : 98-103.

Mayer, R. and Godoy, R. 1989, Seasonal patterns in vesicular arbuscular mycorrhiza in melic beech forest, *Agric. Ecosys. Environ.* 29 : 281-288.

McGongile, T.P. 1987, Vesicular arbuscular mycorrhizas and plant performance in a semi natural grassland, D. Phil Thesis, University of New York.

Meney, K.A., Dixon, K.W., Scheltema, M. and Pate, J.S. 1993, Occurrence of vesicular mycorrhizal fungi in dryland species of Restionaceae and Cyperaceae from South West Western Australia, *Australe J. Bot.* 41 : 733-737.

Miller, R.M. 1979, Some occurrences of vesicular arbuscular mycorrhiza in natural and disturbed ecosystems of the Red Desert, *Can. J. Bot.* 57 : 619-623.

Mishra, R.R., Sharma, G.D. and Gathphon, A.R. 1980, Mycorrhizas in the ferns of North Eastern India, *Proc. Indian Natl. Sci. Acad.* B 46 : 546-551.

Mohan Kumar, V. and Mahadevan, A. 1987, Vesicular arbuscular mycorrhizal association in plants of Kalakad reserve forest India, *Angew. Bot.* 61 : 255-274.

Moorman, T. and Reeves, F.B. 1979, The role of endomycorrhizal in revegetation practices in the semi arid west II, A bioassay to determine the effect of land disturbance on endomycorrhizal populations, *Ame. J. Bot.* 66 : 14-18.

Mosse, B. 1973, Advances in the study of vesicular arbuscular mycorrhizas, *A. Rev. Phytopath.* 11 : 171-196.

Mosse, B. and Hayman, D.S. 1980, Mycorrhizal in agricultural plants, in : *Tropical Mycorrhizal Research,* P. Mikola, ed., Clarendon Press, Oxford, pp.213-230.

Mosse, B. and Hepper, C.M. 1975, Vesicular arbuscular mycorrhizal infection in root organ culture, *Physiol. Plant Pathol.* 5 : 215-223.

Mosse, B. and Philips, J.M. 1971, The influence of phosphate and other nutrients on the development of vesicular arbuscular mycorrhiza in culture, *J. Gen. Microb.* 69 : 157.

Mukerji, K.G. and Dixon, R.K. 1992, Mycorrhizal in Reforestation, in : *Rehabilitation of Tropical Rainforest Ecosystem : Research and Development Priorities*, N.M. Majid, A.K. Malik, Md. Z. Hamzah and K. Tusoff,eds., Salanagar, Malaysia, University Pertanian Malaysia, pp. 62-82.

Mukerji, K.G. and Kochar, B. 1983, Vesicular arbuscular mycorrhiza in rape seed plant, 6th Congress International SUR LE COLZA, Paris, pp. 945-950.

Mukerji,K.G. and Mandeep 1998, Mycorrhizal relationships of wetlands and river associated plants, in : *Ecology of Wetland and Associated System,* S.K. Majumdar, ed., Pennsylvania Academy of Sciences, U.S.A. 240-257.

Mukerji, K.G., Mandeep and Varma, A. 1998, Mycorrhizal microorganisms : screening and evaluation, in : *Mycorrhiza Manual*, A. Varma, ed. Springer Verlag, Berlin, pp.85-99.

Mukerji, K.G. and Sharma, M. 1996, Mycorrhizal relationships in forest ecosystems, in : *Forest : Global Prospective,* S.K. Majumdar, E.W. Miller and F.J. Brenner, eds., Pennsylvania Academy of Science, U.S.A., pp.95-125.

Netzel, P.R. and Vander Walk, A.G. 1995, Vesicular arbuscular mycorrhizae in prairie pathole wetland vegetation in Iowa and North Dakota, *Can. J. Bot.* 74 : 883-890.

Nicolson, T.H. 1955, The mycotrophic habit in grass, Ph.D. Thesis, University of Nottingham, Nottingham, U.K.

Nicolson, T.H. 1960, Mycorrhiza in the Gramineae II, Development of different habitats particularly and dunes, *Trans. Br. Mycol. Soc.* 43 : 132-145.

Pendleton, R.L. and Smith, B.N. 1983, Vesicular arbuscular mycorrhizal of weedy and colonizer plant species at disturbed sites in Utah, *Oecologia* 59 : 296-301.

Perotto, S., Brewin, N.J. and Bonfante, P. 1994, Colonisation of pea roots by arbuscular mycorrhizal fungi and rhizobia : an immunological comparison using monoclonal antibodies as probes for plant cell surface components, *Mol. Plant Microbe Interact.* 7 : 91-98.

Raghupathy, S. and Mahadevan, P. 1991, Ecology of vesicular arbuscular mycorrhizal fungi in a coastal tropical forest, *Ind. J. Microbiol. Ecol.* 2 : 1-9.

Ragupathy, S. and Mahadevan, A. 1993, Distribution of vesicular arbuscular mycorrhizal in the plants and rhizosphere soils of the tropical plains, Tamilnadu, India, *Mycorrhiza* 3 : 123-136.

Read, D.J., Koucheki, H.K. and Hodgson, J. 1976, Vesicular arbuscular mycorrhiza in natural vegetation systems, *New Phytol.* 77 : 641-653.

Redhead, J.F. 1968, Mycorrhizal associations in some Nigerian Forest trees, *Trans. Br. Mycol. Soc.* 51-377.

Reeves, F.B., Wagner, D., Moorman, T. and Kiel, J. 1979, The role of endomycorrhizae in revegetation practices in the semi arid west 1, A comparison of incidence of mycorrhizae in severely disturbed vs. natural environments, *Amer. J. Bot.* 66 : 6-13.

Rickert, D.H., Sancho, E.O. and Ananth, S. 1994, Vesicular arbuscular endomycorrhizal colonization of wetland plants, *J. Environ. Qual.* 23 : 913-916.

Rosendahl, S., Rosendahl, C.N. and Sochting, V. 1990, Distribution of VA mycorrhizal endophytes amongst plants from a Danish grassland community, *Agri. Ecosys. Environ.* 29 : 329-336.

Rozema, J., Arp, W., Van Diggelen, J., Van Esbrock, M. and Brockman, R. 1986, Occurrence and ecological significance of vesicular arbuscular mycorrhizal in the salt marsh environment, *Acta Bot. Nether.* 35 : 457-467.

Sanders, F.E. and Tinker, P.B. 1973, Phosphate flow into mycorrhizal roots, *Pest. Scie.* 4 : 385-395.

Sanders, I.R. and Fitter, A.H. 1992, The ecology and functioning of vesicular arbuscular mycorrhiza in co-existing grassland species, *New Phytol.* 120 : 517-524.

Schwab, S. and Reeves, F.B.1981, The role of endomycorrhiza in revegetation practices in the semi arid west III, Vertical distribution of vesicular arbuscular mycorrhiza inoculum potential, *Amer. J. Bot.* 68 : 1293-1297.

Shamsuddin, M.N. 1979, Mycorrhizas of tropical forest trees, in : *International Symposium on Tropical Ecology,* University of Malaya, Kualalumpur, pp.173.

Simon, L., Bousquet, J., Levesque, R.C. and Lalonde, M. 1993, Origin and diversification of endomycorrhizal fungi and coincidence with vascular land plants, *Nature* 363 : 67-69.

Smith, S.E. and Gianinazzi-Pearson, V. 1998, Physiological interactions between symbionts in vesicular arbuscular mycorrhizal plants, *Ann. Rev. Plant Physiol. Plant Mol. Biol.* 39: 221-244.

Sparling, G.P. and Tinker, P.B. 1975, Mycorrhizas in Pennine grasslands, in : *Endomycorrhiza,* F.E. Sanders, B. Mossee and P.B. Tinker, eds., Academic Press, London, pp.545-560.

Stebbins, L.G. 1965, Discussion of paper by Dr. Stebbins, in : *The Genetics of Colonizing Species,* H.G. Baker and C.L. Stebbins, eds., Academic Press, New York, pp.192-195.

Swanson, G.A., Winter, T.C., Adomaitis, V.A. and Lal Baugh, J.W. 1988, Chemical characteristics of prairie lakes in South Central North Dakota : their potential for influencing use by fish and wild life, U.S. Fish. Nilde. Serv. Tech. Rep. No.18.

Tanner, C.C. and Clayton, J.S. 1985, Effects of vesicular arbuscular mycorrhizas on growth and nutrition of a submerged aquatic plants, *Aquat. Bot.* 22 : 377-386.

Tull, J. 1931, Hoarse hoeing husbandry, Barkshire,M D C C 33.

Van Duin, W.E., Rozema,J. and Ernst, W.H.O. 1989, Seasonal and spatial variations in the occurrence of vesicular arbuscular (VA) mycorrhiza in salt marsh plants, *Agric. Ecosys. Environ.* 29 : 107-110.

Varma, A. 1995, Arbuscular Mycorrhizal Fungi, The State of Art, *Crit. Rev. Biotech.* 15 : 179-199.

Walker, C., Mize, C.W. and McNabb, Jr. H.S. 1982, Populations of endogonaceous fungi at two locations in Central Iowa, *Can. J. Bot.* 60 : 2518-2529.

Wilson, G.W.T. and Hartnett, D.C. 1998, Interspecific variation in plant responses to mycorrhizal colonization in tallgrass praisie, *Am. J. Bot.* 85: 1732-1738.

Woodhouse, W.W. 1982, Coastal sand dunes of the US, in : *Creation and Restoration of Coastal Plant Communities,* R.R. Lewis, ed., CRC Press, Boca Raton, Fl., pp.1-44.

Zaidi,R. and Mukerji, K.G. 1983, Incidence of vesicular arbuscular mycorrhizal in diseased and healthy plants, *Indian J. Plant Pathol.* 1 : 24-26.

ORCHIDACEOUS MYCORRHIZAL FUNGI

Archana Singh and Ajit Varma

School of Life Sciences
Jawaharlal Nehru University
New Delhi-110067, INDIA

1. INTRODUCTION

The underground world harbors one of the most common symbiotic associations between plant root and fungus called 'mycorrhiza' (Smith and Read 1995; Varma 1998, 1999) which is the beneficial association between soil-borne fungus and the roots of about as many as ninety percent of terrestrial plants. Trappe (1996) defined mycorrhizas as 'dual organs of absorption formed when symbiotic fungi inhabit healthy organs of most terrestrial plants'. More than 6000 fungal species are capable of establishing mycorrhizas with about 240,000 plant species, but relatively few anatomical types of plant-fungus interactions result from such impressive biodiversity (Bonfante and Perotto 1995; Varma 1995; 1998). Mycorrhizae are classified on the basis of extent of root penetration, production of external mantle or sheath and the inter- and intracellular structures they form once inside the plant root (Read, 1999) (Fig. 1).

By definition of mycorrhiza, the only true mycorrhizal relationship with orchids is that which occurs between the roots of members of orchidaceae species and symbiotic fungi, but it also includes the association between fungi and the protocorm, the structure intercalated between the embryo and the seedling (Hadley, 1982). The Orchidaceae contains more species than almost any other flowering plant family, something between 12,000 and 30,000 species, being rivaled only by the Compositae. The family is cosmopolitan but with many more species in the tropics than in the temperate regions (Goh *et al.,* 1992). Orchids differ from most other higher plants in that their seeds, which are minute, generally undifferentiated, and with little stored food, are incapable of germination solely in a dilute mineral salts solution but require an external source of organic carbon. It has been shown that in the laboratory, many orchids may be germinated, on media containing inorganic salts and sucrose, but it is generally agreed (Harley, 1969; Peterson and Farquhar, 1994) that in nature orchids germinate in the presence of endophytic fungi that infect germinating seed and later the roots or other absorbing organs of young and adult plants. This symbiotic association is unlike other mycorrhizal associations in that the fungus supplies a source of carbon to a subterranean, achlorophyllous structure.

Mycorrhizal Biology, edited by Mukerji *et al.*
Kluwer Academic/Plenum Publishers, 2000

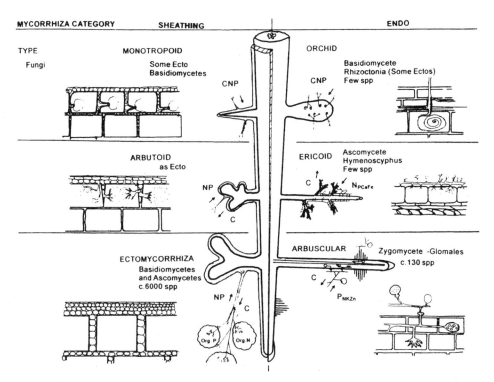

Fig. 1. The structural and major nutrient pathways of six recognised types of mycorrhiza. The defining structures of each type are fungal pegs (Monotropoid), Hartig net and intracellular penetration (Arbutoid), Hartig net, mantle, external mycelial network (Ectomycorrhiza), pelotons (Orchid), hyphal complexes in hair roots (Ericoid) and arbuscules or hyphal coils (Arbuscular). C carbon; N nitrogen; P phosphorus. Arrows indicate direction of flow. c.f. Read, 1999.

Commercial interest in optimizing germination of the tiny orchid seed, either symbiotically or aseptically, has given impetus to the investigation of the precise role of the fungus in the nutrition of the host. The importance of this work is partly because of the commercial value and beauty of the orchid blooms and the need, on the part of growers to raise new varieties including hybrids from seeds. As a result, much more is known in the trade about some aspects of orchid growth and physiology than has been published, and much information gained by amateurs is of extreme practical value but difficult of exact interpretation (Harley and Smith, 1983).

2.   THE ENDOPHYTE: TAXONOMIC STATUS AND SPECIFICITY

Orchid endophytes mostly belong to the form-genus *Rhizoctonia* L. (James *et al.*, 1998; Warcup and Tablot, 1980), some of which have subsequently produced sexual phases (Currah, 1987). This genus, based on mycelial states of certain fungi, is ill-defined, and perfect stages of rhizoctonias are known to occur in both the Basidiomycetes and the Ascomycetes (Warcup and Talbot, 1966, Whitney and Parmeter, 1964). Orchid rhizoctonias are recognized by the general features of the mycelium in culture, the presence in most isolates of short inflated segments which resemble spores (Fig. 2) and the formation of loose aggregates of hyphae regarded as poorly developed sclerotia or resting bodies (Hadley, 1984). In addition to the rhizoctonias, other saprophytic and plant pathogenic fungi including *Marasmius, Xerotus,*

Fomes, Armillaria, Ceratobasidium, Ceratorhiza, Epulorhiza, Thanatephorus and *Tulasnella* are important symbionts (Saunders and Owens, 1998; Terashita and Chuman, 1987). Historically, some isolates, which are often obtained in the sterile, anamorph state, have been difficult to assign to a species, due to the limited morphological features available (Saunders and Owens, 1998). Induction of fructification is occasionally possible (Warcup, 1981), however, sometimes elusive (Ramsay *et al.,* 1986) and, even where possible, can be a time consuming affair. The rhizoctonias include widespread saprophytes and aggressive plant pathogens like *Rhizoctonia solani* (*Thanatephorus cucumeris*) and *R. goodyerae-repentis* (*Ceratobasidium cornigerum*) which can stimulate germination of various orchids, leading to a compatible intracellular infection. Compatible infection normally results in symbiosis, evident as a stimulant to growth of the protocorm (Hadley and Pegg, 1989; Hadley and Williamson, 1971). Deviations from this pattern are of two basic sorts. Occasionally the fungus is contained and eliminated, or, with vigorous and aggressive strains, the compatible phase is succeeded by an incompatible (pathogenic) one in which the fungus spreads into the host cells, including the meristem, and eventually kills the protocorm. *R. solani* and binucleate *Rhizoctonia* spp. are divided into anastomosis groups AG-1 to AG-10, and AG-A to AG-S, respectively. Each anastomosis group has a moderately specialized host range (Adams, 1988; Sneh *et al.,* 1991). Despite the wide distribution of *Rhizoctonia*, no investigation of the colonization of orchid habitats by *Rhizoctonia* spp. other than by mycorrhizal isolates has been carried out.

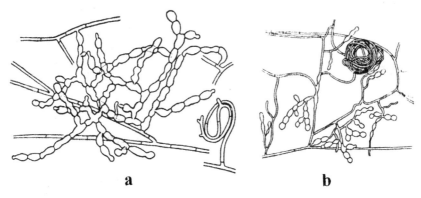

a   **b**

Fig. 2. Two *Rhizoctonia* isolates from orchids. a, *Rhizoctonia mucoroides* (X 160). b, *Rhizoctonia repens* (X 160). c. f. Hadley, 1984.

The degree of host specificity in orchid endophytes is uncertain (Masuhara and Katsuya, 1994), particularly regarding the endophytes of heterotrophic species, since the failure to germinate the seeds of most of these could be due to the specificity of the relationship. Studies of fungi isolated from autotrophic orchids and from roots of other species show a range of specificities in germination trials with orchid seeds (Arditti, 1992; Milligan and Williams, 1988). Warcup (1975) studied the symbiotic germination of seeds of 14 orchids and 17 strains of endophytic fungi and, out of the 229 potential plant-fungus associations tested, found that 88 were beneficial to the plant. Some fungi such as *Tulasnella calospora* had a very wide host range, infecting 11 out of the 14 orchids tested. Similarly, Hadley (1982) who investigated 117 potential interactions between orchid seeds and non-host fungi found that 43 combinations gave infection and stimulation of growth of seeds and a further 29 gave infection without growth stimulation or parasitism. Smreciu and Currah (1989) suggested that in some cases, isolates of adult plants of one species do not necessarily stimulate seed germination or protocorm development of the same species. Some fungal isolates, particularly of *Ceratobasidium* and *Tulasnella*, can stimulate germination of many hosts (Hadley, 1982) but it appears that some orchids have much greater specificities than others do. Xu *et al.* (1989)

proposed that as in *Gastrodia*, the fungi involved in the stimulation of seed germination might be different from those in adult plants.

The identity of the mycorrhizal fungi, their specificity and habitat requirements are, with few exceptions, unknown. The use of genetic fingerprinting and other molecular techniques will greatly facilitate the identification of fungal partners, especially those, which have proved difficult to isolate and grow axenically.

3.   TYPES AND CHARACTERISTICS OF THE SYSTEM

3.1.   Protocorms and Symbiotic Fungi

Most of the information on the colonization process has come from *in vitro* experiments, primarily with terrestrial orchid species. Seeds of many species can be germinated *in vitro* either with various carbon sources, e.g., cellulose with an appropriate fungal species. There is no evidence that embryos/protocorms produce chemotactic substances that attract symbiotic fungal hyphae to them (Williamson and Hadley, 1970) and there is very little information on the first contact of hyphae with the embryo/protocorm. The orchid embryo comprises zones of cells that are functionally differentiated by the presence or absence of mycorrhizal fungi (Clements, 1988) (Fig. 3). There is some variability in the site of hyphal entry with the suspensor end of the embryo being the major entry point of several species (Clements ,1988; Peterson and Currah, 1990; Richardson *et al.,* 1992) and the epidermis hairs in others (Rasmussen, 1990). The best evidence for the latter comes from a time sequence of protocorm colonization using light microscopy in which it was shown that penetration of hairs occurred *via* a single narrow penetration peg (Williamson and Hadley, 1970). Entry of fungal hyphae through the suspensor end of the embryo involves growth into dead cells (Rasmussen, 1990; Richardson *et al.,* 1992). Subsequent to entry of hyphae into developing protocorms, two kinds of infection have been reported.

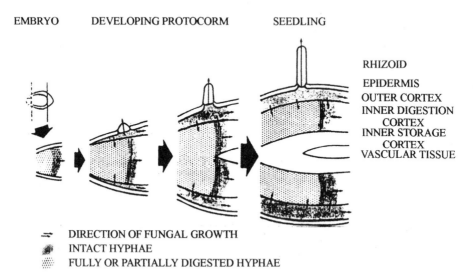

EMBRYO   DEVELOPING PROTOCORM   SEEDLING

RHIZOID

EPIDERMIS
OUTER CORTEX
INNER DIGESTION
    CORTEX
INNER STORAGE
    CORTEX
VASCULAR TISSUE

⇀   DIRECTION OF FUNGAL GROWTH
      INTACT HYPHAE
      FULLY OR PARTIALLY DIGESTED HYPHAE

Fig. 3. A schematic diagram showing the cell layers found in the protocorm and regions of fungal infection. c. f. Clements, 1988.

In the majority of species, fungi form extensive coils within the first few layers of cortical cells but are rarely found in the epidermal layer. Where the fungus infects one or two cells deeper into the cortex, it produces much denser coils, called pelotons (Fig. 4a, b, c), which occupy most of the volume of cells. In these cells, fungus is believed to be digested by the host and eventually sinks into a granular mass, a process that has been termed tolypophagy (Fig. 5a). This kind of infection has been found in *Corallorhiza, Neottia* and *Galeola septentrionalis* (Barmicheva, 1989; Terashita, 1985; Weber, 1981). In the second type of infection which has been reported in *Galeola, Zeuxine purpurascens, Cystorchis aphylla* and *Gastrodia* (Campbell, 1962; 1963), the fungi again form coils and infect the outer cortex of the rhizome and many of the epidermal cells. The coils comprise a less dense growth of hyphae than in the first type and on penetrating the middle cortex, growth is typically limited to a single hypha which does not proliferate or form dense peloton coils but is arrested and then digested by the plant, a process called ptyophagy (Fig. 5b). At the point of penetration of these cells, the wall is often thickened around the fungus forming a curious tubular in-growth, which may prevent the fungus proliferating as in the peloton cells. These pelotons are separated from the protocorm cell cytoplasm by plasma membrane and interfacial matrix material of an unknown chemical nature (Peterson and Currah, 1990; Richardson *et al.*, 1992). Nutrient exchange must occur across this interface, the hyphal wall and the hyphal plasma membrane although this has not been demonstrated experimentally (Peterson and Farquhar, 1994). Pelotons are digested by the host cell shortly after their formation. In some cells, clumps of collapsed hyphae are encased in aniline-blue positive material (Peterson and Currah, 1990; Richardson *et al.*, 1992) which may compartmentalize the hydrolytic enzymes to this region of the cell. Under certain cultural conditions, endophytic fungi become parasitic on the protocorm, ultimately leading to degradation of this structure (Hadley, 1982).

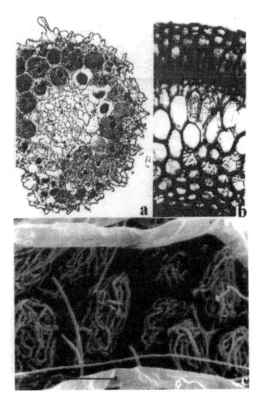

Fig. 4. Organization of mycorrhized roots. a, section of *Dactylorhiza purpurella* root, showing dense infection of the cortex with clumps of digested material and secondary pelotons in some cells (X 100); b, a hand section of *Dendrochilum carnosum* root showing occasional infection of cells of outer cortex (X 200); c, squash preparation of rhizome of *Goodyera repens*, showing infection of cells and development of pelotons (X 300). c. f. Hadley, 1984.

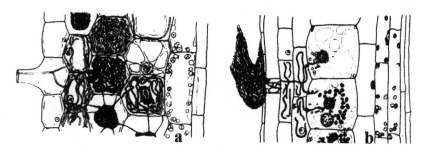

Fig. 5. Mycorrhizal infection in orchid roots. a, tolypophagy in *Platanthera chlorantha*, showing one layer of host cells adjacent to the epidermis and two layers of digestion cells; b, ptyophagy in *Gastrodia callosa*, showing two layers of passage cells and the phagocyte layer with hyphae liberating the cytoplasm, which is absorbed into spherical ptyosomes. c. f. Hadley, 1984.

3.2. Roots and Symbiotic Fungi

There is a paucity of observations on colonization events in roots of orchids. In some species, root hairs may be the sites for hyphal entry (Currah *et al.*, 1988) or epidermal cells may be penetrated following the proliferation of hyphae on the root surface. In *Vanilla* species, the root epidermis is apparently penetrated directly by hyphae, which then enter the short cells of the dimorphic exodermis (Alconero, 1969). In *Spathoglottis plicata* Bl., *Calanthe pulchra* (Bl.) Lindl. and occasionally in *Arundina graminifolia* (Don) Hochr., root hairs adjacent to sites of cortical cell colonization elongated and become distorted although the root hair colonization was not checked (Hadley and Williamson, 1972). As in protocorms, fungal hyphae form pelotons in cortical cells which are eventually digested (Hadley, 1982).

4. MYCORRHIZATION: INFECTION PROCESS AND DEVELOPMENT OF SYMBIOSIS

There are two stages of infection in the life cycle of most orchids: the primary infection of the germinating seedling, and the reinfection of the new roots of the adult. The latter is especially important in those forms which perennate as uninfected tubers of rhizomes and which form a new root system when dormancy is broken. The source of infection may be the soil or sometimes the tuber, where the fungi may persist on the surface or even in the tissues (Harley and Smith, 1983).

In the absence of fungal infection (or exogenous sugars and other essentials), an orchid embryo takes up water, swells, bursts the testa and produces a few epidermal hairs. At this germination stage, the cells may contain a number of starch grains which are hydrolyzed very slowly and do not support the growth of the protocorm (Purves and Hadley, 1975). If a suitable fungus is present, single fungal hyphae penetrate the wall of either the epidermal hairs or epidermal cells near the suspensor of the embryo (Uetake *et al.*, 1997). It is not known whether cellulolytic or pectolytic enzymes are important in this process, but if they are, their action must be very localized because usually minimal disruption of the cells occurs. However, the tubers of both the achlorophyllous *Gastrodia elata* and *Galeola septentrionalis* when attacked by *Armillaria mellea* suffer local tissue breakdown before the symbiosis becomes established and a flowering scape is formed (Harley and Smith, 1983).

As the fungal hypha penetrates the cell of the embryo or root, it progressively permeates the host cell which causes an invagination of the plasmalemma of the host and becomes surrounded by a layer of host cytoplasm which appears to remain healthy and protoplasmic streaming continues. The host cells appear physiologically active and contain numerous mitochondria, well-developed endoplasmic reticulum, dictyosomes and vacuoles of variable size, and few or no starch grains in their plastids. The nuclei of infected orchid cells as well as those in their immediate surroundings are hypertrophied (Fig. 6). The nuclei are much more prominent than those of fungus-free non-infected cells (Goh *et al.*, 1992). Williamson (1970) showed that the nuclei in the cells of infected protocorms had higher DNA contents than those of uninfected ones. The nuclei of the orchids *Dactylorchis purpurella* and *Spathiglottis plicata* in culture with *Tulasnella calospora* were examined by microdensitometry after staining with Feulgen, and also by autoradiography after uptake of tritiated thymidine. The nuclei of infected cells had 2-4 times the stable DNA of the cell of uninfected protocorms. Thus, the DNA level of invaded protocorm cells increases during fungal invasion (Rasmussen, 1990; 1995) so that some of these cells would be expected to have nuclei in S phase.

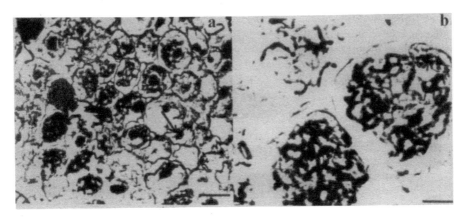

Fig. 6. Root section of *Arachis* Maggie Oei; a, shows the nuclear hypertrophy (enlarged nuclei- arrow) and living pelotons in both outer and inner cortex close to the endodermis (lower left corner), scale = 50 µm; b, root inner cortical cells showing living pelotons, scale = 25 µm. c. f. Goh *et al.,* 1992

Infection spreads from cell to cell so that the basal region of the protocorm becomes extensively infected. Growth and anastomoses of the intracellular hyphae result in the formation of coils (pelotons) (Fig. 4c) which help in increasing the interfacial area between the symbionts.

4.1. Characteristic Features of the Symbiotic Stage

4.1.1. Fungal hyphae

Light micrographs of thin sections of protocorms showed that the intracellular hyphae at an early stage of development were usually filled with cytoplasm and stained heavily. Hyphae in subepidermal parenchyma cells corresponding to the 'host-cell layer' were of this nature. Other hyphae, typically those in the inner cortical parenchyma cells, showed the first stages of digestion and were angular or collapsed and flattened, and contained less cytoplasm.

In electron micrographs, hyphae were seen to be thinly enveloped by host cytoplasm. The cytoplasm of the hyphae was often densely packed with ribosomes, mitochondria (Hadley, 1975), lipid globules, glycogen rosettes, but little ER and nuclei were sometimes seen (Hadley, 1975; Harley and Smith, 1983). Vacuoles were fewer in young than in older hyphae and the tonoplast was sometimes not very clearly defined. The plasmalemma sometimes invaginated to form tubes into the cytoplasm and in one section it appeared continuous with membranes forming a system of tubules or vesicles lying in the central vacuole. Inclusions presumed to be reserve material were sometimes present in the hyphae (Harley and Smith, 1983).

4.1.2. Hyphal walls

The walls of intracellular hyphae usually consist of two layers, the inner electron dense and normally about 60 nm thick, the outer rather granular or flocculose and from 100 to 200 nm thick. Hadley *et al.* (1971) interpreted this outer layer originating from the fungus alone but after further examination of many samples of the material suggested that it may result from the interaction between the fungus and the host. Hyphae grown in culture on agar media do not possess a granular outer wall layer; their walls are similar to the inner wall layer of intracellular hyphae, being 60 to 100 nm thick. They may have an outer electron-transparent layer, not readily stainable and visible only as a gap where adjacent hyphae are in contact. The variations in thickness of the granular outer wall layer and the lack of such a layer in hyphae which are parasitic also suggest that it is a part of the host fungus interface, i.e., an encasement material laid down by the host (Hadley, 1975).

Strullu (1976) proposed the origin of the two layers of the hyphal wall, C1 (inner) and C2 (outer). According to him, the C2 layer within the host cell is most likely to arise in whole or in part by the action of the host. Nieuwdorp (1972), from a study of four chlorophyllous and four achlorophyllous orchids, suggested that the invaginated host plasmalemma continues to synthesize cell wall components, pectins and cellulose, but the young fungal hypha or peloton digests them. As the fungus ages, the protoplasts within the hyphae disappear as a result of the activity of the host indicated by the formation of pinocytotic elaborations of the plasmamembrane and the fungus becomes enveloped in a cellulosic slime layer continuous with the cell wall of the host.

4.1.3. Host-fungus interface

The fungal hyphae after invading the very reduced embryo, form intracellular coils called pelotons, which are separated from the host cytoplasm by an interfacial matrix and the host plasmalemma (Hadley, 1982; Peterson and Currah, 1990; Rasmussen 1995; Uetake *et al.,* 1992). The host-fungus interface comprises of host plasmalemma, the encasement material around the fungal hypha and the wall of the hypha. The encasement may be of uniform thickness but may appear to be extremely thick where hyphae are cut obliquely. The plasmalemma of the host is usually in contact with the encasement layer and there appears to be nothing between them except, for example, where small vesicles are present, or where the plasmalemma was separated from the encasement. Whether the space between plasmalemma and hyphal wall is occupied by an electron transparent material or whether it is an artifact is not known.

Hadley *et al.* (1971) observed that the fungal plasmalemma was invaginated in places to form vesicles and tubules, structures which they believed might be associated with the transfer of substances from one symbiont to the other. Also, the fungal wall outside its plasmlemma bore protuberances on the outer wall to which the host plasmalemma was adpressed. They assumed that these protuberances which increase the contact area by about 15%, might play a part analogous to that believed to be played in transfer cells by their wall protuberances.

Folds or loops in the host plasmalemma occur where it separates from the encasement layer to form vesicle-like bodies extending into the cytoplasm. In some sections, there appeared to be a series of membranes, connected to the plasmalemma, adjacent to healthy hyphae. Where the layer of cytoplasm around a hypha is thin, the loops of plasmalemma may protude into the host cell vacuole, carrying the tonoplast with them. Nieuwdorp (1972) showed a similar arrangement of membranes in later stages in *Corallorhiza trifida* roots, which eventually form groups of pinocytic vesicles around hyphae which are in an advanced stage of disorganization and collapse.

4.1.4. Hyphal senescence

The intracellular hyphal coils have a limited life even in stable mycorrhizal association of orchids. Thus, peloton formation is followed by hyphal collapse resulting in a degenerated hyphal mass surrounded by host plasmamembrane (Uetake *et al.*, 1997). Ultrastructural studies of colonized *Spiranthes sinensis* protocorm cells (Uetake *et al.*, 1992), showed that hyphae forming pelotons and collapsed masses were surrounded by host membrane which expanded quickly to accompany hyphal growth of the cell. In a later study (Uetake and Ishizaka, 1996), it was reported that the membrane hyphae in *S. sinensis* protocorms stained positively as plasma membrane, similar to the membrane adjacent to the host cell wall; however, these membranes showed different activity of adenylate cyclase, suggesting some qualitative differences.

Hadley and Williamson (1971) showed that hyphal senescence could occur within 30-40 hours after peloton or coil formation in *Dactylorhiza purpurella*. After this period of association with active host cells the hyphae degenerate. Strullu and Gourret (1974) classified the process in four stages:

Stage I   : the association of living fungal hyphae and the host cells,

Stage II   : the hyphae still surrounded individually by the plasmalemma of the host, become flattened but the outer layer of the hyphal wall may be very thick. Their contents disappear except for some of the membranes and granules which break up more slowly. The flattened hyphae may now be enclosed in groups in the host plasmalemma rather than singly,

Stage III : consists of the formation of a complicated mass of associated hyphae in which both cell wall layers are apparent, but few hyphal contents exist. The whole is surrounded by host plasmalemma, and

Stage IV : in this last stage, the identity of the hyphae themselves is lost as their walls become diffuse. During the course of this process, reinfection of the cells takes place.

According to Strullu (1976), the reinfection may be a revitalization of hyphae within the cell, a hyphal penetration of the cell from a neighbouring cell, but only very improbably a new infection of the root. The repeated invasion or spread of hyphae in the cells of the host indicates that the degeneration of the fungus is generated by the host, for the host cell remains alive and active, and it receives a new fungal penetration which ends again with the degeneration of the hyphae (Uetake *et al.*, 1997). A single cortical cell outlives several fungal colonization's. Hadley (1975) points out that elaborate membrane systems are present in the cells of the host during the phase of hyphal degeneration (Dorr and Kollman, 1969) (Fig. 7).

The causes of the hyphal collapse are unknown. Generally, it is believed to be caused by the activity of the host cell and to be a manifestation of either a defense reaction against fungal invasion, or a means by which the orchid cell causes a release of nutrients from the fungus. Williamson (1973) showed that the acid phosphatase activity, increases in cells where hyphal collapse is taking place. The localization of the activity was not precise enough to determine whether the enzymes were of orchid or fungal origin, since autolysis of fungal hyphae would result in their own collapse. It may be possible that orchinol, or one of the fungitoxic phenanthrenes which are synthesized by orchids might be important in fungal collapse. Increased activities of oxidase systems (polyphenoloxidase, catalase, ascorbic acid oxidase, etc.) were found by Blakeman *et al.*, (1976) to occur when hyphal collapse was taking place but their precise role in symbiosis is not known.

During peloton collapse, the microtubules surrounding, linking and in circular formation on the pelotons were observed in the cells of *S. sinensis* (Uetake *et al.*, 1997). In the *Zinnia* mesophyll system, Seagull and Falconer (1991) reported that when randomly organized arrays of cortical microtubules are laterally associated they tend to form circular patterns. In the orchid, the formation of new plasmamembrane around hyphae accompanying hyphal growth in the cell might cause disruption of microtubule arrays on the membrane. Upon fungal

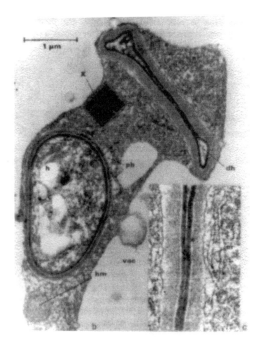

Fig. 7. Ultrastructure of orchid mycorrhiza. Hyphae of *Rhizoctonia* in the host cytoplasm of a cell of *Dactylorhiza (Dactylorchis) purpurella*. Both the living hyphae (h) and the dead collapsed hyphae (dh) are surrounded by an encasement layer (e), which is probably of host origin. Parmural bodies (pb) occur in the interfacial matrix between the host plasmalemma and the encasement layer. Host cytoplasm contains endoplasmic reticulum (er), mitochondria (hm) and a crystal (X). Inset: the host vacuole (vac) is visible, surrounded by the interfacial matrix in greater detail. c. f. Harley and Smith, 1983.

senescence, lateral associations of the microtubules might create the circular structures of the latter observed on hyphae and degenerated masses. Although peloton collapse might be facilitated by host turgor pressure, and ultrastructural observations have shown the appearance of vacuoles in the cortical region of the cells containing collapsing hyphae (Uetake *et al.,* 1992), microtubule arrays might also be involved in the mechanism of peloton condensation. During this process membrane fusion might occur around degenerating hyphae.

Cortical microtubules reappeared in the cells of *S. sinensis* following peloton collapse and before the cells were invaded. It is postulated that cortical microtubules control, among other processes, the orientation of cellulose microfibrils as they are deposited (Giddings and Staehelin, 1991; Gunning and Hardham, 1982). In immunocytochemical studies of orchid protocorm cells (Peterson *et al.,* 1996), cell wall components (cellulose and pectin) were detected in the interface between host plasmamembrane and hyphae which were in the process of collapsing or had collapsed. Therefore, microtubules may be implicated in the deposition of cell wall material around collapsing hyphae. These observations which show changes of microtubule arrays accompanying peloton formation and senescence, suggest that host microtubules are involved in the structural and physiological changes of hyphae (Uetake *et al.,* 1997). To understand the involvement of cytoskeletal arrays in symbiotic orchid protocorms, a correlation of these results with transmission electron microscopy is necessary.

5.   STRUCTURAL FEATURES ASSOCIATED WITH PARASITISM OF THE HOST

Protocorms of *Dactylorhiza purpurella* were infected with a *Rhizoctonia solani* isolate

(RS 16) which was originally obtained from garden soil. In symbiosis tests, this isolate was found to establish a compatible infection but it subsequently turned parasitic (Hadley, 1970). Three weeks after infection, the intracellular hyphae contained in host cells lacked an orderly structure. Some of the hyphae were surrounded by a thin encasement layer characteristic of the compatible structure but others were not encased and were robust, fully turgid and full of cytoplasm. In one instance where a hypha was growing in an intercellular space, no encasement layer was present.

6. THE NUTRIENT FLUX

In the endomycorrhizas of orchids, there is an interchange of carbohydrates between the higher plants and the mycorrhizal fungus (Webster, 1996). Germinating seeds (protocorms) of the host, lacking in food reserves, are supplied through the fungus with carbohydrate and possibly other nutrients. Transfer of carbohydrate from a heterotroph to an autotroph is unusual since in most symbiotic systems the reverse occurs. Although the nutritional requirements of asymbiotic orchids have been studied extensively (Arditti, 1967; Hadley, 1970; Hadley and Harvais, 1968; Harvais, 1972), little comparative investigation of development with and without symbiotic fungi has been carried out. In general, the rate of development of asymbiotic protocorms is much less than that of protocorms infected by suitable endophytes (Hadley, 1970; Hadley and Williamson, 1972).

The cells of the embryo often contain starch grains which remain there in the absence of suitable growth conditions. If soluble carbohydrate is provided to vitamin-dependent seedlings in the absence of suitable vitamins and growth factors, sugar is absorbed and the cells become glutted with starch grains. The embryos of many orchids are unable to form the enzymic systems required to utilize carbohydrates at a sufficient rate to maintain metabolism and to support growth (Harley, 1969). The fungus intervenes in this matter and one of the early results of infection is the disappearance of starch grains from the cells of the embryo. After this the continued growth depends upon the translocation of carbohydrates through the fungal hyphae into the cells of the embryo (Harley and Smith, 1983).

Smith (1966) used cellulose as a carbon source and found that infected seedlings of *D. purpurella* and *D. praetermissa* grew well when cellulose was present in the medium, while seedlings in the absence of cellulose made no growth after seven weeks. The long-term supply of small aliquots of glucose, or alternatively the provision of cellulose, enhanced the growth of populations of infected protocorms. These results indirectly indicate that carbon compounds derived from complex carbohydrates in the surrounding medium are transferred by the fungus into orchid tissue, thereby promoting growth.

6.1. Carbon Nutrition of Mycorrhizal Seedlings

Smith (1966) showed that a fungus, *R. solani* (*Thanetephorus cucumeris*), was not only capable of hydrolyzing cellulose but also translocating them to orchid seedlings in sufficient quantities for growth to occur. The initiation of symbiosis is more certain if the fungus is provided with cellulose than if presented with a medium rich in soluble carbohydrate. In the latter case, destruction of the seedlings is more likely to occur (Hadley, 1969; Harvais and Hadley, 1967).

Inspite of the result of such experiments, the role of the fungus in discharging carbohydrate actually into the seedlings is questionable, and the results of most of the experiments cannot distinguish between direct transport of soluble carbohydrates into the seedling and direct leakage of them from the fungus into the medium followed by uptake by the seedlings (Harley and Smith, 1983). In any population of seedlings on cellulose only a proportion becomes colonised and only such plants show any increase in growth rate over the controls (Purves and Hadley, 1975). Moreover, the direct transport of carbohydrate into the seedlings, which according to these experiments seem improbable, has been confirmed by the use of radioactive substances.

Smith (1967) and Purves and Hadley (1975), using the tracers ^{14}C and ^{32}P, have confirmed the ability of orchid mycorrhizal fungi to translocate and transfer soluble carbohydrates to seedlings. They showed that both tracers could be detected in mycorrhizal seedlings when the fungus alone was fed with ^{14}C D-glucose and ^{32}P orthophosphate. Radioactivity continued to increase in the seedlings for 7 days in experiments with *Dactylorchis purpurella* (Smith, 1967), and for 18 days in experiments with *Goodyera repens* (Purves and Hadley, 1975).

6.2.   Translocation and Transfer of Carbon Compounds

The soluble carbohydrates of uninfected orchid tissues are sucrose, glucose and fructose, whereas those of the mycorrhizal fungi are predominantly trehalose accompanied by glucose and occasionally by mannitol, but not sucrose. Seedlings fed with ^{14}C D-glucose *via* the split plates become labelled not only in the fungal sugars but also in the orchid sugar, sucrose. Changes in the pattern of labelling with time in *Dactylorchis purpurella* indicated that the fungal sugar, trehalose, is the most heavily labelled in the early samples but as the time elapses, sucrose becomes proportionately more heavily labelled as trehalose labelling declines (Harley and Smith, 1983; Smith, 1967). There is thus reasonably good evidence that carbohydrates are translocated in the fungus, and during or following transfer to the orchid cells it is converted to sucrose. Hence, it was established that the form in which carbon compounds move were trehalose and mannitol peculiar to the endophyte (mannitol to certain endophytes only), and sucrose to the orchid (Smith, 1967). Subsequently, Smith (1973) found that seeds of *D. purpurella* could germinate and grow in trehalose as well as in glucose; mannitol was inhibitory. Ernst (1967) reported growth of a *Phalaenopsis* cultivar on mannitol, but did not use carbohydrate-free controls. The fact that no glucose was detected in the medium surrounding germinating *Phalaenopsis* seeds (Ernst *et al.,* 1971), might suggest direct absorption, but a similar result would be obtained if the rate of glucose absorption equalled or could keep pace with that of trehalose hydrolysis.

Mannitol occurs in only a few of the fungi examined and is suitable for the asymbiotic germination of only a few species of orchid. The leaves of *Bletilla hyacintha* absorb and accumulate mannitol but do not seem to metabolize it, nor it is suitable for the germination of the seeds of this species (Smith, 1973; Smith and Smith, 1973). Metabolism of mannitol might possibly provide a basis of crude specificity between fungi producing it and orchids capable of using it.

6.3.   Translocation and Transfer of Mineral Nutrients

Orchidaceous fungi can translocate ^{32}P supplied as orthophosphate both from hyphal tip to older mycelium and *vice versa* (Harley, 1965; Smith, 1966). Such translocation can lead to the accumulation of phosphate in associated seedlings. Mycorrhizal fungi may play a role in mineral nutrition of both seedling and adult orchids in a manner comparable with other mycorrhizal systems. For instance, many orchids have thick magnoloid roots (Baylis, 1975) with a few lateral rootlets and root hairs. The inefficiency of these roots in absorption of nutrients with low movement coefficients in soil would be offset by the fungal hyphae emnating from mycorrhizas.

Mycorrhizal (*Rhizoctonia goodyerae-repentis*) infection of *Goodyera repens*, increases the efficiency of P uptake and under conditions of P stress, this leads to a significant enhancement of growth (Alexander and Hadley, 1985). Inhibition by 1% thiobenzadol of the external mycelium of *R. goodyerae-repentis* attached to plantlets of *Goodyera repens* reduced nitrogen and phosphorus uptake and the growth rate of the plantlets. Also, in *D. purpurella*, evidence of inorganic nutrient transport to the plantlets was found. Mycorrhized plants, mycorrhizal plants treated with the fungicide thiobenzadole, and the non-mycorrhizal plants at the 4-leaf stage were compared. The fungicide did not appear to affect the metabolism of the host. The untreated mycorrhizal plants had higher relative growth rates, and nitrogen and phosphorus contents than the fungicide-treated plants. The problem is of ecological relevance because orchid roots under natural conditions, are usually infected even in most green orchids, and, on comparative grounds, the fungus would be expected to operate as other mycorrhizal fungi.

Besides carbohydrates and inorganic nitrogen sources, orchid seedlings require other substances for continued growth. Nutrients such as amino acids are sometimes stimulatory or even essential. In addition, they require vitamins and growth factors, although the requirement varies with the species. Thiamin and nicotinic acid seem to be required by most (Harley and Smith, 1983). These requirements are presumed to be provided by the activities of the mycorrhizal fungi after infection in nature (Arditti, 1967; 1979).

6.4. Mechanisms of Transfer: Concepts of Biotrophic and Necrotrophic Nutrition

There are three possible ways in which nutrients can be transferred from the fungus to orchid:
(i) digestion of fungal hyphae containing nutrients derived from the external medium
(ii) transfer of substances from fungus to orchid across a living interface
(iii) absorption of substances made available in the external medium by the action of the fungus .

Experiments with *D. purpurella* (Hadley and Williamson, 1971) tend to discredit the third possibility (Harley and Smith, 1983).

Digestion or lysis of the fungus in orchid mycorrhiza was thought to be a mode of nutrient transfer but many have viewed it as a manifestation of defense of the host against invasion. It was believed that nutrient transfer, which appeared to be a continuous process, was more likely to occur across the intact membranes of the host and the fungus. Under such circumstances, control would be operated by the membranes of both the symbionts in uptake, and natural or induced leakiness would be involved in the release of substances. If nutrient transfer occurred following hyphal collapse, selectivity and control would operate at the orchid membrane only. If nutrient uptake were only in one direction, the orchid might possibly be viewed as a necrotrophic parasite of the fungus (Lewis, 1973), but it could be possible that the fungus might be absorbing through its intact hyphal coils material leaking from the orchid while digestion or lysis was taking place elsewhere (Harley and Smith, 1983).

6.4.1. Metabolic changes preceding digestion

The fungal invasion alters the internal metabolism of the orchid. The studies on *D. purpurella* and *G. repens* by Hadley and Williamson (1971) suggest that increase in growth rate of protocorms precede the onset of digestion. On the other hand, the hydrolysis of starch and prevention of excess starch formation in sugar media may be brought about either by infection or by the provision of yeast extract or nicotinic acid in the presence of other vitamins (Harley and Smith, 1983). The fungus may, in this case, relieve an enzymic blockage by providing precursors of NAD and NADP. The different destination of $^{14}CO_2$ fixed by green protocorms of *D. purpurella* into the hexoses: glucose and fructose in infected protocorms, and into sucrose in uninfected ones, is another example of the effect of the fungus in generating conditions for normal metabolism of reserves (Hadley and Purves, 1974). These results argue against an interpretation of the orchid being a parasite or the necrotrophic fungus.

6.5. Reciprocal Movement from Orchid to Fungus

It was presumed that the reciprocal flow from orchid to fungus occurred because the fungus could grow out from infected protocorms onto a carbohydrate-free medium and could cross a diffusion barrier (Hadley and Purves, 1974; Howard, 1978; Smith, 1967). No radioactivity could be detected in the fungal sugars after green infected seedlings of *D. purpurella* had assimilated $^{14}CO_2$ in the light. Nor was there any in the hyphae growing out onto carbohydrate-free medium (Hadley, 1975). However, fungal metabolites sometimes became labelled in similar experiments (Purves and Hadley, 1975) but since the control (uninfected plants) released ^{14}C-labelled nutrients into the medium, carbohydrates released in a similar way by infected plants might have been the source of ^{14}C in the mycelium around infected protocorms. Moreover, dark fixation of $^{14}CO_2$ by the fungus might also have occurred. In experiments with *Goodyera repens*, Hadley and Purves (1974) showed that no ^{14}C-labelled metabolites leaked into the medium and in this case when $^{14}CO_2$ was applied to the green top alone in the light little radioactivity passed to the rhizome and none into the growing hyphae in the medium. When the rhizomes alone were exposed to $^{14}CO_2$, small quantities of radioactivity appeared in the emerging mycelium. It was concluded that carbon movement from the seedlings to the fungus was at most very small in quantity. In contrast, movement into the mycelium from killed $^{14}CO_2$-labelled seedlings was fairly rapid. Thus, a reciprocal movement of carbon compounds into the fungus does not readily occur. This appears to be so, even in those experiments where no carbon source was available to the fungus outside the seedling. According to Harley (1975), since the orchid seedlings in the experiments of Smith (1967) were actually growing, the sink for carbohydrates is the conversion of them into amino acids, proteins and cell walls. In these experiments with *Rhizoctonia solani* and *D. purpurella*, only 30% of the ^{14}C-labelled compounds were soluble carbohydrates in the orchid seedlings; 50% was in insoluble compounds and 15% in amino acids.

It can be assumed from the available evidence that in the adult state, the green orchids examined so far do not provide a carbon source for the associated fungus but that the fungus continues to be self-sufficient.

7. HYDROLYTIC ENZYMES

Orchid fungi, produce cellulases and pectinases which may play a significant part in the saprophytic utilization of insoluble substrates and in obtaining sugars required by both fungus and the orchid (Harley, 1969; Hadley and Ong, 1978). A comparison of the activity of pectinases (endopolygalacturonases, endopolymethylgalacturonase and protopectinase) produced by pathogenic isolates of *R. solani* (*Thanatephorus cucumeris*), *R. repens* (*Tulasnella calospora*) and *R. goodyerae-repentis* (*Ceratobasidium cornigerum*) revealed that the amounts of pectinases produced were not related to their pathogenecity towards orchid seedlings (Perombelon and Hadley, 1965). Some of the most active pathogens may be readily compatible with orchids (Warcup, 1981). Therefore, the activities of pectinases and cellulases must be controlled in some way within the orchid, either by high levels of sugars such as glucose which may inhibit or repress pectinases and cellulases, or by some other means. Starch hydrolysis, which occurs in some newly invaded seedlings might bring about a temporary increase in glucose concentration which might be enough for immediate enzyme repression until photosynthesis starts. Purves and Hadley (1975) observed that the infected roots of *Goodyera repens* have higher hexose levels than uninfected roots. It is not known if mechanism exists in them for maintaining conditions which repress hydrolysis activity by maintaining high soluble sugar concentrations. However, the functioning of such repression in achlorophyllous orchids which have another source of carbohydrate than the fungus, is unexplainable (Harley and Smith, 1983).

The mechanisms such as those favoured by pathogens which are dependent on the copolymerization of proteins of fungus and host, might operate as a mode of enzyme inhibition (Vanderplank, 1978). Selection for such action in evolution would be strong, as it maintains symbiosis.

8.   RESISTANCE TO SYMBIOSIS

8.1.   Phytotoxic Substances and Phytoalexins: An Antifungal Principle

Bernard (1911) suggested that the absence of fungal infection from the tubers of some orchids and the resistance of some seeds to fungal attack were due to the presence within them of an antifungal principle. He showed that the tubers of *Loroglossum* contained a substance which was toxic to many orchid endophytes including *Rhizoctonia repens* (*Tulasnella calospora*) but not to a strain of *Rhizoctonia solani*. The toxic substance was formed after the fungal attack (Nobecourt, 1923), i.e., a phytoalexin. It was shown that orchinol, a dihydroxyphenanthrene, was formed by living tuber tissues of *Orchis militaris* when the tissue was kept in contact with *R. repens*. It was formed not only by the cells immediately in contact , but also, more slowly, by cells upto 12 mm from the contact (Arditti *et al.*, 1975; Fisch *et al.*, 1973; Nuesch, 1963). Many European orchids produce similar phytotoxic compounds and it is likely that the ability is much more widespread (Arditti, 1979).

Three phytoalexins were initially identified- hircinol, loroglossol and orchinol- and others have since been discovered. The dihydroxyphenanthrenes all inhibit mycorrhizal fungi and many other fungi and bacteria. They differ from other phytoalexins in not being formed only around the lesion or the point of attack but also at some distance from it (Wood, 1967). Thus, their synthesis is by no means a simple event but must involve not only a chemical sequence of production but also a signal emnating from the point of attack.

It seems that these substances play a part in controlling or restricting fungal invasion of orchid tissues. To explain as to how mycorrhizal fungi can penetrate some orchid organs such as roots and not others such as tubers, and how are some fungi excluded from both. Gaumann *et al.*, (1960) proposed that the difference in susceptibility of the roots and tubers was mainly due to a difference in the ability of the two kinds of tissue to produce the phytoalexins. Some mycorrhizal fungi, e.g. strains of *R. solani*, will destroy orchinol and it may be that the balance between phytoalexin production and its activation in any combination of the orchid tissue and fungus might determine the stability of the symbiosis. Some other fungi have been found to be resistant to these phytoalexin, and others attack orchid tissues so rapidly that time is not available for their formation before complete parasitization. The site of their production in high concentrations in the tubers for instance, may confer resistance that is nearly complete. At the lower concentrations in the roots, they may be only a coarse control of infection by alien fungi and of the extent of exploitation by mycorrhizal fungi (Harley and Smith, 1983). A fresh impartial examination of this with selected genotypes of fungus and orchid would be worthwhile. Moreover, the effects of low concentrations of these compounds, not only on growth but also on the physiology, including the activity of the exoenzymes, of orchid fungi should be examined.

9. DOUBLE-STRANDED RNA-CONTAINING VIRUSES OF ORCHID MYCORRHIZAL FUNGI

The symbiosis between orchid and the fungus is interesting as *Rhizoctonia* species are known pathogens in species of other plant families. Characteristics of the fungus are of significance in establishment of symbiosis (James *et al.*, 1998).

Rhizoctonia have species containing viruses or unencapsulated double-stranded RNAs (Castanho *et al.*, 1978; Zanzinger *et al.*, 1984). Possession of dsRNA elements can result in hypervirulence as found with *Rhizoctonia solani* isolates (Finkler *et al.*, 1988a,b). The presence of dsRNAs, discovered on examination of symbiotic fungal cells using transmission electron microscopy (James, 1993; James *et al.*, 1996), may therefore, have important implications in the symbiosis.

Fungal isolates can often lose their ability over time to establish a symbiosis with orchids (Mitchell, 1989). This might be due to excision of virus-like particles through culturing as viral particles are found in older parts of the hyphae (Berry and Berry, 1976). Culturing from hyphal tips might result in 'curing' of the fungal virus (James *et al.*, 1998). Thus, the presence of dsRNAs in orchid mycobionts may implicate them in the symbiotic process (James *et al.*, 1996; 1998).

10. A NEW MYCORRHIZA-LIKE FUNGUS, *PIRIFORMOSPORA INDICA*

During routine monosporic mass inoculum preparation of *Glomus mosseae* (Nicol. and Gerd.) Ger. and Trappe isolated from a desert soil in northwest India, a new fungus was discovered as one of the contaminants. It was described as *Piriformospora indica* (Verma *et al.*, 1998). The fungus is cultivable on a wide range of synthetic simple and complex media (Varma *et al.*, 1999; Verma *et al.*, 1998).

Studies on the ultrastructure showed a multilayered hyphal wall and a dolipore septum with continuous parenthosomes, which is a characteristic of some groups of Hymenomycetes (Basidiomycota). In order to find the closest relatives of *P. indica*, the 18S rRNA was partly sequenced and compared to corresponding data from EMBO/Gene Bank. The resulting dendrogram revealed members of the *Rhizoctonia* group (Ceratobasidiales) as the closest relatives. However, the bootstrap value for the probability of *P. indica* belonging to the same branch was quite low. Therefore, the systemic position of the new fungus is not clear, though the molecular data as well as the ultrastructural characters indicate that *P. indica* is related to Ceratobasidiales or Tulasnellales (Varma *et al.*, 1999; Verma *et al.*, 1998).

P. indica forms white to hyaline highly interwoven mycelia with simple septate hyphae (Fig. 8a). In older cultures, the hyphae show irregular constrictions and inflations, which lead to a nodose or coralloid shape. Characteristic pear-shaped chlamydospores are produced singly (Fig. 8b) or in clusters (Fig. 8c) at the tips of hyphae or short side branches.

10.1. Interaction of *P. indica* with Orchids

P. indica is able to grow saprophytically on various media as expected from the systematic relationship of *P. indica* to the species of *Rhizoctonia* group and the members of the Tulasnellales (Verma *et al.*, 1998) which are known as saprophytes, parasites of higher plants and mycorrhizal fungi of orchids.

To test the influence of *P. indica* on the germination of orchid seeds and the growth of developing protocorm, terrestrial orchid species native to the European flora were chosen (Blechart *et al.*, 1999).

The species of *Dactylorrhiza* was chosen as the host because:

(i)   of their low specificity concerning their mycorrhizal partners

(ii)   their ability to interact with species of the *Rhizoctonia* group (Harvais and Hadley, 1967)

(iii) their seeds usually germinate easily and at high rates in axenic and symbiotic culture (Harvais and Hadley, 1967)

(iv)   *Dactylorhiza* species is cultivable *in vitro*.

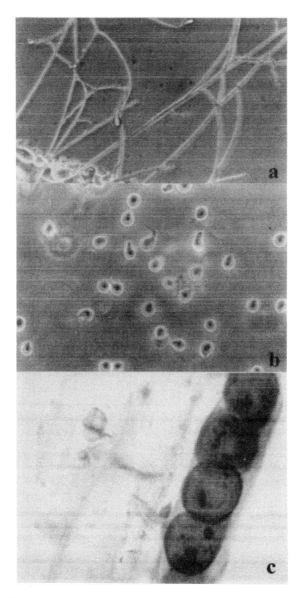

Fig. 8. Morphological features of *Piriformospora indica*. a, hyphae with irregular constrictions and inflations; b, chlamydospores, young thin-walled, mature with thickened walls and refractive content; c, a chain of intracellular spores within the host cell.

10.1.1. Seed germination and symbiotic protocorm development

Seeds of *D. purpurella* (Steph's) Soo and *D. incarnata* (L.) Soo were surface sterilized, placed on a modified oat medium (MOM), (Clements *et al.*, 1985) and inoculated with *P. indica*. After 2 weeks, some seeds of *D. purpurella* began to swell and germinate, while the seeds of *D. incarnata* germinated much slower and only in very small numbers. The germination rate of both orchid species showed no detectable difference between plates inoculated with *P. indica* and the corresponding controls. Light microscopic studies showed that the fungus had already invaded the cells of the testa (Fig. 9a). The hypha also grew on the surface of the embryo, but there was no detectable penetration into or between the embryo cells and no indication for an interaction of the fungus with the seed or an enhancement of the development of the embryo.

Soon after the occurrence of the first rhizoids of the protocorm (Fig. 9b) the rhizoids were penetrated by the fungus. The hyphae formed small appressoria-like structures on the rhizoid surface and penetrated into the cell by a narrow penetration neck (Fig. 9c). Within the cells, the hyphae expanded to a normal diameter and mostly grew towards the protocorm. Within the protocorm, only intercellular hyphae could be detected during this early stage of interaction. No differences in morphology were found between inoculated protocorms and the corresponding controls.

In the later stages of interaction, *P. indica* also penetrated into protocorm cells. The fungus formed dense hyphal coils which almost filled the whole cell. Hyphae growing from one cell into an adjacent cell were not seen. Intercellularly growing hyphae and invaded cells were detectable especially in the basal regions of the protocorm. Differences between the growth of the protocorms inoculated with *P. indica* and the corresponding controls were visible after the beginning of the intracellular interactions. The period of seed germination lasted more than six weeks after placing it on agar plates. All the developmental stages from the germination of seeds to the protocorms, could be observed on the inoculated plates (Fig. 9b) and the controls. The resulting inhomogeneity of the developmental stages of the protocorms hindered the testing of the significance of growth rate differences between inoculated protocorms and the controls. Such differences are expected to appear very distinctly when the plants start to differentiate the first leaves and the roots (Blechart *et al.*, 1999). The mode of interaction described corresponds with the results of Harvais and Hadley (1967).

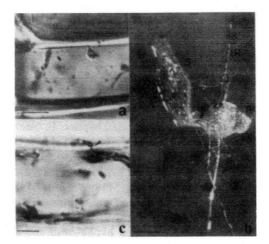

Fig. 9. Interaction of *Piriformospora indica* with *Dactylorhiza purpurella*. A, hyphae penetrating the cell walls of the testa; b, seed (S) and the protocorm (P) with rhizoids (R) covered by a scanty mycelium; c, hypha penetrating the wall (arrow head) of a rhizoid and growing inside towards the protocorm (a, c bar = 4 µm; b, bar 100 µm). c. f. Blechart *et al.*, 1999

10.1.2. *Piriformospora indica* and young asymbiotically grown orchid plants

Plants of *D. maculata* (L.) Soo were grown for two years in the dark axenically on standard medium for asymbiotic orchid cultivation. After transferring to MOM, 12 plants were inoculated with *P. indica* and maintained at a 12-h light/12-h dark regime at room temperature. Controls were grown without the fungus.

Three months after inoculation, the control plants had well-developed leaves but very poorly developed, blackish-brown roots and a differentiated small tuber (Fig 10a). The agar medium had also turned brownish due to the exudation of phenolic compounds by the orchid roots. In contrast, plants inoculated with *P. indica* produced well-developed leaves and root system but the differentiation of a tuber could not be detected (Fig. 10b). The roots were not as dark brown as those from control plants, and the agar was just weakly pigmented.

The light microscopic examination of cross-sections of the roots revealed that hyphae grew on the root surface, intercellularly between the cortical cells as well as intracellularly, forming hyphal coils within the cortical cells (Fig. 10c). To study the structures of orchid mycorrhizas formed in nature, cross-sections of *D. majalis* roots were investigated with the light microscope. The unknown fungal partner of this orchid formed typical pelotons in the cells of the root cortex (Fig. 10d). The structure of the hyphal coils which *P. indica* formed in the protocorms of D. purpurella as well as in the roots of *D. maculata* proved to be similar to the morphology of pelotons from *D. majalis* in nature.

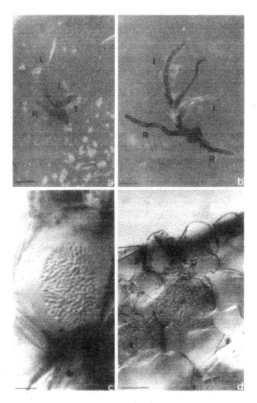

Fig. 10. a-c. Interaction of *Piriformospora indica* with *Dactylorhiza maculata.* a, control plants with short roots (R), leaf (L) and tuber (T); b, symbiotic plant with well-developed roots (R), leaves (L) and no tuber; c, peloton formed in a cell of the root cortex; d, peloton formed in roots of *D. majalis* grown in a natural environment (a,b: bar = 1 cm); c: bar = 4μm; d: bar = 20 μm). c. f. Blechart *et al.,* 1999.

11.   CONSTRAINTS

The relatively massive roots of many orchid species seem to be singularly ill-equipped for soil exploitation and it might be expected that an effective external mycelium system would be essential for survival of mature plants in these circumstances. Many orchids grow in species rich vegetation on water and nutrient-stressed soils, yet we still do not know about their water or mineral nutrient relations, and even less about the possible role of mycorrhizas, in the post-establishment phase of their lives.

12.   FUTURE PROSPECTS

Changes of land-use and increasing collections of bloom by man pose a threat to many orchid species. The progress currently being made on understanding the requirements for symbiotic and asymbiotic germination may enable us effectively to conserve endangered species. Thus, much more work needs to be done to establish the involvement of fungi in, and the environmental conditions required for germination of most these plants and their subsequent developmental stages upto and beyond the emergence of flowering shoots. The conventional methods for the isolation of mycorrhizal fungi and to form successful artificial syntheses with seeds, are of limited value. Improved knowledge of the factors involved in orchid germination could lead to a routine culture even of achlorophyllous species, and this, in turn, if it provides mature plants, would enable us to investigate the carbon and mineral nutrition of orchids more realistic circumstances than has been possible. Much could be achieved by simple experiments in which the seeds are placed in their natural habitats and their subsequent development monitored.

The employment of DNA techniques in the identification of orchid fungal symbionts will complement and supplement data based on morphological characters currently used for taxonomic classification. One major advantage of DNA analysis is that the data is independent of the sexual stage of the fungus and any altered morphology induced by nutritional or environmental factors.

13.   CONCLUSIONS

Orchid seed germination, growth and development is dependent on the symbiotic association with an endomycorrhizal fungus. The fungi of orchid mycorrhizas are mostly basidiomycetes, generally a species from the form genus *Rhizoctonia*, the hyphae of which produce characteristic coils, pelotons, in the cortical tissues of the roots.

The specificity between orchids and mycorrhizal fungi has been the subject of controversy for a long time. It may be necessary to re-evaluate published data on specificity as the current study suggests that no matter what results are obtained from in vitro tests, in situ investigation is still needed before we understand what is occurring in nature.

The mycorrhizal fungi of orchids produce cellulolytic and pectinolytic enzymes, and in contrast to most other mycorrhizal fungi, are active in using complex carbohydrate polymers for growth. It is due to this ability that their success as symbionts of the partially or wholly achlorophyllous orchid depends. Even if the small amounts of the seedlings of green terrestrial orchids could be obtained from the humus and litter layers of the soil, it seems very unlikely that these layers could be the source of the carbon supplies of the large totally achlorophyllous ones. Experiments on the water relations, and on the nitrogen and phosphorus nutrition of mycorrhized and non-mycorrhized orchid plants are urgently needed. Direct evidence of the transfer of vitamins and growth factors synthesized by the fungus, from the latter to the host is lacking, but reasonable circumstantial evidence is available that the requirements of seedlings for vitamins in asymbiotic culture are provided by the fungus when a mycorrhizal association is formed.

The death of the fungus which precedes the death of the host cells in orchid mycorrhizas might be a form of control of fungal exploitation, particularly if transfer occurs across the intact interface. It remains for future research to determine the role of phytoalexins like orchinol in the control of spread of the fungus through orchid tissues.

Subsequent to the discovery of a new fungus, *Piriformospora indica*, from desert soil in northwest India, which was found by 18S rRNA sequencing to be the closest relatives of the members of rhizoctonia group (Ceratobasidiales), its ability to interact with the orchids was tested. *P. indica* was found to interact with orchids as a mycorrhizal partner although further experimentation is required to quantify the promoting effect and to study the interacting structures formed during the early stages of plant development. This fungus promises to open new vistas for understanding the molecular basis of interactions between myco-photosymbiotic partners. This understanding might help to protect orchid seeds from degeneration and lead to mass production of mycorrhized plants for better growth and flowering.

ACKNOWLEDGMENTS

The authors are thankful to UGC and DBT, New Delhi, for partial financial assistance, and Rashmi Mehta and Pragati Tiwari for the preparation of this manuscript.

REFERENCES

Adams, G. C. 1988, *Thanatephorus cucumeris* (*Rhizoctonia solani*), a species complex of wide host range, in: *Advances in Plant Pathology*, Vol.6, D. S. Ingram and P.H. Williams, eds., Academic Press, London, pp. 535-552.

Alconero, R. 1969, Mycorrhizal synthesis and pathology of *Rhizoctonia solani* in vanilla orchid roots, *Phytopathol.* 59: 426-430.

Alexander, C. and Hadley, G. 1985, Phosphorus uptake by mycorrhizas of the orchid *Goodyera repens*. in: *Proceedings of the 6th North American Conference on Mycorrhizae*, R. Molina, ed., Forest Research Laboratory, USA.

Arditti, J. 1967, Factors affecting the germination of orchid seeds, *Bot. Rev.* 33: 1-97.

Arditti, J. 1979, Aspects of physiology of orchids, *Adv. Bot. Res.* 7: 421-655.

Arditti, J. 1992, *Fundamentals of Orchid Biology*, John Wiley and Sons, New York.

Arditti, J., Flick, B.H., Ehmann, A. and Fisch, M.H. 1975, Orchid Phytoalexins II, isolation and characterization of possible sterol companions, *Am. J. Bot.* 62: 738-742.

Barmicheva, K.M. 1989, Ultrastructure of *Neottia nidus-avis* mycorrhizas, *Agric. Ecosystems Environment.* 29: 23-27.

Baylis, G.T.S. 1975, The magnolioid mycorrhiza and mycotrophy in root systems derived from it, in: *Endomycorrhizas*, F.E. Sanders, B. Mosse and P.B. Tinker, eds., Academic Press, London and New York, pp 373-389.

Bernard, N. 1911, Sur la fonction fungicide des bulbes de'ophrydees, *Annls. Sci. Nat. Bot.* 14: 221-224.

Berry, D.R. and Berry E.A. 1976, Nuclei acid and protein systhesis in filamentous fungi, in: *The Filamentous Fungi*, J.E. Smith and D.R. Berry, eds., Edward Arnold, London, pp. 238-291.

Blackeman, J.P., Mokahel, M.A, and Hadley, G. 1976, The effect of mycorrhizal infection on respiration and activity of some oxidase enzymes of orchid protocorms, *New Phytol.* 77: 697-704.

Blechart. O., Kost, G., Hassel, A., Rexer, K.H. and Varma, A. 1999, First remarks on the symbiotic interaction between *Piriformospora indica* and terrestrial orchids, in: *Mycorrhizae: Structure, Function, Molecular Biology and Biotechnology*, A. Varma and B. Hock, eds., second edition, Springer-Verlag, Germany, pp 683-688.

Bonfonte, P. and Perroto, S. 1995, Strategies of arbuscular mycorrhizal fungi when infecting host plants, *New Phytol.* 130: 3-21.

Campbell, E. O. 1962, The mycorrhiza of *Gastrodia cunninghamii* Hook, *Trans. Royal Soc. New Zealand (Botany)* 1: 289-296.

Campbell, E. O. 1963, *Gastrodia minor* Petrie: an epiparasite on manuka, *Trans. Royal Soc. New Zealand (Botany)* 2: 73-81.

Castanho B., Butler, E.E. and Shepherd, R.J. 1978, The association of double-stranded RNA with *Rhizoctonia* decline, *Phytopathol.* 68: 1515-1519.

Clements, M.A., Muir, H. and Cribb, P.J. 1985, A preliminary report on the symbiotic germination of European terrestrial orchids, *Kew Bull.* 41: 437-445.

Clements, M.A. 1988, Orchid mycorrhizal associations, *Lindleyana* 3: 73-86.

Currah, R.S. 1987, *Thanatephorus pennatus* sp. nov. isolated from mycorrhizal roots of *Calypso bulbosa* (Orchidaceae) from Alberta, *Can. J. Bot.* 65: 1957-1960.

Currah, R.S., Hambleton, S. and Smreciu, A. 1988, Mycorrhizae and mycorrhizal fungi of *Calypso bulbosa*, *Amer. J. Bot.* 75: 739-752.

Dorr, I. and Kollman, R. 1969, Fine structure of mycorrhiza in *Neottia nidus avis*, *Planta* 89: 372-375.

Ernst, R. 1967, The effect of carbohydrate selection on the growth rate of freshly germinated *Phalaenopsis* and *Dendrobium* seeds, *Am. Orchid Soc. Bull.* 36: 1068-1073.

Ernst, R., Arditti, J. and Healey, P.L. 1971, Carbohydrate physiology of orchid seedlings, II, Hydrolysis and effects of oligosaccharides, *Am. J. Bot.* 58: 827-835.

Finkler, A., Ben Zvi, B.S. and Koltin, Y. 1988a, dsRNA virus of *Rhizoctonia solani*, in : *Viruses of Fungi and Simple Eucaryotes*, Y. Kolton and M. Leibowitz, eds., Marcel Dekker Inc., New York, pp. 387-409.

Finkler, A., Ben Zvi, B.S. and Koltin, Y. 1988b, Transcription and *in vitro* translation of the dsRNA virus isolated from *Rhizoctonia solani*, *Virus Genes* 1: 205-219.

Fisch, M.H., Flick, B.H. and Arditti, J. 1973, Structure and antifungal activity of hircinol, loroglossol and orchinol, *Phytochem.* 12: 437-441.

Gaumann, E. 1960, Nouvelles donnees sur les reactions chimiques de defense chez les orchidees, *C r hebd Seanc. Acad Sci, Paris* 250: 1944-1947.

Giddings, T.H.J. and Staehelin, A. 1991, Microtubule-mediated control of microfibril deposition: a re-examination of the hypothesis, in: *The Cytoskeletal Basis of Plant Growth and Form*, C.W. Lloyd, ed., Academic Press, London, pp. 85-99.

Goh, C.J., Sim, A.A. and Lim, G. 1992, Mycorrhizal associations in some tropical orchids, *Lindleyana* 7(1): 13-17.

Gunning, B.E.S. and Hardham, A.R. 1982, Microtubules, *Ann. Rev. Plant Physiol.* 33: 651-698.

Hadley, G. 1969, Cellulose as a carbon source for orchid mycorrhiza, *New Phytol.* 68: 933-939.

Hadley, G. 1970, Nonspecificity of symbiotic infection in orchid mycorrhiza, *New Phytol.* 69: 1015-1023.

Hadley, G. 1975, Fine structure of orchid mycorrhiza, in: *Endomycorrhizas*, F.E. Sanders, B. Mosse and P.B. Tinker, eds., Academic Press, London and New York, pp. 335-351.

Hadley, G. 1982, Orchid mycorrhiza, in: *Orchid Biology- Reviews and Perspectives,* Vol. II, J. Arditti, ed., Cornell University Press, Ithaca, pp. 84-118.

Hadley, G. 1984, *Orchid Biology and Perspectives,* Vol.III, J. Arditti, eds., Cornell University Press, Ithaca.

Hadley, G., and Harvais, G. 1968, The effect of certain growth substances on asymbiotic germination and development of *Orchis purpurella, New Phytol.* 67: 447-482.

Hadley, G., Johnson, R.P.C. and John, D.A. 1971, Fine structure of the host-fungus interface in orchid mycorrhiza, *Planta* 100: 191-199.

Hadley, G. and Pegg, G.F. 1989, Host-fungus relationships in orchid mycorrhizal systems, in: *Modern Methods in Orchid Conservation*, H.W. Pritchart, ed., Cambridge University Press, Cambridge, pp. 57-71.

Hadley, G. and Williamson, B. 1971, Analysis of post-infection growth stimulus in orchid mycorrhiza, *New Phytol.* 70: 445-455.

Hadley, G. and Williamson, B. 1972, Features of mycorrhizal infection in some Malayan orchids, *New Phytol.* 71: 1111-1118.

Hadley, G. and Purves, S. 1974, Movement of ^{14}C from host to fungus in orchid mycorrhiza, *New Phytol.* 73: 475-482.

Hadley, G., and Ong., S.H. 1978, Nutritional requirements of orchid endophytes, *New Phytol.* 81: 561-569.

Harley, S.E. 1965, The ecology of orchid mycorrhizal fungi, Ph.D. Thesis, Cambridge University.

Harley, J.L. 1969, *The Biology of Mycorrhiza*, second edition, Leonard Hill, London.

Harley, J.L. 1975, Problems of mycotrophy, in: *Endomycorrhizas*, F.E. Sanders, B. Mosse and P.B. Tinker, eds., Academic press, London and New York, pp. 1-24.

Harley, J.L. and Smith, S.E. 1983, *Mycorrhizal Symbiosis*, Academic Press, London, pp. 268-295.

Harvais, G. 1972, The development and growth requirements of *Dactylorhiza purpurella* in asymbiotic cultures, *Can. J. Bot.* 50: 1223-1229.

Harvais, G. and Hadley, G. 1967, The relation between host and endophyte in orchid mycorrhiza, *New Phytol.* 66: 205-215.

Howard, A. J. 1978, Translocation in fungi, *Trans. Br. Mycol. Soc.* 70: 265-269.

James, J.D. 1993, Ultrastructural, biochemical and molecular biological aspects of the orchid-fungus symbiotic relationship, Ph.D.Thesis, University of Greenwich, London.

James, J.D., Saunders G.C., and Owens, S.J. 1996, Presence of double-stranded RNAs in endomycorrhizal fungi isolated from orchid roots, in: *Mycorrhizas in Integrated Systems from Genes to Plant Development*, C. Azcon-Aguilar and J.M. Barea, eds., Proc 4th Eur. Symp. on Mycorrhizas 1994, European Commission, Brussels, pp. 241-244.

James, J.D., Saunders, G.C. and Owens, S.J. 1998, Isolation and partial characterisation of double-stranded RNA-containing viruses of orchid mycorrhizal fungi, in: *Mycorrhiza Manual*, A. Varma, ed., Springer-Verlag, Berlin, Heidelberg, New York, pp. 413-424.

Lewis, D.H. 1973, Concepts in fungal nutrition and the origin of biotrophy, *Biol. Rev.* 48: 261-278.

Masuhara, G. and Katsuya, K. 1994, *In situ* and *in vitro* specificity between *Rhizoctonia* sp. and *Spiranthes sinesis* (Persoon) Ames var. *amoena* (M Bieberstein) Hara (Orchidaceae), *New Phytol.* 127: 711-718.

Milligan, M.J. and Williams, P.G. 1988, The mycorrhizal relationship of multinucleate rhizoctonias from non-orchids with *Microtis* (Orchidaceae), *New Phytol.* 108: 205-209.

Mitchell, R. 1989, Growing hardy orchids from seeds at Kew, *Plantsman* 11(3): 152-169.

Nieuwdorp, P.J. 1972, Some observations with light and electron microscope on the endotrophic mycorrhiza of orchids, *Acta. Bot. Neerl.* 21: 128-144.

Nobecourt, P. 1923, Sur la production d'anticorps par less tubercles des Ophrydees C r hebd seanc, *Acad Sci. Paris* 177: 1055-1057.

Nuesch, J. 1963, Defense reactions in orchid bulbs, *Symp. Soc. Gen. Microbiol.* 13: 335-343.

Perombelon, M. and Hadley, G. 1965, Production of pectic enzymes by pathogenic and symbiotic *Rhizoctonia* strains, *New Phytol.* 64: 144-151.

Peterson, R.L. and Currah, R.S. 1990, Synthesis of mycorrhizae between protocorms of *Goodyera repens* (Orchidaceae) and *Ceratobasidium cereale, Can. J. Bot.* 68: 1117-1125.

Peterson, R.L. and Farquhar, M.L. 1994, Mycorrhizas - Integrated development between roots and fungi, *Mycologia* 86: 311-326.

Peterson, R.L., Bonfante, P., Faccio, A. and Uetake, Y. 1996, The interface between fungal hyphae and orchid protocorm cells, *Can. J. Bot.* 74: 1861-1870.

Purves, S. and Hadley, G. 1975, Movement of carbon compounds between partners in orchid mycorrhiza. in: *Endomycorrhizas*, F. E. Sanders, B. Mosse and P.B. Tinker, eds., Academic press, London and New York, pp. 173-194.

Ramsay, R.R., Dixon, K.W. and Sivasithamparam, K. 1986, Patterns of infection and endophytes associated with Western Australian orchids, *Lindleyana* 1: 203-214.

Rasmussen, H.N. 1990, Cell differentiation and mycorrhizal infection in *Dactylorhiza majalis* (Rchb. f.) Hunt and Summerh. (Orchidaceae) during germination *in vitro, New Phytol.* 116: 137-147.

Rasmussen, H.N. 1995, Terrestrial orchids from seed to mycotrophic plant, Cambridge University Press, Cambridge.

Read, D.J. 1999, Mycorrhiza- the state of art, in: *Mycorrhizae: Structure, Function, Molecular Biology and Biotechnology*, A. Varma and B. Hock, eds., second edition, Springer-Verlag, Germany, pp. 3-34.

Richardson, K.A., Peterson, R.L. and Currah, R.S. 1992, Seed reserves and early symbiotic protocorm development of *Platanthera hyperborea* (Orchidaceae), *Can. J. Bot.* 70: 291-300.

Saunders, G.C. and Owens, S.J. 1998, RAPD analysis and ITS analysis of orchid mycorrhizal fungi, in: *Mycorrhiza Manual*, A. Varma, eds., Springer-Verlag, Berlin, Heidelberg, New York, pp. 413-424.

Seagull, R.W. and Falconer, M.M. 1991, *In vitro* xylogenesis, in: *The Cytoskeletal Basis of Plant Growth and Form*, C.W. Lloyd, ed., Academic Press, London, pp. 183-194.

Smith, S.E. 1966, Physiology and ecology of orchid mycorrhizal fungi with reference to seedling nutrition, *New Phytol.* 72: 1325-1331.

Smith, S.E. 1967, Carbohydrate translocation in orchid mycorrhizal fungi, *New Phytol.* 72: 371-378.

Smith, S. E. 1973, Asymbiotic germination of orchid seeds on carbohydrates of fungal origin, *New Phytol.* 72: 497-499.

Smith, S.E. and Read, D.J. 1995, *Mycorrhizal Symbiosis*, second edition, Academic Press, London.

Smith, S.E. and Smith, F.A. 1973, Uptake of glucose, trehalose and mannitol by leaf slices of the orchid *Bletilla hyacinthina, New Phytol.* 72: 957-964.

Smreciu, E.A. and Currah, R.S. 1989, Symbiotic germination of seeds of terrestrial orchids of North America and Europe, *Lindleyana* 4: 6-15.

Sneh, B., Burpee, L. and Ogoshi, A. 1991, Characteristics common to a significant number of isolates in anastomosis groups of *Rhizoctonia* spp. in: Identification of *Rhizoctonia* species, B. Sneh, L. Burpee and A. Ogoshi, eds., St. Paul, APS Press, Minnesota, pp. 77-86.

Strullu, D.G. and Gourret, J.P. 1974, Ultrastructure et evolutiondu champignon symbiotique des racines de *dactylorchis maculata, J. Micros.* Paris 2: 285-294.

Strullu, D.G. 1976, Recherches de biologie et de microbiologie forestieres. Etude des relations nutrition, development et cytologie des mycorrhizes chez le douglas (*Pseudotsuga menziesii* Mirb.) et les abietacees, Theses, Univ.Rennes.

Terashita, T. 1985, Fungi inhabiting wild orchids in Japan (III), A symbiotic experiment with *Armillariella mellea* and *Galeola septentrionalis*, *Trans. Mycol. Soc. Japan* 28: 145-154.

Terashita, T. and Chuman, S. 1987, Fungi inhabiting wild Orchids in Japan (IV), *Armillariella tabascens,* a new symbiont of *Galeola septentrionalis*, *Trans. Mycol. Soc. Japan* 28: 145-154.

Trappe, J. M. 1996, What is a mycorrhiza? in : *Proceedings of the Fourth European Symposium on Mycorrhizae,* Granada, Spain, EC Report EUR 16728, pp 3-9.

Uetake, Y., Kobayashi, K. and Ogoshim, A. 1992, Ultrastructural changes during the symbiotic development of *Spiranthes sinensis* (Orchidaceae) protocorms associated with binucleate *Rhizoctonia anastomosis* group C., *Mycol. Res.* 96: 199-209.

Uetake, Y. and Ishizaka, N. 1996, Cytochemical localization of adenylate cyclase activity in the symbiotic protocorms of *Spiranthes sinensis*, *Mycol. Res.* 100: 105-112.

Uetake, Y., Farquhar, M.L. and Peterson, R.L. 1997, Changes in microtubule arrays in symbiotic orchid protocorms during fungal colonization and senescence, *New Phytol.* 135: 701-709.

Vanderplank, J. E. 1978, *Genetic and Molecular Basis of Plant Pathogenesis,* Springer-Verlag, Berlin, Heidelberg, New York.

Varma, A. 1995, Arbuscular mycorrhizal fungi: The state of art, *Crit. Rev. Biotechnol.* 15: 179-199.

Varma, A. 1998, Mycorrhiza- the friendly fungi: what we know, what should we know and how do we know? in: *Mycorrhiza Manual*, A. Varma, ed., Springer-Verlag, Berlin, Heidelberg, New York, pp. 1-24.

Varma, A. 1999, Functions and application of arbuscular mycorrhizal fungi in arid and semi arid soils, in: *Mycorrhizae: Structure, Function, Molecular Biology and Biotechnology*, A. Varma and B. Hock, eds., second edition. Springer-Verlag, Berlin, Heidelberg, New York, pp 521-556.

Varma, A., Singh, A., Sudha, Sahay, N. S., Sharma, J., Roy, A., Kumari, M., Rana, D., Thakran, S., Deka, D., Bharti, K., Franken, P., Hurek, T., Hahn, A., Hock, B., Maier, W., Walter, M., Lemke, P.A. and Strack, D. 1999, *Piriformospoa indica* gen. nov. sp. nov., in: *Mycota* IX, Springer-Verlag, Germany (in press).

Verma, S., Varma, A., Hexer, K.H., Kost, G., Sarbhoy, A., Bisen, P., Butehorn, B. and Franken, P. 1998, *Piriformospora indica* gen. nov. sp. nov., a new root colonizing fungus, *Mycologia* 90: 896-903.

Warcup, J.H. 1975, Factors affecting symbiotic germination of orchid seeds, in: *Endomycorrhizas*, F.E. Sanders, B. Mosse and P.B. Tinker, eds., Academic Press, London, New York, pp. 87-104.

Warcup, J.H. 1981, The mycorrhizal relationship of Australian orchids, *New Phytol.* 87: 371-387.

Warcup, J.H. and Talbot, P.H.B. 1966, Perfect states of some rhizoctonias, *Trans. Br. Mycol. Soc.* 49: 427-435.

Warcup, J.H. and Talbot, P.H.B. 1980, Perfect stages of rhizoctonias associated with orchids, III, *New Phytol.* 86: 267-272.

Weber, H.C. 1981, Orchideen auf dem Weg zum Parasitismus? Uber die Moglichkeit einer phylogenetischen Umbkonstruktion der Infektionsorgane parasitischer Blutenpflanzen, *Bericht der Deutschen Botanischen Gesellschaft* 94: 275-286.

Webster, J. 1996, A century of British mycology, in: *A Century of Mycology*, B.C. Sutton, ed., Cambridge University Press, Cambridge, pp. 1-38.

Whitney, H.S. and Parmeter, J.R. 1964, The perfect stage of *Rhizoctonia hiemalis, Mycologia* 54: 114-118.

Williamson, B. 1970, Induced DNA synthesis in orchid mycorrhiza, *Planta* 92: 347-354.

Williamson, B. 1973, Acid phosphatase and esterase activity in orchid mycorrhiza, *Planta* 112: 149-158.

Williamson, B. and Hadley, G. 1970, Penetration and infection of orchid protocorms *Thanetephorus cucumeris* and other *Rhizoctonia* isolates, *Phytopathol.* 60: 1092-1096.

Wood, R.K.S. 1967, *Physiological Plant Pathology*, Blackwell Scientific Publications, Oxford.

Wood, R.K.S. and Graniti, I. 1976, *Specificity in Plant Diseases*, Plenum Press, New York.

Xu, J., Ran, X. and Guo, S. 1989, Studies on the life cycle of *Gastrodia elata, Acta Acad. Medi. Sinicae* 11: 37-241.

Zanzinger, D.H., Bandy, B.P. and Tavantzis, S.M. 1984, High frequency of finding double stranded RNA in naturally occurring isolates of *Rhizoctonia solani, J. Gen. Virol.* 65: 1601-1605.

EFFECT OF AGROCHEMICALS ON MYCORRHIZAE

S.C. Vyas and Sameer Vyas

Department of Plant Pathology
Jawaharlal Nehru Agricultural University
Indore, M.P., INDIA

1.   INTRODUCTION

The modern crop production techniques require even management of diseases which are important biological determinants of crop productivity. Primary reliance in the control of plant diseases has been placed on fungicidal compounds of intrinsic fungitoxicity, which became standard inputs of crop production. A great number of fungicides (systemic and non-systemic) with selectivity and specificity for the control of seed and soil borne diseases, by seed and soil treatment, respectively and air borne diseases by foliar application are being used. The application of fungicides which ultimately slips into the environment usually results in a diverse array of effects on target and non-target organisms and thereby influence crop productivity, as profoundly or even more so than do the pathogens they are intended to control (Trappe *et al.*, 1984; Vyas, 1984; 1988). Fungicides are biocides, which act upon a great variety of organisms for which they are inteded and thus result in altered biological phenomenon which may be favourable or unfavourable to pathogen activity and disease development (Upadhyaya *et al.*, 1996; 1997; 1998 a,b; 1999 a,b).

Plant roots form mycorrhizae with certain fungi, resulting in a mutually beneficial symbiotic relationship. Most plants have a mycorrhizae rather than roots per se; only few plant families do not form mycorrhizae, The three major types are ectomycorrhizae, endomycorrhizae and ectiendomycorrhizae. Endomycorrhizae on roots of plants in many families are formed in association with fungi belonging to Glomales (Mukerji, 1996) and are commonly known as vesicular-arbuscular mycorrhizae (VAM). These fungi form intramatrical vesicles (terminal hyphal swellings considered to be storage organs and arbuscules (intracellular haustoria-like strucutres). Endomycorrhizal fungi are ubiquitous and they occur on a much larger number of plant species and thus are more common. Endomycorrhizae have received considerable attention in recent years because VAM plants have several advantages over non-mycorrhizal plants (Srivastava *et al.,* 1996). VAM plants grow better in infertile soils largely because of increased uptake of nutrients that are relatively immobile in soils, such as phosphorus; VAM plants enhance water transport in plants (Safir *et al.*, 1971) decreased transplant injury (Menge *et al.*, 1978) and reduce the effect of root infecting fungi (Schenck and Kellam, 1978).

Mycorrhizal Biology, edited by Mukerji *et al.*
Kluwer Academic/Plenum Publishers, 2000

2.   MYCORRHIZAE

Mycorrhiza, the symbiotic association of the mycelium of a fungus with the roots of most plant species, including angiosperms, gymnosperms, pteridophytes, and thallophytes, occurs not only among forest crops but also among shrubs and herbs. The fungi performed the function of root hairs, which were lacking in these much modified dual structures (Mosse *et al.*, 1981). During the prolonged period of physiological interaction, the mycorrhizal association is characterized by a high degree of specificity and specialization of the higher and lower symbionts. Physiologically, mycorrhizae represent a case of symbiosis where:

(i)   There is an increased uptake of nutrients and water transport from soil, particularly in infertile soils, and thus an increase in plant growth.

(ii)   Mycorrhizal roots may afford protection to trees from infection by feeder root pathogens which may reduce growth or kill in plants(biological control).

(iii)   Mycorrhizae make plants drought and frost resistant.

Studies conducted by many workers on the mechanism leading to mycorrhizal development indicate that both the higher and lower symbionts produce a complex of substances: an accumulation of soluble carbohydrates in roots of plants and production of one or more growth-promoting as well as growth-inhibiting metabolites. In addition to thiamine, one or more B-vitamins attract fungal symbionts from the soil to grow on the surface of roots and cause infection. The fungi in turn produce substances related to auxins that result in characteristics root morphogenesis in short roots in which the symbiotic relationship is established (Bakshi, 1974; Mehin, 1963; Slanks, 1971).

The ectomycorrhiza is characteristic of all genera belonging to conifers and hardwood; they can be distinguished from the heterozoic root system, comprising long roots of unlimited length which remain uninfected, and mycorrhizal (short) roots of limited growth which are ephemeral. The infected roots are restricted in growth; they are branched and colored and the root hairs are absent. The mycorrhiza is enveloped completely by a fungal mat or mantle. Endomycorrhizne are characterized by intracellular infection within the root, absence of any organized fungus growth on the root surface, and little, if any, change in morphology. They may be formed by septate or nonseptate fungi; those formed by nonseptate fungi, are referred to as phycomycetous or arbuscular mycorrhizal fungi (AMF), are most widely prevalent not only among conifers in all tree species but also in most agricultural crops throughout the world. The fungi are less abundant in forests though by no means less prevalent (Gardmann, 1971). AMF have been shown to increase plant growth in soil and are increasing utilization of less available forms of phosphorus. They are abundant in forests, and widely distributed in soils in the top 15 cm, and are more numerous and varied in cultivated than in noncultivated soils (Mosse, 1973). The vesicles are intracellular, subglobose, and usually single, although sometimes in groups; they are terminal with open connections and generally in the outer cortex, rarely both outer and inner. The asbuscules are compound and are usually found in the inner cortex (Bakshi, 1974).

3.   FUNGICIDES AND MYCORRHIZAE INTERACTIONS

Fungicides are by definition meant for inhibiting or killing fungi, such compounds might be expected to most profoundly influence mycorrhizal fungi and mycorrhiza formation. The use of fungicides and fumigants to control soil-borne pathogens is common. Recently, concern has developed among agriculturists about the effects of fungicides usage upon beneficial mycosymbiont VAM fungi. Agriculturist often enquire about the benefits of fungicides as increased crop growth via destruction of pathogenic organisms being diluted because they also destroy mycorrhizal fungi and thereby reduce nutrient uptake by crops. A substantial amount of accurate although fragmented data is now available on the interaction between mycorrhizal fungi and fungicides. Earlier reviews (Menge 1982; Nemec, 1980; Trappe *et al.,* 1984) were devoted on the role of fungicides in VAM infection and on spore germination, but in the present review attention has been focussed on the effects of fungicides on the development stages and also mechanisms of fungitoxicity and stimulation. The development of mycorrhizae relationship can be divided into the following four stages (Tommerup and Briggs, 1981) (Table-1).

Table 1 : Influence of systemic and non-systemic fungicides on the establishment of Vesicular Arbuscular Mycorrhizal relationship with host plants

S.No.	Stage	Fungicides
1.	Spore germination or initiation of hyphal growth from the infected root inoculum	Non-systemic fungicides present in soil after seed treatment and foliat application.
2.	Growth of hyphae through the soil to the roots.	Non-systemic fungicides present in soil and around roots.
3.	Penetration and successful initiation of infection in roots	Systemic fungicides movement upward and downward in plant
4.	Spread of infection and development of VAM relationship with roots and spore production	Systemic fungicides moving downward Non-systemic fungicides

Most of the fungicides have been produced from inorganic compounds containg copper and mercury, organic and systemic fungicides. These fungicides simulate the resistance of the host and indirectly affect the VAM association. Sometimes even the VAM fungi can be killed which will indirectly reduce the nuterient uptake of the host. Persistence of fungicides in soil also affects mycorrhization and plant growth (Vyas, 1988).

The effects of agrochemicals on increased plant production have been called a miracle of technological age and these are not mutually exclusive with other methods of crop productivity and protectivity. Environmentalists concern that the substances that enabled conventional agriculture to meet growing demand of food and fibre of ever increasing world population may endanger the safety of the product and the sustainability of the production base led to a search for safer approaches. AMF, as a dominant influence on the soil biota are a part of the solution, as a complement to conventional tehniques such as mineral fertilization and application of biocides. Bethlenfalvay (1992) and Johnson and Pfleger (1992) have reviewed the role of AMF in crop productivity and sustainable agriculture.

Concern over the deterioration of natural bases in agriculture has renewed interest in conservation of tillage (Larsen and Osborne, 1982). Although it is successful in controlling erosion, no tillage necessitates careful, integrated weed, pest and pathogen management (Cornish and Pratley, 1991) and may call for increased biocide use in minimizing crop loss (Phillips *et al.,* 1980). The utilization of biocides, however, is no loss a concern than erosion, because of its impact on the environment and on human health (Huff and Haseman, 1991).

Most of investigations on pesticides and mycorrhizal interactions indicate that pesticides have profound effects on arbuscular mycorrhiza fungi (AMF). However, it is difficult to make simple generalizations from these studies because of variability in pesticide formulations and experimental conditions. A brief review of mycorrhizal responses to some of the more commonly used fumigants, fungicides, herbicides and insecticides, nematicides have been discussed in relation to factors mediating these responses. Trappe *et al.* (1984); Menge (1982); Moorman (1989); Vyas (1988; 1993), Johnson and Pfleger (1992) have reviewed the subject earlier also.

4.   ECTOMYCORRHIZAE

The importance of ectomycorrhizae for the survival and growth of tree seedlings is well documented (Marx and Rowan, 1981). In many parts of the world, special procedures are used in tree nurseries to encourage ectomycorrhizal development (Mikola, 1973). Such procedures include the addition of soil containing these fungi, maintenance of high organic matter and low to moderate fertility, and growing host trees near the nursery to encourage production of fruiting bodies for natural colonization. The ectomycorrhizal fungi include *Pisolithus tinctorius* and *Thelephora terrestris*. Studies indicate that introducing pure cultures of these fungi into nursery soil is more beneficial to tree seedlings after outplanting than are naturally occurring nursery fungi (Marx, 1980).

4.1.   *In Vitro*

General biocides (broad spectrum and nonspecific) will eliminate or kill an organism exposed to them. Hence, it is not surprising that mycorrhizal fungi exposed to fungicides at various concentrations *in vitro* in different media will show variable growth patterns. The available literature is briefly reviewed in the following paragraphs.

4.1.1. Non-systemic fungicides

Non-systemic fungicides are broad spectrum and nonspecific toward mycorrhizal fungi. Captafol (100 to 200 μg g^{-1}), (Kawai and Ogawa, 1977), captian (1 to 15000 μg^{-1}), copper oxychloride (500 to 3000 μg g^{-1}), chlorothalonil (100 to 200 μg g $^{-1}$) (Marx and Rowan, 1984) reduced the growth of fungi in medium. Sobotka (1970) evaluated different dithiocarbamates - thiram, ferbam, zineb, and folpet - using drenched paper strips in medium and found that ferbam and folpet reduced growth in the medium. However, thiram (100 to 200 μg g^{-1}), zineb (500 to 300 μg g^{-1}), and ziram (500 to 300 μg g^{-1}), (Laiho and Mikola, 1964), Mancozeb (1 to 500 μg g^{-1}), (Cudlin *et al.*, 1980) and maneb (100 to 200 μg g^{-1}) (Sobotka, 1968; Laiho and Mikola, 1964) were toxic to fungi in different degrees.

4.1.2. Systemic fungicides

Systemic fungicides were evaluated for their effect on mycorrhizal fungi *in vitro*. The benzimidazoles benomil, fuberidazole, and thiabendazole at 1 to 200 μg g^{-1} (Edington *et al.*, 1971) had no effect on growth *in vitro*. However, benodanil, another systemic fungicides, was toxic to fungi at 1 to 17 μg g^{-1}. Hymexazol still another systemic fungicide, was also toxic at 200 μg g^{-1} (Kawai and Ogawa, 1977; Kelley and Snow, 1978).

The inhibitory or other effects of fungicides on mycorrhizae *in vitro* may not tell us exactly what side effects may occur in nature. Since mycorrhizal associations in forests and fields are the most important, future studies should focus on the beneficial or deleterious effects of fungicides in nautre. Nonetheless, the *in vitro* studies will provide background information for field experiments on fungicide mycorrhizae interactions/associations.

4.2.   *In Vivo*

4.2.1. Non-systemic fungicides

Different fungicides are used to control diseases in the production of nursery seedlings (Peterson and Smith, 1975) and many of them effect ectomycorrhizal development. Bakshi and Dobriyal (1970) found that captan and pentachloronitrobenzene (PCNBP) used to control damping-off in nurseries in India delayed the development of naturally occurring ectomycorrhizae on seedlings of *Pinus patula* until the fungicides had lost their effectiveness. In Malaysia, Hong (1976) observed that ectomycorrhizal development on *P. caribaea* seedlings was suppressed by chlorothanonil and thiram and stimulated by captafol and captan. PCNB inhibited the growth of several ectomycorrhizal fungi in pure cultures but did not alter the development of naturally occurring ectomycorrhizae of pine and spruce seedlings when they were applied to soils in nurseries. In nursery and greenhouse tests in Australia, Theodorou and Skinner (1976) found that seedcoat dressing of captan, zineb, and thiram inhibited ectomycorrhizal development of *P. radiata* seedlings grown from seed inoculated with basidiospores of different fungi, but had no effect on the subsequent development of ectomycorrhizae formed by naturally occurring fungi. Pawuk *et al.* (1980) found that the development of *Pisolithus* ectomycorrhizae on *Pinus palustris* seedlings grown for 14 weeks in a pine bark medium in containers was completely inhibited by PCNB and reduced by captan and fenaminosulf. Because of these known effects, fungicides that are currently used or are being considered for use to control diseases in forest nurseries were evaluated (Marx and Rowan, 1981). A comparison of the results of the laboratory and nursery tests shows that mycelium of *Pisolithus* and *Thelephora* is affected more by fungicides in agar medium than soil (Marx and Rowan, 1981). The limited persistence of these fungicides in soil probably accounts for some of this difference in sensitivity. The lack of leaching, microbial activity by associated microflora, and barriers to fungicide diffusion in agar permits the hyphae to endure long exposure at higher concentrations in agar culture than in soil. Due to the higher nutrient concentrations in agar medium than in soil, the mycelium grows at a faster rate which would increase the metabolic uptake of the fungicides. Also, the hyphae of ectomycorrhizal fungi grown in vermiculite particles as the initial inoculum may be protected to a certain extent from high concentrations of fungicides in soil. These differences between nursery and laboratory studies on fungicides indicate the limited value of studies on agar medium that are not supported by studies carried out in soil.

Arasan, botron, lanstan, and ethazole mixed at 25, 50, 100, and 250 ppm, respectively, captan at 50 and 100 ppm, and preplanting compounds mylone, vapam, and vorlex at 40 ppm restricted mycorrhizal development of *Endogone fasciculata* on corn to some extent, even at the lowest concentration (Nesheim and Linn, 1969). Of the five comparable toxicants, captan was least injurious while vorlex was somewhat less so than mylone and vapam. It was also observed that root volume and branch root development were severely limited by arasan, botran, and ethazole at 100 ppm and moderately so at 50 ppm. There was a slight suppression of both root development and volume by ethozole at 25 ppm. Botran at all concentrations caused some distortion of root hairs. Arasan reduced the number of root hairs, but only at 50 and 100 ppm. The treatment of soil with fungitoxicants is frequently beneficial to crop plants. However, these toxicants may sometimes kill both detrimental and favourable soil microorganisms indiscriminately.

Extreme effects of fungitoxicants found by Nesheim and Linn (1969) in reducing mycorrhizal infections may be attributed to their use of a sterilized soil and mixture. In natural soils, fungal toxicants can be depleted by microbial action, hydrolyses, and chemical reactions, but the time of their disappearance depends on the compound and the physical condition of the soil. This suggests that root hairs and their epidermal cells, which are the potential initials, should be protected from the toxic effects of fungicides and fungitoxicants.

The target organisms of captan are primarily hyphomycetes, ascomycetes, (Bollen and Fuchs, 1970) and certain phycomycetes (Agnihotri, 1971). A lower population of these fungi and other microorganisms may increase the effectiveness of inocula of ectomycorrhizal fungi. A reduction in populations of hyphomycetes and other fungi in the rhizosphere of onion plants by captan has been demonstrated (Bertoldi *et al.,* 1977). Captan temporarily decreased soil populations of *Pythium* of pecan trees and stimulated the development of ectomycorrhizae formed by naturally occurring

Scleroderma bovista. The significane of changes in rhizosphere microbial populations, such as an ectomycorrhizal development of *Pisolithus tinctorius* and *T. terrestris* caused by the indirect action of fungicides, needs further investigation.

In nursery soil infested with mycorrhizal fungi, ectomycorrhizal development on seedlings of *Pinus taeda* and basidiocarp production by *Pisolithus tinctorius* were greater than in fungicide-free controls when the plots were treated with benomyl and captan and less than in those treated with benodanil. PCNB had no effect on the ectomycorrhizal development of *Thelephora terrestris*. However, the development was greater in plots with double doses of benomyl and captan. PCNB and benodanil decreased basidiocarp production. The mycelium of both fungi was affected by the fungicides in agar medium more than in soil (Spokes *et al.*, 1981).

Where maneb, quintozene, captan, thiram, zineb, and ziram were used for the control of damping-off at the rate of 0.7 to 60 kg ha^{-1}, they showed toxicity against mycorrhizal develoment in *Pinus insularis, P. patula*, and *P. radiata* (Bakshi and Dobriyal, 1970). Captafol, chlorothalonil, copper oxychloride, folpet, and triphenyltin acetate at 4.5 to 24.5 kg ha^{-1} showed slight to no toxicity to *P. caribaea* and *P. oocarpa*. Captafol, chlorothalonil, and captan had no effect on mycorrhizal development when applied as foliar fungicides at the operational rate for disease control (Hong, 1976).

4.2.2. Systemic fungicides

Fungicides are being used itensively in the control of diseases in forest nurseries and may affect the development of ectomycorrhizae. Nesheim and Linn (1969) reported that the systemic fungicide ethazole applied at 25, 50 and 100 ppm in soil has a deleterious effect on the ectotrophic fungus *Glomus fasciculatum* on corn but enhances the growth of forest trees. There was a slight suppression of both root development and volume by ethazole at 25 ppm. Benomyl had no effect on the development of ectomycorrhizae on *Pinus caribaea* (Hong, 1976). Benodanil, a systemic fungicide undergoing testing for the control of fusiform rust on pines in the southern U.S., prevented pure culture growth of several ectomycorrhizal fungi. When applied as a drench to nursery soil it delayed the development of naturally occurring ectomycorrhizae on *Pinus taeda* (Kelley, 1980). The development of *Pisolithus* ectomycorrhizae on *Pinus palustris* seedlings was stimulated by ethazole and thiabendazole (Pawuk *et at.*, 1980).

Many factors that influenced the formation of ectomycorrhizae affect either the susceptibility of the host or the survival and infective potential of the fungal symbionts (Bowen and Theodorou, 1973). Although certain fungicides used in this test may have affected the susceptibility of pine seedlings to root infection by fungi, the results of other research suggest that the action of fungicides is more likely to have a direct or indirect effect on the fungi. Benodanil directly inhibits the vegetative growth of symbiotic fungi and causes a decrease in ectomycorrhizal development (Kelley, 1980; Marx and Rowan, 1981). As discussed by Kelley (1980), this type of action by benodanil is an example of its detrimental effect on beneficial, nontarget organisms. However, Marx and Rowan reported that benomyl indirectly stimulated ectomycorrhizal development by specific nontarget fungi. The target organisms of benomyl are primarily hyphomycetes and ascomycetes.

Soil amendments with benomyl, ethazole, metalaxyl, and propamocarb for the control of nursery diseases was assessed on rhododendron mycorrhizae. The foliar application of fungicide had no adverse effect on mycorrhizal formation. Mycorrhizal studies were conducted on the effect of benomyl, chloroneb, and thiophanate-methyl on exotic conifers in west Malaysia (Ivory, 1975). Except for chloroneb, the other fungicides had no effect on the mycorrhizae. A weekly spray of triadimefon at 0.28 kg ha^{-1} for the control of rust on pine (*Pinus radiata*) seedlings was toxic to the mycorrhizae (Cline *et al.*, 1981). This indicates the downward translocation of the fungicide.

When onion seedlings were grown for 6 months in the absnece of pathogens, repeated soil applications of benomyl or captan significantly and simultaneously decreased the diameter and the dry weight by 22 to 25% and 31 to 34%, respectively. The losses were associated with differing effects on soil organisms. Captan had no effect on the occurrence of VAM but appreciably decreased the population of rhizosphere fungi. In contrast, benomyl inhibited mycorrhiza formation but had relatively little effect on rhizosphere fungi (Bertoldi et al., 1977).

These studies indicate that the utmost care should be taken in the selection of fungicides for control of diseases in tree nurseries. The qualitative and quantitative effects of fungicides on the development of ectomycorrhizae should be considered when any fungicide is selected to control specific diseases. Disease-free seedlings lacking adequate ectomycorrhizae may be grown to plantable size in a nursery following fungicide application, but the lack of adequate ectomycorrhizae on such seedlings will often result in substandard survival and growth performance following field outplanting. The stimulating effect of certain fungicides on ectomycorrhizal development, especially those formed by artificially introduced fungi, should be considered an added benefit of these fungicides (Marx, 1980; Marx and Rowan, 1981).

5.   ENDOMYCORRHIZAE

Arbuscular mycorrhizal fungi (AMF) are formed by *Glomus* spp. with the roots of most wild and important agronomic crop plants. Economically important crops with AMF include peas, maize, wheat, soybeans, potatoes, tomatoes, strawberries, apples, citrus, grapes, tobacco, and sugar maple.

AMF, mutually beneficial symbionts, are widely distributed and are physiologically unspecialized. They are so wide-spread in the plant kingdom that there are fewer families in which they do not occur than in which they do. In AMF, there is a nonpathogenic symbiotic association between fungi and the roots of the higher plants. The fungi are members of the endogonaceae. The AMF enhances nutrition uptake, conserve crop plant mineral nutrient resources, and are involved in the protection of the host plant from soil-borne pathogens. Mukerji et al. (1984) reviewed taxonomy, distribution, and other recent developments on AMF. It is becoming increasingly apparent that the AMF have probiotic influence on higher plants. The AMF association is a labile condition; environemntal factors greatly influence the balanced relationship between plant roots and the endophyte (Anderson, 1978; Jalali and Domsch, 1975). This is of particular significance because AMF occur in a large number of commercial crops, and the wide use of toxic materials in normal agricultural practices may therefore upset this balance.

Burpee and Cole (1978) and Smith (1978) have assessed changes in the percentages of infected roots and spore production. Thus, any observed changes result from the response of each of the three preceding stages as well as the stage in which chemicals are added. With the increasing use of fungicides for the control of plant diseases, it is obligatory to review the effects of fungicides on VAM, particularly systemic frungicides which enter the plant system and affect the symbiotic relationship of the host and the AMF.

Diverse groups of fungicides are widely used in the control of plant diseases. The application of these chemicals may, however, result in indiscriminate killing of both pathogenic as well as nonpathogenic or beneficial microorganisms (Domsch, 1964). Several reports have demonstrated the partial or complete inhibition of vesicular-arbuscular endophytes by fungicides. The action of fungicide delays or reduces AMF colonisation. Nonsystemic fungicides such as PCNB, botran, and thiram are highly toxic to mycorrhizal fungi (El-Giahmi et al., 1976).

Some fungicides may increase the infection and sporulation of the AMF fungi. Demosan, daconil, sodium azide, terrazole, captan, and copper sulfate are not very toxic to VAM fungi and may favor VAM activity under certain specific environmental conditions (Vyas, 1988). Systemic fungicides are more damaging than nonsystemic fungicides on AMF fungi. Nonsystemic fungicides can delay colonisation by AM fungi but do not completely eliminate them, whereas systemic fungicides can adversely affect spore germination, infection, and growth within the root.

5.1. Fungicides and VAM Interactions

5.1.1. Sensitivity of VAM species to fungicides

Many of the fungicides directly effect mycorrhizal fungi in the field (Menge, 1982). Different fungicids react differently with different VAM fungi e.g. growth of *Glomus fasciculatum* and *G. constricutum* was suppossed by benzimidazoles and methyl bromude (Benjamin,1979; Dodd and Jefferies, 1989; Menge *et al.,* 1978), Triademifacterrazole chloroneb treatments did not effect colonization by *G.fasciculatum* after 12 weeks, though *G.mosseae* could colonize only 2/3rds of root after fungicides treatment (Spokes and McDonald, 1978). Triademifon reduced VAM formation by *G.microcarpus* to 1/2 and chloroneb increased colonisation to 100% (O'Bannon and Nemec (1978).

Diverse groups of fungicides - systemic and non-systemic (Nene and Thapliyal, 1979; Vyas, 1984) broad spectrum - mostly non-systemic and selective, etc. are used ot reduce soil, seed and airborne pathogens. Irrespective of their application methods, ultimatley the fungicides reach to the soil and affect soil biota. The application of fungicides, however, results in increase (Jababji - Hare and Kendrick, 1985; 1987) or no effect (Nemec, 1980) or in reduction or delaying mycorrhizal infection (Vyas, 1998) but rerely eliminate (Menge, 1982). Most of the fungicides stimulate crop growth because of the elimination of soil borne pathogens. However, since the first use of soil fumigants in 1869 there have been consistent reports of stunting following fumigation with many crops including avocada, citrus, cotton, peach, soybean, white clover and hardweed trees (Powell, 1976; Riffle, 1976). Most to the fumigants induce stunting (Timmer and Leyden, 1978). The stunting of plants includes poor growth and small necrotic leaves which are stiff chlorotic and die prematurely. Stems are thick and roots development is scanty otherwise normal appear; micronutrient deficiency such as Cu, Zn, Fe, Mg and even of K may be implicated in the stunting process *(*Martin *et al.,* 1973). Tucker and Anderson (1972) observed that a decrease in the availability of soil elements as a result of fumigation was not the cause of the stunting problem since the availability of most of the nutrients was unchaged by soil fumigations. For many years the mysterious stunting agent eluded scientists. However, studies of Martin *et al.* (1963) concluded that the stunting was caused by an unidentified substance that are elements or their compounds or soluble salts and, although the stunting problem could be corrected with applications of P, soil deficiency of P was not the cause. The inhibitory condition would last for a few days to several months and most importantly that mixing untreated and treated soil reduced the duration of toxicity (Martin *et al.,* 1963). The solution to the stunting following the fumigation problem was suggested by Clark (1963) and Filer and Tolle (1968) when they demonstrated that mycorrhizal inoculation increased growth of tree seedlings in methyl bromide fumigated soil; they further reported that fumigation reduced VAM population. A mechanism for explaining elevated level of states in some stunted plants was postulated by Ratnayake *et al.* (1978) when they determined that severe P deficiencies induced by lack of mycorrhizae could destroy the selectivity of the phospholipid membranes in root cortical cells, thereby allowing excessive uptake of salts. Mycorrhizae fungi have been shown to provide resistance to salt damage (Hirrell and Gerdemann, 1980).

5.1.2. Fumigants

The fumigants such as chloropicrin, formaldehyde, mylone, methyl bromide, vapam and vorlex consistently reduced mycorrhizal infection; mylon, vapam and vorlex all decompose into methyl isocyanate which is the toxic principle to VAM fungi. Methyl bromide appears to be especially toxic to VAM fungi and many researchers have used this fumigant to eradicate mycorrhizal fungi from experimental soils (Menge, 1982).

5.1.3. Non-systemic fungicides

The fungicides such as PCNB thiram, and botran are consistently toxic to VAM fungi, however, captan, copper fungicides and chlorothalonil were not toxic and somtimes increased VAM infection. The non-systemic fungicides by and large affect stage I and II that is spore germination from plant tissues and hyphal development, respectively. These fungicides accumulate around root soil interface and in rhizoplane in accumulate quantity after foliar, soil and seed application and there toxicity is apparent on these two stages. However, they are being abosrbed partially or in negligible amounts, hence they have no effect on stage III and IV (Tommerup and Briggs, 1981).

5.1.4. Systemic fungicides

Systemic fungicides, which are translocated within the host either in the apoplast (bennomyl, carbendazim, triadimefon, pyrazophos, etc.) or the symplast (metalazyl, fosetyl-Al and chholoroneb) may act either, directly or the VAM fungi or indirectly by altering host root physiology (Edington, 1981). Etridiazole, benomyl, tridemorph, triforine, ethirimol, thiabendazole, thiophanate methyl, triademifom and carboxin all appear capable of adversely affecting infection or development of VAM fungi (Boatman *et al.,* 1978; Jalali and Domsch, 1975; Menge, 1982; Nemec, 1980). Antioomycetes fungicides such as pyroxychlor, fosetyl-Al, Chloroneb, prothiocarb and metalaxyl fungi has been shown to stunt onion and strawberries by reducing mycorrhizal infection and ultimatley P uptake (Boatman *et al.,* 1978). Systemic fungicides are fungistatic hence they have no effect on stage I and II but they are absorbed by the root hairs and present in sufficient quantities in rhizoplane zone of root where they showed toxicity to infection proccess (III) and development stage (IV). Benomyl which decomposes in soil to give carbendazim and butylisocyanate, has been shown to reduce percentage mycorrhizal infection in root samples at growth stage IV (Bailey and Safir, 1978; Boatman *et al.,* 1978; Hayman and Mosse, 1978; Jalali and Domsch, 1975; Sutton and Sheppard, 1976). Suspensions of benomyl are toxic to VAM fungal inoculum on direct immersion and when mixed with irridiated or infected soil, as shown by reduced percentage infection in roots at about 5 weeks, this probably indicates that soil drenched with benomyl was toxic to external mycelium from established infections so that new roots remained uninfected, however, hyphal growth may also have been reduced because the plants did not grow as well in fungicide and possibly aggravated by P deficiency (Boatman *et al.,* 1978). Carbendazim at dose rates comparable to those used for benomyl die not reduce spore germination or hyphal growth on the three VAM fungi *Aculospora laevis, Glomus caledonius* and *G.monosporum* (Tommerup and Briggs, 1981). These two phases of growth stage I and II, are different from hyphal growth from infected roots and their response is not confounded by chemically affected roots. The benomyl group of fungicides such as carbendazim thiophanate mythyl and thizbendazole may prevent penetration and so prevent development beyond growth stage III. They did not reduce spore germination and hyphal growth because most of the systemic fungicides are fungistitic. However, Sutton and Shephard (1976) showed the percentage infection in roots was decreased in soil containing benomyl when spore were used as inoculum. Penetration structures whether developed from germ tubes or hyphae growing from infected roots, are probably morphologically different from other development stages and they may not have same tolerance to benomyl.

Benomyl rapidly hydrolyses to carbendazim and n-butyl isocyanate (BIC), MBC the active moity of the fungicide disrupt microtubes formation affecting nuclear division and hyphal growth (Vyas, 1988; 1993). Benomyl has been found to have deleterious effects on VAM; delaying, diminishing or preventing the formation of symbiosis between fungi and roots (Menge, 1982; Spokes *et al.,* 1981); and disrupting their ability to translocate phosphate (Botman *et al.,* 1978). Tommerup and Briggs (1981) found that a sandy soil containing upto 100 ng⁻¹ carbendazim was not inhibitory to germination or disruptive to hyphal growth of the VA mycorrhizal fungi. Carr and Hinkley (1985) reported that concentration of benomyl greater than 10 ng ml.⁻¹ inhibited germination and hyphal growth of the vesicular-arbuscular mycorrhizal fungus *G.caledonicum* on water agar. This effect was attributed to only the activity of MBC, one of the hydrolysis products; the other product, BIC had not observable effect. MBC disrupts microtubule assembly and such an effect might have caused the responses shown by *G.caledonicum*. However, benomyl has been reported to reduce *de novo* synthesis of RNA in germinating spores of *Fuasarim oxysporum* (Decallonne and Meyer, 1972). If this obtains in *G.caledonicum* then germination could have been inhibited by this mechanism as both Hepper (1979) and Beilby (1983) have provided evidence that *de novo* RNA synthesis is essential for spore germination of the fungus. Carr and Rinkley (1985) observed that the ability of *G.caledonicum* spores to germinate once removed from media containing atleast 1200 ng benomyl mL⁻¹ indicates it to have had a fungistatic effect. Whilst realising that the behaviour of surface decontaminated spores incubated on water agar is unlikely to correspond directly with that of the fungus in roots, in soil a fungisttatic effect of benomyl could help explain (a) the mixed inoculum of spores, hyphae and infected roots used by Boatman *et al.,* (1978) still being able to initiate VAM after being soaked in a 125 g benomyl ml⁻¹ solution for 24 hr. and (b) the non-permanent effects of benomyl soil dencities on the development of VAM (Builey and Safir, 1978; Boatman *et al.,* 1978; Spores *et al.,* 1981).

Table 2 : Effect of nonsystemic fungicides on various developmental stages in establishment of vesicular arbuscular mycorrhizal fungi on host plants

Fungicide	Crop	Mycorrhizal fungi	Effect on develop-mental stages +(positive) or - (negative) or .. (not studied)				References (s)
			I	II	III	IV	
Botran	Maize	GF	..	..	-	-	Nasheim and Linn (1969)
		GM	..	..	-	-	Elgihmi and Linn (1969)
Captan	Maize	GM	..	..	-	-	Elgiahmi *et al.* (1976)
		GM	+	..	..	..	Menge (1982)
	Beans	*Glomus* sp.	..	..	..	+	Sutton and Shephard (1976)
	Citrus	GE	..	..	+	+	Nemec (1980)
	Onion	GC	..	..	-	..	Kough *et al.* (1987)
Captafol	Citrus	GE	+	..	-	-	Nemec (1980)
	Conifers	*Glomus* sp.	+	+	+	+	Ivory (1975)
Chloropicrin	Citrus	GM	-	..	..	..	O'Bannon and Nemec (1978)
Chlorothalonil	Citrus	GM	+	..	..	..	Nemei (1980)
Dicholorfluanid	Wheat	GM	-	-	-	-	Neshim and Linn (1969)
Landstan	Maize	GF	-	-	-	-	Elhinahmi *et al.* (1976)
Maneb	Citrus	GM	+	..	..	..	Nemec (1980)
	Peanut	*Glomus* sp.	..	..	..	+	Backman and Clark (1969)
Quintozene	Maize	GF	-	-	-	-	Elgiahmi *et al.* (1976),Neshim and Linn (1969)
	Wheat	*Glomus* sp.	..	..	-	..	Gray and Gerdman (1969)
	Bentagrass	GF	..	..	-	-	Rhodes and Larson (1981)
Thiram	Maize	GF	..	..	-	-	Neshim and Linn (1969)
	Maize	GM	..	..	-	-	Elgiahmi *et al.* (1976)
	Wheat	GM	..	..	-	-	Jalali (1978)

GM=*Glomus mosseae*, GF=*G.fasciculatum*, GE=*G.etunicatum*, GM=*G.monosporum* GC=*G.caledonicum*,
: stages are given in text

Table 3 : Effect of systemic fungicides on various developmental stages in establishment of vesicular - arbuscular mycorrhizal fungi on host plants

Fungicide	Crop	Mycorrhizal fungi	Effect on developmental stages +(positive) or - (negative) or .. (not studied)				References (s)
			I	II	III	IV	
Benomyl	Cotton	*Glomus* sp	+	..	+	+	Hurlimann (1974)
	Lettuce		-	..	..	-	Spokes *et al.* (1981)
Leriodendron tulipifera	Beans	GF	-	..	..	-	Verkade and Hamilton (1983)
	Wheat	*Glomus* sp.	-	..	-	..	Sutton and Shephard (1976)
		GM	..	..	..	-	Jalali and Domsch (1975)
	Citrus	GM	-	..	-	..	Nemec (1980)
	Clover	*Glomus* sp.	..	-	-	..	Boatman *et al.* (1978)
	Soybean	GF	..	..	-	..	Bailey and Safir (1978)
	In vitro	GC	-	-	..	..	Carr and Hinkley (1985)
		GF	..	..	..	-	Menge (1982)
	Onion	GM	..	..	-	..	Beroldi *et al.* (1977)
		GM	..	..	-	..	Majunath and Bagyraj (1984)
		GI	..	..	-	..	Parvathi *et al.* (1985)
		GI	-	..	..	..	Kough *et al.* (1987)
Carbendazim	Wheat	GM;GM	-	..	..	..	Dodd and Jefferies (1989)
	Clover	AL;GM;GC	+	..	..	..	Tommerup and Briggs (1981)
Carboxin	Beans	*Glomus* sp.	-	..	..	..	Sutton and Shphard (1976)
Fosetyl-AL Metalaxyl	Leek	GI	+	+	+	+	Jabaji - Hare and Kendrick (1987)
	Soybean	GM	+	+	+	+	Vyas *et al.* (1990)
	Leek	GI	-	-	-	-	Jubaji-hese and Kendrick (1987)
	Citrus	GF	+	+	+	+	Nemec (1980)
	Cotton	GI	+	+	+	+	Afek *et al.* (1990; 1991)
	Onion	GI	+	+	+	+	
	Pepper	GI	+	+	+	+	
Prochlorez	Wheat	GM; GM	-	..	..	..	Dodd and Jefferies (1989)
			-	..	..	..	Tommerup and Briggs (1981)
Propiconazole	Wheat	GM; GM	-	..	..	..	Dodd and Jefferies (1989)
Prothiocarob	Barley	*Glomus* sp.	..	..	+	+	Pawak *et al.* (1980)
Pyroxychlor	Pea	GF;GM	+	+	-	..	Steward and Pfleger (1997)
Terrazole	Sorghum	GF	+	..	..	..	Menge *et al.* (1979; 1982)

GM=*Glomus mosseae*,   GF=*G.fasciculatum*,   GE=*G.etunicatum*,   GM=*G.monosporum*
GC=*G.caledonicum*,   GI=*G.intraradices*,   AL=*Aculospora laevis*

Tommerup and Briggs (1981) found that *G. caledonicum* grew normally in a sandy soil amended with MBC. Though zygomycotina and other oomycetes are generally resistant to benomyl (Bollen and Fuchs, 1970; Edington *et al.,* 1971), the inference that *G. caledonicum* is resistant to MBC conflicts with the results of Carr and Hinkley (1985). It is possible that the strain used by Tommerup and Briggs (1981) was less sensitive to MBC than that used by Carr and Hinkley (1985). However, a more plausible explanation is that some of the MBC was absorbed onto sandy soil, decreasing the amounts to which the spores were exposed to levels insufficient to affect either spore germination of hyphal growth (Carr and Hinkley, 1985). Comparing the results of Tommerup and Briggs (1981) on the effects of MBC on hyphae growing in söil free from plants with those already reported on the effects of MBC or benomyl or hyphae growing in association with plant in different soils. Tommerup and Briggs (1981) suggested that distinct growth phases of VAM fungi may tolerate different concentration of MBC and benomyl. This might true but needs substantiating particularly as the persistence of benomyl and MBC in a soil is related to the pH and organic content and will therefore differ between soils.

6.   FUNGICIDES AND ECTOMYCORRHIZAE INTERACTION

Ectomycorrhizae (ECM) improve the survival, promote early growth and development of forest trees especially on low quality or acid coal spoil sites by performing essential physiological functions and help seedlings height and resist infection by certain feeder root pathogens (Marx and Artman, 1979). The various reports of ECM and fungicides interactions demonstract, variability. The same is being reviewed here.

Ectomycorrhizal formation can be delayed when certain fungicides are added to nursery soil (Bakshi and Dobriyal, 1970). On the other hand, ECM formation on pecan increased after several fungicides wre incorporated with dibromochhoropropane (DBCP) into the soil (Powell *et al.*, 1968). *Scleroderma bovista* formed abundant ECM apparently owing to the inhibition of competitive soil organisms (Pawak *et al.*, 1980). Phytoactin applied to western white pine seedlings did not reduce ECM (Harvey, 1967) and cyclohexamide which inhibited growth of several ECM fungi in laboratory had no effect when applied to stems of white pine seedlings (Hacskalo, 1961).

Seedcoat dressing of captan, zineb and thiram inhibited development of ECM on Monterey pines grown from seed inoculated with basidiospores of ECM fungi but ECM formation by naturally occurring symbionts was not affected (Theodoroun and Skinner, 1976). Biweekly fungicide sprays of potted caribbean pine seedlings with capatafal and captan increased both number of seedlings and amount with ECM; chlorothalonil benomyl and thiram reduced ECM development (Hong, 1976).

Kelly (1979) reported that benodanil prevented growth of ECM in laboratory and nursery soils. Benoanyl and be arrot stimulated ECM in pine but triaben degole and truban reduced it (Pawak *et al.*, 1980).

Effect of metalaxyl on growth of ECM of Fraser fir seedling was examined by Kenerly *et al.* (1984) and found that lateral root length number of ECM per cm of lateral root were increased significantly. They observed a marked increase in shoot height and branching and improved needle colour of 5 yr. old seedling for transplants after a spring application of metalaxyl. South and Kelley (1982) also reported similar observations. Triadimefon suppresses development of naturally occurring ECM (Rowen and Kelley, 1986; Show *et al.*, 1979) as well as those formed by *P. tinctorius* (Marx *et al.*, 1986).

6.1.   Seed Treatment

Jalali and Domsch (1975) reported that seed treatment with three systemic fungicides - benomyl, thiabendazole and ethirimol - each at three different concentrations, had an adverse effect on the formation of mycorrhiza on roots of wheat by *Endogone*; they observed that the effect was more pronounced on 9-week-old seedlings than 6-week-old seedlings, with the exception of the thiabendazole treatment (0.6 mg/100g seed). Ethirimol (5.0 mg a.i./100g seed) was most harmful and benomyl was comparatively less toxic than the other two fungicides. They did not observe any toxic effects on the host by seed-treatment fungitoxicants.

Harvey *et al.* (1982) tested the action of a number of fungicides against *Lolium* endophytes and observed that they were resistant to many of the fungicides tested. Propiconazole could reduce the level of the endophytes after foliar application. In culture, the fungus was most sensitive to prochloraz, but showed some sensitivity to propiconazole and imazalil. In seed, the fungus could be completely controlled by prochloraz.

6.2.   Foliar Application

The response of foliar applications with systemic fungitoxicants such as triforine, chloramformethane, tridemorph, thiophanate - methyl, triadimefon, and benomyl and, for the purpose of comparison, with nonsystemic fungitoxicants such as maneb, captan, and dichlofluanid (at recommended rates) showed that most of the treatments had less of an effect on the establishment of mycorrhiza in root tissues than seed treatments. Triforine and benomyl treatments had relatively stronger inhibitory effects; triforine, tridemorph, chloramformethane, and dichlofluanid reduced the number of chlamydospores by 50% or more. Maneb and captan did not have such a pronounced effect on chlamydospore development. Foliar application of maneb at 0.8 kg a.i. ha^{-1} had no effect on mycorrhizal infection and chlamydospore formation (Jalali and Domsch, 1975).

Only a negligible amount of the systemic fungicides used by Jalali and Domsch (1975) translocated. Therefore, inhibitory effects could arise from residues of the fungicides taken up by the germinating seed and translocated passively during root growth. The formation of mycorrhizal chlamydospores was suppressed by foliar application of fungicides. This was brought about by changes in the spectrum of wheat root exudates as a result of the stress caused by the fungicides. It could be further explained by the work of Jalali and Domsch (1975) in which they demonstrated that application of the systemic fungicides triforine and tridemorph on foliage results in changes in root exudations and amino acid metabolism. Alteration of host metabolism is one of the ways in which systemic fungicides may act on fungal invaders. If the concurrent changes in the metabolism of the plant persist long enough, conditions may prevent the fungus from developing normally. According to Ratanayake *et al.* (1978), the root exudates govern mycorrhizal symbiosis. Demonsan, sodium azide, metalaxyl, and terrazole have been found to increase AMF infection and spore production and to stimulate root exudation and thus mycorrhizal infection. Atilano and Vangundy (1979) observed a reduction in the number of mycorrhizal hyperparasites following treatment with fungicides and fumigants, which ultimately led to an increase in mycorrhizal infection.

Trials with foliar application of systemic fungicides (pyroxyclor, prothiocarb, ethyl aluminum phosphate, ethazole, and metalaxyl) effective against phycomycetous fungi have shown that these fungicides are almost ineffective against AMF fungi (Boatman *et al.*, 1978; Menge *et al.*, 1979; Namec, 1980; Stewart and Pflezer, 1977), although these fungicides of their metabolites have been reported by several workers to translocate downward in sufficient quantity to be toxic to pythiacious fungi *in vitro* (Trappe *et al.*, 1984).

6.3.   Soil Application

Because mycorrhizal fungi are memebrs of soil and rhizospheric microbes, the effects of fungicides on the general population may in turn affect either the mycorrhizal fungi or the host and thereby confound cuase-and-effect interpretation (Kawai and Ogawa, 1977; Laliho and Mikola, 1964). The side effects of soil fungicides on soil microbes have been reviewed (Rodriguez-Kabana and Curl, 1980). Trappe *et al.* (1984) reviewed the reactions of mycorrhizal fungi and mycorrhiza formation to soil pesticides.

6.3.1. Nonsystemic fungicides

The effects of soil fumigants and fungicides on VAM fungi (*Glomus etunicatum, G. mosseae, G. fasciculatum, G. macrocarpum* and *G. constrictum*) have been reviewed (Menge, 1982). AMF infection and chlamydospore formation and development was increased after the use of 1,3-D (Telone) in cotton and dibromochloropropane (DBCP) in soybean and sudangrass at the recommended rates of application for plant disease control. However, VAM in crops were reduced by treatment with nonsystemic fungicides such as dicloran (corn), captan (corn, citrus), captafol (citrus), dichlofluanid (wheat), korex (corn), maneb (citrus), quintozene (wheat, corn) sodium azide (peanut), and thiram (corn, bean, wheat, millet). VAM were promoted by captan (bean), captafol (citrus), and sodium azide (citrus).

Copper (up to 22.4 kg ha^{-1}), chlorothalonil (up to 22.4 kg ha^{-1}), and sodium azide (up to 31.4 kg ha^{-1}) were tested against the endomycorrhizal *G. etunicatum* and *G. mosseae* in sour orange. After 8 months, the middle and low rates of copper depressed plant growth, but not infection and sporulation by *G. etunicatum*, as compared to the control. Plant growth and sporulation after the middle and high doses of chlorothalonil were significantly less than the control, whereas plant growth was better and fungus sporulation higher in *G. mosseae* than the control after sodium azide treatment (Bird *et al.*, 1974). Chlorothalonil (1.2 g a.i./m^2), maneb (1.0 g a.i./m^2), and PCNB (0.9 g a.i./m^2) reduced mycorrhizal development of creeping bent grass when applied in the spring to golf course greens or 4 to 8 weeks after bent grass was seeded and inoculated with *G. fasciculatum* in the greenhouse. However, fungicides applied 16 to 20 weeks after bent grass was seeded did not affect mycorrhizal development (Rhodes and Larsen, 1981). The results suggested that spring application of several fungicides on turf may cause reduced mycorrhizal development in bent grass turfs.

Feeder root necrosis of pecans (*Carya illinoensis*) caused by *Pythium* spp. in Georgia was reduced after application of DBCP and other fungicides without the reduction of *Phythium* populations in soil (Powell *et al.*, 1968). There was no apparent overall reduction in nematode population. Since the application of DBCP and the fungicides increased the mycorrhizal fungus *Scleroderma bovista*, it was suggested that a side effect, caused by pesticides, on feeder root necrosis through the mycorrhiza forming fungus, was a possibility.

Mycorrhizal endophytes are normal constituents of most field soils and the application of fungicides may affect this beneficial symbiotic association. Pitcher *et al.* (1966) showed that six soil fumigants improved growth and increased mycorrhizal infection in field-grown apple and cherry trees. It was suggested that this was probably due to reduced microbial competition in the treated soils. Methyl bromide, a general biocide, was found to be an efficient soil fumigant in nursery beds and its application stimulated the growth of sweet gum plants. It most strongly affected soil-inhibiting pathogens and also reduced soil mycorrhizal fungi, but these rapidly recolonized the host roots. In fumigated soils, citrus seedlings may show reduced rates of growth associated with the inhibition of phosphate, absorption by the plant. Such seedlings may be stunted, chlorotic, and nonmycorrhizal (Klein schmidt and Gerdemann, 1972).

El-Giahmi *et al.* (1976) studied the effect of botran, arasan, and ethazole (25 and 50 ppm) and captan (50 ppm) on mycorrhizal maize. Higher levels of fungicides depressed growth; however, preinoculation and fungicidal treatment significantly increased growth. It was observed that the toxicants reduced the growth of mycorrhizae from the indigenous source and that a fairly high level of infection can be produced in the roots if the plants are preinoculated with a selected mycorrhizal variety. Preinoculation may therefore be a way to keep the infection levels high enough to be beneficial to the host.

The influence of carbendazim and captofol *in vitro* on the germination of VA endophyte mycorrhizal spores of *Acaulospora laevis, Glomus calendonius,* and *G. monosporum* was estimated. It was observed that neither germination nor growth of the hyphae from spores was affeted at soil concentrations approximately one half to ten times those likely to be found after field applications (Tommerup and Briggs, 1981). The length of germ tubes of *A. laevis, G. caledonius,* and *G. monosporum* were 8 to 13, 6 to 10, and 5 to 9cm, respectively, for all treatments including the control (Tommerup and Briggs, 1981). Captan (Bertoldi *et al.,* 1977; Nesheim and Linn, 1969; Sutton and Sheppard, 1976) at the recommended rates had no effect or only a slight detrimental effect on the percentage of root infection. Captan may therefore be expected to have little or no effet on stages 1,2 or 3 of the life cycle. Spore germination and hyphal growth of the three VA endophytes tested by Tommerup and Briggs (1981) were unchanged by captafol at rates ten times those of normal field application. If captafol is applied at the rates which do not affect the growth of host plants, then it probably has no adverse effect on any stage of the life cycle of VAM fungi.

Using sour orange seedlings, Nemec (1980) studied the effects of copper, metalaxyl, thiabendozole, captan, captafol, chloroneb, and formaldehyde for their effects on *G. etunicatum* and chlorothanil, sodium azide, benomyl, and maneb for their effects on *G. mosseae*. Captafol, chloroneb, metalaxyl, and captan did not reduce the growth of the mycorrhizal plant, although some adverse effect on the fungus was appparent with all the rates of benomyl and thiabendazole reduced plant growth, but fungal sporulation was higher than that of the mycorrhizal control. In all sodium azide treatment sporulation was higher. Copper at 112 and 224 kg ha^{-1} reduced plant growth significantly but not infection or fungus sporulation. Maneb and chlorothalonil, which had no effect on *G. mosseae* at the low rates of 5.6 kg ha^{-1}, sharply reduced fungal sporulation and mycorrhizal plant growth at 11.2 and 22.4 kg ha^{-1}. Formaldehyde was toxic to the fungus (Nemec, 1980). Kelley and Rodriguez-Kabana (1979) reported that the application of a granular formation of sodium azide to fine nursery beds at the rates of 0,22.4,67.2 and 134.5 kg a.i. ha^{-1} under water seal or plastic seal was compared over 1 year with methyl bromide applied at 650 kg a.i. ha^{-1} to determine their different effects on the development of mycorrhizal roots. An adverse effect of sodium azide was observed. The lack of activity of methyl bromide on mycorrhizal development of pine roots may be attributed to the rapid recolonizaion of the soil by ectomycorrhizal fungi after fumigation; basidiospores of the fungi are produced in copious quantities and are disseminated at the earlier stage by air, insects, and movement in soil (Marx *et al.,* 1976).

The root pathogens that compete with mycorrhizal fungi in the soil and the rhizosphere are adversely affected by fumigants and fungicides; thus the AMF activity increases (Atilano and van Gundy, 1979; Bird *et al.,* 1974; Nemec, 1980; O'Bannon and Nemec, 1978). Some of the soil microorganisms favour VAM activity (Anderson, 1978).

Fungicides can have a negative effect on the development of endomycorrhizal fungi, which in turn may drastically affect the normal development of host plants under nutrient-dificient conditions. Fungicides have been examined for their effects on the inoculum and development of the AMF fungi. Menge *et al.* (1979) reported that soaking inoculata (hyphae, vesicles, arbuscules, and chlamydospores) of *G. fasciculatum* in a suspension of PCNB, ethazole, and DBCP did not impair their viability. The dosages used were 4000, 80, and 240 ppm a.i., respectively. It was further observed that when sudangrass, inoculated with the fungus, was drenched separately with PCNB at 1000 ppm at the time of inoculation, the fungcide restricted spore production by 70% after 104 days, whereas ethazole and DBCP did not significantly change spore production. When these fungicides were applied 60 days after inoculation, PCNB reduced spore production by 100% however, ethazole and DBCP increased spore production by 76 and 63%, respectively. Menge *et al.* (1979) reported that when inocula of *G. fasciculatum* were soaked in a PCNB solution of various concentrations they survived; however, infection and sporulation on sudangrass were considerably and consistently inhibited by soil drenches with this fungicide. Soil drenches with DBCP at 15 to 20 ppm or ethazole 10 to 40 ppm applied 30 to 60 days after inoculation increased infection and sporulation.

6.3.2. Systemic fungicides

Several studies have indicated that benomyl AMF and carbendazim inhibit the formation of AMF belonging to *Endogone* and *Glomus* in various crops such as barley, clover, maize, onion, soybean, and wheat (Bailey and Safir, 1978; Bertoldi *et al.,* 1977; Boatman *et al.,* 1978; Jalali and Somsch, 1975; Spokes *et al.,* 1981; Sutton and Sheppard, 1976). However, other zygomycetes fungi appear to be relatively insensitive to benomyl (Bollen and Fuchs, 1970; Edgington *et al.,* 1971) and some species of *Mortierella*, a genus taxonomically related to theEndogonaceae.

Decrease in the growth of onions in the soil amended with benomyl at 0.6 to 6.0 kg a.i. ha^{-1} were attributed to suppression of the mycorrhizal fungus by the fungicide. Similar growth decreases after captan application appeared to be associated with measured reductions in soil microbes rather than inhibition of the mycorrhizal fungus, because captan did not affect the degree of mycorrhizal colonization of the onion roots. The effects of fungicides on the mycorrhizal fungi and mycorrhiza development on hosts in different soils may partly reflect their effects on differing microbial populations.

Boatman *et al.* (1978) studied the effect of systemic fungicides on VAM infection in clover roots and observed that soil drenches of benomyl and thiophanate-methyl prevented the formation of mycorrhiza and also the spread of established infections. The immersion of a fungal inoculum in a suspension of fungicides reduced infectivity. However, clover plants grown in benomyl-treated soil did not retain enough fungicide to affect the amount of infection after transplanting into benomyl-free soil. These experiments show that suspensions of benomyl and thiophanate-methyl are toxic to *Endogone* inocula on direct immersion and when mixed with irradiated or infested soil. Sutton and Sheppard (1976) reported that benomyl was toxic to VA endophytes in a 3:1 soil-sand mixture, although other zygomycetous fungi appeared to be relatively insensitive to it (Bollen and Fuchs, 1970; Edgington *et al.*, 1971). Boatman *et al.* (1978) observed that longer immersion of the inoculum in a suspension of the fungicides reduced its infectivity, but when different amounts of the fungicide were applied, the lower rates of application were usually as effective as the higher rates, which were occasionally phytotoxic. It was further recorded that as the solubility of both benomyl and thiophanate-methyl is low, the higher rates of application probably did not increase the concentration in the soil solution, but ensured that solid residues capable of going into solution persisted longer (Bollen *et al.*, 1978). Benomyl was not retained in the roots in sufficient quantity to reduce infection when plants were removed from treated to untreated soil. Although the toxic principal of benomyl and carbendazim is known to be rapidly translocated from roots to leaves, it has been suggested that some may be retained on the roots which had been in contact with levels of fungicide high enough to cause considerable phytotoxicity (Fuchs *et al.*, 1972). However, even these roots rapidly become infected when transplanted into untreated soil (Bollen *et al.*, 1978). The objective of Boatman's experiments, to find a way of estimating the efficiency of mycorrhizal endophytes *in situ*, was not achieved because the fungicide drenches, though effective in irradiated soil, were ineffective in the unsterile soils. The workers inhibited two different endophytes in two irradiated soils, but not the native endophytes in the unsterile soils. Whereas the native endophytes could have been more resistant, the more likely explanation is that the fungicides were broken down more rapidly by the general microbial population, or a particular component of it, occurring in the unsterile soil. Although the irradiated soils in open pots were by no means sterile, they may have lacked certain specifically soil-inhibiting microorganisms. The half-life of benomyl in the field is 2.7 to 3.6 months according to pH and soil organic matter content (Austin and Briggs, 1976) or 3 to 6 months in turn and temperatures in greenhouses, breakdown may be more rapid (Boatman *et al.*, 1978).

Benomyl applied at different growth stages has been shown to reduce the percentage of mycorrhizal infection in roots, (Baily and Safir, 1978; Boatman *et al.*, 1978; Sutton and Sheppard, 1976). Suspension of benomyl was toxic to *Endogone* inoculum on direct immersion and when mixed with irradiated or infested soil, as shown by the reduced percentage of infection in roots at about 5 weeks (Boatman *et al.*, 1976). This probably indicates that soil drenched with benomyl was toxic to established infections in the external mycelium so that new roots remained uninfected (Boatman *et al.*, 1978). However, hyphal growth may also have been reduced because the plants did not grow well in fungicide-treated soil due to phytotoxicity caused by the fungicide and possibly aggravated by phosphorus deficiency. Carbendazim at dose rates comparable to those used for benomyl did not reduce spore germination or hyphal growth of three endophytes tested by Tommerup and Briggs (1981). These two phases of growth, stages 1 and 2, are different from hyphal growth from infected roots; their response is not complicated by interactions with chemically affected roots. Benomyl and related fungicides may prevent penetration of the host roots or progression of the initial infection after penetration, and so prevent development beyond growth stage 3. They did not reduce spore germination or hyphal growth. However, Sutton and Sheppard (1976) showed that the percentage of infection in roots was decreased in soil containing benomyl when spores were used as inoculum. Penetration structures, when developed from germ tubes or hyphae growing from infected roots, are probably morphogenetically different from other developmental stages and may not have the same tolerance to benomyl (Tommerup and Brigg, 1981). All rates of thiabendazole (up to 18kg ha^{-1}) reduced plant growth and infection and sporulation decreased as rates increased (Nemec and Bannon, 1979).

Arbuscular mycorrhizal fungi (AMF) are major components of the soil microflora in turfgrasses. Application of recommended rates of benomyl, iprodione, triadimefon, and chloroneb reduced mycorrhizal development of creeping bent grass when applied in the growing season 4 to 8 weeks after bent grass was planted and inoculated with G. fasciculatum. Maximum initiation and growth of turfgrass roots usually occurs in the spring and fungicides applied then apparently prevented the establishment of mycorrhizal fungi in newly produced root tissue (Rhodes and Larson, 1981). However, the fungicides applied 16 to 20 weeks after bent grass was seeded did not affect mycorrhizal development in the greenhouse, as measured by the length of the root colonized by fungi (Rhodes and Larson, 1981).

According to Spokes et al. (1981) different endophyte species vary in infectivity in soils treated with different fungicides. It was reported that the application of chloroneb to inocula of G. microcarpum used with pot-grown lettuce stimulated mycorrhizal development; however, benomyl and triadimefon induced long-term inhibition of G. mosseae on onion (Spokes et al., 1981). It was further reported that the germination of spores of G. epigaeus was completely inhibited by terrazole and PCNB; however, mancozeb did not completely inhibit the germination. Infection by G. fasciculatum was not affected by terrazole triadimefon, or chloroneb, whereas that of G. mosseae was reduced to 2/3 of the control by the same fungicides. It was further observed that infection by G. microsporum was reudced to 1/2 of the control when treated with triadimefon and was decreased 3-fold bychloroneb. Terrazole was found to be ineffective to it (Spokes et al., 1981).

Many fungicides have been used to control pea rot and several new compounds have been made available (Pfleger and Stewart, 1976; Stewart and Pfleger, 1977). The effect of several fungicides on endemic populations of Glomus spp. in association with peas has been studied and the distribution of Glomus spp. in the soil profile was also determined. Pyroxyclor, a systemic fungicide specific for the control of Phytophthora and pythiaceous fungal pathogens (Knauss, 1974) in furrow applications at the rate of 1.1 kg a.i. ha^{-1} significantly reduced the numbers of chlamydospores of G. fasciculatum. Captan, alone and wilthy copper sulfate, pyroxyclor, etridizaole (as Ban-Rot 40 WP), and copper sulfate were applied to seed and furrow at several rates. Afterwards, populationsof G. mosseae and G. fasciculatum were studied. It was observed that the chlamydospores of G. mosseae were significantly more numerous (at the 5% level) in the upper 15-cm soil layer than in the lower, probably because roots of peas proliferate in the upper soil layer. Copper sulfate as a seed treatment and a furrow application had no significant effect on the reduction of chlamydospore numbers. There was a corresponding increase in chlamydospores at higher rates of fungicidal application. However, more chlamydospores of G. mosseae occurred in methyl bromide-treated plots than in controls, although the difference was not significant. Because this chemical is usually considered a soil sterilant, it appears that either tolerant strains were selected or that reinvasion occurred at some time after dissipation of the soil fumigant. No significant effect on the number of G. fasciculatum chlamydospores was observed. Spores of G. fasciculatum were fewer than those of G. mosseae because in this combination of host and soils G. mosseae may have had a competitive advantage. With the increase in the concentration of fungicides, the number of G. mosseae chlaymydospores increased correspondingly. This may be due to the suppression of competitive, non-mycorrhizal soil organisms which permitted greater sporulation of G. mosseae. A positive correlation between spore number and percentage of root infection has been observed after fungicidal application, but not by Stewart and Pfleger, 1977).

Metalaxyl, a systemic fungicide, is highly effective for the control of oomycetes; its incorporation into maize field soil naturally infected with VAM in the greenhouse at 1.4 and 2.9 mg kg^{-1} increased VAM infection from 57% to 62% and 72%, respectively, after 30 days. When *G. fasciculatum* inoculum was added to the soil, VAM infection increased by 16% but metalaxyl had no effect. Metalaxyl at the rate of 9 kg ha^{-1} had no adverse effect on vesicle number and plant growth, but mycelial colonization and chlamydospores were higher than in the *G. etunicatum* control (Nemec and Bannon, 1979).

Studies were conducted on the growth and development of mycorrhizae in Franser fir after soil application of metalaxyl at 1.1 kg a.i. ha^{-1}. Seedling root and shoot dry weights were significantly greater than in controls and mycorrhizal incidence was greater and onset was earlier in the metalaxyl-treated trees. The seedlings from plots that previously had 75 to 100% mycorrhizal incidence decreased to 25%, suggesting that nonmycorrhizal roots may be more suscptible to soil-borne pathogens (Bruck *et al.*, 1982).

Applications of chloroneb (0.8 a.i. /m^2) did not reduce mycorrhizal development in the field (Rhodes and Larsen, 1981). Also, chloroneb resulted in the smallest numerical reduction of mycorrhizal development as compared to benomyl and triadimefon tested by the authors. This indicates that it may therefore be possible to selectively control pythiaceous fungi without substantially reducing mycorrhizal development. It is further suggested that studies involving a wider range of rates and application schedules are needed before the effects of fungicides on mycorrhizal fungi can be compared adequtely.

Application of systemic fungicides in soil at the field rates reduced AMF in soil in different crops: ethazoletridiazole (sudangrass); thiophanate-methly (peas); benomyl (wheat, bean, onion, soybean, citrus, barley); tridmorph, triforine, ethirimol, and chlor - aniformethan (wheat); pyroxychlor (pea); ethazole (lettuce); thiabendzole (wheat, bean, citrus); thophanate (wheat, onion); triadimefon (wheat, lettuce); and carboxin (bean). AMF were promoted by chloroneb (lettuce), efosite-AL (lettuce), metalaxyl (citrus), ethazole (sudangrass), and thiabendazole (citrus) (Menge, 1982).

6.3.3. Fertilizer

Fertilizers have varied effects on the AMF and soil biota (Abbott and Robson, 1991). The effects appear to be mediated by the plant at low and medium levels of soil P, but by the soil at luxury levels. Very high and very low availabilities of phosphorus reduced AMF colonization (Koide and Li, 1990), whereas spore production is generally depressed by P availability and above the levels at which the host plants benefit from AMF colonization. The interaction between phosphorus fertilization and AMF is complex one; high level of P are deleterious to AMF (Abbott and Robson, 1984a). Niterogen can suppress or enhance root colonization and is itself taken up by AMF plants often have depressed K concentrations (Siqueira and Paula, 1986). The world shortage of fossil fuel for chemical fertilizers necessitates the role of biofertilizers and arbuscular mycorrhizal fungi will have great role.

6.3.4. Herbicides

By their nature, herbicides are designed to antagonize plants and not fungi; thus, it is not surprising that many studies report slight to highly toxic effects but some reports indicate no adverse effects of herbicides on VAM fungi. Phenylurea berbicides, diuron and chlorotoluron do not adversely affect sporulation or root colonization (Dodd and Jefferies, 1989; Nemec and Tucker, 1983; Ocampo and Hayman, 1980; Smith *et al.*, 1981). At high application rates diuron actually increased soil densities of AMF spores (Smith *et al.*, 1981).

Dose of herbicides application is an important factor as with fungicides, for mediating effects on mycorrhizae; at low rates, carbamate herbicides like chlorpropham, sulfallate and phenmedipham did not reduce MT of alfalfa or wheat by *G. mosseae*, however, at rates phenmedipham reduced metabolic activity of this mycorrhizae. Smith *et al.* (1981) reported that in field studies, application of paraquat at recommended rates and at 15 times the normal rate did not affect soil densities of VAM fungal spores; similarly, Nemec and Tucker (1983) found that annual field application of paraquat and simazine to a citrus grove for five years had no effect on mycorrhizal colonization; contrary to this, greenhouse studies applying paraquat and simazine at recommended rates as a soil drench caused a significant reduction in root colonization of citrus by *G. etunicatum* (Nemec and Tucker, 1983).

Certain pesticides differentially affect AMF species; thus, pesticides may influence the species composition of VAM fungal communities (Dodd and Jefferies, 1989; Sieverding and Leihener, 1984; Spokes *et al.,* 1981). In a field experiment with Cassava, Sieverding and Leihemer (1984) found that the herbicides oxadiazon selectively depressed populations of *Glomus* spp. while increasing *Gigaspora* and *Sclerocystis* spp. The effects of this altered VAM fungal community on cassava production remain unknown. Future research should be directed toward assessing the ramification of pesticides induced shifts in the species composition of VAM fungal communities.

6.3.5. Nematicides and insecticides

Organophosphate insecticides and nematicides like chlofenviphos, carbaryl, diazinon, ethoprop, malathione and parathion generally have no effect, or a slightly detrimental effet on MI (Backman and Clark, 1977; Burpee and Cole, 1978; Parvathi *et al.,* 1985b; Spokes *et al.,* 1981). The carbamate insecticides aldicarb, carbofuran and oxamyl are reported to have no effect, detrimental effects or beneficial effects on mycorrhizae (Nemec, 1985; Ocampo and Hayaman, 1980; Spokes *et al.,* 1981).

The application rates appear to be a key factor on the toxicity of the pesticides. Carbofuran was shown to reduce MI in field and greenhouse-grown peanut (Backman and Clark, 1977), but at half the recommended rate it increased root colonization and spore production by *G. fasciculatum* (Sreenivasa and Bagyaraj, 1989). DBCP, a nematicide applications enhances MI (Bird *et al.,* 1974; Menge *et al.,* 1979).

6.3.6. Soil fumigants

Soil microorganisms are highly sensitive to the most soil fumigants applied and this is the aim to reduce populations of pathogens and pests, thus stimulating crop production. Soil fumigants besides eradicating harmful organisms also adversely affects beneficial organisms such as AMF (Menge, 1982). This nontarget effects of soil fumigants on AMF cause stunting or poor production, in several crops including celery, onion and pepper (Hass *et al.,* 1987) and clover (Powell, 1976), citrus (Schenck and Tucker, 1974). Methyl bromide is clearly detrimental to AMF but the degree to which it eradicates AMF appears to be determined by the efficiency of the fumigation, and is related to soil type, moisture level, temperature and method of application. The fumigants such as chloropicrin, formaldehyde mylone, methyl bromide, vapam and vorlex consistently reduced mycorrhizal infection; mylon, vapam and vorlex all decompose into methyl isocyanate which is toxic principal to AMF. AMF are more sensitive to methyl bromide than most soilborne fungal pathogens; consequently, it is impractical to reduce methyl bromide application rates to allow AMF to survive (Menge, 1982). Pathogen control measures less lethal to AMF should be applied whenever possible. In sunny climate, soil solarization is a possible alternative to fumigation, because it is generally effective in reducing pathogen population (Katan, 1987) but little or no detrimental effect on AMF (Afek *et al.,* 1991).

7. STUDIES ON FUNGICIDE AND MYCORRHIZAE INTERACTION

(At J.N. Agricultural University, Indore)

7.1. Fungicides and VAM Interactions in Soybean

It is well established that soybean (*Glycine max*) is among the many species of plants that can respond in a positive manner to VAM. Significant increase in growth and yield of soybean reported (Carling and Brown, 1980; Nuffelin and Schenck, 1984; Young *et al.*, 1988). Soybean is also one of the most important oilseeds crop of MP and adjoining states. The crop suffers from several soil and seedborne pathogens which reduce stand and yield (Vyas and Shroff, 1990). This necessitates the use of fungicides for the control of plant pathogens.

(i) Studies were conducted on the effect of fungicides on VAM isolated from black cotton soil cropped with soybean (July through October) and chickpea (October through March) in the cropping season 1989-90. A range of VAM fungi, namely, *Glomus* (five species), *Acaulospora morrowea* and *Gigaspora margirata* were isolated from rhizosphere soil of soybean and chickpea (Table 5). Since carbendazim or benomyl, thiram and metalaxyl are being used intensively for seed treatment. The sensitivity of various VAM fungi isolates was tested in pot culture experiments. *Glomus mosseae* was found to be resistant to carbendazim whereas all the other VAM fungi were sensitive to thiram, other VAM fungi were found resistant and no effect was observed on colonization (Table 2). Metalaxyl was toxic to all the species except *G. epigaeum*, *A. morroweae* and *G. margirata*. These results provide evidence regarding the failure of VAM infection in fugncides applied soil.

(ii) Another study was conducted to evaluate the effect of fungicides carbendazim, thiram and metalaxyl on the *Rhizobium* species, *Glomus* spp. and *Glomus* + *Rhizobium* inoculations of soybean and chickpea in the cultivars JS - 72 - 44 and Ujjain - 21, respectively (Table-3). It was observed that mycorrhizal infection was significantly increased in soybean and chickpea after dual inoculation with *Glomus* spp. and *Rhizobium*. It was also observed that carbendazim was found to be the safest fungicide whereas thiram and metalaxyl were rated second and third, respectively in reducing the mycorrhizal infection (Table 4).

(iii) Effect of dual inoculation in soybean with *Glomus* spp. and *Rhizobium japanicum* and subsequent seed treatment with carbendazim and thiram with regards to germination percentage, mortality percentage, mycorrhizal infection percentage, nodule number and dry weight, plant height and dry weight, grain seed weight per plant and yield kg per ha were recorded and presented in Table 4.

It was observed that mycorrhizal infection was significantly higher when seeds were dually inoculated with *Glomus* spp. and *Rhizobium japonicum*. The *Rhizobium* and VAM inoculation was compatible in the presence of both the fungicides in soybean (Table 4). As evident from mycorrhizal infection, and nodule number and dry weight increased as compared to control indicate growth stimulatory effect with dual inoculations and fungicidal seed treatment. The yield was also constantly significant in dual inoculation. The normal experimental soybean yield is 15 to 17 quintal per hectare with dual inoculation a yield of 22 quintal per hectare was recorded; this indicates mycorrhiza has a great role to play in boosting soybean yield. Our results (Vyas and Shroff, 1990; Vyas *et al.*, 1990) are in agreement with Carling and Brown (1980); Myffelin and Schenck (1984) and Yonng *et al.* (1988).

In each case 1000 root samples were examined for the presence of various VAM fungi in the soil amended with various fungicides; Carbendazim=Carben, thiram and metalaxyl fungicides were used as formulated products at the rate of 1500, 2500 and 2000 ppm in the soil + = No effect; - = Effective in reducing VAM infection.

Table 4 : Vesicular - arbuscular mycorrhizal fungi encountered in black cotton soil cropped with soybean and chickpea and their sensitivity to fungicides

Mycorrhizal fungi	Crop		Fungicides		
	Soybean	Chickpea	Carben	Thiram	Metalayxl
Glomus mosseae	10	08	+	+	-
G. monosporum	40	50	-	-	-
G. constrictum	05	08	-	+	-
G. fasciculatum	25	30	-	+	-
G. eleganum	20	04	-	+	+
Acaulospora morromeae	Rarely	Rarely	-	+	+
Gigaspora margirata	Rarely	Rarely	-	+	+

Table 5 : Effect of dual inoculations with *Glomus* spp. and *Rhizobium japanicum* (soybean) on *Cicer arietinum* (chickpea) and three fungicides on the mycorrhizal infection of soybean and chickepea in cultivars JS-72-44 and Ujjain 21

Treatment	Mycorrhizal infection per cent					
	Carben	Soybean Thiram	Metalaxyl	Carbendazim	Thiram	Chickpea Metalaxyl
Uninoculated	12c*	8c	7c	13c	10c	7c
Rhizobium sp.	20b	10c	14c	18c	12c	11c
Glomus spp.	80a	70b	60b	69b	68b	58b
Glomus spp. + *Rhizobium* sp.	80a	79b	72a	77a	76a	69a

* The data are mean of three replications. The data were analysed as per Duncan's multiple range tests; numbers not followed by same letter are significantly different at the P = 0.05 range.

Table 6 : Effect of dual inoculations in soybean with *Glomus* spp. and *Rhizobium japancum* and subsequent seed treatment with carbendazim and thiram on germination percentage, mortality, mycorrhizal infection, nodule number and dry weight (Cultivar JS-77-44)

Treatment	Germi nation %age	Morta lity %age	Mycorr hizal infec- tion %age	Nodule number	Nodule dry wt.(g)	Plant height (cm)	Plant dry root wt. (g)	Plant dry shoot wt. (g)	Grain seed weight/ plant (g)	yield (kg/ha)
				CARBENDAZIM						
Uninoculated	56c	6a	12c	8c	2.00c	27.89b	4.89c	12.33c	8.6c	1820c
Glomus spp.	62b	8a	80a	5c	1.30c	28.89b	6.12b	17.80b	10.4b	2000b
Rhizobium	64b	7a	20b	28b	7.89b	27.88b	7.34a	16.70b	11.2b	2115b
Glomus spp.+ *Rhizobium*	69a	4a	80a	40a	10.00a	30.89a	6.75ab	20.55b	13.6a	2225a
				THIRAM						
Uninoculated	50c	5a	8c	5c	1.30c	24.73b	3.25c	19.33c	7.9c	1815c
Glomus spp.	64b	8a	70b	0	0	26.55b	5.11a	15.33b	9.4b	1964b
Rhizobium	66b	6a	10c	22b	4.15c	27.05b	3.11a	12.33b	10.2b	1982b
Glomus spp.+ *Rhizobium*	70a	3a	79a	36a	5.00a	28.85a	5.76ab	18.23a	12.8a	2212a

*   The data are mean of three replications
**   Numbers not followed by the same letter are significantly different at the p=0.05 in Duncan's multiple range tests.
***   Soil was infested with 150 sclerotia of *Rhizoctonia batatiocola* per 100 g of soil.

(i)   The experiment was conducted in pot and inoculation, and fungicidal treatment was done in sterilized soil and soybean, and chickpea was planted subsequently, and mycorrhizal infection percentage was recorded after 25 days of germination.

(ii)   The fungicides were used as per their formulation products; carbendazim, thiram and metalaxyl were used at the rate of 1500, 2500 and 2000 ppm.

(iii)   While studying effects of fungicides in the dual inoculation of rhizobia and VAM a stimulatory effect was noticed. Similar observations were also noted by Jabaji - Hare and Kendrick (1985, 1987). Hence this study was undertaken to confirm the stimulatory effect.

The results indicate that metalaxyl and fosetyl-Al were absorbed by soybean roots after seed application, as is evident from the data on the reduction of radial growth diameter (Table 5) of the test fungus *Pythium ultimum* when grown in the PDA medium amended with the extract of the soybean plants raised after seed treatment and applied on the foliage; when seed treatment was carried out the concentration was more in the roots than foliage whereas when foliage applied the concentration was high in leaf than roots; this results suggested ambimobility i.e. upward and downward translocation (Table 5). This confirm the findings of Edigination *et al.* (1980); Erwin (1973) and Vyas (1984) who reported translocation pattern of fungicides in different plant species.

Since the systemic fungicides, metalaxyl and fosetyl-Al both translocated downward through symplast, the experiments were designed to evaluate the effect on the mycorrhizal infection (Table-5). The results indicate that as the concentrations increases there is corresponding increase in the mycorrhizal infection; fosetyl-Al application is significantly superior to metalaxyl in enhancing the infection. Similar results were also observed by Jabaji-Hare and Kendrick (1985, 1987).

Recently attention has been drawn to the possible reduction of VAM infection from fungicidal application (Vyas, 1988). The data on seed treatment and foliar application of metalaxyl and fosetyl-Al and their effects on the germination, mycorrhizal infection, nodule number and weight are presented in Table 4.

Table 7 : Radial growth (mm) of *Pythium ultimum* in potato dextrose agar medium amended with extracts of soybean seedlings and vesicular arbuscular mycorrhizal (VAM) infection percentage in soybean seedlings raised after metalaxyl and fosetyl-Al treatment

Treatment			*Pythium* growth (mm) Root Extract	Shoot Extract	VAM infection
Metalaxyl	Seed		28	70	-
	Spray	0.15mg	60	26	53
		0.2mg	50	18	60
		0.25mg	45	10	70
Fosetyle-Al	Seed		28	50	-
	Spray	0.3mg	70	38	71
		1.0mg	60	41	76
		3.0mg	60	40	80
Untreated seedlings		85	86	17	
Without extracts			87	87	-

* Data are mean of three replications:

The dual inoculation with *Glomus* spp. and *R. japanicum* indicate that mycorrhizal infection and nodule number and weight increased significantly in the presence of fungicides. Yield also significantly increased as compared to control. Metalaxyl and fosetyl-Al showed synergistic effect on dual inoculation between mycorrhiza and rhizobia. Such type of stimulatory effects have been reported for metalaxyl in sour orange (Nemec, 1980) and corn (Groth and Martinson, 1983) and fosetyl-Al increased mycorrhiza in leek (Jabaji-Hare and Kendrick, 1985). However, they also observed toxic effect due to metalaxyl in leek. This seems to be first report where the dual inoculation with mycorrhiza and *Rhizobium* are compatible. The two systemic fungicides can be safely used for soybean for disease control without deleterious effect on VAM and *Rhizobium*. Since VAM and rhizobia symbionts are important in biological control (Jalali and Domsch, 1975) of plant pathogens, the introduction and management of VAM could be used to the advantage of the crop (Table 6).

Table 8: Effect of fungicides on mycorrhizal infection, rooting on mycorrhizal infection, root nodulation of soybean after inoculation with *Glomus* spp. and *Rhizobium*

Treatments	Germi-nation %	Mycorr-hizal infec-tion	Nodule number per plant	Nodule weight (mg) per plant	Plant dry wt. (g)	Yield (kg/ha)
Metalaxyl						
Uninoculated	53c	70c	5c	1.30c	4.75b	1815c
Rhizobium japanicum	60b	14c	22c	4.15b	8.81b	1982b
Glomus sp.	61b	60b	7c	1.61c	11.41b	1984b
R japanicum +						
Glomus sp.	68a	72a	36a	5.00a	13.46a	2212a
Fosetyl-Al						
Uninoculated	61c	10c	8c	0.9c	8.75b	1800c
Rhizobium japanicum	68	12c	21c	1.9b	10.10b	2000b
Glomus sp.	71b	68b	19c	0.7b	11.41a	2100b
R. japanicum +						
Glomus sp.	78a	76a	25a	2.2a	12.77a	2300c

* Number not followed by the same letter are significantly different at the P=0.05 in Duncan's multiple range tests. X-Mean of three replicates.

7.2. Fungicides and Ectomycorrhizae Interaction

Application of chemicals for managements of diseases has become routine in the intensive cultivation of forest. Several products are introduced with growth promoting attributes. This promised use to undertake this study.

Table 9: Colony diameter of *Pisolithus tinctorius* and *Thelephora terestris* after 9 days at 25°C on potato dextrose agar amended with different concentrations of fungicides

Fungicide	Concentration in ppm	Diameter of colony (in mm) of	
		Pisolithus tinctorius	*Thelephora terestris*
Carbendazim	250	85	80
	500	80	79
Benomyl	250	83	82
	500	80	82
Triadimefon	250	79	75
	500	75	74
Metalaxyl	250	-	-
	500	-	-
Fosetyl-Al	250	-	-
	500	-	-
Thiram	250	55	59
	500	40	37
Captan	250	53	50
	500	36	32
Control		89	89

X-Mean of three replicates.

All the systemic fungicides by and large have no fungitoxic effects on both the fungi; among the systemic fungicides benomyl, carbendazim and triadimefon showed some toxic effect compared to metalaxyl and fosetyl-Al. The concentration 500 ppm is more toxic than 250 ppm. Nonsystemic fungicides thiram and captan are toxic and reduction was higher in fungal growth-diameter. *Pisolithus tinctorius* showed less sensitivity to the fungicides as compared to *T. terrestris* (Table 7).

Table 10: Effect of fungicidal treatments on the development of ectomycorrhizae by *Pisolithus tinctorius* and *Thelephora terrestris* on *Leucaena leucocephala* seedlings grown in *Pisolithus tinctorius* and *Thelephora terrestris* infested and non-infested soil.

Fungicides	*Pisolithus* sp. infection		*Thelephora* sp. infection	
	Infested	Non-infested	Infested	Non-infested
Carbendazim	26.21a	0.00	14.95a	0.00
Banomyl	25.00ab	0.00	10.07b	0.00
Triadimefon	19.97bc	0.00	11.84b	0.00
Metalaxyl	14.59d	0.00	9.20b	0.00
Fosetyl-Al	10.65a	0.00	10.23b	0.00
Thiram	6.80e	0.00	0.00	0.00
Captan	0.00b	0.00	0.00	0.00
Control	15.53cd	0.00	6.20c	0.00

Note: Values are expressed as a percentage of short roots forming ectomycorrhizae; Means followed by similar number are not different at the 0.05 level.

Pisolithus sp. did not form ectomycorrhizae on seedlings grown in uninfested soil. In infested soil, *P. tinctorius* formed ectomycorrhizae (ECM) on seedlings, in all fungicide treatment except captan (Table 8). The degree of ECM development differed signigicantly among fungicial treatments. Seedlings drenched with carbendazim benomyl had triadimefon, metalaxyl and fosetyl- Al had greater development by *Pisolithus* than the control were also stimulatory out thiram and captan drench depressed ECM. Ectomycorrhizae formed by *T. terrestris* were found in all treatments and formed about 95% of these. The behaviour of the fungicidal drenching was a almost identical in both the fungi.

Table 11: Effect of fungicidal dranch on mean fresh top and root weight and average stem diameters of *Leucaena leucocephala* seedlings grown in *Pisolithus tinctorius* infested and non-infested soil.

Fungicides	Conc. in %age	Top Weight (g)		Root Weight (g)		Stem weight (g)	
		Infested	Non-Infested	Infested	Non-Infested	Infested	Non-Infested
Carbendazim	0.15	1.30b	1.20a	1.80c	0.97b	2.85b	2.64a
Benomyl	0.15	1.48ab	1.23a	1.21b	1.04a	2.87b	2.68a
Triadimefon	0.15	1.34b	1.15b	1.14b	1.08a	3.09a	2.65a
Metalaxyl	0.20	1.53a	1.25a	1.16b	0.86c	2.81b	2.35
Fosetyl-Al	0.20	1.41ab	0.96c	1.40a	1.00b	3.07a	2.07
Thiram	0.30	1.51a	0.97c	1.04c	0.85c	2.54c	2.32
Captan	0.25	1.25c	0.37d	0.96c	0.73d	2.34d	2.28
Control	-	0.65d	0.37d	0.79d	0.60e	2.54c	2.07

Means in vertical columns followed by similar letters are not significantly different at the 0.05 level.

Inoculation with the ECM fungi did increase seedlings growth in this 5 week study. Systemic fungicides carbendazim, benomyl, tridionefon, metalaxyl and fosetyl-Al drench increased all the parameters responsible for *Leucaena* growth in the presence of *Pisolithius tinctorius* and in non infested soil the growth of comparatively less but significantly higher as compared to control. Root growth in fungicide treated soil was significantly, affected and higher as compared to control. Systemic fungicides has stimulatory effect by increasing growth significantly. Stem diameters of seedlings in systemic fungicial drenches were higher as compared to non systemetic fungicides thiram and captan (Table 9).

Fungicides used to control damping off of seedlings were chosen carefully so development of ECM is not inhibited. Systemic fungicides by and large increased ECM activity. Benzimidazoles are generally innocuous to phycomycetes and many basidiomycetes but at low concentrations inhibit many soil fungi such as species of *Penicillium, Trichoderma* and *Fusarium* (Vyas, 1994). Antioomycetes fungicides metalaxyl and fosetyl-Al were also found to be stimulatory. This suggest that the origin of mycorrhizae is not from oomycetes and hence their relationship needs to reinvestigated. In the present study captan and thiram inhibited development of ECM. Obviously-further testing of other fungicide-symbiont combinations is needed to identify fungicides compatible.

A comparison of the results of the laboratory and nursery tests shows that mycelium of *Pisolithus tinctorius* and *Thelephora* sp. is affected more by fungicides in potato dextrose agar (PDA) medium than in soil (Tables 6,7, 8). The limited persistence of these fungicides in soil probably accounts for some of this difference in sensitivity. Lack of leaching, microbial activity by associated microflora and barriers to fungicide diffusion in agar permits longer exposure at higher concentrations to the hyphae in agar culture than soil (Max and Rowan, 1981). Due to higher concentrations in agar medium than in soil, the mycelium grows at a faster rate which would increase the metabolic uptake of the fungicides. Also, hyphae of ECM fungi grown in vermiculite particles as initial inoculum may be protected to a certain extent from higher concentration of fungicides in soil. These differences between nursery and laboratory studies on fungicides indicate the limited value of studies on agar medium that are not supported by studies carried out in soil.

Most of the factors influencing the formation of ectomycorrhizae affect either the susceptibility of the host or the survival and ineffective potential of the fungal symbionts. Although certain fungicides used in this test may have affected the susceptibility of the *Leucaena* seedlings to root infection by the fungi, results of other research suggest that the action of fungicides is more likely a direct or indirect effect on the fungi, results of other research suggest that the action of fungicides is likely a direct. Our results and results of other workers indicate that fungicide directly inhibits vegetative growth on the symbiotic fungi and causes increase or decrease in ECM development (Marx and Artman, 1979; Marx *et al.,* 1979).

These studies show that care must be taken in selection of fungicides used to control diseases on *Leucaena* seedlings in nurseries. The qualitative and quantitative effect of fungicides on the development of ECM should be considered when any fungicide is selected to control specific diseases. The stimulating effect of certain fungicides on ECM development, especially those formed by artificially introduced fungi should be considered an added benefit to the merits of these fungicides (Marx *et al.,* 1986).

Many fungicides tested are commonly used in forest tree nurseries. Although nursery application are usually discontinued a few weeks after seed germination. Systemic fungicides should be used in tree nurseries because of their non-toxic nature to ECM and stimulating effect.

8.   *IN VIVO* ARBUSCULAR MYCORRHIZAL FUNGI AND PESTICIDES INTERACTION

The advantages of AMF in crop productivity raises concern for their survival and preservation in soils. The conservation in soils covers many aspects of agricultural crop production and management programmes. Some of these concerns include preventing reduction in infection through excessive use of fertilizers high phosphorus and nitrogen and inhibition of infection and reduced population caused by pesticides and agrochemicals (Trappe *et al.,* 1984; Vyas, 1988; 1993). Some chemical treatments are of particular interest because of their highly toxic effects on nontarget organisms such as AMF (Vyas, 1988) but others can actually enhance colonization and sporulation (Jabaji-Hare and Kendrick, 1987). A study was undertaken on the influence of pesticides on AMF and their formation to elucidate the effects of recently developed fungicides, insecticides and herbicides (Tables 12,13,14). The soil was inoculated with root fragments of maize which contained *Glomus fasciculatum, Glomus* spp., *Gigaspora margirata* and *Acaulospora* sp. Chlamydospores number was counted as per method of Gerdmann and Nicolson (1963) and mycorrhizal infection by Phillips and Hayman (1970).

Carbendazim, fosetyl-Al and mancozeb possess a growth promoting effects significatly which is evident from the data and similarly they significantly enhanced both mycorrhizal infection (MI) and chlamydospores number. So there is some in relationship in healthy growth of plants and higher mycorrhizal activity.

Table 12: Growth of soybean and mycorrhizal infection and chlamydospore production in inoculated soil with *Glomus* spp. and treated with field dose of fungicides

Fungicides	Rate/ha in kg	Plant		Mycorrhizal infection (% age)	Chlamy- dospore 25g
		Height (cm)	Weight (g)		
Thiram (Thiride)	15.00	40.00d	25.00c	50c	177b
Carbendazim (Bavistin)	10.00	45.00a	35.00a	65a	180b
Triadimefon (Bayleton)	10.00	28.00d	30.00b	10d	62c
Metalaxyl (Ridomil)	25.00	35.00c	28.00b	65b	185b
Fosetyl-Al (Aliette)	25.00	40.00b	32.00a	80a	251a
Mancozeb	25.00	38.00b	26.00b	53c	170b
Control-1	-	24.00d	15.00d	80a	185b
Control-2	-	20.00e	10.00e	-	-

(i)   The data are mean of three replicates;

(ii) The application rates of fungicides are field rates used for control of soilborne plant pathogens;

(iii) Number not followed by the same letter are signigicantly different at P=0.05 in Duncan's multiple range tests;

(iv) Control - 1 is inoculated but not treated and Control -2 is uninoculated and untreated.

Triadimefon, a fungicide which inhibits sterol biosynthesis in plant and fungi and thus results in the stunting of plant growth. This also adversely affect the growth of mycorrhiza. By and large fungicides application enhanced the growth of mycorrhizal infection and chlamydospore number. The results also indicate that metalaxyl and fosetyl-Al have stimulatory effect on mycorrhizal development after soil application; this should be exploited for harvesting more mycorrhizal spores.

Table 13: Growth of soybean and mycorrhizal infection and chlamdospore production in inoculated soil with *Glomus* spp. and treated with insecticides

Insecticides	Rate/ha in kg	Plant Height (cm)	Weight (g)	Mycorrhizal infection (% age)	Chlamy-dospore 25g
Quinolphos Gr (Ekalux)	12	35.00b	25.00d	80a	86b
Carbaryl (Sevin)	30	45.00a	36.00a	68b	90b
Thimet (Phorate Gr)	12	30.00c	29.00c	65b	85b
Methyl prathion (Folidol)	30	35.00b	32.00b	70.b	90.b
Chlorophyriphos (Ruben)	1.51t	40.00a	30.00b	75.a	90.b
Control - 1	-	25.00d	18.00e	80.a	170.a
Control - 2	-	20.00d	15.00e	-	-

1. The data are mean of three replicates;
2. The application rates of insecticides are field rates used for control of soilborne plant pathogens;
3. Number not followed by the same letter are signigicantly different at P=0.05 in Duncan's multiple range tests;
4. Control - 1 is inoculated but not treated and Control - 2 is uninoculated and untreated.

Among the pesticides, insecticides were least toxic to the MI and chlamydospores production. The data indicate that a definite trend has been established based on their toxicity toward mycorrhizal is : thimet > carbasyl > methyl-parnthione > chloropyriphos > quinolphos. The results are as per reports made by Parvathi *et al.* (1985b) and Spokes *et al.* (1981).

Herbicides were found to be highly toxic than other pesticides. The MI and chlamydospores number is significantly reduced in various tretments ; this gives us a caution that their use should be judicious to harnessed the benefits of the mycorrhiza in crop productivity.

Table 14 : Growth of soybean and mycorrhizal infection and chlamydospore production in inoculated soil with *Glomus* spp. and treated with herbicides

Insecticides	Rate/ha in kg	Plant Height (cm)	Weight (g)	Mycorrhizal infection (% age)	Chlamy-dospore 25g
Fluchloralin (Basalin)	Pre-pl One kg	40.00a	25.00a	30b	50b
Metolachlor (Dual)	Pre-E One kg	35.00b	20.00b	20e	40d
Metribuzin (Sencor)	Pre-E 0.05kg	38.00a	21.00b	20e	60b
Alachlor (Lasso)	Pre-E Two kg	40.00a	27.00a	50b	50c
Nitrofen (Tok)	Pre-E Two kg	30.50b	20.00b	35c	40d
Control-1	-	22.00c	72.00c	80a	175a
Control-2	-	20.00c	10.00c	-	-

1. The data are mean of three replicates;
2. The application rates of herbicides are field rates used for control of soilborne plant pathogens;
3. Number not followed by the same letter are signigicantly different at P=0.05 in Duncan's multiple range tests;
4. Control - 1 is inoculated but not treated and Control - 2 is uninoculated and untreated.

This study concludes that pesticides and mycorrhizal interactions studied through this technique showed varied effects hence the compatibility between these should be predetermined; among the pesticides, herbicides are highly toxic to MI hence their use should be restricted and judicious. Such study should also be conducted under different soil, variety, cropping pattern and climatic conditions to understand the pattern of toxicity because pesticides and biofertilizing agents such as AMF are indispensable in the present day commercial farming system and those showed toxicity towards the AMF should be either discarded or should be used judiciously.

Information to predict results of most crop, pesticides and mycorrhiza interactions is scanty and future requirements to be directed toward developing mechanistic understanding of pesticide effects on mycorrhizae. A standardized research protocol should be defined and used to allow comparison between studies. This protocol would define factors that may mediate outcomes of pesticides and mycorrhizal interactions such as field or greenhouse conditions, pesticide dosage, application method, crop species, soil type and species of AMF. Because of different AMF taxa vary in their sensitivities to pesticides, studies designed to screen AMF responses to pesticides should examine a variety of species and genera. Hence, it is suggested that *in vivo* trials should be conducted in lieu of, or concurrently with, greenhouse (or growth chamber) studies because it is virtually impossible to recreate or simulate in a greenhouse the conditions of AMF systems that exists *in vivo*.

9. MECHANISMS OF PESTICIDE ACTION

9.1. Fungicide Toxicity

The fungicides are per se toxic to the VAM spores and mycelium and which is consistent for many species. Because a large proportion of fungus hyphae are inside roots, it would appear that most of the nonsystemic fungicides can affect VAM fungi primarily by inhibiting the spore germination and infection processes; hence considered probably less damaging to VAM symbiosis than systemic fungicides. Most of the nonsystemic fungicides may postpone infection, but not eliminate completely. Quintozene and other nonsystemic fungicides inhibited germination of *Glomus epigaeus* chlamydospores while mancozeb did not completely inhibit germination (Daniels and Menge, 1980). Menge (1982) postulated that these fungicides may be fungistatic but once the mycorrhizal fungus gains entry to the root these fungicides have little influence and may actually stimulate the spread within the host root. Sutton and Shephard (1976) reported that benomyl was toxic to VAM in a soil and mixture 3:1 ratio, although other group of zygomycetes are innocuous (Edgington *et al.* (1971). Jabaji-Hare and Kendrick (1987) reported that metalaxyl reduced the growth of mycorrhizal infection and colonization in leek root segments by *Glomus intraradices*. The mechanism by VAM fungi may be a direct action or may be mediated through its phytotoxic effect on root function and growth, though toxicities occurred on symptoms were not seen (Table 15).

9.2. Stimulation

On the other hand some systemic fungicides actually stimulate root colonization by VAM fungi. Menge *et al.* (1979) showed that terrazole when applied as soil drench caused a signigicant increase in root colonization and spore production by *Glomus fasciculatum* in *Sorghum vulgare* while Groth and Martison (1983) reported that incorporation of metalaxyl into soil containing different VAM fungi increased root colonization; Jabaji-Hare and Kendrick (1987) found inhibitory effect on VAM in *Allium porrum* by metalaxyl. Spokes *et al.* (1991) found stimulatory effect of chloroneb in citrus and lettuce. Clarke (1987) and Jabaji-Hare and Kendrick (1985, 1987) found stimulatory effect of-fosetyl-Al in lettuce seedlings on VAM infection by *G. microcarpum* and leek (*A. porrum* L.) respectively. Stewart and Pfleger (1977) reported that pyroxychlor and etridizole (ETMT) stimulate *Glomus fasciculatum* and *G. mosseae* infection in field grown peas. It is interesting to note that the systemic fungicides mentioned above are all active against oomycetes. It may therefore be possible to selectively control downy mildew, damping off and other diseases caused by oomycetes, without substantially interfering with VAM colonization. Benzimidazoles such as benomyl, carbendazim and thiabendazole are particularly suppressive towards the zygomycotina but less effective towards most members of Ascomycotina and Basidiomycotina (Dodd and Jeffries, 1989). The VAM fungi (Endogonaceae) have been placed in the Zygomycotina as a separate order Endogonales (Benjamin, 1979) and the differential effects of thiazoles and antioomycetes supports this taxonomic placement (Dodd and Jeffries, 1989).

Table 15: Mechanism of action of fungicides on VAM

1. Fungicides *per se* toxic	Adversely affect spore germination and infection process.	Daniels and Menge (1980, 1980); Menge (1982)
2. Stimulation		
a. Root exudation such as sugars and amino acid	Infection process	Jabaji-Hare and Kendrick (1985)
b. Fumigants and fungicides Sterilise soil and make soil free of pathogens	Infection process	Atilono and Van Gundy (1979); Bird *et al.* (1964); Menge *et al.* (1979)

Attempts have been made to investigate the stimulatory effects of systemic fungicides such as pyroxychlor (Stewart and Pflegar, 1977) and fosetyl - Al (Jabaji-Hare and Kendrick, 1985) on VAM.

Systemic fungicides, which translocated within the host either in the apoplast or the symplast, may act either directly on the VAM fungus, or indirectly by altering host root physiology; the specifically anti-oomycetes fungicides, metalaxyl and fosetyl are both systemic. However, the former is translocated mainly in the transpiration stream (Bruin and Edgington, 1984) while the latter is translocated in the symplast and can move downwards in plants (Bertrand *et al.,* 1977). Partial downward translocation has also been reported for metalaxyl by Staub *et al.* (1978) and Vyas (1984, 1988). Jabaji-Hare and Kendrick (1985, 1987) studied the factors for stimulation by fungicides. They selected metalaxyl and fosetyl-Al on colonization of leek (*Allium porrum*) roots by the VAM fungus *Glomus intraradices*; they selected these fungicides by two further considerations (i) it has been suggested by Pirozynski and Malloch (1975) and Benjamin (1979) that VAM fungi have zygomycotinous affinities and (ii) since both fungicides are systemic with different translocations pathway, their effects on the VAM fungus may well be different.

Jabaji-Hare and Kendrick (1985) reported that a single foliar applications of fosetyl-Al at rates of 0.3 - 3.0 mg a.i. ml^{-1} effectively increased colonization of leek roots by a *Glomus* sp. and increased the growth of plants; these results confirm the finding of Clarke (1978) who found that lettuce seedlings colonized by a mixture of VAM fungi showed a 10% increase in colonization after being sprayed with fosetyl - Al. It is suggested that increases in plant growth strongly paralleled increases in percent colonization. This also suggests that increase in dry weight could be partly due to increase in mycorrhizal colonization (Abbot and Robson, 1977). Similar response of growth due to fungicide application have been reported for sour orange plants inoculated with *G. mosseae* (Nemec, 1980) and for soybean plants inoculated with *G. fasciculatum* and subject to different concentrations of metalaxyl (Gorth and Martinson, 1983). This increase in colonization of fungicide treated leek plants by *Glomus* sp. cannot be attributed to decrease in organisms pathogenic to leeks, since the experiments were conducted in a controlled environment growth chamber, mycorrhizal cultures were established from surface sterilized root inoculum and there was no evidence of disease on roots, or evidence of pathogenic fungal propogules in the potting medium and lastly fosetyl - Al was applied only to the foliage (Jabaji-Hare and Kendrick, 1985).

Root exudation is thought to be one of many factors which govern mycorrhizal development (Glaham *et al.,* 1981; Ratnayake *et al.,* 1978). It has been conclusively demonstrated that application of certain fungicides to leaves affects the kind and amount of root exudates (Rovira, 1969). Jalali and Domsch (1977) reported that foliar application of certain systemic fungicides increased the exudation of amino acids from wheat root. This led Jabaji-Hare and Kendrick (1985) to explore the stimulatory effects of fosetyl - Al on leek VAM fungi. These studies indicated that the fungicide caused a significant increase in soluble sugars exuded from roots of mycorrhizal (M) and nonmycorrhizal (NM) plants depending on the fungicide concentration and harvesting time. Fosetyl - Al treated M plants had greater amounts of soluble sugars in their root exudation than similarly treated NM plants; this indicates synergistic effect of fungicide and VAM on plant metabolism (Jabaji-Hare and Kendrick, 1985). The root extract studies revealed a quantitative direct relationship between fosetyl - Al and total soluble sugars and free amino acids. It has been suggested that the activity of fosetyl - Al against its target pathogenic fungi results from stimulation of host defence reactions and the synthesis of phytoalexins (Bompeix *et al.,* 1980), but recently Feen and Coffey (1984, 1985) presented evidence that foliar spray of either fosetyl - Al or its breakdown product phosphorus which is present in fungicide treated plant tissue (Vo-Thi-Hi *et al.,* 1979).

Jabaji-Hare and Kendrick (1985) observed that roots of M plants had lower soluble sugars that control NM plants at 2,7 and 27 days after treatment. It seems probable that these metabolites are utilized by the VAM fungus or transformed by it into the lipids which are present in large quantities in VAM fungus structures (Jabaji-Hare *et al.,* 1984; Nemec, 1981). Significantly more lipid has been extracted from VAM fungus colonized roots than from uncolonized roots (Cooper and Losel, 1978). Similar results were also obtained by Janaji-Hare and Kendrick (1985). In conclusion, Jabaji-Hare and Kendrick (1985) presented evidence that fosetyl-Al increase colonization and development in roots of leeks and directly or indirectly, growth of the leek plants themselves; they have further demonstrated that the fungicide has an effect on both host plant and VAM fungus.

With fosetyl - Al, plant growth and proportion of root segments (i) containing more than one kind of fungal structure (FS); (ii) with intramatrical vesicles only, (iii) with arbuscules only (AR) and (iv) with extramatrical hyphae only (EH), were singnificantly increased with increasing concentration and this effect did not diminish (Jabaji-Hare and Kendrick, 1987). Similar results were also reported by Clarke (1978); he found the lettuce seedlings colonized by a mixture of VAM fungi showed a 10% increase in colonization after being sprayed with fosetyl - Al. It is of interest that the increase in growth of mycorrhizal plant strongly parallel increases in percent colonization by the fungicide fosetyl - Al not only increased the extent of colonization by *Glomus intraradices* in leek roots, but also increased the production of intramatrical vesicles. Bird *et al.* (1974) and Nemec and O,Bannon (1979) observed similar increase in vesicle formation of *Endogone* sp. in cotton roots and of *Glomus etinucatum* in roots of sour orange, respectively, when DBCP was applied to the soil. The stimulation of *Glomus intraradices* by fosetyl-Al as observed by Jabaji-Hare and Kendrick (1985, 1987) may be a direct effect of the fungicide on the plant, or the mycorrhizal fungus or both (Jabaji-Hare and Kendrick, 1987). Fosetyl - Al causes a significant increase in exudation of soluble sugars from roots of leek, the magnitude of which is maximal during the first few days after treatment (Jabaji-Hare and Kendrick, 1985). While these results do not prove that root exudation is responsible for increased *G. intraradices* colonization, they do provide support for the hypothesis that root exudates are predisposing factors for colonization and spread of mycorrhizal fungi (Graham *et al.,* 1981; Ratnayake *et al.,* 1978).

The fungicides which are weak in their action usually stimulate VAM colonization. The systemic fungicides chloroneb, metalaxyl and sodium azide and etridiazole and ethazole (ETMT) have been found to stimulate mycorrhizal colonization (Vyas, 1988). Fungicides may affect soil and rhizosphere populations of microorganisms (Vyas, 1988). These microorganisms may interact with germination of infection process by VAM. Mosse (1962) showed that certain bacteria could significantly increase infection by VAM. This may be due to reduction of number of mycorrhizal hyperparasites or predators (Atilano and Van Gundy, 1979; Daniels and Menge, 1980) which could ultimately increase VAM infection. Several workers (Atilano and Van Gundy, 1979; Bailey and Safir, 1978; Bird *et al.,* 1974; O'Bannon and Nemec, 1978; Stewart and Pfleger, 1977) suggested that fungicides reduce root pathogens that compete with mycorrhizal fungi for root nutrients. The best evidence for this hypothesis is presented by Bird *et al.* (1974) who found that DBCP reduced plant parasitic nematode population and increased mycorrhizal infection. However, Menge *et al.* (1979) obtained similar increases in VAM infection DBCP autoclaved soil without root pathogens. It is suggested that another hypothesis, perhabs root exudation induced by DBCP could better explain the increased mycorrhizal infection caused by DBCP. Soil incorporation of metalaxyl increased VAM infection (*Glomus fasciculatum*) in maize and soybean (Groth and Martinson, 1985) and observed that increase in VAM infection is due to decreased competition from other root infecting microorganisms rather than affecting the VAM fungi directly.

10. METHOD FOR PREDICTING FUNGICIDAL EFFECT

Application of either benomyl or captan significantly decreased the growth of VAM fungi *Glomus intraradices* and *G. caledonicum* in onion (*Allium cepa* L. cv Hyper) plants four weeks after treatment at the rate of 100 µg but non-VAM plants are not affected (Kough *et al.,* 1987). The growth depression of VAM plants with fungicides at the concentrations used Kough *et al.* (1987) were also earlier reported by Bertoldi *et al.* (1977), Nemec (1980), Manjunath and Bagyaraj (1984) and Parvathi *et al.* (1985). Captan is considered less harmful to VAM fungi than benomyl (Vyas, 1988) but Kough *et al.* (1987) found captan more harmful than benomyl; both captan and benomyl significantly decreased the growth of VAM onion plants four weeks after fungicide treatment, irrespective of the species of VAM fungus colonizing the roots (Kough *et al.* 1987). Kough *et al.* (1987) explored the possibilities of easily detecting early affects of fungicides on the metabolic activity of VAM fungi by screening chemicals and for predicting their long term effects on the growth of mycorrhiza dependent plants. Succinate dehydrogenase (SDH), a tricarboxylic acid cycle (TCA) enzyme which has been reported to be an indicator of metabolically active fungal tissue in VAM (MacDonald, 1980; MacDonald and Lewis, 1978), can be examined directly in tissue using a chlorogenic assay (Pearse, 1968). Kough *et al.* (1987) using SDH assays revealed significant decreases in the level of metabolically active VAM tissue as soon as 3 days after fungicide application, whilst the non-vital staining technique employing CBE did not reflect effects to a chemical treatment illustrates how slowly physical changes in both the structures of the VAM fungus and plant growth lag behind the physiological processes involved. Carr and Hinkely (1985) suggested benomyl to be fungistatic to *G. caledonicum*; on the contrary the rapid decrease in SDH activity detected in the present study when plants are treated with a comparable concentration of benomyl indicates, a fungicidal effect on the internal mycelium developing within roots (Kough *et al.,* 1987). This support the suggestion of Tommerup and Briggs (1981) that distinct growth phases of VAM fungi may differ in their tolerance to benomyl. This may well also apply to captan, since these authors reported that spore germination and hyphal growth of *G. caledonicum* are similarly unaffected by captafol (Kough *et al.,* 1987). They have also reported that arbuscules showed a decreased SDH activity following fungicide application. These structures are belived to be main site of nutrient exchange between the host plant and the VAM fungus.

11.   EFFECT OF FUNGICIDES ON PHOSPHATE ACCUMULATION BY MYCORRHIZAE

Arbuscular mycorrhizae fungi (AMF) have been demonstrated to increase plant growth in soil and utilization of less available forms of phosphorus. Gray and Gerdemann (1969) observed a 16-fold reduction of P uptake in 12-week-old mycorrhizal onions when PCNB was applied 48 hr before P application. Similarly, Hirrel and Gerdemann (1979) found that in 10-week-old onion plants PCHB (applied 2 and 5 days before C-glucose infjection into the soil) inhibited C translocation in maize by *Endogone fasciculata* and greatly reduced phosphate uptake and accumulation by mycorrhizal roots, but did not significantly affect phosphate accumulation by mycorrhizal roots.

Benomyl and thiophanate-methyl soil drenches the phosphate uptake of inoculated onion and strawberry plants grown in irradiated soil, but not in unsterile soil (Boatman *et al.*, 1978). Although the results were variable, the possibility of measuring short-term P uptake by perfusing parts of the soil using a muslin wick with a radioactive solution of approximately the same concentrations of phosphate as the soil solution seems viable. Its success depends on a reasonably uniform root distribution in the pot and the retention of an equilibrium P concentration little changed from that normally present in the soil solution (Boutman *et al.*, 1978). Since it is capable only of establishing what proportion of P uptake is lost when endoyphyte activity is inhibited, and not obtaining an accurate measurement of total uptake, however, the technique may be useful but obviously it requires further work.

Studies on the effects of soil fungitoxicants on the development of AMF and phosphate uptake in wheat have been carried out (Jalali, 1979). The results clearly showed that AMF infection is considerably influenced by the soil application of fungicides. The greatest effects were observed with PCNB (100 ppm) and thiram (100 ppm) which also interfered, in varying degrees, with the growth and phosphate uptake of the plants. It was further observed that the toxic effects of fungitoxicants, particularly PCNB and thiram, persist long enough to upset the delicate symbiotic balance. Benomyl (25 and 50 ppm) and captan (50 and 100 ppm) also showed toxic effects in varying degrees over controls.

12.   FUTURE OUTLOOK

The review was engendered by two apparently contradictory observations concerning vesicular-arbuscular mycorrhizal (VAM) fungi and fungicides; on the one hand, it is expected that fungicides will have detrimental effects on VAM and other hand, some fungicides actually stimulate root colonization by VAM fungi. Further studies relating the effects of fungicides to VAM activity at the root surface are of growing interest because it is in this zone of intense microbial activity that pathogens and/or mycosymbiont of either disease of symbiosis must penetrate before establishing infection. Apparently, side effects on VAM are more permanent than similar effects of ectomycorrhizal fungi, perhaps because recolonization of amended soil by ectomycorrhizal fungi is higher than soil with VAM. Also, information is scarce on the importance of the side effects of fungicides on mycorrhizae with respect to disease development or in yield.

The fungicides are an important segment of productivity and their continuous use in farming systems merits consideration. Some fungicides are reported to induce instability of fungal pathogens. Since VAM mycosymbionts may become important in the biological control of plant/pathogens (Schonbeck, 1979), the introduction and management of VAM could be used to the advantage of the crop. This is of practical significance, since any interference with the VAM development may ultimately have a depressive effect on plant growth and development.

Diverse group of fungicides are being used for the control of plant diseases. The broad spectrum soil biocides application induce mineral nutrient deficiency symptoms on plants have also been observed (Hirrel and Gerdemann, 1980; Jalali, 1978). Their use may also reduce mycorrhizal infection from the indigenous soil population. Bertoldi *et al.* (1977) showed that repeated applications of toxicants decreased the growth of onion seedlings and these losses were also associated with deffering effects on soil biota.

Plants in fungicide-treated soil in the field would probably develop eventually a mycorrhizal association, depending to a great extent on the biological properties of the toxicant, the degree of volatility, the speed of diffusion, and the method of application. VAM fungi resume growth quickly following the degradation of fungicides such as botran or terrachlor, but might not do so after exposure to biocides that are strongly fungicidal. Following ban application of toxicants, the roots could grow both horizontally and vertically from the treated zone into untreated soil and mycorrhizal development might be delayed longer, since the roots could reach untreated soil only by vertical growth. Thus, the least interference with mycorrhizae theoretically would follow narrow-band application of non-volatile fungistats that are relatively non-specific of endotrophic fungi.

Since VAM and ectomycorrhizal symbionts application may become important as biofertilizers and biological control agents of plant-pathogens, the introduction and management of these could be used to the advantage of the crop and tree. This is of practical significance in their activities by the fungicides may ultimately have a depressive effect on the tripartite relationship of mycorrhizae, rhizobia and plant.

Also, chloroneb resulted in the smallest numerical reduction of mycorrhizal development as compared to benomyl and triadimefon. This indicates that it may therefore be possible to selectively control pythiaceous fungi without substantially reducing mycorrhizal development. It is further suggested that studies involving a wider range of rates and application schedules are needed before the effects of fungicides on mycorrhizal fungi can be compared adequately.

Application of systemic fungicides in soil at the field rates reduced AMF in soil in different crops: ethazoletridiazole (sudangrass): thiophanate-methyl (peas); benomyl (wheat, bean, onion, soybean, citrus, barley); tridmorph, triforine, ethirmol, and chlor-aniformethan (wheat); pyroxychor (pea); ethazole (lettuce); thiabendzole (wheat, bean, citrus); thophanate (wheat, onion); triadimefon (wheat, lettuce); and carboxin (bean). AMF were promoted by chloroneb (lettuce), fosetyl-AL (lettuce), metalaxyl (citrus), ethazole (sudangrass), and thiabendazole (citrus) (Menge, 1982).

REFERENCES

Abbot, L.K. and Robson, A.D. 1977, Growth stimulation of subterranean clover with vesicular-arbuscular mycorrhizas, *Aust. J. Agric. Res.* 28: 639-649.

Afek, U., Menge J.A. and Johnson, E.L.V. 1990, The effect of *Pythium ultimum* and metalaxyl treatments on root length and mycorrhizal colonization of cotton, onion and pepper, *Plant Dis.* 74 : 117-120.

Afek, U., Menge J.A. and Johnson, E.L.V. 1991, Interaction among mycorrhizae, soil solarization, metalaxyl and plants in the field, *Plant Dis.* 75 : 665-671.

Agnohotri, V.P. 1971, Persistence of captan and its effect on microflora respiration and nitrification of a forest nursesy soil, *Can. J. Microbiol.* 13: 377.

Anderson, J. R. 1978, Pesticides effects on nontarget soil microorganisms, in : *Pesticides Microbiology*, I.R. Hill, and S.J.L. Wright, eds., Academic Press, New York, pp. 313-345.

Atiloano, R.A. and Vangundy, S.D. 1979, Effects of some systemic fumigant and nematicides on grape mycorrhizal fungi and citrus nematode, *Plant Dis. Rep.* 63: 729-733.

Austin, D.J. and Briggs, G.G. 1976, A new extraction method for benomyl residues in soil and its application in movement and persistence studies, *Pestic. Sci.* 7: 201.

Azcon, R. and Barea, J.M. 1992, Nodulation, N_2 fixation (N^{15}) and N-nutritional relationship in the mycorrhizal or phosphate amended alfalfa plants, *Symbiosis* 12: 33-44.

Backman, P.A. and Clark, M. 1979, Effect of corbofuran and other pesticides and vesicular arbuscular mycrrohizae in peanuts, *Nematropica* 7:14-18.

Baude, F.J., Pease, H.L. and Holt, R.F. 1974, Fate of benomyl on field soil and turf, *J. Agric. Food Chem.* 22: 413, 1974.

Bailey, J.E. and Safir, G.R. 1978, Effect of benomyl on soybean endomycorrhizae, *Phytopath.* 68 I. 1810-1812.

Bakshi, B.K. and Dobriyal, N.D. 1970, Effect of fungicides on control damping off on development of mycorrhiza, *Indian Forester* 96(9): 701-703.

Beilby, J.P. 1983, Effect of inhibitors on early protein , RNA and lipid synthesis in germinating vesicular-arbuscular mycorrhizal fungal spores of *Glomus calendonicum, Can J. Microbiol.* 29:596-601.

Benjamin, R.K. 1979, Zygomycetes and their spores, in: *The whole Fungus,* W.B. Kendrick, ed., National Museums of Canada, Ottawa, pp. 256-573.

Bertold, M. De, Giovanneh, M. and Rambelli, A. 1977, Effects of soil applications of benomyl and captan on the growth of onion and the occurrence of endoyphytic mycorrhizae and rhizosphere microbes, *Ann. Appl. Biol.* 86: 111-115.

Bertrand, A., Ducret, J., Debourge J.C. and Horriere, D. 1977, Etude des properties d'une nouvelle famille de fungicides: Les monethyl phospnites metalliques. Caracteristiques physicochemiques et properties biologigues phytiatrrie, *Phytopharamacie* 26: 3-17.

Bethlenfalvay, G.J. 1992, Mycorrhizae and crop productivity, in: *Mycorrhizae in Substainable Agriculture,* ASA Special Publication No. 54, 677 S. Segoe Rd., Madison, USA, pp. 1-27.

Bird, J.W., Rich, J.R. and Clover, S.V. 1974, Increased endomycorrhizae of cotton in soil treated with nematicides, *Phytopath.* 64:48-51.

Boatman, N.D., Paget, D.K. Hayman, D.S. and Moses, B. 1978, Effects of systemic fungicides on vesicular-arbuscular mycorrhizal infection and plant phosphate uptake, *Trans. Br. Mycol. Soc.* 70:443-450.

Bodmer, M., Sun Chengm G. and Scheupp, H. 1986, Monitoring of vesicular arbuscular mycorrhiza in field plots treated with selected fungicides, in: *Mycorrhizae : Physiology and Genetics,* Ist ESM, Dijon, 1-5 July 1985, INRA, Paris, pp. 700-706.

Bollen, G.J. 1972, A comparison of the *in vitro* antifungal spectra of thiophanates and benomyl, *Neth. J.Pathol.* 78:55.

Bollen, G.J. and Fuchs, A. 1970, On the specificity of the *in vitro* and *in vivo* antifungal activity of benomyl, *Neth. J.Pl. Path.* 76:299-312.

Bompeix, G., Ravise, A., Raynal, C., Fettouche F. and Durang, M.C. 1980, Modalite de l'obtention des necroses bloquantes sur feuilles detachees detomate pe l'action durtris-o-ethyl phosphontae d'aluminum (phosethyl d'aluminum) hypothesis sur son mode d'action *in vivo, Ann. Phytopathol.* 12:337-351.

Bowen, G.J. and Theodorou, C. 1973, Growth of ectomycorrhizal fungi around seeds and roots, in: *Ectomycorrhizae - Their Ecology and Physiology,* D.C. Marks and T.T. Kozlowski, eds., Academic Press, New York, pp. 107.

Bruck, R.I., Kenerley, C.M. and Grand, L.F. 1982, The effects of metalaxyl on growth and mycorrhiza incidence Fraser Fir, *Phytapath.* 72: 335.

Bruin, C.G.A. and Edgington, L.V. 1984, The chemical control of plant disease caused by zoosporic plant pathogens, in: *Zoosporic Plant Pathogens,* S.T. Buczacki, ed., Academic Press, New York, pp. 193-232.

Burpee, L. and Cole, H. Jr. 1978, The influence of alachlor, trifulrain and diazinon on the development of endogenous mycorrhizae in soyabean, *Bull. Enviro. Cont. Toxicol.* 19 : 191-197.

Carling, D.E. and Brown, M.F. 1980, Relative effect of vesicular arbuscular mycorrhizal fungi of the growth and yield of greenhouse soybean, *Soil. Sci. Soc. Am.J.* 44 : 528-532.

Caron, M. 1989, Potential use of mycorrhizal in control of soilborne disease, *Can. J. Plant Pathol.* 11: 177-179.

Carr, G.R. and Hinkley, M.A. 1985, Germination and hyphal growth of *Glomus caladonium* on water agar containing benomyl, *Soil Biol. Biochem.* 17: 313-316.

Clark, C.A. 1978, Effect of pesticides on vesicular-arbuscular mycorrhizae, in: *Rothamsed Exp. Sta. Res. Rep. for 1978,* Part 2, pp. 236-237.

Clark, F.B. 1963, Endotrophic mycorrhizae influence yellow poplar seedling growth, *Science* 140: 1220-1221.

Cooper, K.M. and Losel, D.M. 1978, Lipid physiology of vesicular arbuscular mycorrhizae, I, Composion of lipids in root of onion, clover and ryegrass infected with *Glomus mosseae, New Phytol.* 80: 143-151.

Cornish, P.S. and Pratley, J.E. 1991, Tillage practices in substainable farming system, in: *Dryland Farming: A Systems Approach,* L.White, eds., Sydney Univ. Press, South Melborne, Australia, pp. 76-101.

Cudlin, P., Mejstrick, V. and Sasek, V. 1980, The effect of fungicide ditane M-45 and the herbicide gramoxone on the growth of mycorrhizal fungi *in vitro, Ceska Mykol.* 34: 191.

Decallonne, J.R. and Meyer, J.A. 1972, Effect of benomyl on spores of *Fusarium oxysporum, Phytopath.* 11: 2155-2160.

Despatie, S., Furlan V. and Fortin, J.A. 1989, Effect of successive applications of fosetyl-Al on growth of *Allium cepa* associated with endomycorrhizal fungi, *Plant Soil* 113: 175-180.

Dodd, J.C. and Jaffries, P. 1989, Effect of fungicides on three vesicular-arbuscular mycorrhizal fungi associated with winter wheat *Triticum aestivum* L., *Biol. Fertil. Soil* 7: 120-128.

Domsch, K.H. 1964, Soil fungicides, *An.. Rev. Phytopath.* 2: 293-320.

322

Edginton, L.V. 1981, Structural requirements of systemic fungicides, *An. Rev. Phytopathol.* 19: 107-124.

Edginton, L.V., Khew K.L. and Barrn, G.L. 1971, Fungitoxic spectrum of benimidazol compounds, *Phytopath.* 61: 42-44.

Edgington, L.V., Martin, R.A., Briun, G.C. and Parsons, I.M. 1980, Systemic fungicides : a perspective after 10 years, *Plant Dis.* 64: 1-29.

El-Giahmi, A.S., Nicolson, T.H. and Daft, M.J. 1976, Effects of fungal toxicants on mycorrhizal maize, *Trans. Br. Mycol. Soc.* 67: 172-173.

Erwin, D.C. 1973, Systemic fungicides : disease control, translocation and mode of action, *Ann. Rev. Phytopathol.* 11: 389-402.

Fenn, M.E. and Coffey, M.D. 1984, Studies on the *in vitro* and *in vivo* antifungal activity of fosetyl-Al and phosphorus acid, *Phytopath.* 74: 606-611..

Fenn, M.E. and Coffey, M.D. 1985, Further evidence for the direct mode of action of fosetyl-Al and phosphorus acid, *Phytopath.* 75: 1064-1068.

Filer, T.H. Jr. and Toole, E.R. 1968, Effect of methyl bromide on mycorrhizae and growth of sweetgum seedlings, *Plant Dis. Rep.* 52: 483-485.

Filter, A.H. 1986, Effect of Benlate on leaf phosphorus concentration in alpine grasslands: a test of mycorrhizal benefit, *New Phytol.* 103: 767-776.

Fitter, A.H. and Nichols. R. 1988, The use of Benlate to control infection by vesicular-arbuscular mycorhizal fungi, *New Phytol.* 110: 201-206.

Fuchs, A., Van Den Verg, G.A. and Davidse, L.C. 1972, A comparison of benomyl and thiophanates with respect to some chemical and systemic fungitoxic characteristics, *Pesic. Biochem. Physiol.* 2: 191.

Furlan, V. and Bernier Cardou, M. 1989, Effects of N.P.K. on formation of vesicualr arbuscular mycorrhizae, growth and mineral content of onion, *Plant Soil* 113: 167-174.

Gerdemann, J.W. 1971, Fungi that form veiscular-arbuscular type of mycorrhiza, in : *Mycorrhizae*, E. Hacskaylo, ed., Forest Service Misc. Publ., 1189, U.S. Department of Agriculture, Washington, D.C., pp 9-30.

Gerdemann, J.W. and Nicolson, T.H. 1963, Spores of mycorrhizal *Endogone* species extracted from soil by wet sieving and decanting, *Trans. Br. Mycol. Soc.* 46: 235-244.

Grahmm J.H., Leonard, R.T. and Menge, J.A. 1981, Membrane mediated decrease in root exudation responsible for phosphorus inhibition of vesicular-arbuscular mycorrhiza formation, *Plant Physiol.* 68: 548-552.

Gray, L.E. and Gerdemann, G.W. 1969, Uptake of phosphorus ^{32}P by vesicular-arbuscular mycorrhizae, *Plant Soil* 30: 415-423.

Groth, D.E. and Martinson, C.A. 1983, Increased endomycorrhizal infection in maize and soybean after soil treatment with metalaxyl, *Plant Dis.* 67: 1337-78.

Habte, N., Aziz T.L. and Yuen, J.E. 1990, Residual effect of chlorothalonil on vesicular arbuscular mycorrhizal symbiosis in *Leucaena leucocephala*, in: *VIII North Am. Conf. Mycorrhiza*, W.Y. Jackson, ed., 5-8 September 1990, pp. 130-134.

Hacskaylo, E. 1961, Influence of cycloheximide on growth of mycorrhizal fungi and mycorrhiza of pine, *For. Sci.* 7: 377-379.

Hanz, A., Snow, G.A. and Marx, D.II. 1981, The effects of benomyl and *Pisolithus tinctorius* ectomycorrhizae on survival and growth of longleaf pine seedlings, *South J. Appl. For.* 5(4): 189-194.

Harvey, A.E. 1967, Effect of phytoaction treatments on mycorrhizae-root association in western white pine, *Plant Dis. Rep.* 51: 1012-1013.

Harvey, I.C., Fletcher, L.R. and Emme, L.M. 1982, Effects of several fungicides on the *Lolium* endophytes in ryegrass plants, seeds in culture, *N.Z. J.Agric. Res.* 25: 601.

Hatrick, B.A.D. and Wilson, G.W.T. 1991, Effect of mycorrhizal fungus species and metalaxyl application of microbial suppression of mycorrhizae symbiosis, *Mycologia* 83: 97-102.

Hayman, D.S. 1970, *Endogone*, spore numbers in soil and vesicular-arbuscular mycorrhizae in heat as influenced by season and soil treatment, *Trans. Br. Mycol. Soc.* 54: 53.

Hepper, C.M. 1979, Germination and growth of *Glomus caledonium* spores: the effect of inhibitors and nutrients, *Soil Biol. Biochem.* 11. 269-277.

Hirrel, M.C. and Gerdemann, J.M. 1980, Improved growth of onion and bell pepper in saline soil by two vesicular-arbuscular mycorrhizal fungi, *J. Soil Sci. Soc. Am.* 44 : 654-655.

Hong, L.T. 1976, Mycorrhizal short root development on *Pinus caribea* seedlings after fungicide treatment, *Malays. For.* 39: 147-156.

Huff, J.E. and Haseman, J.K. 1991, Exposure to certain pesticides may pose real carciogenic risk, *Chem. Engg. News* 69: 32-37.

Hwang, S.F. 1988, Effects of VA mycorrhizae and metalaxyl on growth of alfala seedlings in soils from fields with "Alalfa Sickness" in alberta, *Plant Dis.* 72: 448-452.

Ivory, M.H. 1975, Mycorrhizal studies on exotic conifers in West Malaysia, *Malays. For.* 38: 149.

Jabaji-Hare, S.H. and Kendrick, W.B. 1985, Effects of fosetyl-Al on root exudation and composition of extracts of mycorrhizal and non-mycorrhizal leek roots, *Can. J.Pl. Path.* 7: 118-126.

Jabaji-Hare, S.H. and Kendrick, W.B. 1987, Response of an endomycorrhizal fungus in *Allium porrum* L. to different concentrations of systemic fungicides, metalaxyl (Redomi) and fosetyl-Al (Aliette), *Soil Biol. Biochem.* 19: 95-99.

Jalali, B.L. 1978, Response of soil fungitoxicants on the development of mycorrhiza and phosphate uptake in cereals, Abs., 4th International Congres of Plant Pathology, Munchen, 182 p.

Jalali, B.L. and Domsch, K.H. 1975, Effects of systemic fungitoxicants on the development of endotrophic mycorrhiza, in: *Endomycorrhizas,* E.E. Sanders, B. Mosse and P.B. Tinker, eds., Academic Press, London, pp. 619-626.

Jalali, B.L. and Domsch, K.H. 1977, Effect of some fungitoxicants on amino acid spectrum of wheat root exudataes, *Phytopathol. Z.* 90: 22-26.

Johnson, N. C. and Efleger, F.L. 1992, Vesicular-arbuscular mycorrhizae and cultural stresses, in: *Mycorrhizae in Sustainable Agriculture,* ASA Special Publication No. 54, 677, S.Segoe Rd., Medison, USA, pp. 71-92.

Katan, J. 1987, Soil solarization, in: *Innovative Approaches to Plant Disease Control,* L. Chet, ed., John Wiley & Sons, New York, pp. 77-105.

Kellye, W.D. 1978, Control of *Cronartium usiforme* on loblolly pine seedlings with experimental systemic fungicides, benodanil, *Plant Dis. Rep.* 62: 595-598.

Kelley, W.D. 1979, Status of fusiform rust control with systemic fungicides, Proc. 1978. Southern Nurserymen's. Conf. Tech. Publ. SE-TA6. pp 168-170, USDA Forest Serv. State and Prov., Aatlantta, Ga. 170 p.

Kelley, W.D. and Snow, G.A. 1978, Non-target effect of benodanil on mycorrhizal fungi of loblolly pine, *Int. Congr. Plant Pathol.* (Abstr.), Munich, 176.

Kenerley, C.M., Bruck, R.I. and Cornl, L.F. 1984, Effects of metalaxyl on growth and ectomycorrhizae of fraser for seedlings, *Plant Dis.* 68: 32-35.

Kleinschidt, G.D. and Gerdemann, J.W. 1972, Stunting of citrus seedlings in fumigated nursery soils related to the absence of endomycorrhizae, *Phytpath.* 52: 1447-1452.

Knauss, J.F. 1974, Pyroxychlor, a new systemic fungicide for control of *Phytophthnora patmicora, Plant Dis, Rep.* 58: 1100.

Koide, R.T. and Li, M. 1990, On host regulation of the vesicular arbuscular mycorrhizal symbiosis, *New Phytol.* 114: 59-65.

Kough, J.L., Gianinazzi - Pearson, V. and Gianinazzi, S. 1987, Depressed metabolic activity of vesciular-arbuscular mycorrhizal fungi after fungicide applications, *New Phytol,* 106: 707-715.

Laiho, O. and Mikola, P. 1964, Studies on the effect of some eradicants on mycorrhizal development in forest nurseries, *Acta For. Fenn.,* 77: 1.

Larson, W.F. and Osborne, G.J. 1982, Tillage accomplishment and potential, in: *Predicting Lillage Effects on Soil Physical Properties and Processes,* P.W. Unger and D.M. van Doren, eds., ASA Spec. Publ. 44. ASA and SSSA, Madison, WI pp. 1-11.

Mac Donald, R.M. and Lewis, M. 1978, The occurrence of some acid phosphatase and dehydrogenases in the vesicualr - arbuscular mycorrhizal fungus *Glomus mosseae, New Phytol.* 80: 135-141.

Manjunath, A. and Bagyaraj, D.J. 1984, Effects of fungicides on mycorrhizal colonization and growth of onion. *Plant Soil* 80: 147-150.

Martin, J.P., Baines, R.C. and Page, A.L. 1963, Observations on the occasional temporary growth inhibition of citrus seedlings following heat or fumigation treatment of the soil, *Soil Sci.* 95: 175-185.

Martin, J.P., Farmer, W.J. and Ervin, J.O. 1973, Influence of steam treatment and fumigation of soil on growth and elemental composition of avocado seedlings, *Proc. Soil Sci. Soc. Am.* 37: 56-60.

Marx, D.H. 1980, Ectomycorrhizal fungus inoculations: a tool for improving forestation practices, in: *Tropical Mycorrhizae Research,* P., Mikola, ed., Oxford University Press, London, pp 13-30.

Marx, D.H. and Artman, D. 1979, *Pisolithus tinctorius* ectomycorrhizae improve survival and growth of pine seedlings on acid coal spoils in kentucky and Virginia, *Reclam Rev.* 2: 23-31.

Marx, D.H. and Rowan, S.J. 1981, Fungicides influence growth and development of specific extomycorrhizae on loblolly pine seedlings, *For. Sci.* 27(i) 167-176.

Marx, D.H., Cordell, C.E. and France, R.C. 1986, Effects of triandimefon on growth and ectomycorrhizal development of loblloly and slash pines in nurseries, *Phytopath.* 76: 824-831.

Melin, E. 1963, Some effects of forest tree roots on mycorrhizal basidiomycets, in: *Symbiotic Association,* P.S. Nutman and B. Mosse, eds., The University Press, Cambridge, p. 125.

Menge, J.A. 1982, Effect of soil fumigents and fungicides on vesicular arbuscular mycorrhizal fungi, *Phytopath.* 72: 1125-1132.

Menge, J.A., Johnson, E.L.V. and Minassian, V. 1979, Effect of heat treatment and three pesticides upon the growth and production fo the mycorrhizal fungus *Glomus fasciculatum, New Phytol.* 82: 473-480.

Menge, J.A., Munnecke, D.E., Johnson, E.L.V. and Carnes, D.W. 1978, Dosage response of the vesicualr-arbuscular mycorrhizal fungus *Glomus fasciculatum* and *Glomus constrictum* to methyl bromide, *Phytopath.* 68: 1368-1372.

Mikola, P. 1973, Application of mycorrhizal symbiosis in forestry practice, in: *Ectomycorrhizae - Their Ecology and Physiology,* G.C. Marks and T.T. Kozlowski, eds., Academic Press, New York, pp. 383-395.

Moorman, T.B. 1989, A review of pesticide effects in microorganisms and microbial processed related to soil fetility, *J. Prod. Agric.* 2: 14-23.

Mukerji, K.G. 1996, Taxonomy of Endomycorhizal fungi, in: *Advances in Botany*, K.G. Mukerji, B. Mathur, B.P. Chamola and P. Chitralekha, eds., A.P.H. Publihsing Corp., New Delhi, pp. 213-222.

Mosse, B. 1962, The establishment of vesicualr-arbuscular mycorrhizae under aseptic conditions, *J. Geb. Microbiol.* 27: 509-520.

Mosse, B., Stribley, D.P. and Letacon, F., 1981, Ecology of vesicular asbuscular mycorrhizal fungi, *Adv. Microb. Ecol.* 5: 137.

Nemec, S. 1980, Effect of 11 fungicides on endomycorrhizal development in sour orange, *Can. J. Bot.* 11: 171-196.

Nemec, S. and O'Bannon, H.J. 1979, Response of *Citrus aurantium* to *Glomus etunicatum* and *G. mosseae* after soil treatment with selected fumigants, *Plant Soil* 53: 351-359.

Nene, Y.L. and Thapliyal, P.N. 1979, *Fungicides in Plant Disease Control*, Oxford and IBH Publisher, New Delhi, 461 p.

Nesheim, O.N. and Linn, M.B. 1969, Deleterious effect of certain fungi-toxicants on the formation of mycorrhiza on corn by *Endogone fasciculata* and on cron-root development, *Phytpath.* 59: 297-300.

Nuffelen, Van H. and Schenck, N.C. 1984, Spore germination, penetration and root colonization of six species of VAM fungi in soybean, *Gen.J.Bot.* 62: 624-628.

Ocampo, J.A. and Hayaman, D.S. 1980, Effects of pesticides on mycorrhiza in field grown barley, maize and potatoes, *Trans. Br. Mycol. Soc.* 74: 413-416.

O'Bannon, J.H. and Nemec, S. 1978, Influence of soil pesticides on vesicular-arbuscular mycorrhizae in a citrus soil, *Nematropics* 8: 56-61.

Parvathi, K.K., Venkateswarlu, K. and Rao, A.S. 1985a, Toxicity of soil applied fungicides to the vesicular-arbuscular mycorrhizal fungus *Glomus mosseae* in groundnut, *Can.J. Bot.* 63: 1673-1675.

Parvathi, K., Venkateswarlu, K. and Rao, A.S. 1985b, Effects of pesticides on development of *Glomus mosseae* in groundnut, *Trans. Br. Mycol. Soc.* 84: 29-33.

Pawak, W.H., Ruehle, J.L. and Marx, D.H. 1980, Fungicide drenches affect ectomycorrhizal development on container grown *Pinus palustris* seedlings, *Can.J. For. Res.* 10: 61.

Pearse, A.G.E. 1968, *Histochemistry: Theoretical and Applied* Vol. I, Churchill Livingstone, Edinburgh, 63: 1673-1673.

Peterson, G.W. and Smith, R.S. 1975, *Forest Nursery Diseases in the United States*, Agric,. Handb. No. 470, U.S. Department of Agriculture, Washington, D.C.

Pfleger, F.L. and Stewart, E.L. 1976, The Influence of fungicides on endemic population of *Glomus* spp. in association with field grown peas, *Proc. Phytopathol. Soc.* 3: 274.

Phillips, J.M. and Hayman, D.S. 1970, Improved procedures for clearing root and vesicular arbuscular mycorrhizal fungi for rapid assessment of infection, *Trans. Br. Mycol. Soc.* 55: 158-161.

Phillips, R.E., Blevins, R.L., Thomas, G.W., Frye W.W. and Phillips, S.H. 1980, No-till agriculture, *Science* (Washington, D.C.) 208: 1108-1113.

Pirozynski, K.A. and Malloch, D.W. 1975, The origin of land plants: a matter of mycotrophison, *Biosystems* 6: 153-164.

Powell, C.L. 1976, Development of mycorrhizal infections from *Endogone* spores and infected root segments, *Trans. Br. Mycol. Soc.* 66: 439-445.

Powell, W.M., Hendrix F.F. and Marx, D.H. 1968, Chemical control of feeder root necrosis of pecans caused by *Pythium* species and nematodes, *Plant Dis. Rep.* 52: 577-578.

Ratnayake, M., Leonard, R. and Menge, J.A. 1978, Root exudation in relation to supply of phosphorus and its possibel relevance to mycorrhizal formation, *New Phytol.* 81: 543-552.

Redhead, J.F., 1961, Mycorrhizae association in some Nigerian forest trees, *Trans. Br. Mycol. Soc.* 51: 377.

Rhodes, L.H. and Larsen, P.O. 1981, Effect of fungicides on mycorrhizal development of creeping bentagras, *Plant Dis.* 65: 145-150.

Riffle, J.W. 1976, Effects of fumigation on growth of hard wood seedlings in northern plains nursery, *Proc. Am. Phytopath. Soc.* 3: 215 pp

Rodriguez-Kabana, R. and Curl, E.A. 1980, Nontarget effects of pesticides on soilborne pathogens and disease, *Annu. Rev. Phytopathol.* 18: 311.

Rowan, S.J. and Kelley, W.D. 1986, Survival and growth of out planted pine seedlings after mycorrhizae were inhibited by use of triadimefon in the nursery, *South J. App. For.* 10:21-23.

Schenck, N.C. and Tucker, D.P.H. 1974, Endomycorrhizal fungi and the development of citrus seedlings in Florida fumigated soils, *J.Am. Soc. Hortic. Sci.* 99: 284-287.

Schenck, N.C. and Kellam, M.K. 1978, The influence of vesicular-arbuscular mycorrhizae on disease development, *Florida Tech. Bull.* 798: 66.

Scheupp, H. and Bodmer, M. 1991, Complex response of VAMycorrhizae to xenobiotics substances, *Toxicol. Environ. Chem.* 30: 193-199.

325

Schonbeck, F. 1979, Endomycorrhiza in relation to plant disease, In: *Soilborne Plant Pathogens*, B. Schippers and W. Gams, eds., Academic Press, New York, pp. 271-280.

Schwab, S.M., Johnson, E.L.V. and Menge, J.A. 1982, Influence of simazine on formation of vesicualr arbuscular mycorrhizae in *Chenopodium qunona, Plant Soil* 64: 283-287.

Sieverding, E. and Leihner, D.E.1984, Effect of herbicides on population dynamics of VA-mycorrhiza with cassava, *Angew. Bot.* 58: 283-284.

Siqueira, J.O., Colozzi-Filho, A. and Oliveira, E. 1989, Occurrencia de micrrizas vesiculo-rbusculares em agro ecossistemas naturaise de esttado de minas gerais, *Pesqui Agropecu Bras.* 24: 1499-1506.

Slankis, V. 1971, Formation of ectomycorrhizae of forest trees in relation to light, carbohydrates and auxinsm, in: *Mycorrhizae*, E. Hacskaylo, ed., Forest Service Misc. Publ. 1189, U.S. Department of Agriculture, Washington, D.C. p. 151.

Smith, T.F. 1978, Some effects of crop protection chemicals on the distribution and abundance of vesicular arbuscular endomycorrhizae, *J.Austr. Inst,. Agric. Sci.* 44: 82-85.

Smith, T.F., Noack, A.J. and Cosh, S.M. 1981, The effect of some herbicides on vesicular arbuscular endophyte abundance in soil and on infection of host roots, *Pestic. Sci.* 12: 91-97.

Sobotka, A. 1970, Die Testuns des Einflues von Pestiziden auf die Mykorrhiza-Pilze in Waldboden. Zentrabl. Backeteriol. Parasitenkd, *Infektionskr, Hyg.* Abt. 2, 152, 723.

South, D.B. and Kelley, W.D. 1982, The effect of selected pesticides on short root development of greenhouse grown *Pinus taeda* seedlings, *Can. J. For. Res.* 12: 237-241.

Spokes, J.R., Hayman, D.S. and Kandasamy, D. 1989, The effects of fungicide coated seeds on the establishment of VA mycorrhizal infection, *Ann. Appl. Biol.* 115: 237-241.

Spokes, J.R. and Mac Donald, R.M. 1978, Effects of pesticides on VA mycorrhiza, Rothamsted Exp. Stan. Rep. for 1978, Part-I, pp. 236-350.

Spokes, J.R., Mac Donald, R.M. and Hayman, D.S. 1981, Effects of plant protection chemicals on vesicular arbuscular mycorrhizae, *Pestic. Sci.* 12: 346-350.

Snow, G.A., Rowan, S.J., Jones, J.P., Kelley W.D. and Mexal, J.G. 1978, Using bayteton to control fusiform rust in pine trees nurseries, U.S. Dep. Agric. Res. Note 50: 253-5 pp.

Sreenivasa, M.N. and Bagyaraj, D.J. 1989, Use of pesticides for mass production of vesicular arbuscular mycorrhizas, *Plant Soil* 119: 127-132.

Srivastava, D., Kappor, R., Srivastava, S.K. and Mukerji, K.G. 1996, Vesicular arbuseular mycorrhiza - an overview, in: *Concepts in Mycorrhizal Research*, K.G. Mukerji, ed., Kluwer Academic Publishers, Dordrecht, The Netherlands, pp. 1-40.

Staub, T.H., Dabmen and Schwinn, F.J. 1978, Effects of Ridomil on the development of target pathogens on their host, *Proc. 3rd Int. Conf. Plant Path.* (16-23 August, 1978), Munich, pp. 356.

Stewart, E.L. and Pfleger, P.L. 1977, Influence of fungicides on extant *Glomus* species in association with field grown peas, *Trans. Br. Mycol. Soc.* 69: 318-319.

Sukarno, N., Smith, S.E. and Scott, E.S. 1993, The effect of fungicides on vesicular-arbuscular mycorrhizal symbiosis, *New Phytol.* 25: 139-147.

Sutton, J.C. and Sheppard, B.R. 1976, Aggregation of sand dune soil by endomycorrhizal fungi, *Can.J. Bot.* 54: 326-333.

Theodorou, G. and Skinner, M.F. 1976, Effects of fungicides on seed inocula of basidiospores of mycorrhizal fungi, *Aust. For. Res.* 7: 53-58.

Tiefenbrunner, F. 1972, Mycelgewichtszunahme and Mycorrhizapilzen under Einwirkung von Fungiziden *in vitro, Z. Pilzkd.* 38: 105.

Timmer, L.W. and Leyden, R.F. 1978, Relationship of seedbed fertilization and fumigation to infection of sour orange seedlings by mycorrhizal fungi and *Phytophthora parasitica, J.Am. Hortic. Soc.* 103: 537-541.

Tommerup, J.C. and Briggs, G.G. 1981, Influence of agricultural chemicals on germination of vesicular-arbuscular endophyte spores, *Trans. Br. Mycol. Soc.* 76: 326-328.

Traffe, J.M., Molina, R. and Casallano, M. 1984, Reactions of mycorrhizal fungi and mycorrhiza formation to pesticides, *Ann. Rev. Phytopath.* 22: 331-355.

Tucker, D.P.H. and Anderson, C.A. 1972, Correction of citrus seedling stunting of fumigated soil by phosphate application, *Proc. Fla. State Hortic. Soc.* 85: 10-12.

VO-Thi-Hi, G., Bompeix and Ravise, A. 1979, Role dutris-o-ethyl phosphonate d'aluminium dans la stimulation des reaction de defense detomate contree le *Phytopathora capsicie*, C.R. Acad. Sci., Paris, D. 288: 1171-1174.

Vyas, S.C. 1984, *Systemic fungicides*, Tata McGraw Hill Publishing Co., New Delhi, 360 p.

Vyas, S.C. 1988, *Nontarget Effects of Agricultural Fungicides*, CRS Press, Florida, USA, 272 p.

Vyas, S.C. 1992, Effect of systemic fungicides metalaxl and fosetyl-Al on mycorrhizal infection in soybean, Abstr. 10th International Symposium on Modern Fungicides and Antifungal Compounds, Castle Reinhardsbrunn, Germany, p. 48.

Vyas, S.C. 1992, Vesicular arbuscular mycorrhizal fungi, *Mycorrhizae : Asian Overview*, Sujan Singh, ed., Tata Energy Research Institute, New Delhi, pp. 36-42.

Vyas, S.C. 1993, *Handbook of Systemic Fungicides*, Vol. I,II and III, Tata McGraw Hill Publihsing Co., New Delhi, pp. 400-436.

Vyas, S.C. 1992, Role of systemic and nonsystemic fungicides in formation of vesicualr arbuscular mycorrhizal fungi in plants, in : *Proc. 10th International Symposium on Systemic Fungicides and Antifungal Compounds*, H. Lyr and C. Polter, eds., Castle of Rheinhardsbrunn, Thuringia, Germany, May 3-8, 1992, pp. 19-32.

Vyas, S.C. 1992, Effect of systemic fungicides metalaxyl and fosetly-Al on mycorrhizal formation in soybean, in : *Proc. 10th International Symposium on Systemic Fungicides and Antifungal Compounds*, H. Lyr and C.Polter, eds., Rheinhards, Thiringia, Germany (May 3-8, 1992) pp. 449-454.

Vyas, S.C. 1993, *Handbook of Systemic Fungicides*, Tata McGraw Hill Publishing Co., New Delhi, Vol. I,II,III pp. 430, 389, 410.

Vyas, S.C. and Shroff, V.N. 1990, Interactions between fungi and fungicides, in : *National Con. Mycorrhizae*, Department of Plant Pathology, Haryana Agricultural University, Hissar (Feb. 14-16, 1990), pp. 175-177.

Vyas, S.C., Vyas, A. and Shroff, V.N. 1990, Interactions between fungi and fungicides, in : *National Con. Mycorrhizae*, Department of Plant Pathology, Haryana Agricultural Universtiy, Hissar (Feb. 14-16, 1990), pp. 175-177.

Vyas, S.C., Vyas, A., Mahajan K.C. and Shroff, V.N. 1990, Effect of seed treatment fungicides on mycorrhizal and rhizobial development in soybean, in : *Natl. Seminar on Mycorrhizae*, Department of Plant Pathology, Haryana Agric. University, Hissar, (Feb. 14-16, 1990), pp. 188-189.

Young, C.C., Juang, T.C. and Chao, C.C. 1988, Effects of *Rhizobium* and vesicular arbuscular mycorrhiza inoculations on nodulations, symbiotic nitrogen fixation and soybean yield in subtropical - tropical field, *Boil. Fertil. Soils* 6: 165-169.

INDEX

Cajanus cajan, 175	Epulorhiza sp., 267
Calamagrostis canadensis, 257	Eragrostis ciliaris, 253
Calanthe pulchra, 270	Eragrostis japonica, 252
Capsicum frutescens, 175	Eucalyptus camaldulensis, 70
Carex atherodes, 257	Eucalyptus globulus, 37, 47, 128, 129
Carex stricta, 257	Eucalyptus pilularis, 189
Carex vesicaria, 257	Eucalyptus sp., 80, 102, 106, 107, 112, 128, 135, 189
Carica papaya, 159	Eupatorium odoratum, 75
Carrizo citrange, 68	Euphorbia hirta, 252
Carya sp., 115	Euphorbia pulcherima, 175
Cassia tora, 175	Fagus sp., 105, 135
Casuarina sp., 135	Festuca rubra, 253
Chamaecyparis lawsoniana, 175	Fragaria ananossa, 175
Chenopodium album, 252	Fragaria vesica, 182, 246
Chenopodium quinoa, 157	Fragaria vesica var. alpina, 176
Chloridium sp., 45	Fragaria sp., 176
Chloris dolichostachya, 253	Fraxinus excelsior, 220
Chloris sp., 253	Fraxinus pennsylvanica, 75
Chrysanthemum morifolium, 157	Galeola septentrionalis, 269, 270
Cicer arietinum, 175, 309	Galeola sp., 269
Ciriconemella sp., 179	Galinsoga parviflora, 256
Cirsium discolor, 90	Gastrodia callosa, 270
Cirsium geophilum, 92, 95	Gastrodia elata, 270
Citrus jambhiri, 178	Gastrodia sp., 268, 269
Citrus limon, 178	Gerbera jamesonii, 220, 221, 240
Citrus sp., 148, 175, 178	Gliricidia sepium, 238
Commelina benghalensis, 252	Glycine max, 76, 176, 178, 197, 308
Convolvulus arvensis, 252	Gnaphalium purpureum, 256
Convolvulus sepium, 206	Goodyera repens, 269, 276, 277, 278
Corallorhiza trifida, 272	Gossypium hirsutum, 75, 179, 184
Corallorhiza sp., 269	Gossypium sp., 176
Corchorus olitorius, 175	Hemarthria compressa, 253
Corchorus tridens, 252	Hevea brasiliensis, 176
Corylus americana, 90	Hibiscus lobatus, 252
Corylus cornuta, 90	Holcus lanatus, 253
Cucumis melo, 175, 178, 182	Hordeum jubatum, 257
Cucumis sativus, 204	Hordeum vulgare, 76, 176
Cuminum cyminum, 175	Imperata cylindrica, 252
Cydonia oblonga, 178	Ipomea batatas, 179, 204, 205
Cynodon dactylon, 252, 253	Lathyrus pratensis, 253
Cyperus bulbosus, 252	Launaea nudicaulis, 256
Cyperus compressus, 252	Lectochola panicea, 253
Cyphomendra betacea, 178, 183	Lepidosperma gracile, 255
Cystorchis aphylla, 269	Leptochola penicea, 252
Dactylolectinium aegypticum, 252, 253	Leucaena leucocephala, 70, 71, 73, 74, 147, 148, 220,
Dactylorchis purpurella, 271, 276, 277, 278	221, 222, 224, 238, 241, 312
Dactylorhiza praetermissa, 275	Leucaena sp., 313
Dactylorrhiza incarnata, 282	Linum usitatissimum, 176
Dactylorrhiza maculata, 283	Lolium sp., 300
Dactylorrhiza majalis, 75, 283	Lucerne sp., 176
Dactylorrhiza purpurella, 282	Lupinus albus, 160
Dactylorrhiza sp., 275, 269, 273, 274, 275, 278, 281	Lupinus cosentinii, 160
Dalbergia sisoo, 80	Lupinus sp., 160
Daucus carota, 178, 200, 206	Lycopersicon esculentum, 176, 179
Diervilla lonicera, 91	Lyginia barbata, 255
Distichlis spicata, 58	Meconopsis simplicifolia, 238
Distichlis stricta, 257	Medicago indica, 252
Echinochola colonum, 252	Medicago sativa, 179, 187
Eclipta alba, 252, 256	Medicago truncatula, 187
Elettaria cardomomum, 178, 183	Musa acuminata, 179, 183
Elusine indica, 252, 253	Nephrolepis exaltata, 240

332

Parasitic Plants

Other Plants

Plant Pathogens
Fungal Pathogens

Cylindrocladium scoparium, 177
Gliocladium virens, 144, 184
Gliocladium sp., 189
Humicola sp., 145, 148
Mortierella sp., 303
Paecilomyces lilacinum, 124
Palaeomyces sp., 10
Penicillium sp., 145, 148, 313
Phanerochaete chrysosporium, 125
Phialophora sp., 45
Stachybotyris sp., 145
Trichoderma sp., 145, 148, 189, 313

Fungal Structure
Dolipore, 280
Haustoria, 147
Haustorium, 165
Rhizomorphs, 16, 29, 105, 107, 113

Mycorrhiza
Ectomycorrhiza
Ectomycorrhizal Basidiomycetes
Agaricus angustus, 110, 111
Agaricus silvaticus, 109
Agaricus trisulphuratus, 109
Amanita berkeleyi, 110
Amanita emilii, 109, 110
Amanita flavoconia, 109
Amanita fulva, 109
Amanita gemmata, 109, 110
Amanita hemibapha, 109
Amanita inaurata, 109
Amanita muscaria, 107, 113
Amanita muscaria var. fluavivolvata, 111
Amanita pantherina, 109
Amanita rubescens, 109, 110
Amanita umbonata, 110
Amanita vaginata, 109, 110, 111
Amanita verna, 109
Amanita sp., 30, 45, 106, 109, 111, 113, 114, 136
Arctostaphylos sp., 31
Armillaria ostoyae, 123
Armillaria sp., 106, 267
Astraeeus hygrometricus, 109, 111
Astraceus sp., 106, 114
Boletellus sp., 109
Boletinus caripes, 113
Boletinus sp., 48
Boletopsis sp., 107, 109, 111
Boletus erythropus, 111
Boletus gertrudiae, 110
Boletus hoarkii, 107
Boletus vermiculosoides, 110
Boletus sp., 30, 45, 106, 109, 110, 114, 136, 139
Calvatia sp., 109
Cantharellus cibarius, 48, 107, 109, 111
Cleome viscosa, 252
Clitocybe clavipes, 110
Clitocybe dealbata, 109, 110

Clitocybe infundibuliformis, 111
Clitopilus sp., 30
Collybia fusipes, 110
Collybia sp., 111
Coprinus cinereus, 47, 125, 126
Corticium bicolor, 114
Cortinarius camphoratus, 49
Cortinarius cinnabarinus, 109
Cortinarius hemitrichus, 107
Cortinarius obtusus, 49
Cortinarius sp., 47, 48, 50, 106, 109, 114
Cystoderma amianthianum, 109
Dermocybe sp., 48, 50
Fomes sp., 267
Fuscoboletinus sp., 48, 107, 114, 184
Ganoderma pseudoferreum, 183
Ganoderma sp., 174
Gautieria sp., 109
Geastrum fimbriatum, 109
Gomphidius vinicolor, 107
Gomphidius sp., 30, 124
Gomphus clavatus, 110
Hebeloma circinans, 36, 126
Hebeloma crustuliniforme, 79
Hebeloma cylindrosporum, 35, 47, 49, 50, 113, 123
Hebeloma sp., 30, 34, 47, 93, 95, 106, 114, 122, 123
Heterobasidium annosum, 79
Hydnangium sp., 34, 124
Hygrocybe conica, 110
Hygrocybe sp., 110
Hygrophorus chrysodon, 110
Hygrophorus pudorinus, 110
Hygrophorus subalpinus, 110
Hymenogaster sp., 114
Hymenoscyphus ericae, 45
Hysterangium inflatum, 183
Hysterangium sp., 102, 109
Inocybe fastigata, 109
Inocybe sp., 93, 106, 136
Laccaria amethystina, 47, 49, 110, 123
Laccaria laccata, 47, 48, 49, 73, 93, 95, 107, 110, 111, 123, 124, 126, 183
Laccaria proxima, 47, 50, 79, 123
Laccaria tortilis, 123
Laccaria sp., 30, 34, 45, 46, 93, 95, 106, 107, 122, 123, 124, 136
Lactarius bicolor, 34, 122
Lactarius camphoratus, 111
Lactarius deliciosus, 110, 111
Lactarius deterrimus, 136
Lactarius hygrophoroides, 110
Lactarius indicus sp., 110
Lactarius laccata, 33, 34, 35, 121, 122
Lactarius obseuratus-Alnus, 107
Lactarius piperatus, 110, 111
Lactarius proxima, 34
Lactarius sanguifluus, 110, 111
Lactarius scorbiculatus, 109
Lactarius zonarius, 110, 111
Lactarius sp., 30, 106, 113, 120
Leccinum luteum, 110

Heterotrophic, 148
Heterotrophs, 2
Hydroponics, 29, 200, 203
Micropropagation, 217, 218, 219, 221, 222, 225, 235
Mycotrophic, 2, 3, 4, 256
Necrotrophic, 5, 7, 277
Necrotrophs, 18
Necrotrophy, 6
Recognition, 16
Solarization, 307
Stress, 27, 62, 68, 70, 71, 72, 73, 74
Translocation, 5, 7, 76

Interactions and Genetic Manipulation
Antagonistic, 186, 257
Antibiosis, 186
Biocide, 302
Biocontrol, 144, 145, 146, 149, 174, 184, 189, 225
Compatibility, 16, 36, 201, 225, 316
Compatible, 107, 204, 267
Complementation, 125
Endomycobiont, 3, 4
Incompatibility, 121
Mutualism, 10
Mutualistic, 5
Mycoparasitism, 145
Phycobiont, 1
Phytoalexins, 163, 187, 279
Phytobiont, 3, 4, 17, 31
Phytotoxic, 279

Ptyophagy, 270
Symbiosis, 1, 2, 3, 4, 5, 6, 7, 8, 10, 17, 30, 36, 37, 38, 45, 59, 76, 127, 128, 129, 147, 153, 166, 167, 174, 187, 206, 221, 223, 238, 258, 267, 275, 279, 280, 316
Transformation, 35, 36, 125, 126, 127, 129
Transgenic, 146
Virulent, 147

Microbial Ecology
Biomass, 77, 87, 88, 92, 136, 139
Biosphere, 87, 88
Biotic, 46
Epigeous, 30
Hypogeous, 17, 30
Mycorrhizosphere, 143, 149
Niche, 67
Propagules, 89, 91, 92, 94, 95, 116, 200, 201, 209, 251
Rhizoplane, 184, 258
Rhizosphere, 32, 62, 87, 88, 89, 95, 96, 143, 145, 147, 148, 149, 153, 155, 165, 166, 167, 184, 186, 217, 255, 258, 260, 303

Fossils
Asteroxylon sp., 3
Fossil, 1, 3, 11, 18, 85
Rhynia sp., 3, 13

Forestation
Reforestation, 57, 218, 224

DATE DUE

MAY 0 6 2002	
JAN 1 3 2004	